AUSTRALIAN MINE VENTILATION CONFERENCE 2019

26–28 AUGUST 2019
PERTH, AUSTRALIA

The Australasian Institute of Mining and Metallurgy
Publication Series No 4/2019

≋ AusIMM

Edited by Bharath Belle MAusIMM(CP)

Published by:
The Australasian Institute of Mining and Metallurgy
Ground Floor, 204 Lygon Street, Carlton Victoria 3053, Australia

ISBN 978 1 925100 90 7

ORGANISING COMMITTEE

Duncan Chalmers MAusIMM
(Conference Chairperson)

Guang Xu MAusIMM (Deputy Chairperson)

Bharath Belle MAusIMM(CP)
(Technical Editor)

Basil Beamish MAusIMM(CP)

Rick Brake FAusIMM(CP)

Yingjiazi Cao AAusIMM

Wendy Harris MAusIMM

Johannes Holtzhausen MAusIMM

Katie Manns MAusIMM

Ting Ren MAusIMM

John Rowland

Michael Shearer MAusIMM

Guangyao Si MAusIMM

Craig Stewart MAusIMM

Jerry Tien MAusIMM

Mick Tuck MAusIMM

Leon van den Berg MAusIMM

AUSIMM

Kirsty Grimwade (Senior Manager, Events)

Alice Angley (Coordinator, Events)

Claire Stuart (Coordinator, Event Publishing)

TECHNICAL REVIEW MEMBERS

We would like to thank the following people for their contribution towards enhancing the quality of papers included in this volume:

Ms Cheryl Allen, Vale North Atlantic, Canada

Dr Rao Balusu, CSIRO Energy Flagship, Australia MAusIMM

Dr Basil Beamish, CB3 Mine Services Pty Ltd, Australia MAusIMM(CP)

Dr Rick Brake, Mine Ventilation Australia FAusIMM(CP)

Dr Jurgen Brune, CSM, USA

Dr Yingjiazi Cao, Monash University, Australia AAusIMM

David Carey, Mines Rescue Services, Australia MAusIMM

Mr Duncan Chalmers, UNSW, Australia MAusIMM

Prof David Cliff, University of Queensland, Australia MAusIMM

Mr Andrew Derrington, Ozvent Consulting Pty Ltd, Australia MAusIMM(CP)

Dr Gerrit Goodman, Office of Mine Safety and Health Research, USA

Dr Adrian Halim, Lulea University of Technology, Sweden MAusIMM(CP)

Dr Gerald Joy, National Institute for Occupational Safety and Health, USA

Dr Gareth Kennedy, SIMTARS, Australia

Ass Prof Mehmet Kizil, The University of Queensland, USA MAusIMM

Mr Bob Leeming, Health and Safety Executive, UK

Mr Kevin Lownie, Australia

Dr Pierre Mousset-Jones, Professor Emeritus, USA MAusIMM

Prof Huw Phillips, Professor Emeritus, South Africa

Dr Ting Ren, University of Wollongong, Australia MAusIMM

Mr John Rowland, Dallas Mining Services Pty Ltd, Australia

Dr Guangyao Si, UNSW, Australia MAusIMM

Prof Jerry Tien, Monash University, Australia MAusIMM

Dr Michael Tuck, Federation University, Australia MAusIMM

Mr Leon van den Berg, BBE, Australia MAusIMM

Dr Hsin Wei Wu, Gillies Wu Mining Technology Pty Ltd, Australia MAusIMM

Prof Pawel Wrona, Silesian University of Technology, Poland

Dr Guang Xu, Western Australian School of Mines, Australia MAusIMM

FOREWORD

Firstly, greetings to you all; welcome to Perth, Australia.

In this space, it is the ordinary duty of the Technical Editor to highlight the elements of the conference proceedings. Fair dinkum! Let me sincerely thank you all for participating in this biennial Australian mine ventilation conference. Our aim of this gathering is to improve the health and safety of Australian mine workers and to contribute towards our understanding and willingly share the wide range of expertise that resides amongst our engineering professionals, researchers and academics represented herein.

The proceedings of the 2019 Australian Mine Ventilation Conference 'Mine Vent 2019' focuses on the theme of 'Managing Ventilation Risk'. The organising committee, chaired by Mr. Duncan Chalmers, aims to provide an opportunity to present the latest developments so that the Australian and global mining industry can benefit from this collective knowledge. The main topics of this conference encapsulate various direct and indirect elements of metal and coal mine ventilation planning and case studies; main and booster fans; numerical modelling and integration with planning and remote monitoring; occupational health hazard monitoring and management (mine gases, dust, DPM, radioactive dust); heat and refrigeration; prevention of methane, coal and sulfide dust explosions; prevention and fighting of mine fires; detection and control of spontaneous combustion; ventilation monitoring and control; mining regulations, including principle hazard management plans; understanding radiation; and seal-up management.

The conference is supported by featured keynote speakers from local and international regulators, industry, research and academia, namely, Ms. Cheryl Allen (Canada), Dr. Heidi Roberts (QLD, Australia), Mr. Andrew Chaplyn (WA, Australia), Mr. Bob Leeming (UK), Prof. Shugang Li (China), Prof. Fubao Zhou (China) and Prof Qin Botao (China).

At most of our surface and underground mines, we are often faced with crossroads on safety and health-related engineering design decisions. Most of our preferences or choices have a consequential effect. Some would have multi-fatality risks and some would have loss of assets. Last year, both QLD and NSW coal mines were faced with heating and sponcom events, resulting in significant operational loss and production. In addition, there were a significant number of alarming cases of silicosis amongst the stone cutting industry. In situations like these, where do you lean to for guidance and solutions or advice? Mine ventilation engineering is a critical safety and health function in various spheres. It is hoped that the MVS technical symposium proceedings presented herein will be a valuable reference material.

In addition, as an Australian representative of the IMVC Committee, it is with great pleasure to inform you that Australia (AusIMM/UNSW) has won the bid to host the Sydney 12th International Mine Ventilation Congress in 2021. Winning the bid to host the next IMVC is a significant recognition of the Australian mine ventilation engineering community of mine ventilation engineering professionals, academics, researchers, suppliers and regulatory bodies.

I trust that you will enjoy the next three days and hope that you will use this occasion to get to know people from various specialist areas of the mine ventilation discipline, mining operations and from other parts of the world. A number of suggestions, practices and recommendations made during the conference should be scrutinised with further discussions. We should debate their intentions and the logic behind innovative operational practices to benefit all to improve mine health and safety.

Finally, thanks and appreciation go to all the technical peer reviewers (local and overseas) cited below for their precious time and assistance in making these proceedings a reality. Also, special mention goes to Claire Stuart who has provided a kind and constant line between the technical committee and the authors.

If you are still with me and you want to know the latest in mine ventilation engineering – keep reading the proceedings. That's enough of the editorial introduction. Finally, best wishes to you all for a successful 2019 AusVent conference, and I look forward to meeting you all at the 12th International Mine Ventilation Congress in Sydney in 2021.

Dr B Belle, MAusIMM(CP)

Proceedings Editor

Australian Representative, IMVC

SPONSORS

Platinum Sponsor

Howden

Gold Sponsors

gordon

zitron
AUSTRALIA

Silver Sponsors

ALETEK
Advanced Exhaust Technologies

MTV
Mine & Tunnel Ventilation

MST

MECANICAD
PLASTIC · VENTILATION · EXPERTS

Welcome Reception Sponsor

BBE
GROUP

Coffee Cart Sponsor

GT PLASTICS

Technical Session Sponsors

HATCH

Dräger

Name badge and lanyard

Howden

Conference App Sponsor

MINOVA

Notepads and Pens

AirEng

Destination Partner

PERTH
CONVENTION
BUREAU

Media Supporters

Mining

mine

MINING CHRONICLE

RESOURCES

IQ
INDUSTRY
QUEENSLAND

EXHIBITORS

ABC VENTILATION SYSTEMS

AC INDUSTRIES

Accutron

aggreko

Air Eng

ALETEK Advanced Exhaust Technologies

APACservices

BBE GROUP

ClemCorp Australia

encore monitoring

FREUDENBERG INNOVATING TOGETHER

gordon

Howden

Johnson Controls

MacLean Performance. Reliability. Innovation.

Maestro Digital Mine

MECANICAD PLASTIC · VENTILATION · EXPERTS

MINETEK

MINOVA

MST

MTV Mine & Tunnel Ventilation

OHMS HYGIENE

PINSSAR Air Monitoring Technology

PLASCORP MINEMASTER Mine Ventilation

Polyline Industries Tough Plastics Fabrication

TLT-Turbo

VumA-3D

WILSHAW

zitrön AUSTRALIA

CONTENTS

Health and Safety Hazard Management

Health Hazard Management-Respirable Dust / Diesel Emissions

Heat and Refrigeration

Methane and Coal Dust Explosions

Ventilation Planning and Management

Keynotes

Ventilation and workers health: a regulator's perspective

A Chaplyn[1]

1. State mining engineer, Director Mines Safety, Department of Mines, Industry Regulation and Safety, Dept. Mines, Industry Regulation & Safety 1 Adelaide Tce Perth Western Australia

ABSTRACT

Western Australian underground mines over the decades have increased in productivity and depth requiring more sophisticated controls. Concurrently, airborne contaminant exposure standards have become more stringent. Mine operators are required to review and adapt the controls for these changing risks.

Contemporary engineering practices and an informed workforce are key aspects enshrined in the objects of the Mines Safety and Inspection Act 1994.

The key note will highlight:-

- Ventilation initiatives undertaken by the department to engage with mine operators which led to issuing Mine Safety Bulletin 151

- The progress towards the development of a ventilation code of practice

- Requirements for a Health and Hygiene Management Plan to control health hazards to miners

- Projects undertaken to address emerging issues in diesel exhaust emissions to characterise nano Diesel Particulate Matter (nDPM), through the Mining Industry Advisory Committee (MIAC) working group, sponsored by DMIRS and Minerals Research Institute of Western Australia (MRIWA)

- What the future holds with respect to new legislation (Work Health and Safety) and competency requirements for the role of ventilation officers in WA.

KEYWORDS Improving the health of miners through ventilation and other controls by minimizing atmospheric contaminant exposure.

(risk assessment, design standards, training, monitoring)

Avoiding repeat incidents – learning the lessons of the past

J R Leeming[1]

1.HM Chief inspector of Mines, Health and Safety Executive, Sheffield, S3 8NH, UK
Bob.leeming@hse.gov.uk

ABSTRACT

From the advent of large scale mining in the UK from the 1750's, major incidents and disasters started to occur. With time, and after some repeats, lessons were learned and control measures developed to reduce the numbers and severity of incidents to a minimum.

With the passage of time, most of these events are now beyond the memory of the current workforce and management. The reasons for certain control measures and practices have been lost and there is a danger that complacency might set in, despite the fact that the hazards largely still exist.

It is important that the lessons are not lost. This paper will seek to examine a number of past methane ignition, coal dust explosion and fire incidents, where the lessons learned are still relevant.

INTRODUCTION

As mines have been developed to access deeper and further reserves, firstly artificial light then ventilation and latterly mechanisation have all had to be introduced to allow that development. Each brought their own hazards.

Artificial light in the early days was by unguarded open flame: a clear ignition source but one which took time and effort to make safe, by the Davy Lamp. Ventilation was difficult and costly to produce, and a lack of adequate ventilation caused flammable gas to build up into the explosive range, where it was readily ignited by open flames. Mechanisation has introduced numerous ignition sources, from friction at bearings, brakes or misaligned conveyors, to cutter picks in stone, to hot engine or exhaust parts.

Ignitions of methane have been the vast majority of initiators of coal dust explosions. Although the demise of the large scale coal industry in the UK has virtually eliminated this hazard, the risk remains and manifests itself regularly around the world. With mechanisation, underground fires have been a more significant problem, not from the fire itself but from the products of combustion transported by the mine ventilation and the resulting difficulties with escape.

Table 1 shows the major events under these topic areas that have occurred in the UK since 1960, and their casualties. Most are from coal mines, although it is recognised that other mines can generate flammable atmospheres, and other dusts apart from coal can be explosive. Any mine can potentially have an underground fire.

METHANE IGNITIONS

Ignitions of methane occur when two components are present: a flammable atmosphere, and an ignition source. As I always argue, controlling every ignition source is difficult, so it is imperative to avoid the flammable atmosphere. (Leeming, 2018). Prevention of a flammable atmosphere occurring depends primarily on adequate ventilation. The primary ventilation should be sufficient to dilute, render harmless and remove, the gas produced by the mining operation. The quantity of air required will have to be derived from the predicted or known gas make. It is also important to provide sufficient air velocity to encourage turbulence to prevent layering. (National Coal Board, 1979). Airflows and gas levels must be regularly monitored, and the ventilation adjusted if necessary to achieve the primary aim.

Ignition locations are analysed into three areas: headings while unventilated or being degassed; headings during cutting, and coal faces. I am including the Gresford colliery ignition in the paper.

The incident does not fit within the causes / locations analysed below as, unusually, gas was ignited outbye. However this demonstrates just how poor the ventilation must have been. It was a very significant event and the last of the >100 fatality underground mining events in the UK. It smacks right at the heart of what colliery management must be about: ensuring that the mine is ventilated adequately.

Table 1. Significant UK fires and ignitions since 1960.

Date	Mine	Type	Killed/injured	Comments
1960	Six Bells	Ignition	45 killed	u/mgr, 3 deps
1962	Hapton Valley	Ignition	19 killed	2 deps.
1962	Tower	Ignition	9 killed	2 deps.
1965	Cambrian	Ignition	31 killed	Mgr and u/mgr
1967	Michael	Fire	9 killed	Dep and o/man
1975	Houghton Main	Ignition	5 killed	2 deps.
1979	Golborne	Ignition	10 killed	1 dep.
1982	Cardowan	Ignition	40 serious	
1982	Coventry	Ignition	8 serious	
1989	Bevercotes	Ignition	No injuries	
1992	Bevercotes	Ignition	No injuries	
1992	Shirebrook	Fire	No injuries	Nearly closed mine
1997	Maltby	Ignition	No injuries	
1998	Prince of Wales	Ignition	No injuries	
2010	Kellingley	Ignition	No injuries	218 evacuated
2012	Maltby	Ignition	No injuries	
2013	Daw Mill	Fire	No injuries	Closed mine
2016	Boulby	Fire	9 heat illness	

Gresford Colliery, 1934. 265 killed

This was a gassy mine, and poorly ventilated. The workforce was used to working in gas: layers and accumulations were common, and flame safety lamps were often extinguished by the concentrations of firedamp and having to be re-lighted underground. This incident was an ignition of firedamp: there was no evidence of coal dust being involved. A fire subsequently developed which could not be contained. That, and further ignitions, resulted in that section of the mine being sealed a few days later, with not all the bodies being recovered. It was concluded that flammable gas was ignited outbye of the working places by the use of a telephone – likely being used to raise the alarm of gas, but with the area sealed off this could not be conclusively proven. The death toll includes three members of a mines rescue team, and one person killed on the surface three days later by a subsequent ignition. (HMSO, (1937)).

Headings – unventilated / degassing

Unventilated headings are a clear opportunity for a flammable atmosphere to build up, and a clear argument for continuous ventilation to be provided. If this fails and a heading becomes gas-fast, then carefully controlled degassing must take place before any potential ignition sources are introduced into the heading. Degassing has to be carefully controlled because if control is lost there is a high risk of a flammable atmosphere being discharged into the main ventilation. To maintain adequate control, a purpose made degassing device should be incorporated into the ducting outbye of the heading, whereby the flow into the heading can be controlled by shutters and flaps to expel the gas at a rate slow enough for the main ventilation stream to cope with.

Tower Colliery, 1962. 9 killed.

To enable an extension to the electrical supply in the heading, power was disconnected which also resulted in power being removed from the fan ventilating the heading. As a result, the heading became gas fast. When power was reapplied to equipment in the heading, a short circuit fault on a newly installed cable caused an ignition. From that incident, all headings within National Coal Board (NCB) mines were fitted with interlocks (known as Tower interlocks) that prevent power being applied to a heading if the fan is not running, and also to trip power to the heading if the fan stops. As a consequence, there has been no repeat of this event. (HMSO (1962)).

Cambrian Colliery, 1965. 31 killed.

This was not known as a gassy mine, but ventilation to the face was severely deranged by the opening of access holes in an air crossing. This reduced face ventilation to a minimum, whereby gas layering developed. The gas was ignited by sparking from an electrical panel which was being worked on by electricians with the panel door open and the power on. Even a relatively low gas make will be a problem if the ventilation is not sufficient to dilute and remove it. (HMSO (1965)).

Houghton Main Colliery, 1975. 5 killed.

A fan exhaust ventilating from a heading had been stopped because of a defect which caused the blades to foul the casing, causing incendive sparking. The heading became gas fast. The fan was switched back on after nine days because the reason for its being off was not appreciated. An explosive mixture was drawn through the fan and ignited. A fire developed which resulted in this section of the mine being sealed off. Following this incident, mining fan design was amended to increase the clearance between the fan blades and the casing. In addition, the manager of every NCB mine had to make auxiliary ventilation rules, and a 'defective equipment lock-off' system was established to prevent accidental or unauthorised operation of defective equipment. There has been no repeat of this event. (HSE (1976)).

Golbourne Colliery, 1979. 10 killed.

Two unconnected jobs were taking place simultaneously, one involving the degassing of a heading, and the second involving electricians working on panels just downstream of the heading. The panels were open with power on while fault-finding was taking place. Door interlocks had been defeated. These jobs were uncoordinated, and uncontrolled degassing resulted in a flammable gas plug passing over the live and open panels, which was ignited. Figure 1 shows the aftermath. This incident demonstrates the importance of co-ordinating work to avoid potential conflict, and reinforces the practice of not working on live equipment if at all avoidable. (HSE (1979)).

Cardowan Colliery, 1982. 40 serious injuries.

A flammable gas plug was blown out of a gas fast heading when disconnected ventilation duct in the heading was reconnected. The gas plug passed onto an adjacent working face, where it was ignited by the shearer picks that were cutting into floor stone at the time. From this incident, methane detectors were installed at the approach to any machine – cut working place where there was the potential for methane to be present in the intake air and arranged to trip the power to the cutting machine if 1.25% methane was detected. As a consequence, there has been no repeat of this event. (HSE (1982)).

Headings – while cutting

Bevercotes Colliery, 1989(x2), 1992. No injuries.

In 1989 a boom type roadheader was operating in the presence of a flammable mixture, known to the official on site. Gas was ignited by the cutter picks resulting in a double ignition. The second was believed ignited by a hanging flame and was powerful enough to dislodge the stonedust barriers. Agreement was reached with the NCB that air movers would be provided on machines to ventilate the cut, with power not being available to the cutting head if correct pressures and flows were not achieved. In addition, methane detection would be provided in the roof at the face of headings with an automatic trip facility to cut power to the heading machine if 1.25% was detected.

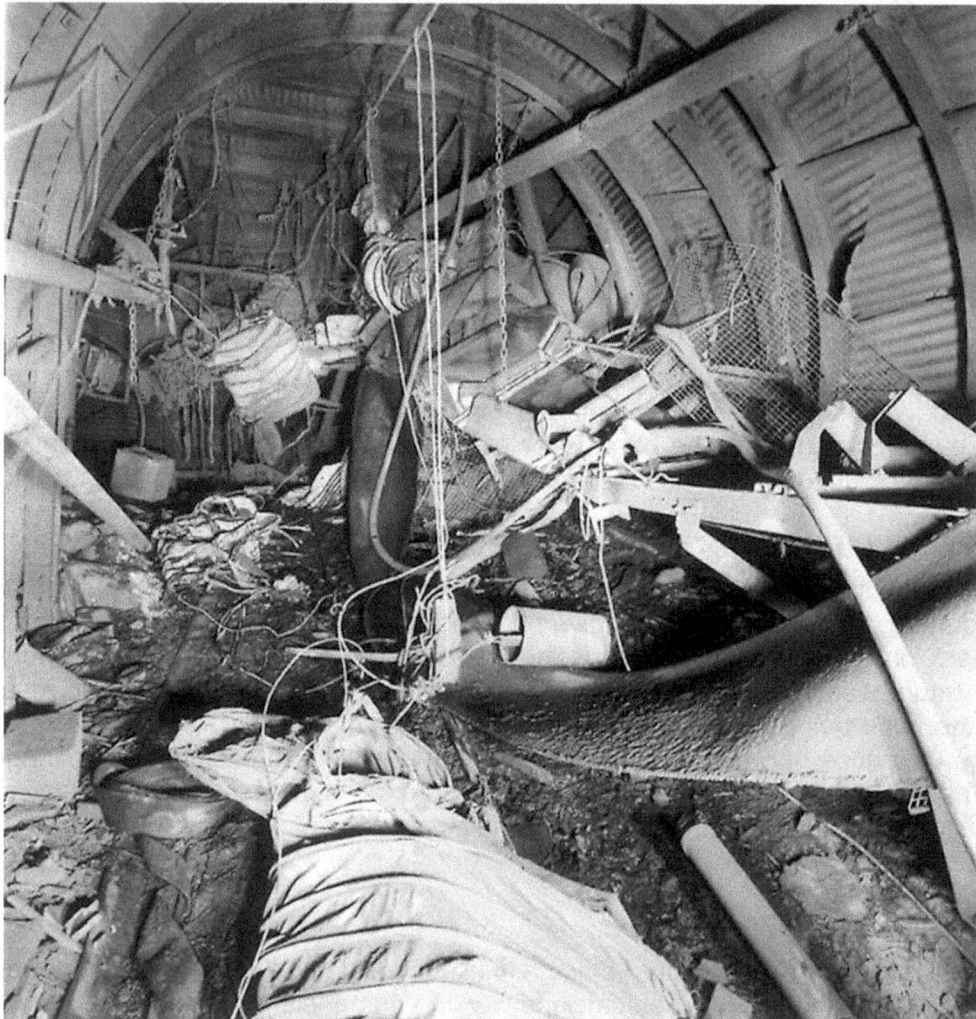

Figure 1. Golbourne colliery following the ignition.

There have been two ignitions in the UK at the face of a heading since this was implemented. The first was a very minor event, also at Bevercotes, in 1992. The small scale of the incident gave reassurance that the measures in place, while not guaranteeing no repeats, at least minimised the effects. A further event occurred at Maltby Colliery in 2012, detailed below:

Maltby Colliery, 2012. No injuries.

An ignition of firedamp occurred at the face of a coal development heading. No injuries or damage resulted, however the flame burned for the full width of the heading in front of the continuous miner cutter head. Ignition prevention sprays failed to quench the flames, which were dowsed by a wander hose after approximately two minutes.

The main contributory cause to the extent of the incident was the presence of a significant gas feeder in the face. The igniting source was a very hard quartzitic rock in the face. Other factors were that the main ventilation had been compromised by damage to the forcing duct (although the quantity delivered was still above the manager's minimum); and the cutting pattern adopted.

Despite the occurrence of this event the effects were contained to a non-violent ignition with no damage or injuries resulting. This again demonstrated that the control measures introduced were effective.

Coal Faces

Following a number of minor ignitions on the coal face on the 1960's, attempts were made to ventilate the back of the cut to dilute and remove any gas from the cut face. Initially using hollow shaft ventilators (HSV) through the centre of the drum, these were later superseded by the more efficient, multi-tube Rotary Air Curtain (RAC) drums, as shown in Figure 2.

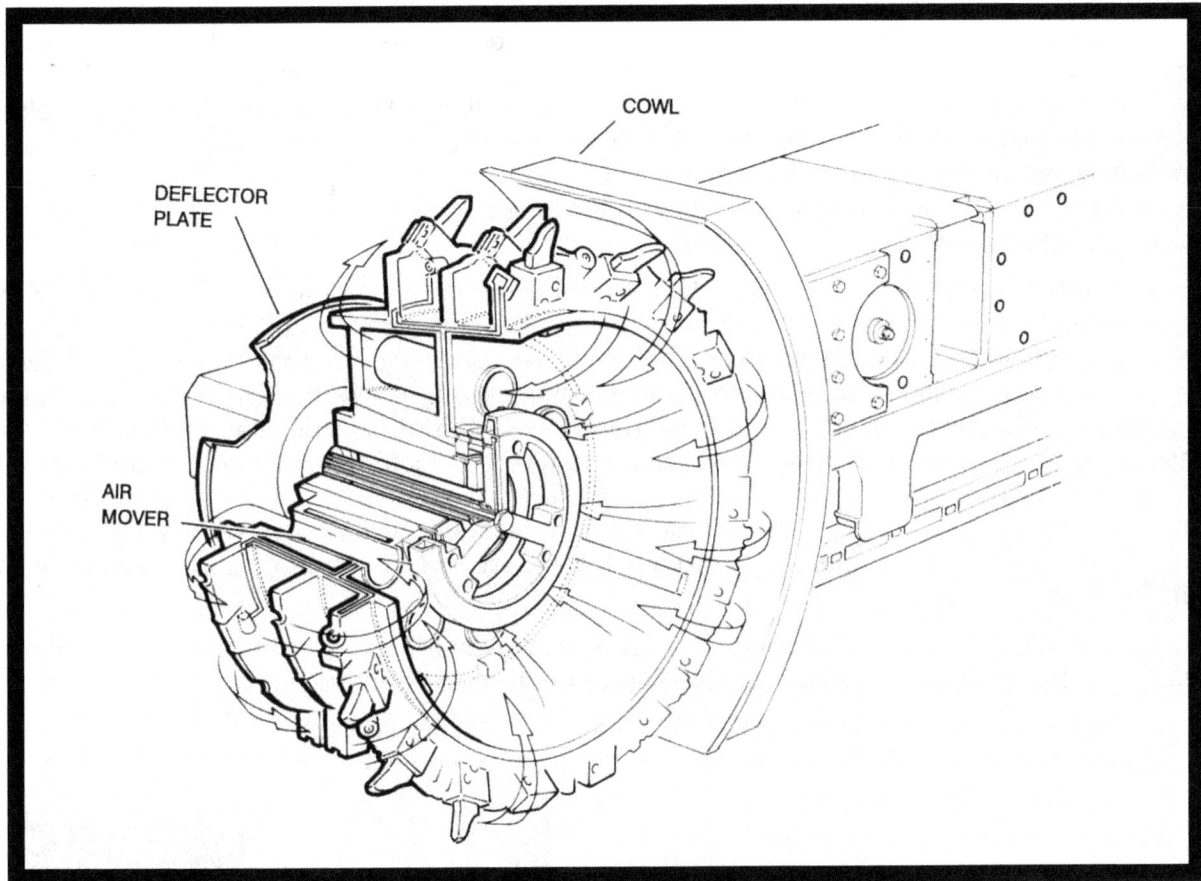

Figure 2. The RAC drum.

Ignitions at coal faces have been rare, mainly because of good faceline ventilation. It has long been recognised, however, that on retreat faces the gas fringe in the waste can encroach onto the faceline. This can occur in times of falling barometer, reduced airflow, or when the ventilation is deflected into the waste by the presence of the shearer near the tailgate end, so flushing out higher concentrations.

A particular problem with retreat mining is that most of the methane emitted from the waste is emitted at the tailgate end of the face, where high concentrations can be seen. To provide a margin of safety to the lower explosive limit at 5%, UK mining law requires that significant potential ignition sources such as electricity and diesel engines are turned off, and the use of explosives discontinued at 1.25%; and that personnel are withdrawn at 2%.

Methane is evolved at very high (near 100%) purity and is diluted when mixed with air. An area known as the 'fringe' is present. This is the region where the methane concentration is unacceptably high, usually taken to be 2% or above. A problem with the fringe encroaching into the working area at the tailgate was identified soon after retreat mining evolved. A solution was adopted to force the fringe back in to the waste to ensure that the fringe, and more importantly the area where an explosive concentration is present, is maintained away from the working area on the face and away from potential ignition sources.

This movement of the fringe is accomplished by forcing most of the ventilating air coming along the face into the waste at the tailgate end of the face. In the early days of retreat mining, this was achieved by leaving a pillar of coal next to the tailgate road; with the pillar being breached at intervals some way behind the face so that the ventilating air had to turn back inbye before coming around the pillar and into the tailgate. Thus the mixing and dilution of the high purity methane took place behind the face. This system was known as the 'back return' system.

In modern retreat longwall districts, the leaving of a pillar of valuable coal, and the additional manpower required to drive the snickets through the pillars and to support the back return have been all been eliminated by the development of the Sherwood curtain system.

A Sherwood curtain is a preformed partition along the tailgate from a point some distance outbye of the face, typically 30 – 50m, extending into the waste behind the face. The partition is usually formed by nailing lengths of air-proof brattice cloth to wooden supports set down the centre of the roadway. The function of the curtain is to divert a major portion of the ventilation flow coming along the coal face into the waste at the tailgate end, so flushing and diluting any gas from the area immediately behind the powered supports at that end of the face. Airflow is so encouraged by forming an airlock on one side of the curtain with a number of doors to allow access. The system is depicted in Figure 3 below.

The curtain does not alter the general body gas level in the tailgate – that is only a function of the gas make, and the amount of clean air being supplied for its dilution. The effects of the curtain are local to the tailgate end, and solely to keep the explosive gas fringe back into the waste and away from any potential ignition sources at the tail end of the face. Typically, the 'clean' or doors side of the curtain would contain perhaps 0.4% methane, while the 'dirty' side may contain over 1%. These flows would combine outbye of the curtain to give a tailgate general body level of perhaps 0.8% methane. The lack of a back return, or significant damage to one, would result in the ventilation taking the shortest and easiest route off the face, effectively hugging the coal side. This would allow the gas in the waste to move forward, and over potential ignition sources in the vicinity of the faceline.

One face ignition occurred at Kellingley Colliery in November 2010, which while causing no injuries, resulted in the mine being evacuated and 218 men safely accounted for.

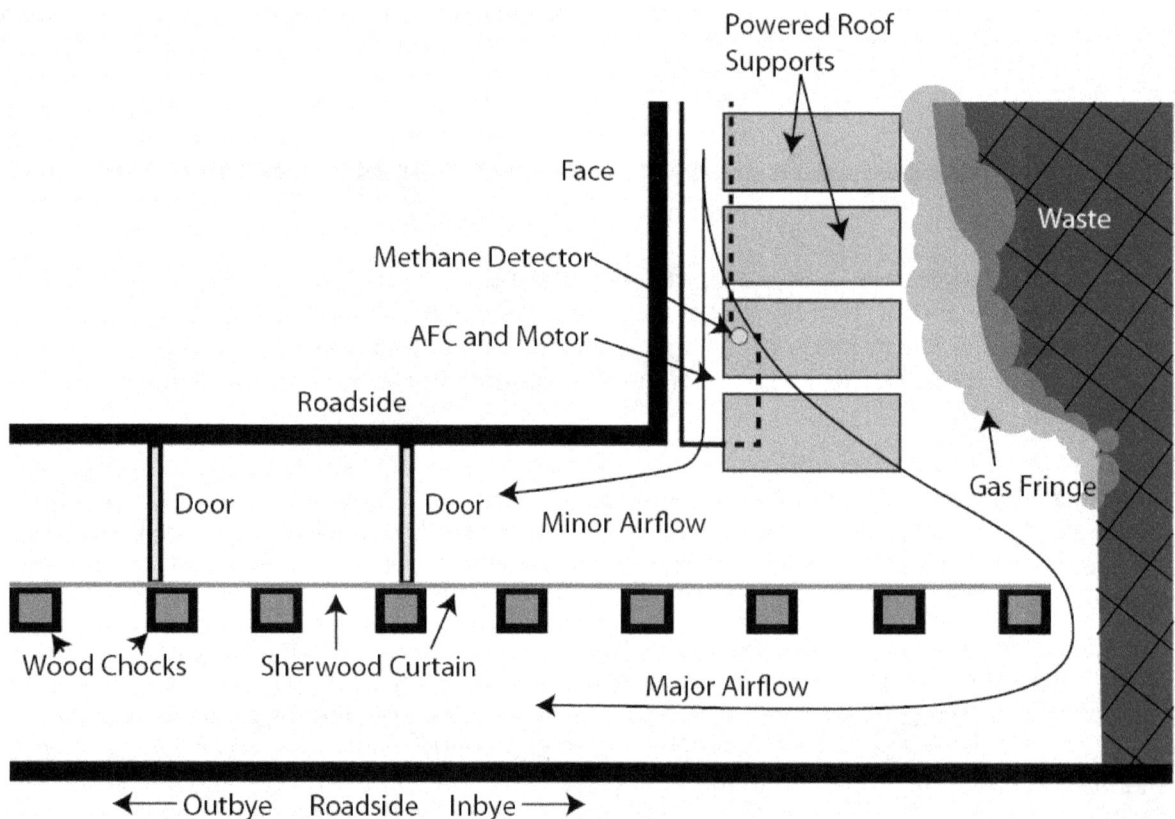

Figure 3. Schematic representation of a Sherwood Curtain.

Kellingley Colliery, 2010. No injuries.

At a retreat face, the shearer had completed cutting into the tailgate end and was some 40m down the face cutting to the maingate. Gas fringe control at the tailgate was by Sherwood curtain. A section had been deliberately cut out to allow timber to be passed through, resulting in it becoming totally ineffective. The gas fringe came forward and was ignited by a fall of sandstone laminations forming the roof at the tailgate, just behind the powered supports. It was determined that the workforce and officials had no appreciation of the reason for the Sherwood curtain, nor how critical its maintenance was. There was little in the way of management oversight.

This is an example of where a lesson had been learned, and very effective control measures introduced, only to have been forgotten with the passage of time. It may be that the lack of direct experience of ignitions had allowed some complacency to set in, throughout the management structure. The ignition was quite violent and extensive, destroying part of the curtain and resulting in material being burned / melted a few metres down the face and 10s of metres outbye in the tailgate. It was fortunate that no-one was close to the seat of the ignition, otherwise very serious injuries would have resulted.

COAL DUST EXPLOSIONS

By the late 1800s, research had identified that coal dust was explosive when mixed into a cloud in air at a concentration of about 600 mg/m^3. This is a concentration that does not normally occur, so to reach this concentration standing dust has to be raised into the air. The dust must then be ignited. Coal dust is not easy to ignite and thus, apart from a few instances where friction, electricity or the use of explosives have been thought to be the initiators, all coal dust explosions have been initiated from methane ignitions. These have provided the energy both to raise sufficient dust then to ignite it. Once initiated, a coal dust explosion is self-propagating and can travel right through the mine so long as there is a supply of combustible dust that can be raised into the air.

Various methods have been adopted to prevent a cloud being raised, including consolidation. The roadways could also be wetted, but this control measure is unsatisfactory because excessive water causes problems, from floor heave to difficulty walking, and may also have to be pumped. Water also evaporates and the treatment must be repeated at frequent intervals for it to remain effective.

A far more effective and efficient method of explosion prevention was to ensure that any cloud of dust lifted was not ignitable. Studies determined that coal dust of different ranks could be rendered safe if mixed with particular amounts of non-flammable stonedust. This is shown in table 2.

Table 2: Minimum percentage of incombustible matter.

Coal volatile matter (%)	Min.%-age incombustible matter required
20	50
22	55
25	60
27	65
30	68
32	70
35	72
>35	75

The widespread adoption of these measures almost eliminated the large coal dust explosions in the UK within the first couple of decades of the 20th century.

One area where the above control measures are not appropriate and cannot be employed is where coal is being transported by belt conveyor. The raw product needs to be kept clean, so stone dust cannot be added. Instead, barriers to the extension of flame must be erected at strategic points in the mine to suppress any explosion on the conveyors. These were traditionally wooden boards covered in stonedust and arranged so they would readily tip if subjected to an explosion blast wave, so filling the road with inert dust before the flame front arrived. These are highly effective but require some maintenance resource. A modern development is the stonedust bag barrier which once installed is virtually maintenance free.

Water trough barriers have been used but require a very high maintenance commitment, and are not recommended in the UK.

1960, Six Bells Colliery. 45 killed including the u/mgr, and 3 deputies.

An accumulation of flammable gas was allowed to develop at the maingate ripping of an advancing face, because of derangement of the district ventilation by leaving ventilation doors open. Ignition was caused by the frictional heat produced by the impact of quartzitic rock falling from the roof exposed by shotfiring, onto a steel girder forming part of a conveyor canopy. The ignition developed into a coal dust explosion, which was allowed to propagate because of the lack of stonedust barriers. These did not become a legal requirement in the UK until October 1960; the explosion occurring in the June of that year. Nevertheless, such barriers had been required by company (NCB) instruction since 1953. (HMSO (1961)).

This is an example of lessons learned but not put into practice.

1992, Westray, Canada. All 26 underground killed.

McPherson (2001) described how this was a gassy mine, where again accumulations were allowed to develop through poor ventilation control. In addition, there was little in the way of dust suppression, removal, or inertisation through the application of stonedust. There were no stonedust barriers. A layer of firedamp from a leaking stopping was ignited by continuous miner cutter picks, and there was nothing to stop the resulting coal dust explosion which killed everybody underground.

Again, widely known lessons were not put into practice.

2010, Upper Big Branch, USA. 29 killed

Page, *et al*, (2010) described the incident in the MSHA report. Poor ventilation, exacerbated by a roof fall that restricted airflow, allowed the gas make to build up on the face and not be diluted. Worn picks and defective anti-ignition water sprays caused an ignition to be triggered at the shearer. A failure to maintain adequate levels of incombustible matter in the roads allowed the ignition to propagate into a coal dust explosion which travelled more than two miles, killing 29 and injuring two. Evidence suggests a concerted effort by management not to comply with the law.

FIRES

No fire underground can be thought of as trivial, as all large fires start off as small fires. The Great Fire of London in 1666 started as an insignificant fire in a bakery, but developed into a conflagration that destroyed 70,000 homes. Historically in the UK, belt conveyors have been the major cause of underground fires, although since the advent of fire resistant conveyor belting, fires have been limited in extent and have not resulted in any deaths. This is not the case in some other countries.

I have included details of the Michael Colliery fire here, as despite it not being on a conveyor or a machine, it was again a significant event that really brought in the widespread use of self-rescuers.

Michael Colliery, 1967. 9 killed.

A fire developed from spontaneous combustion, but ignited polyurethane foam previously sprayed onto the roadway sides as a seal. This material gave off thick smoke and toxic fumes. The incident led to the banning by the NCB of the use of polyurethane foam and precipitated the adoption throughout the NCB mines of filter self-rescuers. As a consequence, there has been no repeat of this event, and the use of self-rescuers has allowed many men to safely leave mine workings in conditions where before they may not have. (HMSO (1968)).

Conveyors

Creswell Colliery 1950. 80 killed.

A fire developed in an outbye conveyor belt transfer point chute, by frictional heating on a strip of belting torn from the edge (approx. 125 mm x 79 m). The initial fuel was the rubber and cotton conveyor belt, but the fire quickly transferred to the timber roadway supports. Problems with the water supply severely hampered fire-fighting operations. All fatalities were from CO_2 poisoning at locations remote from the actual fire site, and the last six bodies were not recovered for 11 months. From this accident, fire-proof construction of roadways at belt drives and transfer points was recommended, along with amongst other things, fire resistant (FR) conveyor belting being developed and introduced. As a consequence, there has been no repeat of this event. In addition, work started to explore self-rescuer provision. (HMSO (1952)).

I cannot emphasise the dangers of rubber belting enough – Figure 4 is a photograph of a rubber belt on fire from end to end – in this case on a surface gantry but it takes no imagination to see what effect this fire would have in the confines of an underground roadway, and how escape would be compromised.

Figure 4. Fire on a rubber conveyor belt.

By way of contrast, Figure 5 shows the damage to a section of FR conveyor belt that was subject to an intense, oxygen-rich fire from an oxygen self-rescuer that had been ignited when it passed through a crusher onto the conveyor.

Palabora Copper mine, RSA. 2018, 6 killed, 48 injured.

This is almost a repeat of the Creswell colliery fire. A rubber belt was ignited either by hot working or belt slip. It burned for about three hours, consuming about 1400m of belt. Again inadequate water supplies hampered firefighting efforts. Criticism was levelled at the inadequate state of the refuge chamber, in which three of the deceased were found, and its air supply, but the root cause here was clearly the use of non-FR belt. This is a good case of a lesson long known, not being heeded. Von Glehn (February 2019, personal communication).

Figure 5. Damage to FR belt after an O_2 self-rescuer fire.

Mobile Plant

The second major cause of fires in the UK has been mobile plant, particularly but not exclusively those powered by diesel engines. With the use of very large plant this is of significant concern as the fire loading on a large loading shovel or dump truck, including the tyres, is huge. Consequently, a fire that gets out of hand could burn for several hours. Although the fire itself does not pose a major risk as there is limited opportunity for the fire to spread (not the case in a coal mine, of course), the products of combustion can create an irrespirable and toxic atmosphere which the ventilation can carry quickly through the mine, filling the ventilation network downstream with toxic gases and an oxygen deficiency. The lesson here is 'don't let a fire start,' but if it does, put it out while you still can, and make sure it stays out.

In the UK we have developed a 'no leaks' policy whereby any leak of fuel or hydraulic oil is immediately rectified. Machines are switched off immediately and not used until repaired. This has been tightened up recently, following two incidents where fires have occurred on vehicles being driven back to the workshops for repair.

Tara mine, Republic of Ireland, 2014. 1 injured from smoke inhalation.

A fire developed on a shotcrete rig when tramp metal in a roadway became wrapped around the driveshaft, causing the rupture of hydraulic and fuel hoses. Fluids were sprayed onto hot surfaces, resulting in a fire that the on-board suppression system and the manual actions of the operator failed to extinguish. The whole of the vehicle was engulfed, including the tyres. (Figure 6). The fire was reported at 1135, stench gas was released into the mine at 1137 to warn the workforce. By 1210 all personnel were accounted for in refuge chambers apart from the machine operator who was later found by the mine's rescue team at 1407, treated and brought to the surface. The workforce stayed in the refuge chambers until the fire had been extinguished / burned itself out and were returned to surface by 2103. When the vehicle was subsequently salvaged the oil and diesel tanks were empty and aluminium components had melted indicating that the fire had exceeded a temperature of 660°C. Holmes (February 2019, personal communication)

Figure 6. Tara shotcrete rig after recovery to surface.

It is vital that plant with a large fire load, including small personnel carriers, are equipped with a fire detection and suppression system that not only extinguishes the initial fire but then quenches the area to cool it and prevent re-ignition – which then cannot be suppressed because the system has already been discharged. This incident also highlights the need for good housekeeping standards and to keep roadways free from scrap. It also demonstrates the benefit of properly designed, sited and maintained refuge chambers.

CONCLUSIONS

There are a number of lessons apparent from these short case studies. One is that all these incidents are entirely preventable, but if we ignore the lessons (the learned control measures) then they will continue to repeat, as the Palabora belt fire demonstrates. Great store is often put on mitigation after an event – stonedust barriers, even the explosion-proofing of fan installations has been advocated. The author's message is prevention is better than cure. Yes, mitigation is important, but far better to plan to avoid the incident in the first place.

These are the main lessons: clearly there is a lot of detail behind them but their adoption should minimise the effects of any future incidents:

1. Establish and constantly maintain adequate ventilation throughout the mine, but particularly in working places where firedamp is likely to be emitted. If you don't have a flammable mixture, you can't light it.

2. Do not allow places to become gas fast, but if you do, have a safe method of work established and the correct equipment pre-installed, to affect a safe and controlled recovery.

3. This degassing method of work must include the isolation of potential ignition sources downstream of degassing operations.

4. Ensure adequate ventilation of the cut.

5. Monitor gas levels and remove power by tripping when necessary: take the decision away from the workforce.

6. The primary preventative measure for coal dust explosions is not to have a firedamp ignition.

7. Ensure that appropriate levels of incombustible matter are maintained in all roadways by ensuring that sufficient stonedust is spread.

8. Maintain adequate explosion barriers at appropriate positions.

9. Use only fire resistant conveyor belt.

10. Properly maintain diesel vehicles, including shrouding hot parts. Shroud oil and fuel lines to prevent sprays in cases of bursts, and have a 'no leaks' policy.

11. Ensure plant with a fire loading is equipped with a 'extinguish and quench' fire suppression system.

REFERENCES

Health and Safety Executive, 1976. Report on Explosion at Houghton Main Colliery Yorkshire, June 1975. HSE: London. ISBN 0-11-880328 X.

Health and Safety Executive, 1979. The explosion at Golborne Colliery, Greater Manchester County, 18 March 1979. HSE: London. ISBN 0-11-883288-3.

Health and Safety Executive, 1982. The Explosion at Cardowan Colliery, Stepps, Strathclyde Region, 27 January 1982., HSE: London. ISBN 0-11-883644-7.

Her Majesty's Stationary Office, 1937. Explosion at Gresford Colliery, Denbighshire. HMSO: London. Cmnd 5358.

Her Majesty's Stationary Office, 1952. Accident at Creswell Colliery, Derbyshire. HMSO: London. Cmnd. 8574.

Her Majesty's Stationary Office, 1961. Explosion at Six Bells Colliery, Monmouthshire. HMSO: London. Cmnd. 1272.

Her Majesty's Stationary Office, 1962. Explosion at Tower Colliery, Glamorgan. HMSO: London. Cmnd. 1850.

Her Majesty's Stationary Office, 1965. Explosion at Cambrian Colliery, Glamorgan. HMSO: London. Cmnd. 2813.

Her Majesty's Stationary Office, 1968. Fire at Michael Colliery, Fife. HMSO: London. Cmnd. 3657

Leeming, 2018. Avoiding Catastrophe – the Importance of Ventilation, paper presented to 11th International Mine Ventilation Congress, Xia'an, China, 16-18 September.

McPherson, MJ. 2001. The Westray Mine Explosion, in Proceedings of 7th International Mine Ventilation Congress (ed: S Wasilewski), pp 557-564. (Research and Development Center for Electrical Engineering and Automation in Mining: Crakow).

National Coal Board, 1979. Ventilation in Coal Mines: A Handbook for Colliery Ventilation Engineers, 159 p. (National Coal Board: London).

Page, N G, et al. Mine Safety and Health Administration. Report of Investigation, Fatal Underground Mine Explosion April 5, 2010. US Department of Labor. Available from <https://arlweb.msha.gov/Fatals/2010/UBB/PerformanceCoalUBB.asp.>. [Accessed 27 February 2019].

The reasonable location of high drainage roadway to control gas exceeding and spontaneous combustion in goaf

S Li[1], P Xu[2], Y Ding[3] and H Lin[4]

1. Ph.D., Professor, College of Safety Science and Engineering, Xi'an University of science and Technology, Xi'an 710054, China. E-mail: lisg@xust.edu.cn
2. Ph.D. College of Safety Science and Engineering, Xi'an University of science and Technology, Xi'an 710054, China. E-mail: xupeiyun1994@outlook.com
3. Ph.D. College of Safety Science and Engineering, Xi'an University of science and Technology, Xi'an 710054, China. E-mail : dingyang@xust.edu.cn
4. Ph.D., Professor. College of Safety and Engineering, Xi'an University of Science and Technology, Xi'an 710054, P.R.China. E-mail: lhaifei@163.com

ABSTRACT

In order to study the problem of the layout of the high drainage roadway in the fully mechanized caving face of high gassy and easy spontaneous combustion seam to achieve the synergistic prevention of gas and coal spontaneous combustion, this paper will firstly use the physical similar simulation method to determine the distribution of overburden fracture, and the factors influencing the gas drainage effect and air leakage of coal mining face are obtained by analyzing the parameters of high drainage roadway. Based on the above influencing factors, the numerical simulation method is adopted to obtain the layout level of the high drainage roadway and the reasonable range of the suction negative pressure. Research indicates that the vertical horizon location of high drainage roadway and the negative suction pressure are the main controlling factors affecting the gas drainage effect and the spontaneous combustion of coal in the goaf. Under the action of suction negative pressure, the arrangement of the high drainage roadway at a lower position will cause a large number of fissure air leakage passages in the rock mass near the pumping section, resulting in the expansion of the oxidization and heat accumulation zone of the coal body in the goaf. The higher layer position can not solve the problem of gas accumulation in the upper corner. When the high drainage roadway is arranged at a cutting height of 2.8~3.2 times from the roof of the coal seam and 0.4~0.5 times the width of the crack belt of the return airway, the gas concentration of the upper corner together with the oxidization and heat accumulation zone of the coal body in the goaf can be controlled within the safe range, the safe and efficient recovery of the working face provides effective protection.

Key words: U+I-shape ventilation, high drainage roadway, coal spontaneous combustion, Synergistic prevention and control

INTRODUCTION

As the main part of primary energy in China, coal has accounted for more than 60% in primary energy for a long time. With the continuous progress of high-intensity and intensive mining of coal resources and mining technology, the coal resources that are easy to be mined in shallow burial have been nearly exhausted, and most of the well mining mines have entered the deep mining stage one after another [1,2]. With the increase in mining depth, a series of new problems, such as ground stress increases, coal broken degree increase, coal seam gas content and gas pressure, gas emission and poured out of strength increasing, the higher temperature, the gas and coal spontaneous combustion is highlighted increasingly symbiotic disasters, for coal mine safety and efficient production brings huge hidden trouble. According to incomplete statistics, in state-owned key mines, high-gas mines account for more than 48%, and mines with the tendency of spontaneous combustion account for more than 56%, among which the mine with the tendency of spontaneous combustion with high-gas accounts for 32.2%.

In the actual production process of high gas spontaneous combustion mine, gas and coal spontaneous combustion are two kinds of disasters that interact. Many studies show that the coupling of fracture field, goaf gas migration field and temperature field is a necessary and sufficient condition for spontaneous combustion of gas and coal. The decrease of oxygen concentration and the increase of gas concentration can reduce the risk of spontaneous combustion of coal to some extent. When CO and gas are mixed after coal spontaneous combustion, the upper and lower limits of gas explosion concentration will be greatly increased [3-5]. At the same time, the spontaneous combustion of coal will intensify the air leakage in the goaf and increase the intensity of gas emission In the process of gas and coal spontaneous combustion disaster prevention and control, gas control demands to reduce the working face advancing speed, increased the strength of extraction, etc., but the decrease in working face advancing speed provides conditions for the regenerative heating of residual coal in goaf. The increment of the strength of extraction can greatly increase the air leakage intensity of gob, which provides the oxygen environment for coal spontaneous combustion. Therefore, gas control and coal spontaneous combustion prevention and control in coal mining face are contradictory [6,7]. To find the balance between gas control and coal spontaneous combustion control is the key point of gas and coal spontaneous combustion combined disaster control. Taking the 401102 working face of Hujiahe coal mine in the Binchang coal mine in Shaanxi province as the entry point, this paper will discusses the influence of the layout of high drainage roadway and the setting of various extraction parameters on the combined disaster of gas and coal spontaneous combustion, which can provide a basis for the layout of gas extraction system in this type of working face.

GAS AND COAL SPONTANEOUS COMBUSTION COUPLING DISASTER MECHANISM

Characteristics of spontaneous combustion of gas and coal

The spontaneous combustion of residual coal in goaf is closely related to oxygen concentration distribution and air leakage intensity in that area. In the process of gas extraction using high drainage roadway, air leakage intensity and high oxygen concentration area in goaf will increase[8-10]. Therefore, there are problems that the gas oxidation caused by the gas is accelerated, the risk of spontaneous combustion of residual coal is reduced, and the gas extraction requirement cannot be met, which make the imbalance of two disasters easy to occur. The coupling disaster relationship between gas and coal spontaneous combustion is shown in figure 1.

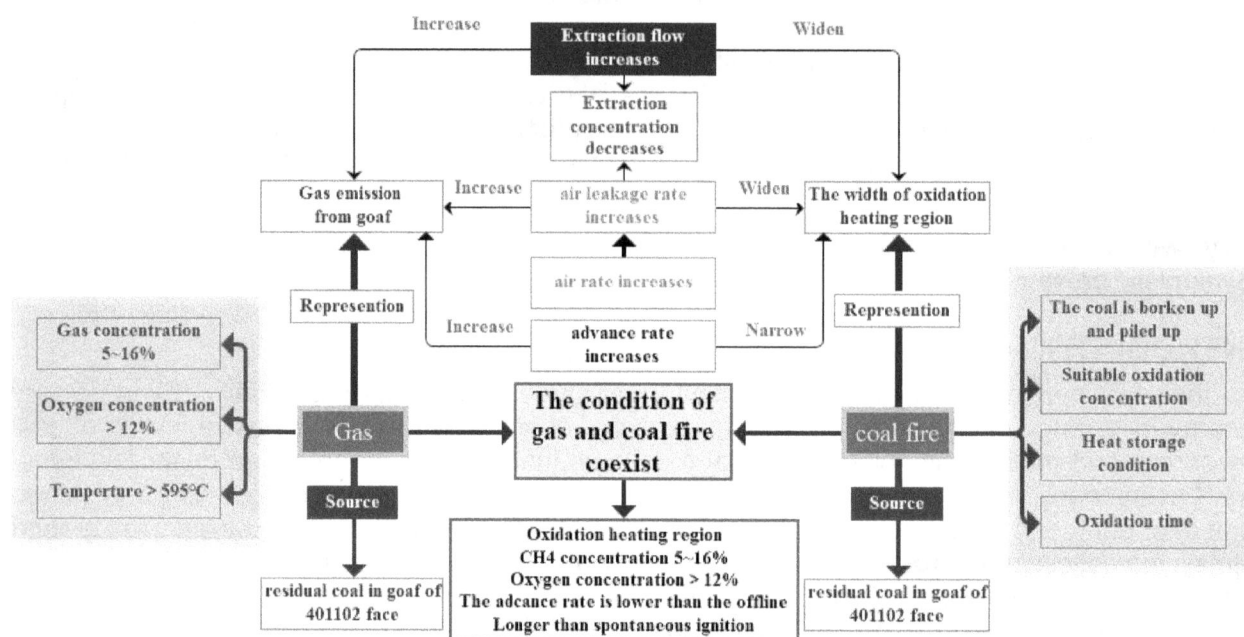

Fig 1 The relationship between gas and coal spontaneous combustion in the goaf

From the conditions of the two disasters, the occurrence of gas disasters requires the joint action of gas, oxygen and ignition source, among which gas and oxygen exist objectively in the goaf, while the ignition source in the goaf mainly comes from coal spontaneous combustion. When the area corresponding to the gas explosion coincides with the oxidative heating zone, and the continuous propulsive degree of the working face is less than the lower limit and the time is longer than the spontaneous ignition period, spontaneous combustion disasters of gas and coal may occur in the goaf area of the test working face[11].

The coupling mechanism of gas and coal spontaneous combustion

With the development of coal mining activities, a large number of secondary fractures are generated . Complex fracture network structure provides space for disaster occurrence. Oxygen and gas flow fields and temperature fields exist in the fracture field. The flow field and temperature field are coupled within the range of fracture field. In the intersection area of fracture field, flow field and temperature field, the coupling conditions of natural combustion of gas and coal are as follows[1,12]:

$$S_e(\tau,x,y,z)=S_{O_2}(\tau,x,y,z)\cap S_T(\tau,x,y,z)\cap S_l(\tau,x,y,z)\cap S_{CH_4}(\tau,x,y,z) \qquad (1)$$

Where, S_e is the coexisting area of spontaneous combustion disasters of gas and coal; S_l is the crack field where the disaster occurs; S_{O2} is to meet the oxygen concentration area of disasters; S_{CH4} is the concentration area of combustible or explosive gas. S_T is the temperature region where gas is combustible or explosive.

Conditions to be met when a gas disaster is triggered by spontaneous combustion in a fracture field

$$\begin{cases} S_{l_c} < S_l(x,y,z) < S_{l_d} \\ S_e \Big|_{(C_{O_2}>C^*_{O_2},T>T_s,C_1<C_{CH_4}<C_2)} > 0 \end{cases} \qquad (2)$$

Where: l_c, l_d are the upper and lower limits corresponding to the gas disaster caused by spontaneous combustion of coal in the fracture field; C'_{O2} is the lower limit of gas concentration of gas combustion (explosion); T_s is the critical temperature of ignition; C1 and C2 are the upper and lower limits of gas concentration at the time of disaster occurrence..

According to the conservation of energy, when the heat generated by the oxidation of the floating coal in the goaf is greater than the heat lost (including the loss of the top and bottom plates and the heat taken away by the wind), the coal can continue to heat up, eventually causing the spontaneous combustion of the floating coal, which can be obtained by formula (3):

$$div[\lambda_e grad(T_m)]+q_0(T)-div(n\rho_g c_g \bar{U}T_m)\geq 0 \qquad (3)$$

Where: λ_e is the thermal conductivity of the floating coal,(kJ·s^{-1}·m^{-1}); $q_0(T)$ is the experimental determination of the coal heat release intensity, kJ·m^{-3}·s^{-1}; ρ_g represents the wind density of the working face , kg·m^{-3}; c_g is air heat capacity (J·g^{-1}·K^{-1}); $\bar{U}$ is the air leakage speed in the goaf, m/s, T_m is the highest temperature in the coal body, °C.

Therefore, the self-ignition of the floating coal requires sufficient floating coal thickness to facilitate the accumulation of heat generated by the oxidation temperature, and at the same time, sufficient oxygen required to ensure the temperature rise of the oxidation is required. Although the air leakage in the goaf can take away a certain amount of heat, it also provides an oxygen environment for coal spontaneous combustion. When all the above conditions are met, sufficient time is needed [13].

In the high gassy self-igniting mine, the mining method of fully mechanized top coal caving is generally adopted for thick coal seams, which also causes more floating coal in the goaf. At the same time, the gas drainage in the goaf is carried out by using the high drainage roadway. Although the area where gas disasters can occur in the fracture field is reduced, the risk of spontaneous combustion of the floating coal in the goaf is also increased. By adjusting the arrangement position of the high drainage roadway and the key parameters such as the suction negative pressure, the balance between the gas and coal spontaneous combustion combined disaster prevention and control can be realized.

ENGINEERING PRACTICE BACKGROUND

Binchang Hujiahe Coal Mine mainly collects 401102 working face, the working face design length is 1643 m, the recoverable length is 1493 m, the inclined length is 180 m, and the area is 268740m^2. The geological reserves of the working face are 9.694 million tons, the recoverable reserves are 5.453 million tons, the gas reserves are 45.367 million m^3, the working surface predicts the gas emission amount of 4 3m³/min, the design wind discharge gas volume is 8 m³/min, and the rest of the gas gush will be solved by the extraction system. The gas control of the working face is mainly composed of pre-drainage pre-extraction, pre-mining drainage, high drainage roadway extraction and upper-level corner pipe extraction. It has been identified that the spontaneous combustion tendency of the 4# coal seam is Class I, which belongs to the easy spontaneous combustion coal seam. The shortest time for spontaneous combustion on site is 20 days.

Therefore, during the recovery of the working face, the danger of gas and coal spontaneous combustion combined disaster is a key factor that restricts the safe and efficient production of the working face.

HIGH DRAINAGE ROADWAY ROADWAY LAYOUT HORIZON PREDICTION

4.1 Principle of position selection for high drainage roadway layout

Based on the coal mining technology of the test working face, the floated coal in the goaf mainly consists of the bottom coal with lower layers (solid coal without mining) and the top coal without effective mining, accounting for about 40% of the thickness of this coal seam, and the minimum period of natural combustion of floated coal is only 20 days. Therefore, there is a large amount of residual coal and gas emission in the goaf and a high risk of spontaneous combustion of residual coal in the goaf. For the layout of high drainage roadway, comprehensive consideration should be given to relieved pressure gas extraction and prevention of spontaneous combustion of residual coal in the goaf.

Affected by mining, the coal and rock mass in front of the working face has experienced multiple stress concentration and pressure relief, resulting in the development of fractures in the coal and rock mass and connection with high drainage roadway. In the process of advancing the working face, the rock layer where the high pumping roadway is located fails to collapse in time, which will make the high drainage roadway extend to the goaf. In addition, the negative pressure of pumping will cause the formation of air leakage channel and power at the goaf and pumping section. Therefore, the position of the high drainage roadway and its extraction parameters are the key to solving the problem. Generally, the layout of high drainage roadway is in the "O" shaped ring at the side of return air lane on the working face [14]. At this time, the width of "O" shaped ring determines the horizontal distance between high drainage roadway and return airway. Under the influence of rising and floating, the discharged gas migrated and accumulated along the fracture zone[15,16].When the horizontal distance remains unchanged, the concentration of drainage and pressure relief gas in the high drainage roadway will increase with the increase of the vertical distance (the high drainage roadway is arranged within the mining-induced fracture zone). However, in the actual production process, the layout of high drainage roadway is too high, which has little impact on solving the working face and upper corner gas concentration. Therefore, it is very important to find the reasonable layout of high drainage roadway for gas extraction effect.

From the perspective of preventing the spontaneous combustion of residual coal in goaf, the layout level of high drainage roadway is relatively low (in the caving zone), and the surrounding rock fractures near high drainage roadway are highly developed, forming a large number of air leakage channels. This will lead to a substantial increase in the air leakage intensity in goaf, thereby increasing the risk of spontaneous combustion of residual coal in goaf. Therefore, the arrangement of high drainage roadway at the junction of caving zone and fissure zone can reduce the influence of high negative pressure extraction of high drainage roadway on air leakage in goaf to some extent. In summary, under the condition of high gas spontaneous combustion coal seam, high drainage roadway should be arranged in the "O" shape circle of return air roadway above the regular caving zone to extract the relieved pressure gas in goaf.

According to the theory of pressure relief gas extraction and the characteristics of upper corner gas easy to exceed the limit, the roadway for high gas extraction is suitable to be arranged near the return wind side, and the approximate positions of the vertical distance and horizontal distance of the high drainage roadway can be expressed as follows[17]:

$$h = h_1 \cos \beta + h_B \qquad (4)$$

$$S = h\cos(\alpha - \beta)/\sin\alpha + S_p \qquad (5)$$

Where, h is the vertical distance from the high drainage roadway to the roof of coal seam ; h_1 is the height of the rule fall-zone , m; β is the dip Angle of coal seam, °; ;h_B is the safe height to prevent high drainage roadway from damaging safety, m; S is the horizontal distance between the high drainage roadway and the return air roadway , m; α is the mining pressure relief Angle, S_p is the horizontal distance between the high drainage roadway and the boundary of the mining-induced fracture zone, which is about 0.4~0.6 times the width of the return wind side fracture zone. The above parameters can be obtained by simulating the physical similarity of the working face.

Physical simulation experiment of mining-induced fracture in overburden rock on working face 401102

The experiment design

The experiment is divided into two stages: model building and model mining. Stress sensors are laid at the bottom of the test bed in advance, and the model is laid according to the requirements of similarity ratio. After the model is dried naturally, corresponding loads are loaded on the top of the model according to the working face depth, and displacement measuring points are arranged on the model surface.

Physical analog simulation test bed size is 3000 mm * 200 mm * 1500 mm (length x width x height), the model for slicing, the upper layer is 13.5 cm, the bottom layer is 10 cm. 30 cm boundary pillars are set at both ends of the model. The length of the simulated working face is 240m, the average mining depth of the working face is 650m, and the simulated height of the model is 120cm. The rest overlying strata are loaded in accordance with gravity compensation load. The layout and installation of model sensors are shown in FIG. 2~3.

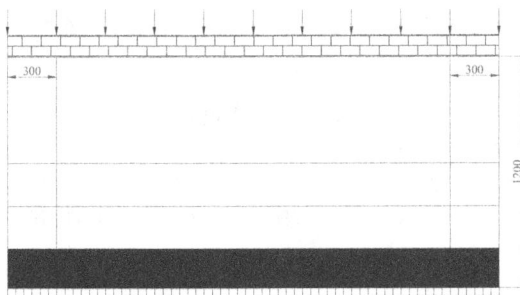

FIG. 2 Schematic diagram of simulated layout

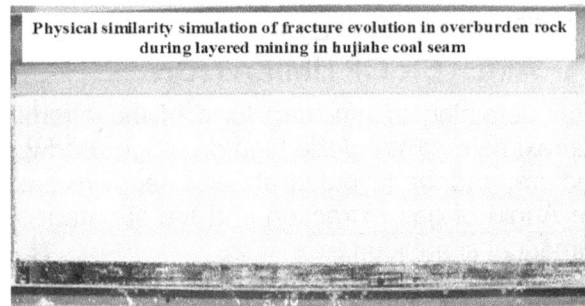

FIG. 3 Plane model

The experimental results

After the completion of the working face, the fracture line of the rock formation forms a certain Angle with the coal seam, which is the mining pressure relief Angle. According to the experimental results, the mining pressure relief Angle at the working face is 61°, the width of the fracture area is about 25m, the mining pressure relief Angle at the open-off cut is 52°, and the width of the fracture area is about 23m. At the same time, after mining, the lower rock strata are in the irregular accumulation state, which is irregular caving zone; while the upper rock strata are in a regular accumulation state due to the small caving space, which is regular caving zone. According to the experimental results, the irregular caving zone is 44.5m, and the height of the regular caving zone is 98m. The final form of overburden movement is shown in FIG. 4.

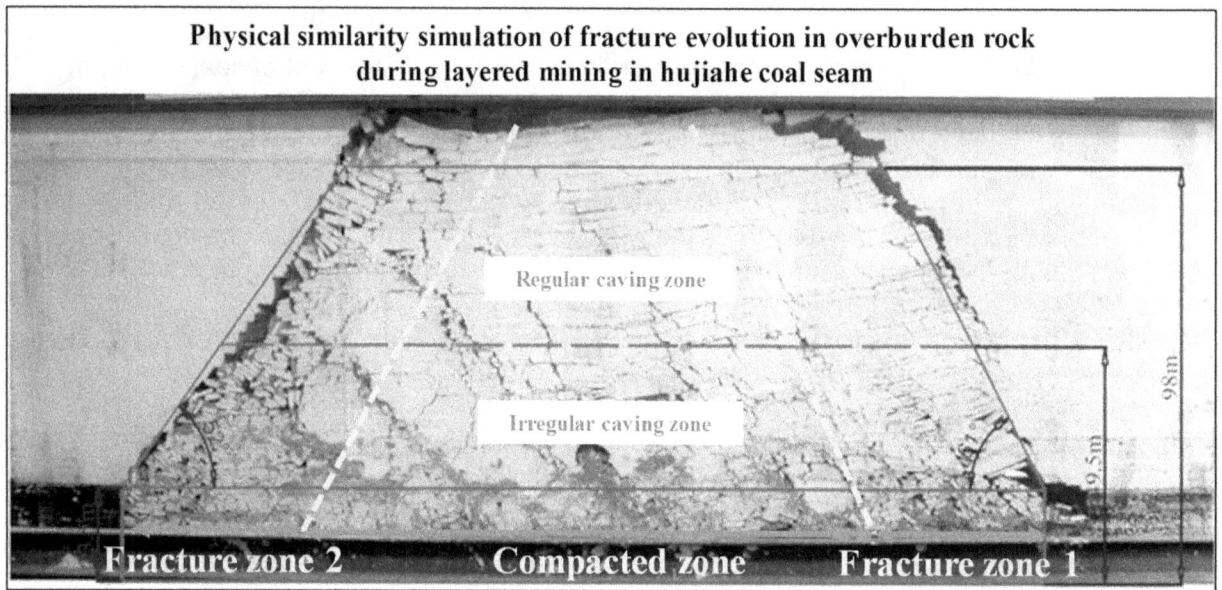

FIG. 4 Final movement form of overlying strata

By substituting the simulation results of physical similarity into equations (4) and (5), it can be obtained that the offset h=24.96 m, horizontal distance S=25 m. The expansion of the longitudinal through crack in this position can make the high drainage roadway give full play to the role of high concentration gas extraction in the gas active area, improve the gas extraction rate in the stope and reduce the gas emission to the working face in the goaf.

NUMERICAL SIMULATION STUDY ON LAYOUT LAYER AND EXTRACTION PARAMETER OPTIMIZATION

After obtaining the moving form of the overburden in the goaf through physical simulation, the approximate strata of the high drainage roadway are determined from the Angle of mining-induced fracture, and the layout strata and negative pressure of the drainage can be further optimized from the Angle of gas extraction and the spontaneous combustion of residual coal in the goaf through numerical simulation.

Modeling scheme

Buildup of model

According to the simulation results of physical similarity, it is simplified into the engineering model of ladder platform belt, and the physical model of goaf is established accordingly, as shown in FIG. 5. The geometric parameters of the physical model are shown in table 1.

Table 1 model parameters

Definition	
Length Unit	Meters
Bounding Box	
Length X	190.0 m
Length Y	276.0 m
Length Z	103.5 m
Properties	
Volume	3.2746e+006 m³
Statistics	
Bodies	6
Nodes	440550
Elements	413994

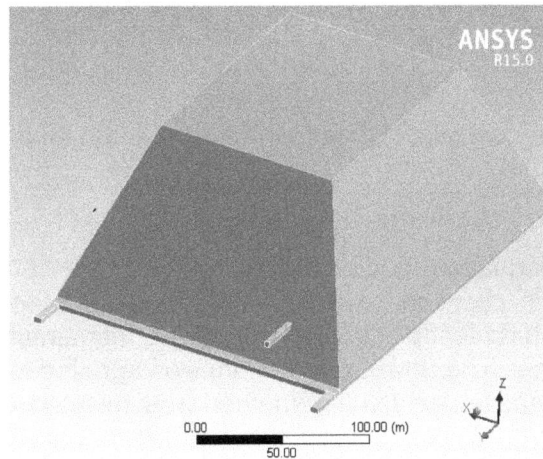

Fig 5 goaf model

Setting of boundary conditions

The inlet boundary of air inlet in the roadway is Velocity-inlet and air volume is 1450 m³/min, then inlet velocity is :

$$V = \frac{Q}{S} = \frac{1450}{60 \times 3.5 \times 5.4} = 1.28 \text{ m/s} \qquad (6)$$

Airflow turbulence =0.05, then the hydraulic radius

$$L = \frac{4d}{S} = \frac{4 \times (3.5 + 5.4) \times 2}{3.5 \times 5.4} = 3.78 \text{ m} \qquad (7)$$

Where Q is roadway wind volume, m³/min; L is perimeter of roadway, m; S is section area of roadway, m2;

Lane airflow exit is set as outflow; Air flow outlet of high drainage roadway was set as Pressure-outlet, and the negative Pressure value was 2.5kpa respectively.

The mining-induced fracture zone is characterized by porous media, so the porosity of the mining-induced zone can be simulated by establishing a porous media model. The permeability (k) and viscous resistance coefficient (R) are:

$$k = \frac{n^3 d_m^2}{150(1-n)^2} \qquad (8)$$

$$R = \frac{1}{k} = \frac{150(1-n)^2}{n^3 d_m^2} \qquad (9)$$

Where, n is porosity, $n=1-1/K_p$; dm is the mean harmonic particle size, m; K_p is the crushing expansion coefficient of a certain point in the mining-induced fracture zone, which can be determined by similar simulation experiments.

The gas quality source term in the goaf can be expressed as

$$Q_s = \frac{Q_g \cdot \rho_g}{V_g} \qquad (10)$$

Where, Q_s is the mass source term of model gas, kg/(m3•s); Q_g is the quantity of gas emission, m³/s; ρ_g is the gas density; pg=0.7167kg/m³;Vg is the total volume of the gas quality source term, m³.

When there is no gas extraction on 401012 fully mechanized caving face, the average daily advance speed is 4m/d, and the absolute gas emission is 40.93m³/min; When gas extraction is carried out, considering the influence of gas extraction (generally, its influence coefficient is 1.1~1.4, and 1.2 is taken here), the absolute gas emission amount at this time is 51.16m³/min. However, according to the law of gas attenuation, the emission amount of waste coal gas in goaf varies with the concentration in the goaf. The attenuation coefficient can be expressed as

$$q(x) = 0.00572\left(1 + \frac{x}{10}\right)^{-0.5} \qquad (11)$$

Where, q(x) is the attenuation coefficient of gas emission; x is the length from goaf to working face, m.

The specific parameter value is calculated using UDF function input.

The standard k-epsilon two-equation model was selected for the solution of the model, and the composition transport and no chemical reaction mode was adopted for the gas transport. The turbulent kinetic energy and turbulent dissipation rate of the momentum equation were discretized by first-order windward scheme. The SIMPLE algorithm was adopted for the pressure coupling, and the uncoupled and implicit steady-state solution method was selected.

Simulation scheme

In order to determine the optimal location of high drainage roadway, on the basis of physical simulation and theoretical calculation, different horizontal and vertical strata of high drainage roadway are taken as variables under the same negative drainage pressure (2.5kpa) to conduct numerical simulation on the drainage effect of high drainage roadway and the width of oxidation and temperature rise zone in goaf. The specific simulation scheme is shown in table 2.

Table 2 layout scheme of high drainage roadway

Serial number	Simulation conditions	Simulation purpose
1	Air volume : 1450m³/min The offset : 25m, Horizontal distance : 20m、25m、30m	The best horizontal distance
2	Air volume 1450m³/min, Horizontal distance : Determine the best horizontal distance from the first step The offset : 20m、25m、30m	Determine the best offset

Simulation Results Analysis

Effects of horizontal horizon on the prevention and control of spontaneous combustion of gas and coal

Fig 6 shows the extraction effect of high drainage roadway obtained through the numerical simulation study of scheme 1.

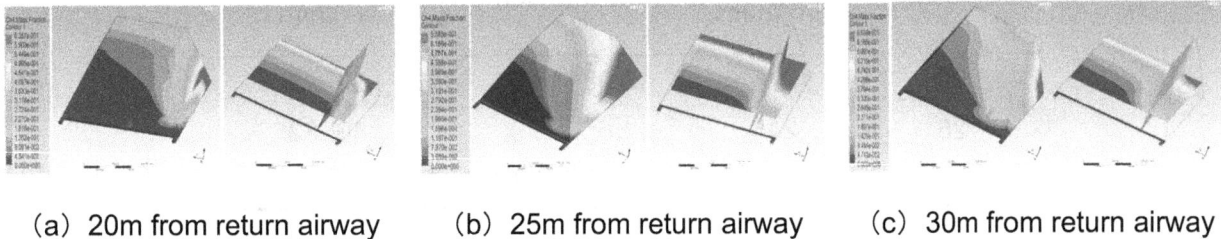

(a) 20m from return airway   (b) 25m from return airway   (c) 30m from return airway

Fig 6 distribution of gas concentration in high drainage roadway
under different vertical spacing conditions

Specific effects of gas extraction in high drainage roadway with different horizontal distances obtained through simulation results are shown in table 3.

Table 3 comparison of extraction effects of high drainage roadway at different levels

The level of distance /m	Mixed extraction /m³/min	Extraction concentration /%	Extraction of pure quantity /m³/min	Concentration in upper corner /%
20	245	13~16	31.85~39.20	0.76~1.15
25	245	19~23	46.55~56.35	0.35~0.62
30	245	15~18	36.75~44.10	0.68~0.95

According to the simulation results, when the horizontal distance of the high drainage roadway is 25m, the extraction concentration and pure amount remain at a high level, and the gas concentration at the upper corner is also far lower than the stipulated maximum value of 1%.The increase of the horizontal distance will lead to high alley pumping horizon from outside of the fracture zone return air lane, the degree of fracture zone gas is poorer, gas migration channel is relatively out-of-the-way, floating active gas migration degree is low, led to a decline in gas extraction effect. Meanwhile , the increase of the horizontal distance, is not conducive to reduce gas concentration in upper corner. If the horizontal distance decreases, the position of the high drainage roadway will be closer to the edge of mining-induced fractures. Similarly, the degree of fracture development is low, the gas extraction concentration and pure quantity are small, and the gas extraction effect in goaf is not good.

The influence of the horizontal distance of high drainage roadway on the spontaneous combustion oxidation temperature rise zone of coal in goaf is shown in Fig 7.

(a) 20m from return airway   (b) 25m from return airway   (c) 30m from return airway

Fig 7 oxygen concentration distribution in goaf at different plane distances

Table 4 shows the width of spontaneous combustion oxidation temperature rise zone of coal in goaf under different horizontal distance conditions obtained by numerical simulation.

Table 4 comparison of natural band widths at different horizontal layers

Horizontal distance/m	Mixed extraction /m³/min	Extraction of oxygen concentration /%	Spontaneous combustion zone width/m
20	245	13.98~16.3	39~89
25	245	14.65~16.9	40~82
30	245	13.69~16.1	46~85

It can be seen from the numerical simulation results that the horizontal position of high drainage roadway has little influence on the width of coal spontaneous combustion zone in goaf, while it has a great influence on the gas extraction effect. When the horizontal distance of high drainage roadway is 25m, the pure quantity of gas extraction reaches the maximum, which is 46.55m³/min~56.35m³/min, and the gas concentration in the upper corner is the lowest, which is 0.35-0.62%. At the same time, the width of spontaneous combustion zone in goaf also reaches the minimum value. Therefore, it can be determined that the reasonable range of the level distance of high drainage roadway is 23~28m.

Influence of vertical distance on prevention and control of spontaneous combustion of gas and coal

After determining a reasonable range of horizontal distance, the distribution of gas in stope is simulated when the horizontal distance is about 25m and the vertical distance is 20m, 25m and 30m respectively, as shown in FIG.8 and table5.

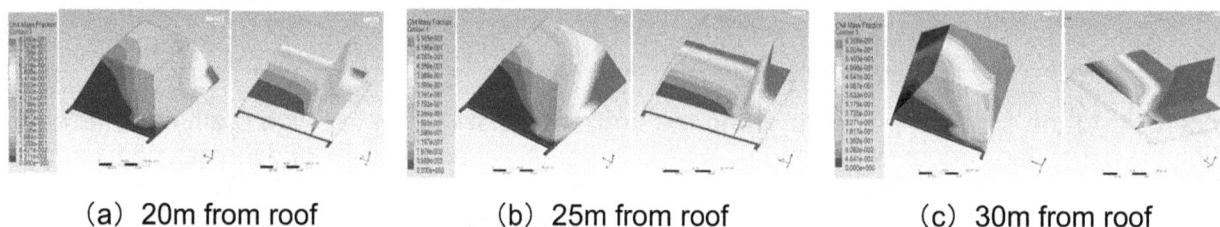

(a) 20m from roof　　(b) 25m from roof　　(c) 30m from roof

FIG. 8 Distribution of gas concentration in stope under different vertical distances

Table 5 Comparison of gas extraction effects in high drainage roadway of different vertical distance

vertical distance /m	Mixed extraction/m³/min	Extraction concentration /%	Extraction of pure quantity/m³/min	Upper corner gas concentration/%
20	300	12~15	36.00~45.00	0.85~1.12
25	245	19~23	46.55~56.35	0.35~0.62
30	120	36~40	43.20~48.00	0.45~0.75

With the increase of the vertical level of high drainage roadway, the drainage flow gradually decreases, but the extraction concentration will gradually increase. The main reason for this change is that the higher the level of mining-induced fracture development is lower, and the higher the level is, the lower the drainage flow is. The pressure relief gas in the goaf rises to the higher level through the longitudinal permeability crack and finally concentrates in the area with less permeability. Therefore, the extraction concentration will increase with the increase of the vertical layer arranged in the high drainage roadway. When the vertical distance is 25m, the pure quantity of gas extracted by high drainage roadway reaches the maximum. At this time, the upper corner gas concentration is controlled at 0.35~0.62%, which can ensure the safe and efficient recovery of the working face.

The influence of the horizontal distance of high drainage roadway on the spontaneous combustion oxidation temperature rise zone of coal in goaf is shown in FIG.9.

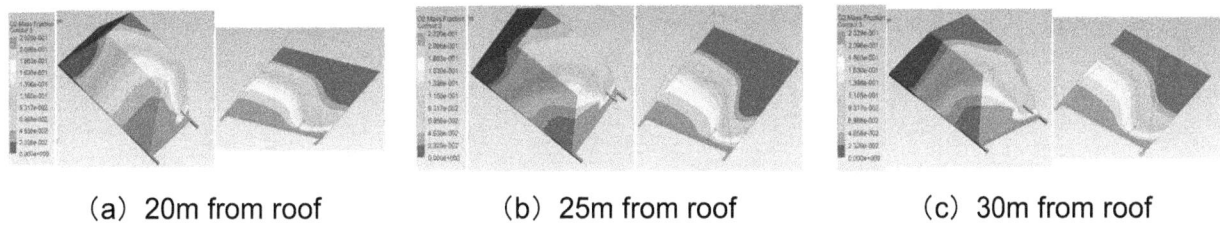

| (a) 20m from roof | (b) 25m from roof | (c) 30m from roof |

FIG.9 Oxygen concentration distribution in goaf with different vertical distances

Under different horizontal distances obtained by numerical simulation, the width of spontaneous combustion oxidation temperature rise zone in goaf is shown in table 6.

Table 6 Comparison of extraction effects of different vertical level high drainage roadway

vertical distance/m	Mixed extraction /m³/min	Oxygen concentration at extraction time/%	Spontaneous combustion zone width/m
20	300	13.98~16.3	43~85
25	245	10.65~12.9	45~81
30	120	6.99~9.31	40~82

With the increase of vertical horizon, the influence of high negative pressure extraction in high drainage roadway on air leakage in goaf gradually decreases, the oxygen concentration in extraction gas in high drainage roadway gradually decreases, and the width of coal spontaneous combustion zone in goaf gradually decreases.

According to the requirements of gas control and coal spontaneous combustion prevention, it is the best to arrange high drainage roadway 25m away from the roof of coal seam, and the reasonable range of vertical distance of high drainage roadway is 23m~28m.

Influence of negative pressure on spontaneous combustion of gas and coal in high drainage roadway

After determining the layout layer of high drainage roadway, gas drainage effects under 4 conditions, 0.5, 1.5, 2.5 and 3.5 kPa, were simulated respectively, so as to study and analyze the influence of different negative drainage pressures on gas volume fraction distribution and drainage effect in goaf. Simulation results of gas volume fraction distribution in high roadway stope are shown in FIG.10 and 11.

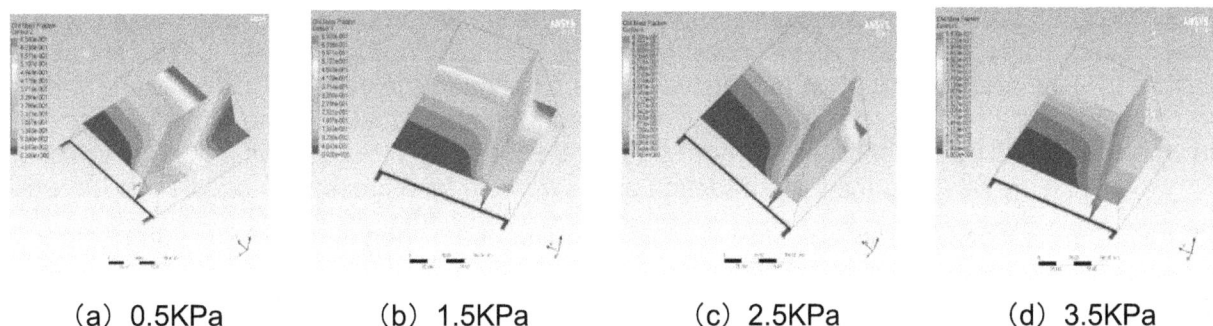

| (a) 0.5KPa | (b) 1.5KPa | (c) 2.5KPa | (d) 3.5KPa |

FIG.10 Distribution of gas concentration in stope under different negative extraction pressure

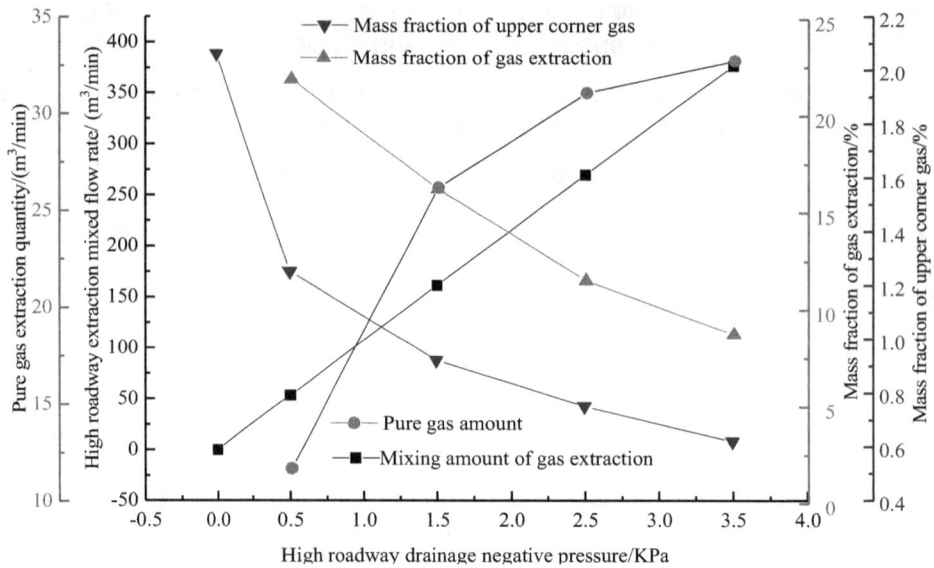

FIG.11 Influence of negative extraction pressure on extraction effect

According to the numerical simulation results, with the increase of negative pressure of drainage, the overall gas drainage effect gradually increases, with the increase range gradually decreases. However, with the increase of negative pressure of extraction, the degree of air leakage in goaf will be more serious, which increases the risk of spontaneous combustion of coal in goaf, making the prevention and control of spontaneous combustion of coal in goaf more difficult.

In order to study the influence of negative extraction pressure on the distribution of "three belts" of spontaneous combustion in goaf, the distribution of "three belts" of spontaneous combustion under negative extraction pressure of 1, 2 and 3kPa was simulated, and the simulation results were shown in FIG.12.

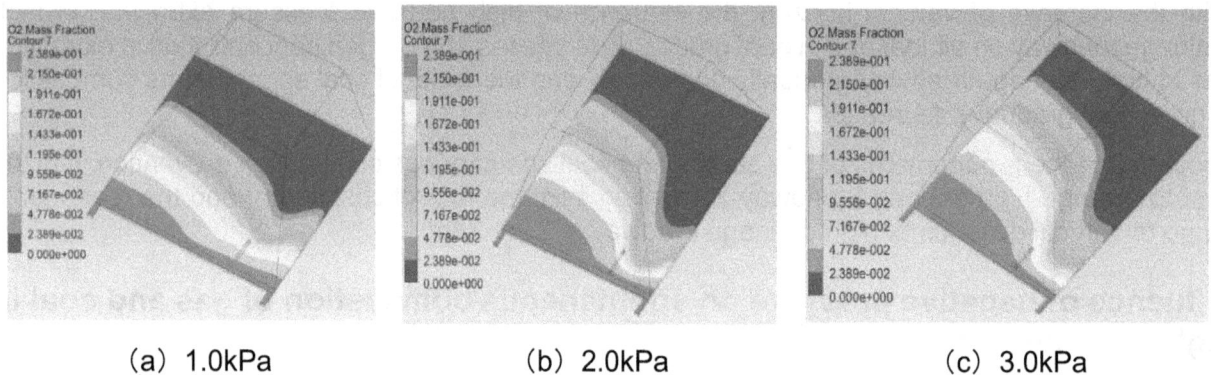

(a) 1.0kPa   (b) 2.0kPa   (c) 3.0kPa

FIG.12 Oxygen concentration distribution in goaf under different negative extraction pressure

The negative pressure of high drainage roadway extraction increases, the dividing line between the heat dissipation zone on the air inlet side and the oxidation and heating zone extends to the deep part of the goaf, the range of the oxidation and heating zone widens, and the range of the choking zone moves to the deep part of the goaf. Moreover, the dividing line between the central heat dissipation zone and the oxidation and heating zone extends to the depth of the goaf, the width of the oxidation and heating zone shrinks, and the range of the suffocation zone moves forward. On the whole, the dividing line between the heat dissipation zone and the oxidation and heating zone in fully mechanized goaf along the trend direction of the working face presents a saddle-like pattern.

According to the above numerical simulation of the influence of negative pressure on gas control and coal spontaneous combustion, the value of the optimal negative pressure of extraction to balance them is shown in FIG.13.

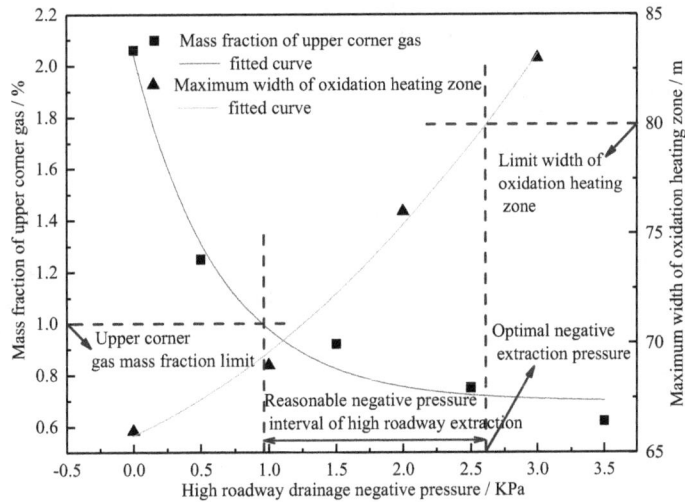

FIG.13 Value basis of reasonable extraction negative pressure

Considering the gas control(upper corner gas concentration<1%),the shortest spontaneous combustion period of floating coal（T_{min}=20d）and the average advance speed of working face（V=4m/d），the reasonable negative pressure range of gas extraction in high drainage roadway is [0.9516, 2.558], the optimal negative pressure point of gas extraction is 2.558 kPa, and the corresponding drainage flow is 275.78 m³/min. In theory, this negative pressure can achieve the maximum efficiency of gas extraction and prevent the occurrence of spontaneous combustion of coal in fully mechanized caving goaf, so as to ensure the safe and efficient production of the working face.

Effect of working face air supply amount on gas and coal spontaneous combustion in pumping conditions

Theoretical calculation of air supply amount in working face

There is a direct intrinsic connection between working face air supply amount and gas emission from goaf , the spontaneous ignition of the floating coal. When the air volume changes, it will inevitably affect the change of the wind flow in the goaf, which will lead to the redistribution of gas and oxygen concentration in the goaf. Especially under the conditions of fully mechanized caving mining, due to the high-intensity mining, the increase of air volume will lead to a sharp increase in the amount of gas emission. On the other hand, it will expand the range of spontaneous combustion zone in the goaf, increasing the spontaneous combustion risk.

The relationship between the air supply amount and the gas emission from the goaf to the working face and the spontaneous combustion zone width in the goaf has been obtained by numerical simulation, and it has shown as follows [18]:

$$\begin{cases} Q_{CH_4} = a_0 \cdot Q^{a_1} \\ L_m = b_1 - b_2 \cdot e^{-\lambda \cdot Q} \end{cases} \qquad (12)$$

Where Q is the air supply amount in working face, m³/min; Q_{CH4} is absolute gas emission from goaf to working face, m³/min; L_m is the width of spontaneous combustion zone, m. a_0, a_1, b_1, b_2 are parameter, which can be got by analyzing the results of numerical simulation using regression analysis

$$\begin{cases} Q_1 = 100(Q'_{CH_4} + a_0 \cdot Q^{a_1}) \\ Q_2 = -\dfrac{1}{\lambda}\ln\dfrac{b_1 - v\tau}{b_2} \end{cases} \qquad (13)$$

Where Q_1 is the air volume limit (upper and lower limits) required for gas control, m³/min; Q_2 is the maximum allowable air volume for coal spontaneous combustion control, m³/min; Q_{CH4} is absolute gas emission from goaf to working face, m³/min; v is advanced speed of working face, m/d; τ is the shortest period of spontaneous combustion of the coal in the working face, d.

Therefore, considering gas and spontaneous combustion control, the range of air supply amount in working face is $Q= [Q_1,Q_1]\cap[0,Q_2]$. The upper limits of air supply amount in 401102 working face in Hujiahe coal mine is about 1600m³/min, which can be obtained by theoretical calculation.

The numerical simulation of rational air supply amount

After determining the various pumping parameters of the high pumping lane, numerical simulation of the air supply amount in working face under extraction conditions has been carried. Based on calculation, air supply amount can be set up as 1400m³/min、1500m³/min、1600m³/min. the results have been shown as follows:

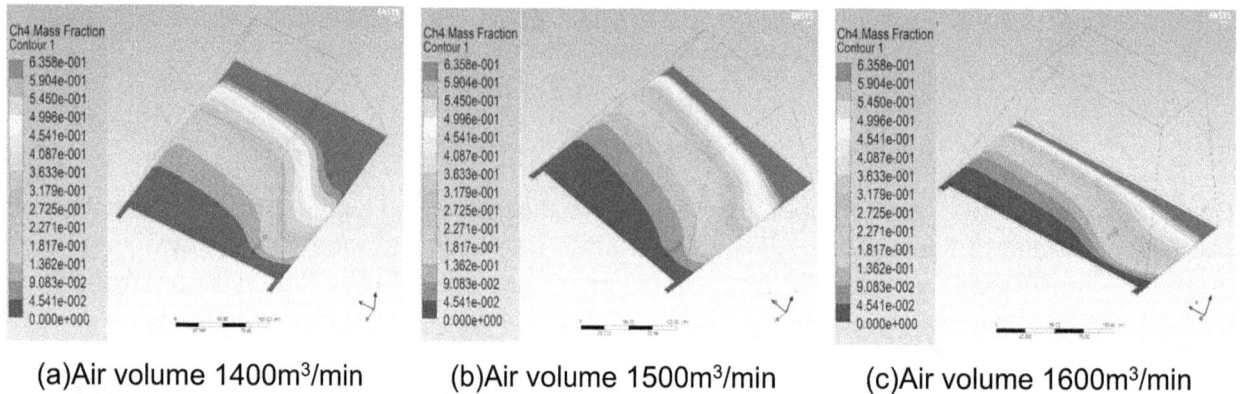

| (a)Air volume 1400m³/min | (b)Air volume 1500m³/min | (c)Air volume 1600m³/min |

Fig 14 Gas distribution law with different air volume

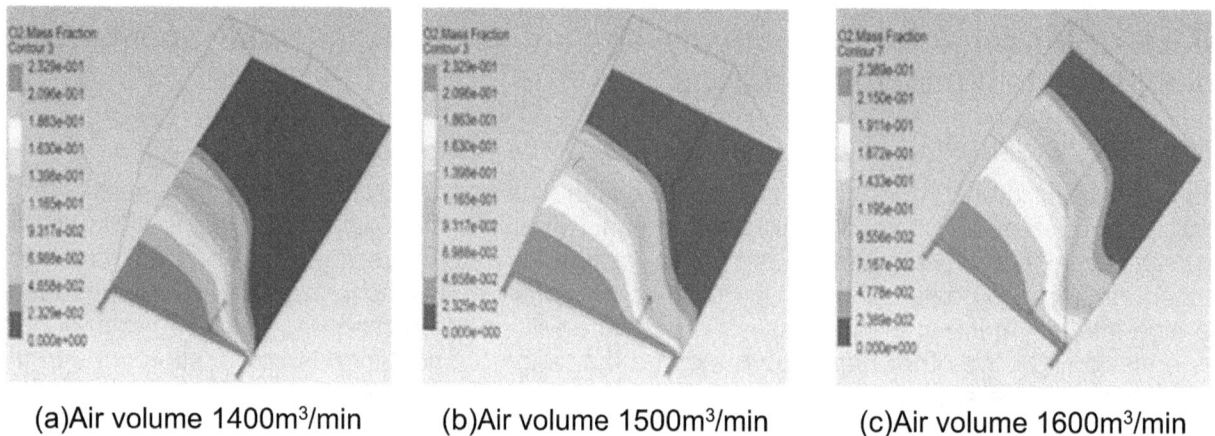

| (a)Air volume 1400m³/min | (b)Air volume 1500m³/min | (c)Air volume 1600m³/min |

Fig 15 Oxygen concentration distribution law with different air volume

Table 7 Gas concentration in upper corner and the width of oxidative zone with different air volume

Air volume/m³/min	Upper corner gas concentration /%	Oxidative heating zone in goaf		
		Initial position of maximum width/m	final position of maximum width /m	maximum width /m
1400	1.235	25.8	82.8	57
1500	0.7562	31.1	108.9	77.8
1600	0.5436	38.9	127.1	88.2

Shown form the simulation results, when the air volume increases from 1400m³/min to 1500m³/min, the gas concentration near the upper corner decrease the most obvious, and the gas concentration near the upper corner decreases from 1.235% to 0.5436%. when the air volume increases continuously, the decreasing range of gas concentration near upper corner reduces. Therefore, it shows that increasing the air volume in working face can prevent gas accumulation in upper corner. However, when the air volume increases beyond a certain limit, the effect will be weakened. On the other hand, when the air volume in working face increases, the air leakage intensity of the goaf increases, so that the high concentration of gas in the goaf is brought to the working surface, thereby breaking the original gas adsorption and desorption equilibrium state in the goaf, resulting in gas emission increase from the goaf. Therefore, there are certain limitations to reduce the gas in return airflow only by increasing air volume in working face.

In different air volume conditions, the initial distance and the longest distance of the oxidative heating zone appearing in the goaf increase with the increase of the inlet air volume. Meanwhile, the larger the air volume is , the wider the oxidative heating zone width will be. Therefore, when the air volume is small in intake airway, it is beneficial to prevent coal spontaneous combustion in the goaf.

To summarize, when air volume is 1500m³/min in working face, the gas concentration near the upper corner is less than 1.0%. Meanwhile the coal spontaneous combustion oxidation zone width in the goaf is less than 80m, which meets the basic requirement of daily gas control and coal spontaneous combustion prevention. Therefore, the air volume range in working face of 401102 should be controlled around 1500m³/min.

ANALYSIS OF ENGINEERING PRACTICE EFFECT

FIG.16 shows the relationship between the extraction concentration and the extraction amount of high drainage roadway with the advance distance and the layout layer of high drainage roadway. In the early stage of mining, the mining-induced fracture is not connected with the high pumping roadway, and the extraction concentration and extraction purity are relatively low. With the emergence of periodic overburden pressure, fractures in overburden rock gradually develop to a higher level, and the extraction volume and concentration gradually increase. When the vertical distance is about 25m, the high drainage roadway has the best drainage effect, which verifies the reliability of theoretical calculation and numerical simulation.

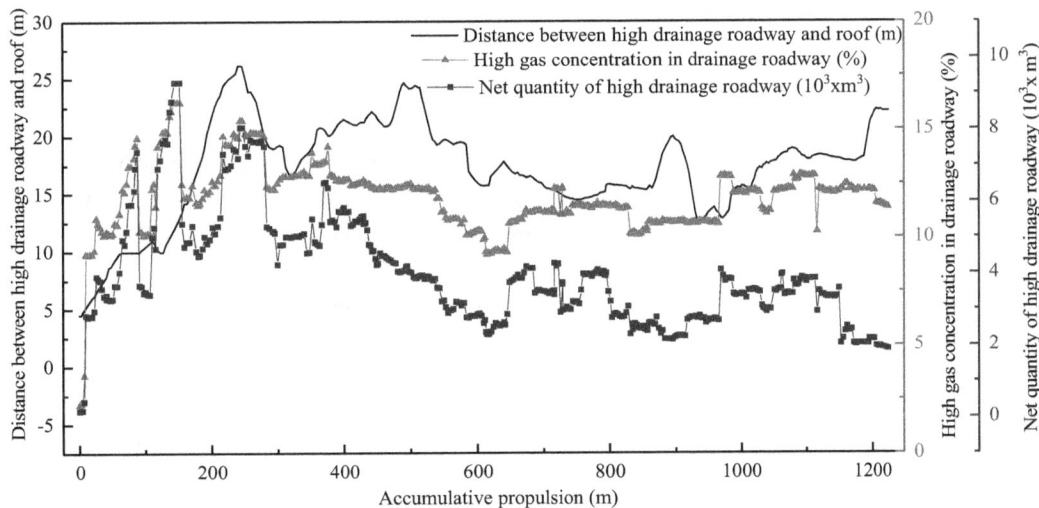

FIG.16 Relationship between drainage effect of high drainage roadway and change of horizon

Through real-time observation, the gas concentration at the corner of the test work surface and in the return air roadway is shown in FIG. 17. During the period when high drainage roadway is adopted to extract pressure relief gas from goaf, the average gas concentration in the upper corner and the return air roadway is lower than 0.6% and 0.45% respectively, which is far below the specified maximum value of 1.0% , ensuring the safe production during the working face mining.

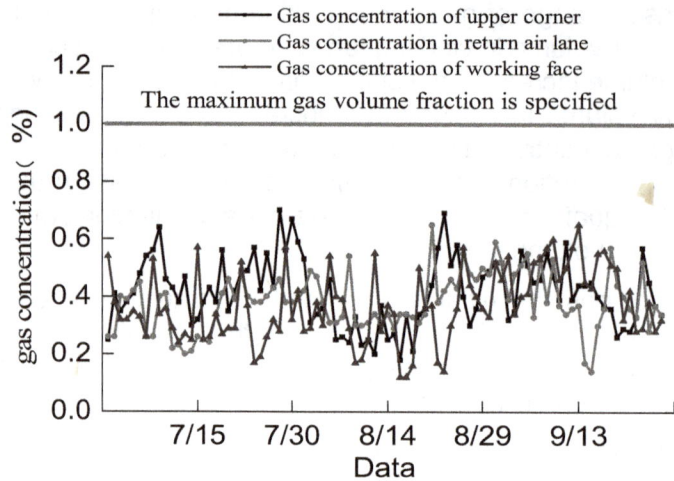

FIG.17 Gas control effect on test working face

CONCLUSION

1）The selection of reasonable level of high drainage roadway and the determination of negative pressure of drainage should comprehensively consider the two factors of gas control effect and coal spontaneous combustion risk in goaf, and find the balance point between them is the key to solve the problem.

2）The vertical layout of high drainage roadway is the leading factor of gas extraction effect and coal spontaneous combustion prevention. With the increase of vertical stratum, the gas extraction concentration increases, the gas extraction flow decreases, the air leakage intensity in goaf decreases, and the risk of coal spontaneous combustion decreases.

3）Combining the requirements of gas control and coal spontaneous combustion control, the high drainage roadway should be arranged 2.8~3.2 times the mining height from the roof of the coal seam and 0.4~0.8 times the width of the crack zone from the return air roadway. The engineering practice shows that by arranging the high drainage roadway in this area, the gas drainage effect is good, which can ensure the safe and efficient mining of the face.

4)Under the condition of gas extraction by high-extraction roadway, the adjustment of air volume in working face should not be arbitrarily. Due to the irreconcilability between the gas overrun and the coal spontaneous combustion in the goaf, if there is no intersection between the air volume required for gas control in working face and coal spontaneous combustion control air volume, under the premise of meeting gas control, measures such as inerting the goaf should be adopted to increase the maximum air volume allowed by coal spontaneous combustion control.

ACKNOWLEDGEMENT

This study was financially supported by the National Natural Science Foundation of China (No. 51734007 , 51704227 and 51874236).

REFERENCE

[1]Zhou Fubao, Xia Tongqiang, Shi bobo. Coexistence of gas and coal spontaneous combustion(Ⅱ)：new prevention and control technologies[J]. Journal of China coal society, 2013,38(05),843~849

[2]Qin Botao, Lu Yi, Yin Shaoju, Prevention and control technique of complex disaster caused by gas and spontaneous combustion for fully-mechanized sublevel caving face in close-distance seams[J]. Journal of Mining & Safety Engineering,2013,30(02),311~316

[3] Mao J D, Schimmelmann A, Mastalerz M, et al. Structural Features of a Bituminous Coal and Their Changes during Low-Temperature Oxidation and Loss of Volatiles Investigated by Advanced Solid-State NMR Spectroscopy [J]. Energ. Fuel, 2010, 24: 2 536-2 544

[4] Qin Botao, Zhang Leilin, Wang Deming, et al. Mechanism and restraining technology on spontaneous combustion of coal detonation gas in goaf [J] . Journal of China Coal Society, 2009, 34(12) : 1655－1659.

[5] LI Gui-he. Control of gas and coal spontaneous combustion for the large mining height and fully mechanized mining face[J]. Journal of Anhui Institute of Architecture & Industry : Natural Science Edition, 2010, 18(3) 38-41.

[6]REN Wan-xing, WU Bin-wei, WANG De-ming, et al. Fire prevention technology for high-gas-easy- spontaneous combustion super large underhand working face [J]. Journal of Mining & Safety Engineering, 2009 26(2) : 199-202.

[7] JIANG Jun-cheng , WANG Xing-shen .Method of artificial neural network for the prediction of coal spontaneous combustion[J] .Journal of China Mining & Technology, 1997, 26(1):19-22.

[8] Hao Yu, Liu Jie, Wang Changyuan, et al. Application of feeding nitrogen in extra—thick coal seam of comprehensive mechanized and breaking props coal mining [J]. Safety in Coal Mines, 2008, 39 (7) : 41—44.

[9] Li Zongxiang. Numerical simulation of nitrogen injection process for fire prevention and extinguishment in fully mechanized longwall top coal goaf [J]. China Safety Science Journal, 2003, 13(5) : 53—57.

[10] WU Ren-lun, XU Jia-lin, QIN Wei. Numerical simulation of gas control in fully mechanized top coal caving face with roof pre-splitting at initial mining period[J]. Journal of Mining & Safety Engineering, 2011, 28(2) : 319-322.

[11] DAI Guang-long .Study on change law of gas product of coal oxidation in low temperature [J] .Safety in Coal Mines, 2007, 38(1):1-4.

[12] Zhou Fubao, Coexistence of gas and coal spontaneous combustion(I) : disaster combustion [J]. Journal of China coal society, 2013,38(05),843~849

[13] Media-Struminska B. Correlation between methane and fire hazards in abandoned workings of long wall mining [A]. Proceedings and Monographs in Engineering, Water and Earth Sciences [C]. 2006: 325—330.

[14] Skotniczny Przemyslaw. Three-dimensional distribution of temperature and gas concentration in long wall drifts accompanying the phenomenon of self-combustion of coal deposited in long wall goafs [J]. Archives of Mining Sciences, 2008, 53(2) :235—255

[15] LI Shugang, LIN Haifei, ZHAO Pengxiang, et al. Dynamic evolution of mining fissure elliptic paraboloid zone and extraction coal and gas [J]. Journal of China Coal Society, 2014, 39 (8) :1455—1462.

[16]LI Shugang, XU Peiyun, ZHAO Pengxiang, et al. Analysis and application on the mining height effect of evolving law of compaction area at fully mechanized face [J]. Journal of China Coal Society, 2018, 43(S1) : 112—120

[17] LI Shugang, XU Peiyun, ZHAO Pengxiang, et al. Aging induced effect of elliptic paraboloid zone in mining cracks and pressure released gas drainage technique [J]. Coal Science and Technology, 2018, 46(9) : 146—152.

Case Studies

Study on turbulent dispersion simulation of concentrative methane emission in mine ventilation network

K Y Chen[1], S G Li[2], F B Zhou[3], L J Wei[4] and T Q Xia5

1. Professor, China University of Mining & Technology, Xuzhou, Jiangsu, China
Email: kychen109@126.com
2.Master, China University of Mining & Technology, Xuzhou, Jiangsu, China
Email: mm19833@126.com
3.Professor, China University of Mining & Technology, Xuzhou, Jiangsu, China
Email: zfbcumt@163.com
4.Associate Professor, China University of Mining & Technology, Xuzhou, Jiangsu, China
Email: cumtvb@qq.com
5.Associate Professor, China University of Mining & Technology, Xuzhou, Jiangsu, China
Email: xtq09cumt@163.com

ABSTRACT

The methane transmission with ventilation airflow in the mine ventilation network can be considered as a problem of one dimensional longitudinal turbulent dispersion. Based on the one dimensional longitudinal turbulent dispersion equations, we educe the analytical solution of temporal and spatial distribution of methane concentration for continuous point source methane emission in the ventilation roadway theoretically. The methane concentration relationship formula between the two sections is derived in the roadway. Combining with the law of conservation of methane quality, the mathematical model of methane transmission in the mine ventilation network is established, and the simulation program of methane transmission is compiled. Examples analysis show that the methane peak spread in the ventilation roadway is the decay process, and its decay rate depends on the methane peak height and its duration time when the distance between the two airflow sections is fixed. The time of methane peak appearing in the downstream section lags behind the upstream section. If the length or the section area of the roadway increases, the methane peak will appear later in the downstream section, and the peak height will be lower. The smaller the roadway air quantity is, the later the peak appears at the downstream section and the lower the peak value is. If the coefficient of friction resistance increases, the time of peak appearing remains unchanged and the peak value decreases. Furthermore, the fundamentals of the methane peak spread in the ventilation network are analyzed by numerical simulation. Finally, Application results show that the proposed model and algorithm of methane spread in ventilation networks are effective. The researching findings in this paper can contribute to quantitative analysis of the degree of pollution and containment spread caused by abnormal gas emission in the underground for the future research.

Key words: one dimensional longitudinal turbulent dispersion, methane emission, ventilation network, spread law, numerical simulation

INTRODUCTION

While the concentrative methane emission transfers in mine ventilation roadway, the methane concentration can be increased in airflow direction, and seriously endanger the safety production of coal mine.

The methane, which emits continuously from coal or rock in coal mine, can mix and transfer with the airflow in ventilation roadways. In the turbulent flow field, the effects of molecular diffusion, advection diffusion and turbulent diffusion are coexistent. The coefficient of molecular diffusion can be ignored as it is much smaller than the turbulent diffusion. Because the longitudinal size of roadways is larger than the lateral one. Therefore, the turbulent flow of roadway macrocosmically occurs almost only in the longitudinal direction, and the longitudinal pulse velocity is also much larger than the lateral one. In the turbulent transfer process of mine roadway, the longitudinal dispersion phenomenon caused by uneven fluid velocity distributed in the station of roadway is called dispersion of shear flow. It is essentially different from the diffusion caused by molecular motion or fluid points pulse. The turbulent diffusion is caused only by turbulent velocity pulse, and methane mixture is in the lateral direction. But the longitudinal turbulent dispersion is the diffusion in the airflow direction, which is caused by the uneven distribution of airflow velocity on the cross section and the velocity pulse in the airflow direction. The dispersion can be described as the varied process of mean values on the cross section in the longitudinal direction (Wang, 1994). The phenomena of concentrative methane emission transfer belongs to the longitudinal turbulent dispersion problem.

The model of longitudinal turbulent dispersion, which is based on instantaneous emission source, has been established (Lu et al 2006). The transfer distributed process of the instantaneous methane emission source was researched by this model in ventilation network, and the transfer law of the methane peak was obtained. In general case, the characteristics of methane emission are continuous and unsteady, e.g. abnormal emission or coal gas outburst. How does the methane transfer in mine ventilation networks should be investigated. In order to provide a theoretical analysis method for determining the influence extent of the abnormal methane emission, the transfer law of continuous and concentrative methane emission in ventilation roadway would be researched in this paper.

THE MODE OF LONGITUDINAL TURBULENT DISPERSION IN VENTILATION ROADWAY

The mode of longitudinal dispersion

In the turbulent transfer process of ventilation roadways, turbulent diffusion and longitudinal turbulent dispersion are concurring. And the effect of the longitudinal turbulent dispersion is larger than the diffusion in lateral direction. Because the lateral mixing, pulse and exchange among levels of fluid of turbulence flow, the methane concentration tends to equality while gas leaves the emission source behind a certain distance (Chen K.Y. et al 2008), that is to say, the diffusion in lateral direction may be ignored. Assuming that methane emission is not enough to change the density and velocity of the airflow, then methane transfer process in the airflow can be described by the model of longitudinal dispersion based on the parameter's average values on cross sections (Li, E.L.1985 & Wang, Y.M. 1994). The model shows as follow:

$$\frac{\partial C}{\partial t} + U\frac{\partial C}{\partial x} = E_x \frac{\partial^2 C}{\partial x^2} \tag{1}$$

Where, C is the average methane concentration of some cross section in ventilation roadway; U is the average airflow velocity of the cross section, m/s; E_x is the longitudinal turbulent dispersion coefficient, m²/s.

According to the mass transfer model of the circular roadway, the velocity distribute function in the roadway, and the analogy principle between momentum and quality transfer (Li, E.L. 1986), the formula of the coefficient E_x can be derived as follow:

$$E_x = 65.47r\sqrt{\alpha}U \tag{2}$$

Where, r is the equivalent radius of ventilation roadway, m; α is friction resistance coefficient in standard condition, kg/m³; 65.47 is a constant with the unit, (m³/kg)¹ᐟ².

The methane transfer model of instantaneous point source in ventilation roadways

Assuming that M is the release amount of an instantaneous source in ventilation roadway, if the total amount of methane is an invariant constant in the transfer process, then $\int_0^\infty \left(\int_0^\infty C(x,t)\mathrm{d}x\right)\mathrm{d}t = M$. When the average velocity is a constant, the concentration distribution of an instantaneous point source can be obtained by equation (1) as follows (Yu C.S. 1992):

$$C(x,t) = \frac{M}{\sqrt{4\pi E_x t}}\exp\left[-\frac{(x-Ut)^2}{4E_x t}\right] \tag{3}$$

Where, $C(x, t)$ is methane concentration; x is distance from an instantaneous point source in airflow direction, m; t is the time releasing methane from the point source.

The methane transfer model of continuous point source in ventilation roadways

For the concentrative and continuous methane emission source, the longitudinal turbulent dispersion process can be regard as the superposition of methane dispersion from numerous instantaneous point sources in space-time. Assuming the adding diffusion mass increment is $\Delta M = f(\tau)\Delta\tau$ in a micro-period of time $\Delta\tau$ at time τ, the concentration change caused by dispersion, which keep on going for the time of $(t-\tau)$, has the following form:

$$\Delta C(x,t) = \frac{f(\tau)\Delta\tau}{\sqrt{4\pi E_x(t-\tau)}} \exp\left[-\frac{[x-U(t-\tau)]^2}{4E_x(t-\tau)}\right] \tag{4}$$

then, the methane concentration in longitudinal direction can be expressed as followings:

$$C(x,t) = \int_0^t \frac{f(\tau)}{\sqrt{4\pi E_x(t-\tau)}} \exp\left[-\frac{[x-U(t-\tau)]^2}{4E_x(t-\tau)}\right] d\tau \tag{5}$$

METHANE TRANSFER MODEL IN MINE VENTILATION NETWORK

To solve the problem of methane turbulent dispersion in mine ventilation networks, except for the two fundamental sets of equations determined by Kirchhoff's first and second laws, and law of ventilation resistance, the relational formulas are also required for calculating methane flow and concentration.

Methane flow conversation law of nodes

The methane volume flow entering a junction equals the volume flow leaving that junction. For the ventilation network with n branches and m nodes, the law is expressed as following:

$$\sum_{j=1}^{n} a_{ij}G_j = 0, \quad i = 1,2,\cdots,m\text{-}1 \tag{6}$$

Where, G_j is the methane volume flow of Branch j, m³/s ; a_{ij} is the fundamental incidence matrix element of ventilation networks.

The formula for calculating methane concentration

Assuming methane is instantaneously and fully mixed with air in a junction, when airflows enter the junction from several branches, the junction's mixing concentration equals the average value of methane concentration in the branch airflows, and methane concentration of airflows leaving that junction are all equal nearby that junction. The junction methane mixed concentration is expressed as following:

$$C_i = \frac{G_i}{Q_i} = \frac{\sum_{j\in J_i} C_j q_j}{\sum_{j\in J_i} q_j} \tag{7}$$

Where, Q_i is the sum of airflows entering the ith junction; q_j is the airflow of Branch j entering the ith junction; J_i is the branch set entering the ith junction.

THE SOLUTION OF THE PROBLEM OF METHANE TURBULENT DISPERSION IN VENTILATION ROADWAYS

Figure 1 illustrates which there are two stations on this ventilation roadway.

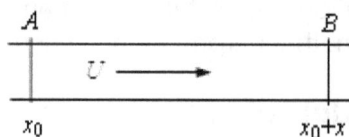

FIG 1 – Ventilation roadway

Assuming that the distance from station A to B is x, the airflow rate is U, the average gas concentration is $C_0(x_0,\tau)$ of station A at an arbitrary time τ, then the average gas concentration is $C(x_0+x,t)$ of station B at time t. Since that Station A is the source position of gas emission, the average concentration of station B can be regarded as the superposition of the dispersion processes generated by numerous different intensity instantaneous source in the same position at the same time. The micro instantaneous source intensity is $\Delta M = UC_0(x_0,\tau)\Delta\tau$ in station A in the period of $\Delta\tau$. According to equation (5), the concentration increment of station B at time t, which caused by the micro instantaneous source, has the following expression (Yu, C.S. 1992).

$$\Delta C(x_0+x,t) = \frac{UC_0(x_0,\tau)}{\sqrt{4\pi E_x(t-\tau)}} \exp\left\{-\frac{\left[x-U(t-\tau)\right]^2}{4E_x(t-\tau)}\right\}\Delta\tau \tag{8}$$

The concentration of station B at time t can be obtained by using the superposition principle.

$$C(x_0+x,t) = \int_0^t \frac{UC_0(x_0,\tau)}{\sqrt{4\pi E_x(t-\tau)}} \exp\left\{-\frac{\left[x-U(t-\tau)\right]^2}{4E_x(t-\tau)}\right\}d\tau \tag{9}$$

Above formula's discrete expression as following:

$$C(x_0+x,t) = \frac{1}{2}\sum_{i=0}^{\left[\frac{t}{\Delta\tau}\right]-2} C_0(x_0,i\Delta\tau)\{\text{erf}(X)-\text{erf}(Y)\} + \frac{1}{2}C_0(x_0,t-\Delta\tau)\{1-\text{erf}(Z)\} \tag{10}$$

$$X = \frac{(Ui\Delta\tau+U\Delta\tau-Ut+x)}{\sqrt{4E_x(t-i\Delta\tau-\Delta\tau)}}, \quad Y = \frac{(Ui\Delta\tau-Ut+x)}{\sqrt{4E_x(t-i\Delta\tau)}}, \quad Z = \frac{(x-U\Delta\tau)}{\sqrt{4E_x\Delta\tau}}$$

Where, $\Delta\tau$ is discrete time span; $[t/\Delta\tau]$ is the rounding integer of $t/\Delta\tau$; erf(X) is the error function.

The error function expression is $\text{erf}(X) = \frac{2}{\sqrt{\pi}}\int_0^X e^{-t^2}dt$. In order to speed up the calculation of error function erf(X), the best fitting function is obtained by using the known error function erf (X) values:

$$\text{erf}(X) = \begin{cases} 1 & X \geq 3 \\ 1.128\sin(0.7121X)+0.1352\sin(2.185X) & -3 < X < 3 \\ -1 & X \leq -3 \end{cases}$$

Formula (10) is called the relationship of methane concentration change between two cross section of the branches.

THE CALCULATION OF METHANE TURBULENT DISPERSION IN VENTILATION NETWORKS

After air quantities of all braches are obtained by ventilation network solution, beginning from the source node of the mine ventilation network with one source and one sink, all nodes are visited by Breadth First Search method, according to the methane data at monitoring emission points, the methane concentration at the end node of each branch leaving the node with known methane concentration is calculated by equation (10) and (7). This procedure continues until the methane concentration of all nodes have been computed. Finally, The methane concentration of proportional sub-sites in the branches are also calculated by equation (10).

Setting up mine ventilation network is the network with one total inlet point and one total outlet point, methane emission source points are at nodes in mine ventilation network, and the methane emission in the branch is uniformly distributed longitudinally. The specific algorithm is as follows:

(1) Ventilation network solution

Assuming that the air flow state of mine ventilation network is stable and not affected by methane concentrated emission, all branch air volume can be obtained by solving ventilation network.

(2) Node parameter initialization

Node parameter values include *ok*, *r*, *in*, *tin*, w[*i*], *q*. Where, *ok* denotes whether all branches entering the node have been computed, yes *ok* = true, no *ok* = false; *r* indicates whether all branches leaving the node have been computed, yes *r* = true, no *r* = false; *in* record the number of branches that have been computed into the node; *tin* record the in-degree of the node; w[*i*] array record node methane concentration value; *q* represents the sum of all branch airflows into the node. Node parameter initialization settings: *q*=0, *ok*=false, *r*=false, *in*=0, *tin*=0, w[*i*]=0.

(3) Node in-degree calculation

The in-degree of a node is equal to the number of branches flowing into the node. Traversing through all branches, the *tin* value of the node parameter corresponding to the end node number of each branch adds 1.

(4) Node airflow calculation

Node air quantity is defined as the sum of all branch air quantity entering the node. Traversing through all branches, the air quantity of each branch is added to the air quantity *q* of the node corresponding to the number of the last node of the branch.

(5) The methane concentration treatment of monitoring node

According to the read methane concentration value of the monitoring node, the parameter w[*i*] value of the node is modified, and the parameter ok value of the node is set to true.

(6) The treatment of methane emission branch

When there is a methane emission source in the branch, according to the incidence relation between the branch and the node, methane concentration increment caused by methane emission in the branch is added to the value of w[i] at the end node of the branch.

(7) Setting the parameter value of total inlet and outlet nodes in the ventilation network

In the ventilation network, the parameters of total outlet node are set as follows: *r* = true, *ok* = true, and the parameter *ok* value of total inlet node is set as true.

(8) Node methane concentration calculation

According to the known node methane concentration in the ventilation network and formulas (10) and (7), the methane concentration values of unknown methane concentration nodes in the ventilation network are calculated, and the calculation process is as follows:

① Traversing all nodes in the network, for the nodes with parameter values *ok* = true and *r* = false, according to formulas (10) and (7), the methane concentration values of end nodes of all branch outflow from the node are calculated, and the parameter r of the node is set as true, and the parameter in value of the end node adds 1.

② Traversing all nodes in the ventilation network, if node parameter *ok*=false and *in*=*tin*, then the node parameter *ok* value is modified as true.

③ Determine whether there is a node with *r* = false in the ventilation network. If there is such a node, return ① until all nodes have *r* = true.

(9) Calculation of methane concentration at branch quantile point

The methane concentration at any position in the branch can be calculated by using branch quantile point. The calculation process is as follows:

① Traversing all branches in the ventilation network, according to the methane concentration value of the branch start node and formula (10), the methane concentration value at the branch quantile point is calculated.

② For the branch with methane emission, the increment of methane concentration caused by the methane emission from the start point to the quantile point is calculated, and this increment is added to the methane concentration value at the sub-point of branches calculated by step ①.

(10) Save the result and end the program

The calculated methane concentration value at the quantile point is added to the branch gas table to update the information in the database and end the program.

The above calculating process have been achieved by computer programming.

ANALYSIS OF APPLICATION EXAMPLES

The methane turbulent dispersion laws of single airway

Assuming that the methane concentration in the upstream station A is known, and the distance from A to B is 1000m, the equivalent radius of the roadway is 2m, the coefficient of friction resistance is $0.02\,\mathrm{N}\cdot\mathrm{s}^2/\mathrm{m}^4$, the average airflow velocity is 3m/s, the longitudinal dispersion coefficient is 55.553m²/s. Figure 2 shows the sampling data of methane concentration in Station A. There are 1441 data samples, the interval is 5s, so sampling time is from 0 to 1440×5s. According to equation (10), the methane concentration in Station B can be calculated by Station A. Then draw the graph of methane concentration change in the two stations (Figure 2). Ordinal numbers 1'-8' represent the peak or valley of station B, which corresponding with ordinal numbers 1-8 of station A.

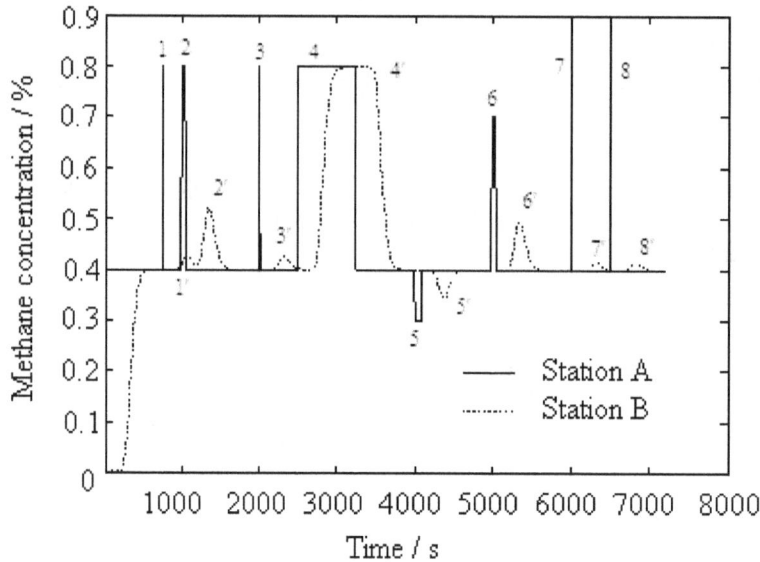

FIG 2 – Methane concentration change in Station A and B

From Figure 2, we can know: The peak and valley of methane concentration in the downstream station are not greater than the corresponding values in the upstream one. The decay rate of the methane concentration in the downstream station depends on the duration of the peak of methane in the upstream one if the diffusion distance is fixed. When the highest Peaks 7,8 with a short time at station A transfer to arrived at station B, and its Peaks 7', 8' are the lowest, so its decay rate is the biggest; The height of Peak 4, which has the longest duration, is equal to the one of Peak 4'. The depth of Valley 5' is smaller than Valley 5 too. Comparing Peaks 2,6 with Peaks 2',6', we can know: When the duration of two peeks are equal in the upstream station and their height are different, so the two decay rate values of peak height are different in the downstream one , which depend on height and duration of the peak in the upstream one. Generally speaking, the methane dispersion is a decaying process in roadway airflow.

If Peak 4 of station A have a long duration, then does corresponding Peak 4' of station B, and the slope of the peak is steeper; contrariwise, if the duration of Peaks 1,2,3,6,7,8 is shorter at station A, the slope of Peaks 1',2',3',6',7',8' will be slower at station B, and the peak height is smaller. Generally speaking, the shorter the duration of the peak in the upstream station is, the slower the peak in the downstream station is, and even could disappear.

The time appearing Peaks 1'-8' of station B are lag behind Peaks 1-8 of station A. which mainly depends on the time that air flow from station A to B.

The Influence of airway parameter change on methane dispersion

Airway parameters mainly include length, cross section area, friction resistance coefficient and air quantity. Analyzing the influence of airway parameters on methane dispersion can reveal the basic law of methane dispersion in roadways.

In the series airway shown in Figure 3, the node 2 is the source of methane emission, and the variation of methane concentration with time as shown in Figure 4. The elevation of each node is 0, the air density is 1.2 kg/m^3, and the parameters of each branch are length is L=1000 m, sectional area is S=10 m^2, friction resistance coefficient is 0.01 N·s^2/m^4, sectional shape is rectangular, and the quantile points of branches are 0.5.

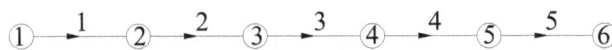

FIG 3 – Branch series connection

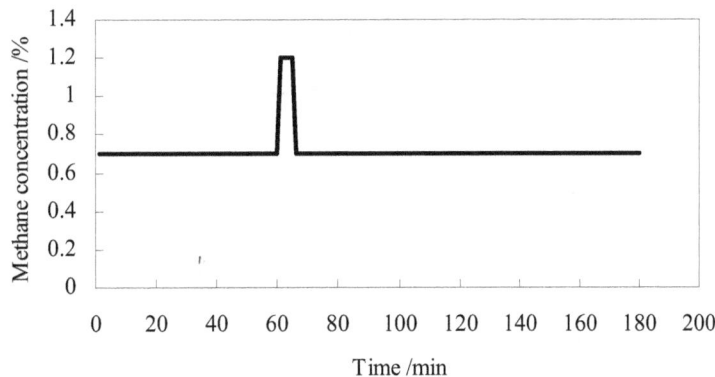

FIG 4 – the variation of methane concentration with time at the node 2

Effect of airway length

The methane dispersion are simulated respectively for the cases that the length of Branch 3 is L=1000m, 2000m, 4000m, 8000m and other parameters are unchanged. The variation law of methane concentration at the centre location of branch 4 is shown in Figure 5. When the length of branch 3 increases, the occurrence time of methane peak in branch 4 will be delayed and the peak value will be reduced.

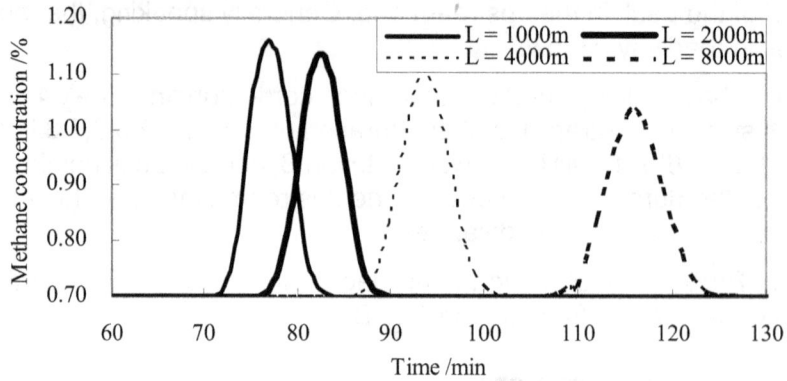

FIG 5 – Influence of airway length on methane dispersion

Effect of branch section area

The methane dispersion simulations are carried out respectively for the cases that the length of Branch 3 is $S=5m^2$, $10m^2$, $20m^2$, $40m^2$ and other parameters are unchanged. The variation law of methane concentration at the centre location of Branch 4 is shown in Figure 6. When the section area of Branch 3 increases, the occurrence time of methane peak in branch 4 will be delayed and the peak value will be reduced.

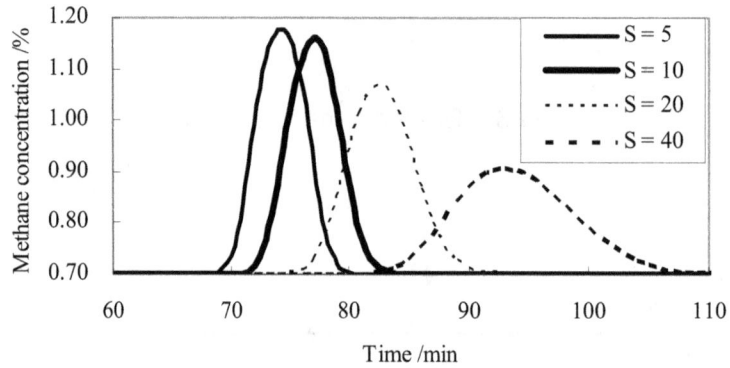

FIG 6 – Influence of section area on methane dispersion

Effect of branch friction resistance coefficient

The methane dispersion simulations are carried out respectively for the cases that the friction resistance coefficient of branch 3 is α =0.01N·s^2/m^4, 0.04 N·s^2/m^4, 0.2 N·s^2/m^4, 1 N·s^2/m^4, and other parameters are unchanged. The variation law of methane concentration at the centre location of Branch 4 is shown in Figure 7. When the friction resistance coefficient of Branch 3 increases, the occurrence time of methane peak in Branch 4 does not change, and the methane peak value will decrease.

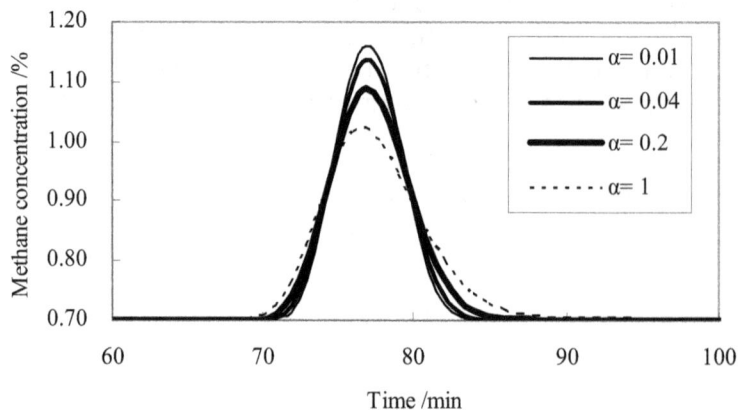

FIG 7 – Influence of friction resistance coefficient on methane dispersion

Effect of branch air quantity

The methane dispersion was simulated respectively for the branch 3 with air quantity q=30 m³/s, 50 m³/s, 70 m³/s and other parameters unchanged. The methane concentration variation law at the centre location of branch 4 was obtained as shown in Fig. 8. When the air quantity of branch 3 increases, the methane peak in branch 4 will appear earlier, and the methane peak will rise, and its value will not exceed the peak value at node 2.

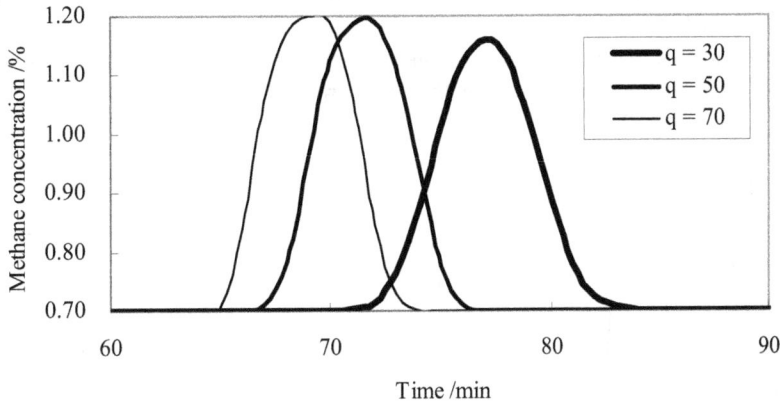

FIG 8 – Influence of air quantity on methane dispersion

Analysis of methane turbulent dispersion in the ventilation network

Fig.9 shows the parallel ventilation network. Where, Node 1 is general entering junction, Node 6 is general returning junction, Node 2 is the source of methane emission. Fig.4 show that the methane concentration of Node 2 changes with time. The elevation and air density of each junction is the same, air density is 1.2 kg/m³. The parameters of each branch are the same: cross section area 10m², friction resistance coefficient 0.01N·s²/m⁴, the shape of cross section is rectangle, the calculation position point sets in the middle of branches. The length of Branch 1, 2, 7 and 8 is 1000 m, respectively. Total air quantity is 30 m³/s in the ventilation network. The quantile point of each branch is 0.5.

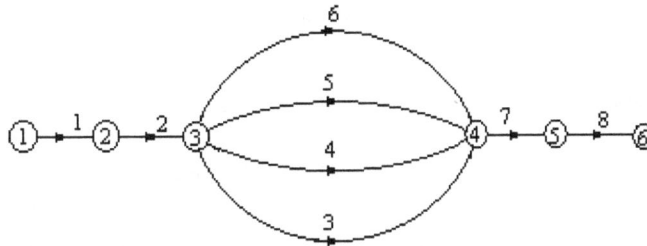

FIG 9 – parallel ventilation network

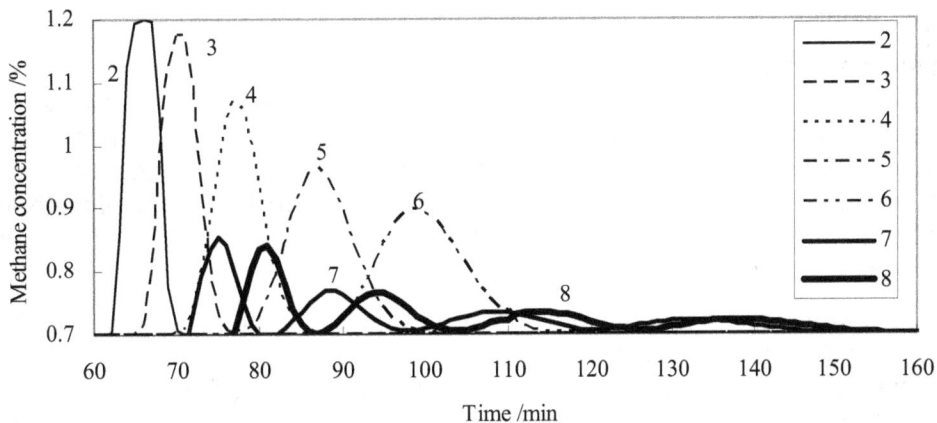

FIG 10 – Four methane peaks shown in Branch 7

When the length of Branch 3,4,5,6 is 250m, 750m, 1250m, 1750m respectively, the corresponding average airflow velocity calculated from the ventilation network is 1.249m/s, 0.721m/s, 0.558m/s, 0.472m/s. Through the methane dispersion simulation in the ventilation network, the results show that four methane peak appear in Branch 7 as shown in Figure 10. This case is the result of the different time it takes for methane to flow through the four paths from Node 2 to the middle point of Branch 7.

When all basic parameters of Branch 3,4,5,6 are the same, the corresponding average airflow velocity obtained by ventilation network calculation is 0.75m/s, the simulation results of methane dispersion show that only single methane peak appear in Branch 7 as shown in Figure 11. this is due to having the same migration time that the methane flows through the four paths from Node 2 to the middle point of Branch 7.

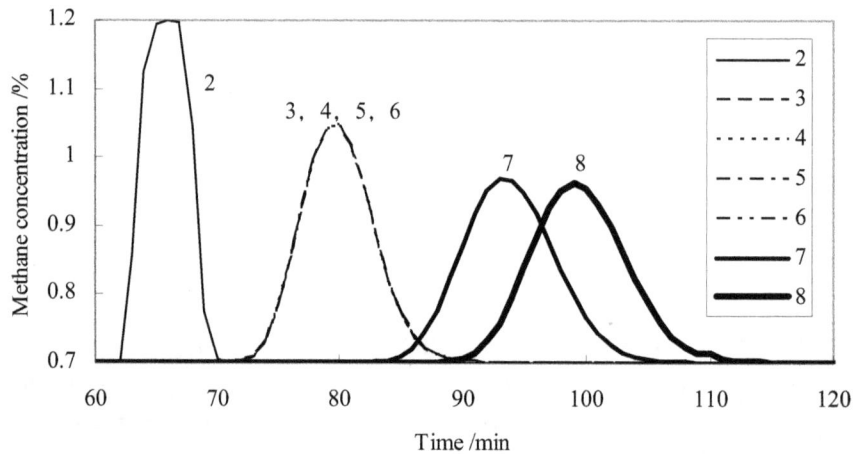

FIG 11 – Only one peak appear in Branch 7

CONCLUSIONS

Essentially speaking, the continuous methane emission transfers in mine ventilation networks is a process of longitudinal turbulent dispersion in airflow direction. Because of influences by the ventilation network topology structure and ventilation parameters, the transfer process is very complicated.

Generally, there is a decay process appeared in the transfer of methane peaks in mine ventilation roadway. Methane concentration distribution is a function that depends on the time and travel distance, the peak value of methane concentration would gradually down to averages with the time period of transfer and travel distance. If the transfer distance is fixed, its decay rate depends on the maximum value and duration time of peak in the upstream station. The time of methane peak appearing in the downstream station lags behind the upstream station, and the extent of lagging time depends on the travel time that the airflow needs flowing from the upstream station to the downstream one. If the interval time between two peaks is too short, they would be combined together. If propagation distance is long enough, the peaks would disappear eventually.

when methane peak propagates in series airway, it is related to the length, cross-section area, friction resistance coefficient and air quantity of airway. When the length or cross-section area of the airway increases, the occurrence time of methane peak will delayed and the peak value will reduce; when the air quantity increases, the occurrence time of methane peak will be advanced and the peak value will be increased; the friction resistance coefficient only affects the size of methane peak value, but does not affect the occurrence time of methane peak; when the friction resistance coefficient increases, the peak value will decrease.

In a complicated ventilation network, when there are several paths linking the methane emission sites with a site in downstream, at this point, the number of peaks and the duration time observed are largely dependent on the different travel paths, travel time and emission peak durations. If the peak duration is long, the number of peaks appeared at downstream point should be equal or less than the product of the number of pathways and peaks observed at the methane emission site. If the peak duration is too small, there may not be seen any peaks.

Finally, Application results show that the developed model and algorithm of methane propagation in ventilation networks are effective. The researching findings in this paper can contribute to quantitative analysis of the degree of pollution and containment spread caused by abnormal gas emission in the underground for the future research.

ACKNOWLEDGMENTS

This financial supports from the National key research and development program of China (Grant No.2018YFC0808100) is deeply appreciated.

REFERENCES

Chen, K.Y.; Li, S.G.; Zhang, Z.H. et al. 2008. Theoretic research related to the method of measuring air quantity by means of tracing gas. Journal of China University of Mining & Technology 37(1): pp10-14.

Li, E.L. & Wang, B.Q. 1985. Mathematical model of turbulent mass transfer and coefficient of turbulent dispersion for mine ventilation. Journal of Northeast Institute of Technology 6(3): pp41-47.

Li, E.L. & Wang, B.Q. 1986. Study on diffusion coefficient of the transverse turbulence of pollutant in mine tunnel. Journal of Northeast Institute of Technology 5(1):pp91-96.

Li, E.L.; Wang, B.Q. & Wang, Z.C. 1986. Experimental research on turbulent diffusion and dispersion for mine ventilation. Journal of Northeast Institute of Technology 5(2):pp38-43.

Lu, G.L.; Li, C.S. & Xin, S. 2006. Study on the law of concentrative harmful gas spreading in ventilation networks. Journal of Shandong University of Science and Technology, (6):pp120-122.

Wang, Y.M. 1994 Mine Aerodynamics and Mine Ventilation System. Beijing: Metallurgical Industry Press.

Wen, D.S.; Li, Z.N. & Huang, Z.H. 2004. Engineering Hydrodynamics. Beijing: Higher Education Press.

Yu, C.S. 1992. Introduction to Environmental Hydrodynamics. Beijing: Tsinghua University Press.

Risk assessment of spontaneous combustion of sulfide ores based on thermal analysis and AHP method

Z L Chen[1], and H M Zhou[2]

1. Safety engineer, Sinosteel corporation Wuhan safety & Environmental Protection Research Institute, No. 122, Luoshi Road, Hongshan district, Wuhan City, Hubei Province, CN 430070. Email:13627111370@163.com.
2. Senior Engineer, Sinosteel corporation Wuhan safety & Environmental Protection Research Institute, No. 122, Luoshi Road, Hongshan district, Wuhan City, Hubei Province , CN 430070. Email:2363311844@qq.com.

ABSTRACT

The sulfide ores is easily oxidized in air during the mining and processing. A large quantity of heat by the oxidation reaction resulted from spontaneous combustion of sulfide ores. It would not only threaten the life of employees, but also cause economic losses and environmental issues. Therefore, the spontaneous combustion of sulfide ores is one of the major safety problems in mining at present. There are many factors affecting the spontaneous combustion of sulfide ores. However, the spontaneous combustion of sulfide ores is a complex process, it's necessary to consider more about the influence of various factors on the spontaneous combustion tendency of sulfide ores. Analyzing the influence degree of factors on the spontaneous combustion of sulfide ores is conductive to effectively controlling the risk of the spontaneous combustion.

In this paper， based on thermal analysis and AHP method， the risk of spontaneous combustion of sulfide ores was evaluated. The TG-DSC curves under different experimental parameters were obtained by using the thermal analyzer. The activation energy was calculated by the Coats-Redfern method. The activation energy and the initial decomposition temperature were used to determine spontaneous combustion tendency. The risk assessment index system of spontaneous combustion including the target layer, criterion layer and the index layer was established by using AHP method. The judgment matrices were constructed according to the results of thermal analysis. The risk weight of each factor was obtained.

The research shows that the risk weight of the factors following big to small is particle size, weight and heating rate. The quantitative results of safety assessment are obtained based on thermal analysis and AHP method. It provides reference for mine enterprises to assess the risk, so as to reach the goal to prevent and control the risk of the spontaneous combustion.

Key words: risk assessment; sulfide ores; thermal analysis; AHP method;

INTRODUCTION

The spontaneous combustion of sulfide ores is one of the major safety problems in the process of metal mine at present (Yang, 2013)(Wang, 2013). Sulfide ores are easy to be oxidized by air and to cause a temperature increasing, due to the large amount of heat released. The spontaneous combustion may occur when the temperature reaches the self-ignition point. The fire caused by spontaneous combustion results in great economic loss and environmental pollution (Wang, 2014) (Shao, 2014). In order to ensure the mine safety in the process of production, storage and transportation, it is necessary to evaluate the risk of spontaneous combustion of sulfide ores. Therefore the thermal disaster can be effectively controlled by risk control measures.

Thermal analysis, as a test method, has been widely used to study the thermo-oxidation behavior of sulfide ores. The activation energy and the initial decomposition temperature have been recognized as important indexes to determine spontaneous combustion tendency (Lu, 2006) (Wang, 2015). Zhao (2009) studied the sulfide ores absorbability by TPO experiments, and the total quantity of absorption oxygen and the self heating starting temperature were regarded as the identification indexes of sulfide ores spontaneous combustion process. Liu (2006) studied the spontaneous combustion tendency of coal on different heating rate by using thermogravimetric analysis instrument. Yang (2009) (2011) obtained dynamic parameters of spontaneous combustion of sulfide ores using Coats-Redfern method and FWO method, and the activation energy was used as an index to evaluate spontaneous combustion tendency of sulfide ores. However, few scholars quantitatively analyze the influence degree of various factors on spontaneous combustion tendency, so as to establish spontaneous combustion risk assessment system.

Analytic Hierarchy Process (AHP) combining qualitative and quantitative analysis, has been successfully applied to risk assessment in municipal, construction, mining and other fields. Researches show that the risk assessment results obtained by AHP method have certain reliability. Sun (2015) established a complete bridge fire risk assessment system and assessed the fire risk by using the Analytic Hierarchy Process (AHP) and the fuzzy comprehensive evaluation(FCE). Hao (2015) assessed the fire risk of high-rise building construction site by using the Analytic Hierarchy Process (AHP) and put forward the improvement countermeasures. Wang (2017) established the evaluation index of coal and gas outburst danger degree and determine the risk level using the AHP method. However, Analytic Hierarchy Process (AHP) has certain subjectivity in the construction of the judgment matrix, which can be combined with other methods to reduce errors.

In this paper, the Analytic Hierarchy Process method is combined with the thermal analysis method to assess the risk of spontaneous combustion of sulfide ores. The activation energy is calculated by using Coats-Redfern method and the initial decomposition temperature is obtained by TG-DSC curves. Both parameters are used to determine spontaneous combustion tendency. The risk assessment index system is established by the AHP method. The system includes 1 target layer, 2 criterion layers and 3 index layers. The criterion layer includes activation energy and initial decomposition temperature. The index layer consists of the heating rate, weight and particle size. The weights of indexes are obtained to study the influence of various indexes on the spontaneous combustion of sulfide ores, which provide a reference for enterprises to improve the risk management level.

EXPERIMENTAL

Sample preparation

The pyrite used was obtained from the street of Tongling, Anhui Province, China. The chemical compositions of pyrite were obtained by Axios X-ray fluorescence spectrometer, with an effective content of more than 78.8%, as shown in TABLE 1. The X-ray diffraction result (FIG 1) shows that the non-activated pyrite consists of cubic pyrite as a predominant component. The pyrite contains a few impurities, such as, SiO_2 and metal oxides.

TABLE 1 – The chemical composition of pyrite

Elements	Content(%)
Si	2.97
Al	1.33
Fe	42.75
Ca	0.10
Mg	0.12
K	0.097
Na	0.086
S	42.02
O	10.27
Ignition loss	0.25

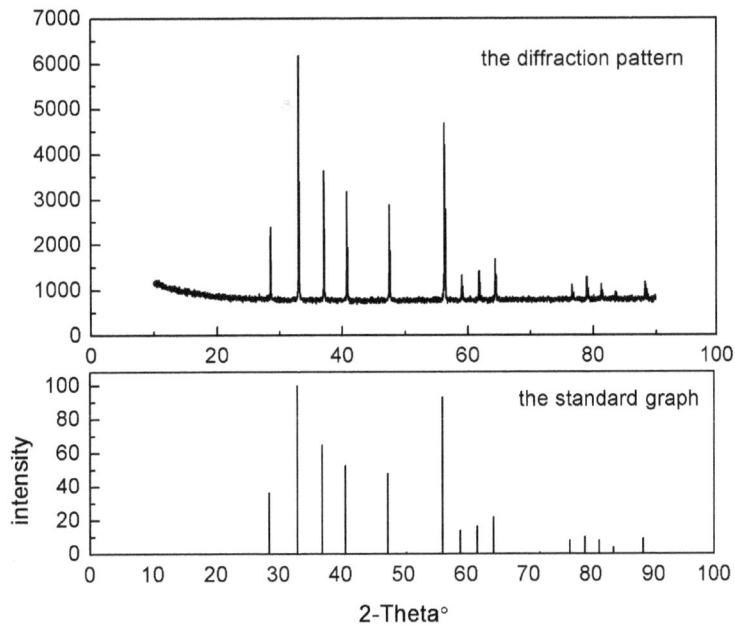

FIG 1 – The diffraction pattern and standard graph of pyrite

Firstly, the pyrite in a mortar was crushed into a particle size less than 1mm in diameter. And then, three samples with different particle sizes of 30-45μm (sample 1), 45-90μm (sample 2) and 90-150μm (sample 3) were separated by 500 mesh sieve, 325 mesh sieve and 100 mesh sieve. The particle size distribution of the samples was measured by laser particle sizer (Mastersizer 2000) in the liquid mode, and the mean particle diameter was obtained (FIG 2). The mean particle diameter of sample 1, Sample 2 and sample 3 is 30.19μm, 74.32μm and 131.10μm respectively.

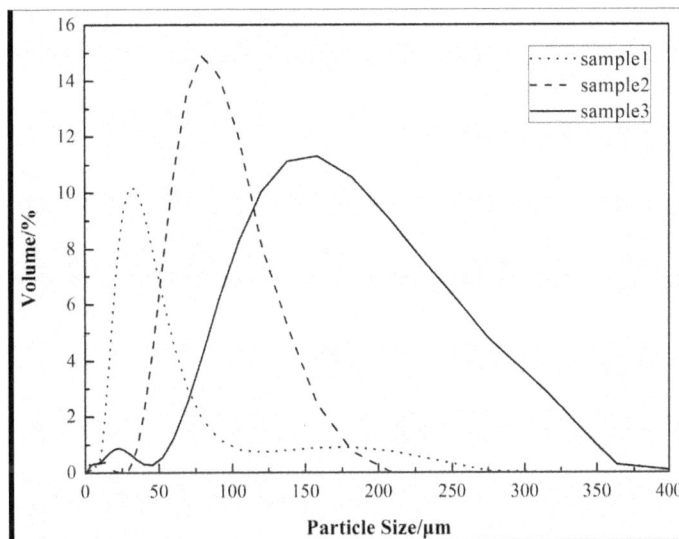

FIG 2 – The particle size distribution of the samples

Experimental process

Thermal analysis was carried out using a simultaneous PerkinElmer STA-6000. The samples in Al_2O_3 crucible were subjected to a temperature scanning programmer, and was heated from 50℃ to 800℃ at the different heating rates of 10 K·min^{-1}、15 K·min^{-1} and 20 K·min^{-1} in a dynamic nitrogen-atmosphere (flow rate : 20cm^3/min). The TG-DSC curves of samples under different conditions were obtained, as shown in Figure 3-5.

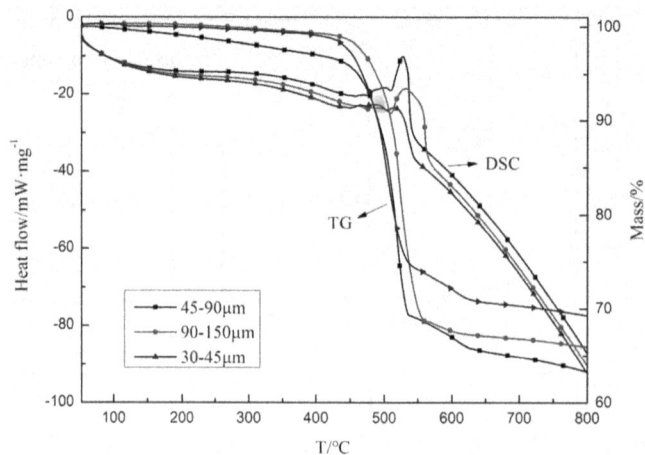

FIG 3 –   TG-DSC curves of samples with different particle size

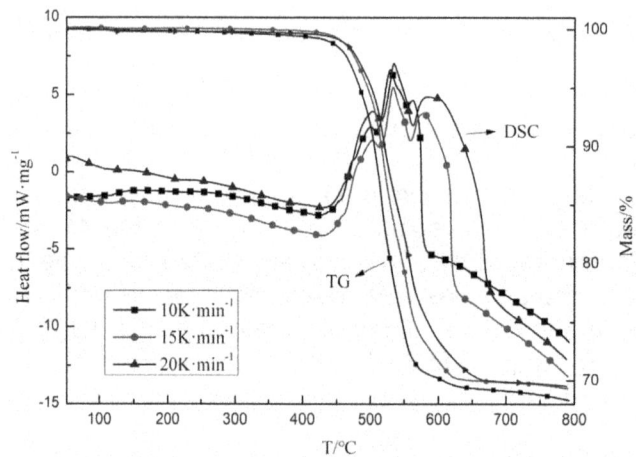

FIG 4 –   TG-DSC curves of samples under different heating rates

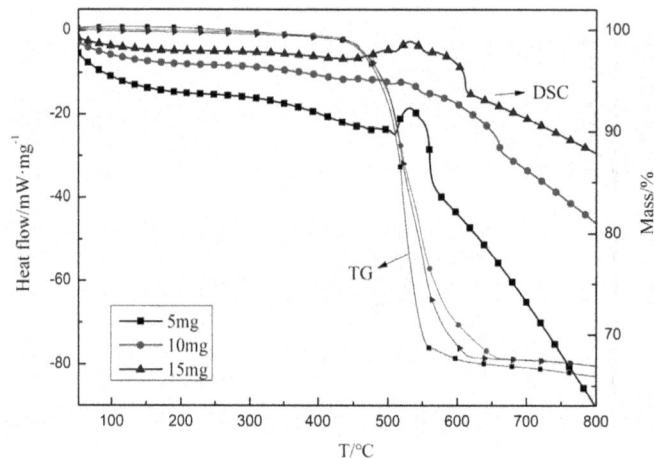

FIG 5 –   TG-DSC curves of samples with different weight

THERMAL ANALYSIS RESULTS

Coats-Redfern method (Ebrahimi-Kahrizsangi, 2008) is used to analyze the samples' spontaneous combustion dynamic characteristics. The reaction process of the samples is approximately considered as the second-order reaction. The corresponding reaction mechanism function is as follows.

$$G(\alpha) = (1 - \alpha)^{-1} - 1 \qquad (1)$$

Where α is the oxidation degree of the reduced samples. α equals $(m_t-m_0)/(m_\infty-m_0)$, where m_t is an actual weight of the sample at time t in the oxidation stage, m_0 is the initial weight of the samples, and m_∞ is the weight after complete samples oxidation.

The kinetic equation in non-isothermal proceed is integrated into the integral equation, as follows.

$$\ln\left(\frac{G(\alpha)}{T^2}\right) = \ln\left(\frac{AR}{\beta E}\right) - \frac{E}{RT} \qquad (2)$$

where R is the universal gas constant, 8.314J/ (mol·K), E is activation energy, A is pre-exponential factor, β is the heating rate, T is an actual temperature of sample oxidation, and T_0 is the initial reaction temperature.

The experimental data and Eqs. (1) are brought into Eqs. (2).The relationships between $\ln(G(\alpha)/T^2)$ and $1/T$ is linear. The results show that the Second-order equation can give a good description for the experimental data. The activation energy could be obtained by the fitting line slope. Therefore, the activation energy (E) and the initial decomposition temperature (T_0) under different conditions are obtained, as shown in TABLE 2.

Table 2 shows that the initial temperature of the sample decreases with the decreasing of the particle size and the temperature-decreasing rate increases. The initial temperature of the sample decreases with the decreasing of the heating rate and the temperature-decreasing rate just has a little change. However, the initial temperature has no change with the increasing of weight. The influence of three factors on the initial temperature can be judged by the change extent and velocity of the initial temperature. Therefore, the influence of three factors on the initial temperature is as follows: particle size > heating rate > weight.

Furthermore, table 2 shows that the activation energy of the sample also decreases with the decreasing of the particle size and temperature-decreasing rate decreases. With the increasing of heating rate，the activation energy increases gradually. However, the activation energy decreases with the increasing of weight. The influence of three factors on activation energy can be judged by the change extent and velocity of activation energy, Therefore, the influence of three factors on activation energy is as follows: particle size > weight > heating rate.

The researchers found that the lower initial temperature and activation energy indicate that the sample has a higher spontaneous combustion tendency. Therefore, the spontaneous combustion tendency decreases with the increasing of the heating rate and particle size. And the spontaneous combustion tendency increases with the increasing of the weight.

TABLE 2 – the activation energy and the initial decomposition temperature under different conditions

	Heating rate(β)/ K·min^{-1} (D=48~50μm,M=10mg)			Weight(M)/mg (D=98~150μm,β=10 K·min^{-1})			particle size(D)/μm （M=5mg, β=10 K·min^{-1}）		
	10	15	20	5	10	15	45~90	90~150	30~45
T_0	489	491	495	491	491	491	483	487	455
E	212	213	222	255	244	232	244	318	206

APPLICATION OF THE AHP METHOD

The Analytic Hierarchy Process (AHP) is a systematic analysis method proposed by the American renowned operation scientist, to solve complex muti-criteria decision problems (Saaty,2007). The basic principle of AHP is to decompose the system into target layer, criterion layer and index layer, and establish an analytic hierarchy model. The factor's relative importance in each layer is determined by comparing the factors at the same layer, so as to construct the judgment matrix by standard comparison table. The weight vector of judgment matrix is obtained by the mathematical method. Finally, The consistency of judgment matrix is checked by calculating consistency ratio. The AHP method provides reliable quantitative basis for analyzing and predicting the risk of target.

Establishing the AHP Assessment Model

The AHP model established in this study is a three-layer tree in which the risk assessment of spontaneous combustion of pyrite is regarded as the target layer. The criterion layer that influences the target layer include the activation energy (E) and the initial decomposition temperature(T_0).E and T_0 are related to the index layer consisted of heating rate, weight and particle size. The system structure is as shown in FIG 6.

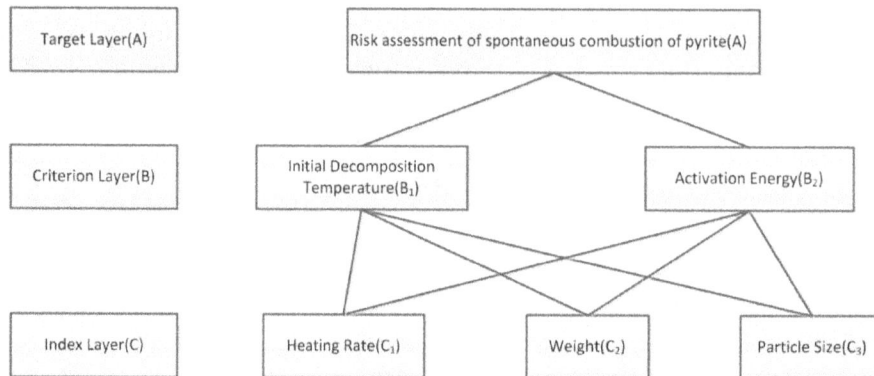

FIG 6 –AHP structural chart of the spontaneous combustion of pyrite

Performing the comparative judgment matrix

The judgment matrix can be constructed by comparing their relative importance with respect to the parent element in the adjacent upper layer. The element x_{ij} of the matrix X-Y represents the relative importance of factor i and factor j. X represents upper layer relative to Y. The value of a_{ij} is determined by using Saaty's 1-9 scale (TABLE 3) (Dong, 2008).The X-Y judgment matrix is as follows.

$$X\text{-}Y = \begin{pmatrix} a_{11} & \cdots & a_{1n} \\ \vdots & \ddots & \vdots \\ a_{n1} & \cdots & a_{nn} \end{pmatrix}$$

In the paper, combining the thermal analysis results (TABLE 2) and the standard comparison table (TABLE 3), the judgment matrix can be obtained. Taking B_1-C as an example, the influence of three factors on the initial temperature is as follows: particle size > heating rate > weight. Therefore, the judgment matrix(B_1-C) can be obtained combined the TABLE 3.

$$B_1\text{-}C = \begin{pmatrix} 1 & 5 & 1/3 \\ 1/5 & 1 & 1/7 \\ 3 & 7 & 1 \end{pmatrix}$$

TABLE 3 – Standard comparison table

Standar a_{ij}	Definition
1	Factor i is as important as factor j.
3	Factor i is as a bit more important as factor j.
5	Factor i is as more important as factor j.
7	Factor i is as much more important as factor j.
9	Factor i is as most more important as factor j.
2、4、6、8	Between two standards above
Count Down	If the factor j is compared with factor i, the value is $a_{ji}=1/a_{ij}$, $a_{ii}=1$

Calculating the Weight of Each Factor

The square root method (Shang, 2008) is utilized to calculate the weight in this context. The calculation steps of the method are as follows. First, the new vector is obtained by multiplying all elements of each row of the judgment matrix. Then, the cube root of every element of new vector is calculated. Finally, the weight vector can be obtained by normalization. The calculation formula is Eqs. (3). Once the weight of single-layer ranking is calculated, the weight of total hierarchy ordering can be calculated by Eqs. (4). The weights of the total hierarchy ordering express the influence of each factor to the target of the AHP hierarchy. The weight values are shown in TABLE 4 - TABLE 7.

$$W_i = \frac{\left(\prod_{j=1}^{n} a_{ij}\right)^{\frac{1}{n}}}{\sum_{k=1}^{n}\left(\prod_{j=1}^{n} a_{kj}\right)^{\frac{1}{n}}} (i = 1,2,3 \dots, n) \tag{3}$$

where W_i is the weight vector of judgment matrix, n is the order of judgment matrix.

$$W = \sum_{j=1}^{m} a_j\, b_{ij} \tag{4}$$

where W is the weight vector of the total hierarchy ordering, a_j is the weight vector of judgment matrix (A-B), b_{ij} is the weight vector of judgment matrix (B-C).

TABLE 4 – The weights of judgment matrix (A-B)

A	B_1	B_2	W_0
B_1	1	0.5	0.333
B_2	2	1	0.667

TABLE 5 – The weights of judgment matrix (B_1-C)

B_1	C_1	C_2	C_3	W_1
C_1	1	5	1/3	0.278
C_2	1/5	1	1/7	0.072
C_3	3	7	1	0.650

TABLE 6 – The weights of judgment matrix (B_2-C)

B_2	C_1	C_2	C_3	W_2
C_1	1	1/3	1/5	0.101
C_2	3	1	1/4	0.226
C_3	5.	4	1	0.673

TABLE 7 – The weights of the total hierarchy ordering

C	A		W
	B1	B2	
	0.333	0.667	
C_1	0.278	0.101	0.160
C_2	0.072	0.226	0.175
C_3	0.650	0.673	0.665

Checking consistency

The judgment matrix (A-B) is a second order operator matrix, which has complete consistency. However, when n>2, it is necessary to check the consistency of the judgment matrix by calculating the consistency ratio(CR). When CR<0.1, the judgment matrix is to be regarded as consistent.

$$CR = CI/RI \qquad (5)$$

where CR is the consistency ratio, RI is the random index which is taken as in TABLE 8, CI the consistency index which is obtained by Eqs. (6).

$$CI = \frac{\lambda_{max} - n}{n - 1} \qquad (6)$$

where λ_{max} is the largest eigenvalue which is obtained by Eqs. (7).

$$\lambda_{max} = \sum_{i=1}^{n} \frac{AW_i^T}{nW_i} \qquad (7)$$

TABLE 8 – Random index values for matrices

n	1	2	3	4	5	6	7	8	9	10
RI	0	0	0.58	0.90	1.12	1.24	1.32	1.41	1.45	1.49

The weight vector of judgment matrix is brought into Eqs. (7) to obtain the largest eigenvalue, and the consistency ratio is obtained by Eqs. (5) and Eqs. (6). By calculating, CR value of the judgment matrix (B$_1$-C) is 0.0559. Therefore, we think the consistency of judgment matrix (B$_1$-C) can be be acceptable. Likewise, the consistency of judgment matrix (B$_2$-C) and can be be acceptable.

CONCLUSION

（1） The activation energy and the initial temperature are obtained by using the thermal analysis, so as to determine the influence of each factor of the target layer on the criterion layer, and to reasonably construct the judgment matrix. Combining the thermal analysis and the AHP method, the subjective error of the traditional AHP method is reduced. Therefore, risk assessment results are accurate and reliable.

（2） By calculating, the weight values of heating rate, weight and particle size on the spontaneous combustion of pyrite were 0.160, 0.175 and 0.665 respectively. The three factors sorted according to the influence on spontaneous combustion of pyrite are particle size, weight and heating rate. The results can provide a reference for mine to assess the risk of spontaneous combustion.

ACKNOWLEDGEMENTS

This research was financially supported by National Key R&D Program of China (No.2018YFC0808100).

REFERENCES

Dong, Y C, Xu, Y F, Li, H Y, Dai, M, 2008. A comparative study of the numerical scales and the prioritization methods in AHP, Eur J Oper Res, 186:229-242.

Ebrahimi-Kahrizsangi, R, Abbasi, M H, 2008. Evaluation of reliability of Coats–Redfern Method for kinetic analysis of non-isothermal TGA, Transactions of Nonferrous Metals Society of China, 18(1):217-221.

Gao S, 2007. Three calculating weights methods in analytic hierarchy process, Science Technology and Engineering, 20(7):1-4.

Hao L, 2015. Fuzzy evaluation on fire risk assessment of high-rise building construction site based on AHP, Hebei Journal of Industrial Science and Technology, 32(3):224-229.

Liu, J, Zhao F J, 2006. Study on effect of heating rate on spontaneous combustion tendency characterization of coal, Safety in Coal Mines, 5(2): 4-6.

Lu, W, Wang, D M, Zhong, X X, Zhou, F B, 2006.. Tendency of spontaneous combustion of coal based on activation energy, Journal of China University of Mining & Technology, 35(2): 201-205.

Saaty, T L, Tran, L T, 2007. On the invalidity of fuzzifying numerical judgments in the Analytic Hierarchy Processs, Math Comput Model, 46:962-975.

Saaty, T L, Shang, J S, 2007. Group decision-making:Head-count versus intensity of preference, Socio Econ Plan Sci, 41:22-37.

Shao, H, Jiang, S G, Wu, Z Y , Zhang, W Q, Wang, K, 2014. Comparative research on the influence of dioxide carbon and nitrogen on performance of coal spontaneous combustion, Journal of china coal society, 39(11):2244-2249.

Sun, B, Xiao, R C, 2015. Bridge fire risk assessment system based on analytic hierarchy Process-Fuzzy Comprehensive Evaluation Method, Journal of Tongji University(Natural Science), 43 (11):1619-1625.

Wang, H J, Xu, C S, Wu A S, Ai, C M, Li, X W, Miao, X X, 2013. Inhibition of spontaneous combustion of sulfide ores by thermopile sulfide oxidation, Minerals Engineering, 49: 61-67.

Wang, D M, Qi, G S, Qi, X Y, Xin, H H, Zhong, X X, Dou, G L, 2014. Quick test method for the experimental period minimum of coal to spontaneous combustion, Journal of China Coal Society，39(11):2239-2243.

Wang, Y, Wang, H H, 2015. Physical nature of the indexes for ranking self-heating tendency of coal based on the conventional crossing-point temperature technique, Journal of China Coal Society, 40(2):377-382.

Wang, J R,2017. Evaluation of coal and gas outburst hazard grade based on analytic hierarchy process and gray relational analysis, Coal, 9:1-4.

Yang, F Q, Wu, C, Shi Y, 2009. Thermal properties of sulfide ores Determined by using Thermogravimetric and Differential Scanning Calorimetric Method, Science & Technology Review, 27(22): 66-71.

Yang, F Q, Wu, C, Liu, H, Pan, W, Cui Y, 2011. Thermal analysis kinetics of sulfide ores for spontaneous combustion, Journal of Central South University(Science and Technology), 42(8):2469-2474.

Yang, F Q, Wu, C, 2013. Mechanism of mechanical activation for spontaneous combustion of sulfide minerals, Transactions of Nonferrous Metals Society of China, (23):276-282.

Zhao, J, Zhang, X K, Wang Y H, 2009. Study on Appraisal Technique of Sulf-ores spontaneous combustion tendency, Journal of Safety Science and Technology, 5(6):105-109.

Study on distribution law of air leakage in gas drainage borehole

J Gao[1], J Liu[2], J Zhang[3] and M Yang[4]

1. Professor, School of Safety Science and Engineering, Henan Polytechnic University, Jiaozuo, China, 454000. Email: gao@hpu.edu.cn
2. Ph.D, School of Safety Science and Engineering, Henan Polytechnic University, Jiaozuo, China, 454000. Email: liujiajia@hpu.edu.cn
3. Master, Division of Exploration Engineering, Pingdingshan Tian'an Coal Industry Co., Ltd., Pingdingshan, China, 467000. Email: 857708604@qq.com
4. Associate Professor, The collaborative Innovation center of coal safety production of Henan Province, Jiaozuo, China, 454000. Email: yming@163.com

ABSTRACT

Gas drainage is one of the important technical means to control gas disaster in coal mine. Air leakage around boreholes has great influence on gas extraction effects and reduces extraction efficiency. In this paper, Gas drainage mathematical model considering the air leakage of boreholes has been constructed and COMSOL Multiphysics software is used to simulate the coal seam gas drainage process. The distribution and quantity of the air leakage under different borehole sealing length are studied. The results show that with the increase of pumping time, the range of air leakage around the borehole expands gradually, and increase reduces after 30 days of extraction. With the increase of sealing length of the borehole, both the amount and the scope of air leakage around boreholes decreases gradually. The decreasing trend of air leakage area around boreholes is obviously slowed down if the borehole sealing depth is close to the depth of surrounding rock fragmentation zone. The negative pressure of pumping has little influence on the amount and scope of gas leakage in boreholes. Air leakage in boreholes increases a little with the increase of negative pressure of pumping. When negative pressure of pumping increased by 10 kPa, the air leakage increases by only 0.5 L/min. With the increase of suction negative pressure, the area of air leakage near roadway side remains basically unchanged, while the range of air leakage around borehole expands radially and axially. The research result of this paper provides a theoretical basis to develop specific measures to reduce air leakage and improving the effect of gas extraction.

Keywords: Gas drainage, Air leakage, Pumping negative pressure, Pumping time

INTRODUCTION

Drilling and extracting coal seam gas is the main measure to control coal mine gas outburst and utilize gas resources. However, air leakage in actual gas drainage borehole often leads to poor gas extraction effect. Therefore, it is necessary to study the effect of borehole leakage on the drainage effect. The conclusion of the study is of great significance for solving the problem of low gas drainage quality and controlling coal gas accidents.

Zhou S, Lin B et al. analyzed borehole sealing mechanism and hole sealing technology, and derived the calculation formula of air leakage in sealing section. Wang Z et al. determined the calculation formula for air leakage in air leakage ring of the borehole, and studied the new grouting sealing method according to the actual situation of a mine. The effectiveness of the sealing was verified on site. Hu S et al. analyzed loose rings of the roadway and borehole, and proposed the mechanism of powder plugging, which played the role of sealing the fracture. Zhang X et al. calculated the range of air leakage ring of the borehole and derived the formula for calculating air leakage of the borehole. Based on this, variation law of the borehole air leakage was analyzed under the parameters of different borehole radius, drainage system suction pressure, sealing length and coal physical properties, which provided a theoretical reference for determination of suitable sealing parameters. According to the multi-media properties of coal body. Xia T derived the gas extraction quality evaluation equation considering gas diffusion in coal matrix system and gas-air seepage in coal seam fissure system, and analyzed gas drainage effect under different fracture opening and different air leakage rates. Lin B et al. improved the technology of gas drainage and sealing of coal seams with high gas and low permeability, and tested the new sealing technology in a working face of Changcun Mine, and the verification results were ideal. To seal the fracture surrounding the borehole, Sun Y et al. combined the borehole air leakage ring model to divide hole sealing into two steps of combination technical approach, namely, the first sealing with pressure grouting and the second sealing with air leakage treatment. Wang G et al. discussed the drainage radius considering drilling interaction based on three-dimensional monitoring of coal seam gas pressure, and provided reference data for numerical simulation to offer practical guidance for the project. Si G et al. studied the sensitivity of drilling slotting performance to various field and operating parameters. In a series of numerical simulations, a large number of geomechanical properties, geostress conditions, groove geometry and spacing of a plurality of grooves were considered. Zhang C et al. obtained experimentally that in the initial stage of coal seam gas extraction, permeability near the borehole decreased rapidly, and the later permeability was positively correlated with the degree of closeness to the borehole. Based on on-site monitoring of the amount of gas emission from the borehole (q(t)). Ti Z et al. obtained the function of the amount variation of gas emission from the borehole over time using multivariate statistical regression method. Zhao D et al. used COMSOL Multiphysics software to carry out three-dimensional numerical simulation of gas seepage around gas drainage boreholes. The gas pressure distribution, gas seepage velocity distribution and permeability change between the two boreholes and around the borehole were given.

Based on theoretical analysis and numerical simulation method, the mathematical control equations of gas drainage considering rock mass deformation, dynamic permeability changes and coal seam gas-air coupled movement was established. The simulation of physical process of gas drainage considering borehole leakage was made using COMSOL numerical simulation software, and the distribution law of air leakage pressure in coal seam was obtained. Through the numerical simulation, the influence law of extraction time, sealing depth, drainage system suction pressure on borehole air leakage range, air leakage and gas drainage quantity, extraction concentration was obtained.

ESTABLISHMENT OF GAS-SOLID COUPLING MODEL FOR GAS DRAINAGE BOREHOLE

Extraction mathematical model

In the actual production process, gas drainage borehole is subject to very complicated influencing factors, and gas migration is controlled by factors like structure, geostress, temperature and pressure of the coal. This paper simplifies the handling of some complex motion processes, and proposes the following hypotheses:

1. With coal body as the elastic pore fissure medium, it is considered that the free gas and the leaked air exist only in the fracture system of coal, and the deformation of the coal body conforms to the small deformation hypothesis under the mining action;

2. It is considered that the coal body is homogenous, the porosity of the coal body is evenly distributed, the coal body only adsorbs gas and has no adsorption effect on other gases;

3. The gas in the coal seam is regarded as an ideal saturated gas, ignoring the heat dissipated by the gas during the adsorption and desorption process, and the coal body is considered to be in a constant temperature state;

4. The gas flow in the coal body meets Darcy's law;

5. Excluding the extreme phenomena such as collapse and plugging of gas drainage boreholes, it is assumed that the borehole is always intact during the extraction cycle.

By borehole air leakage model considering stress deformation of coal and rock mass, gas adsorption-desorption process, gas and air seepage in coal, it is possible to obtain the control equation describing intercoupling state of coal-rock deformation and coal gas-air seepage migration:

$$Eu_{i,jj} + \frac{E}{1-2\theta} u_{j,jj} - \alpha(P_1 + P_2)_i + F_i = 0$$

$$\left(n + P_1 \cdot \frac{1-n}{k_s} + \frac{abcp_1p_n}{(1+bp_1)^2}\right)\frac{\partial P_1}{\partial t} + P_1 \cdot (1-n) \cdot \frac{\partial \varepsilon_v}{\partial t} - \nabla \cdot \left(\frac{k}{\mu} \cdot P_1 \cdot \nabla (P_1 + P_2)\right) = 0$$

$$\left(n + P_2 \cdot \frac{1-n}{k_s}\right)\frac{\partial p_2}{\partial t} + P_2 \cdot (1-n) \cdot \frac{\partial \varepsilon_v}{\partial t} - \nabla \cdot \left(\frac{k}{\mu} \cdot P_2 \cdot \nabla (P_1 + P_2)\right) = 0$$

$$n = 1 - \frac{(1-n_0)}{1+\varepsilon_v}\left(1 - \frac{\Delta (P_1 + P_2)}{k_s}\right)$$

$$k = \frac{k_0}{1+\varepsilon_v}\left[1 + \frac{\varepsilon_v + \Delta (P_1 + P_2)(1-n_0)/k_s}{n_0}\right]^3$$

(1)

Physical meaning of the symbols in the formula：

K is the permeability of the gas in the fracture system, μ is the dynamic viscosity coefficient of the gas, $P = P_1 + P_2$, P_1 is the gas pressure, P_2 is the air pressure. E is the elastic modulus of coal rock mass, θ is the coal body Poisson's ratio. Δp is the coal seam gas, air pressure change. a, b, c are langmuir adsorption constants, n is the coal porosity.

Physical model of bedding gas drainage borehole

This paper mainly studies the effect of air leakage from bedding gas drainage borehole on the gas extraction result. The punching and drilling is generally from the coal roadway to the coal seam. The drilling hole is located between the top and bottom boards of the coal seam. The gas around borehole moves radially around the center of the borehole under the pressure difference between the coal seam and the borehole. The air in the roadway also performs the same migration under pressure difference through the fracture around the borehole. Therefore, vertical cross section of the borehole is selected for two-dimensional geometric model. As shown in Fig.1, the two-dimensional geometric model can also explain the research problem, and reduce the modeling workload while lowering the requirements on computer.

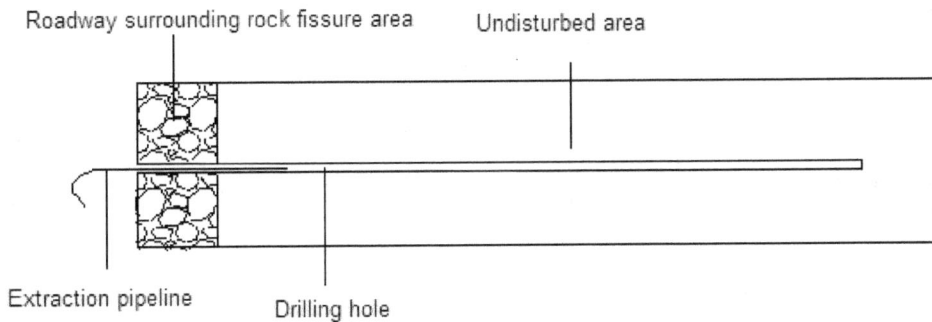

FIG 1– 2D extraction gas drilling section

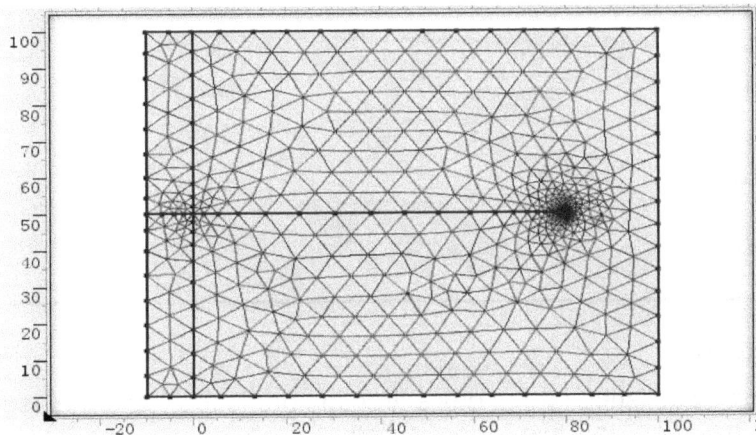

FIG 2 – Schematic diagram of geometric model.

The geometric model is built in the simulation software and meshed as shown in Fig.2. With 100*110m rectangle as coal body, borehole diameter is 100mm, geostress is 8.5 MPa, sealing material support force is 0.2MPa, coal body cohesion force is 100MPa, internal friction angle is 300, roadway width is 5m, coal body compression resistance strength is 7.3 MPa, the Poisson's ratio of the coal rock mass is 0.339, and the residual gas intensity is 1.75 MPa. It can be obtained that the distance of fracture area under roadway extraction effect is 9.25 m, and the section line is located at the coordinates (0,0)~(0,100). The calculation by formula shows that the range of borehole air leakage ring is only 0.169 m, and the relative range of roadway fracture can be ignored.

The section line in parallel direction of the model is the borehole. The meshing method adopts free-subdivision triangle mesh. The left side of the vertical section line in the figure is the pressure relief zone due to roadway excavation. The right side of the section line is the area without coal seam disturbance, and permeability is the original permeability. According to the actual situation of a mine, physical parameters selected by the model are as follows:

TABLE 1– Physical parameters table

Parameter	Expression formula	Value	Unit
The maximum adsorbed gas volume per unit mass of coal	a	28.8436	m^3/t
Coal adsorption constant	b	0.494	MPa^{-1}
Gas density under standard condition	ρ_n	0.717	kg/m^3
Standard atmospheric pressure	p_n	101325	Pa
Initial coal seam porosity	n_0	0.04	-
Initial coal seam permeability	k_0	2.8×10^{-17}	mD
Initial permeability of coal seam in fracture zone	K_{20}	2.8×10^{-15}	mD
Coal moisture	M	0.014	-
Coal ash content	A	0.127	-
Coal density	ρ_s	1380	kg/m^3
Convective mass transfer coefficient	D_q	5.924×10^{-7}	m/s
Coal particle radius	r_0	1.875	mm
Coal matrix diffusion coefficient	D	5.599×10^{-11}	m^2/s
Geostress	p_d	8.5	MPa
Poisson's ratio of coal	θ	0.339	-
Elastic modulus of coal body	E	2863	MPa
Gas dynamic viscosity	μ	1.08×10^{-5}	Pa·s
Gas drainage borehole radius	r	0.05	m
Initial gas pressure of coal seam	p_0	1.57	MPa
Drainage system suction pressure	p_c	15	kPa

Definite conditions

The control equation established for the gas extraction process by gas drainage borehole concerns a solid mechanics equation and two partial differential equations. The solid mechanics equation mainly controls coal body deformation, and one partial differential equation controls the gas migration of the entire coal seam while the other partial differential equation controls air migration within the coal seam. The physical forces include coal body skeleton displacement u, coal seam gas pressure P_1, air pressure P_2 and adsorption gas content of the coal seam. The following is the analysis of the involved definite conditions.

1) Initial conditions

Ignoring the coal seam gas dissipation during roadway excavation and drilling, it is assumed that the coal seam gas pressure is the initial coal seam gas pressure at t=0, and the initial displacement amount is assumed as u=0. That is, the coal body does not undergo displacement deformation, and the specific expression formula is as follows:

$$\left.\begin{array}{l} P_1 \big|_{t=0} = P_0 \\ P_2 \big|_{t=0} = 0 \\ u \big|_{t=0} = 0 \end{array}\right\} \tag{2}$$

Where, P_1 is the gas pressure; P_0 is the initial gas pressure; P_2 is air pressure; U is the displacement.

2) Boundary conditions

The upper and lower boundary of coal body stress is defined as a fixed constraint, that is, no displacement occurs, while the left and right of the coal body is defined as stick support, and a certain body load is set inside the coal body. The boundary load is set for the upper boundary of the coal body to represent the geostress. The specific expression is as follows:

$$\left.\begin{array}{l} F_A = pd \\ F_V = gamma \end{array}\right\} \tag{3}$$

Where, F_A is the boundary load; p_d is the geostress magnitude set as 8MPa. For details, refer to Table 1 of the physical parameters; F_V is body load of the coal body; $gamma$ is a variable.

All the boundary conditions of the flow field adopt the Dick boundary condition, that is, the first type of boundary condition. Assume that the top and bottom boards of the coal seam are airtight and the gas pressure at infinity maintains the initial gas pressure, and the model roadway boundary is set under atmospheric pressure, as follows:

$$\left.\begin{array}{l} Left\ border: \qquad P_1 = 0, P_2 = P_n \\ Up, down\ and\ right\ borders: P_1 = P_0, P_2 = 0 \\ Pumping\ negative\ pressure: \quad P_1 + P_2 = P_C \end{array}\right\} \tag{4}$$

Where, p_c is the gas extraction pressure; P_n is the standard atmospheric pressure, and the specific values are shown in Table 1.

SIMULATION RESULTS AND ANALYSIS

Analysis of the influence of extraction time on borehole air leakage

The sealing depth is set to 4m, the permeability of the roadway fracture zone is $2.8e^{-15}$ mD, the initial permeability of the undisturbed zone is $2.8e^{-17}$ mD, the drainage system suction pressure is 15 kPa, and the simulated extraction time is 60days. The variation of borehole air leakage range and the variation law of borehole air leakage with the extraction time are obtained at 10days, 30days and 60days' extraction. The results are shown in the figure below.

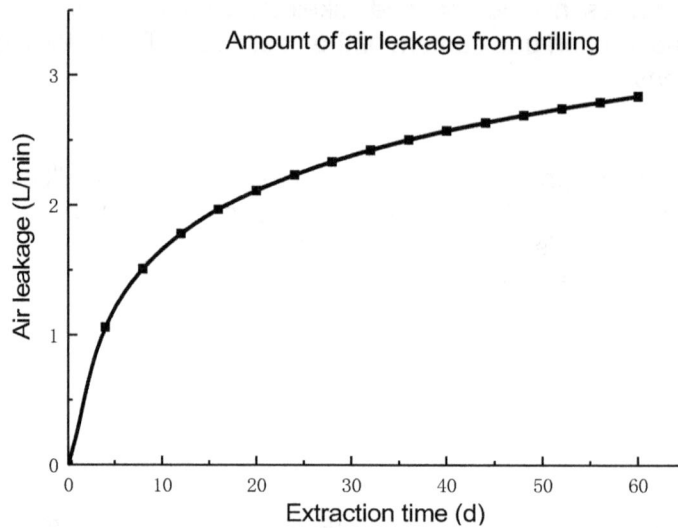

FIG 3 – Drilling air leakage changes over time

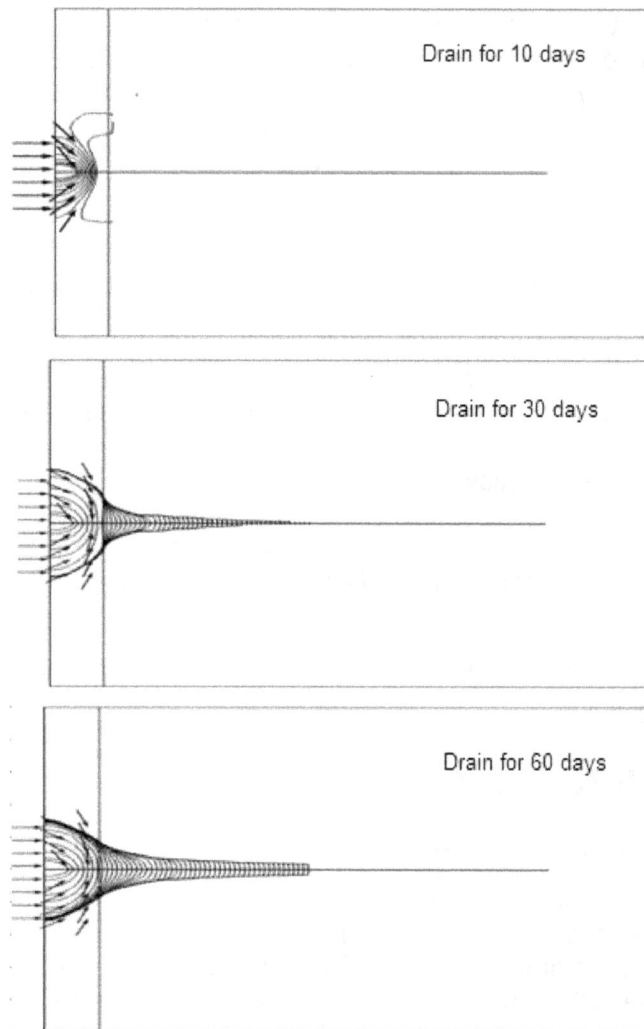

FIG 4 – Different extraction time borehole leakage flow chart

It can be seen from the figure that under the action of drainage system suction pressure, the air around the roadway enters the air drainage borehole through the fracture zone and a small area around the borehole in radial direction. It can be seen from Fig.3 that as the extraction time increases, the borehole air leakage increases, but the increase amplitude will gradually decrease; as can be seen from Fig.4, as the extraction time increases, the borehole air leakage area is gradually enlarged along axial and radial directions of the borehole. The air leakage is mainly concentrated in the fracture area around the roadway, and for an area farther away from the periphery of the roadway, air leakage range around the borehole is smaller.

Analysis of the influence of sealing depth on borehole air leakage

The determination of sealing depth is one of the key parameters to ensure gas drainage quality. Under different sealing depths of 4m, 6m, 8m, and 10m, the variation law of the borehole air leakage is simulated as shown in the figure below.

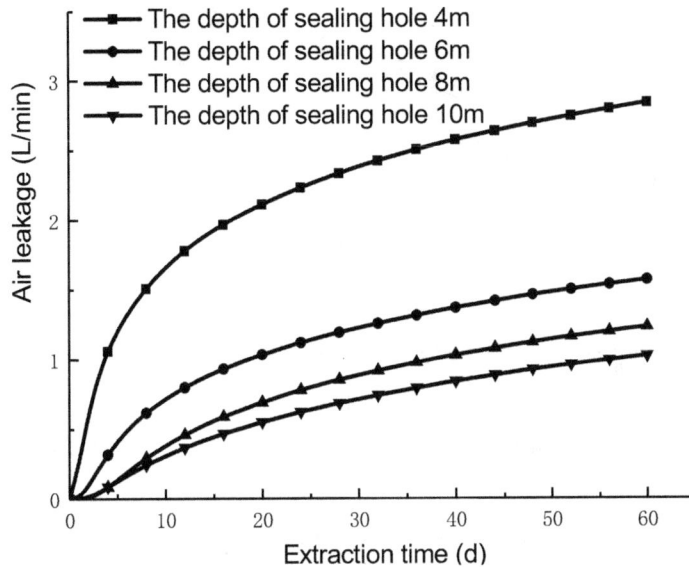

FIG 5 – Different depth of the hole drilling air leakage

It can be seen from Fig.5 that the borehole air leakage gradually decreases with the deepening depth of the hole sealing. When the hole sealing is 4m deep, air leakage is about three times that of 10m-deep hole sealing. As can be seen from Fig.6, as the hole sealing depth is deepened, the circumference of the borehole is gradually lowered. When the hole sealing depth is close to the depth of fragmentation circle of surrounding rock of the roadway, borehole air leakage range narrowing is obviously slowed down; it shows that deepening the hole sealing depth can effectively increase difficulty of borehole air leakage. As the hole sealing depth is deepened, the distance for the roadway air to enter the borehole is extended, and then it becomes more difficult for the air to enter the borehole.

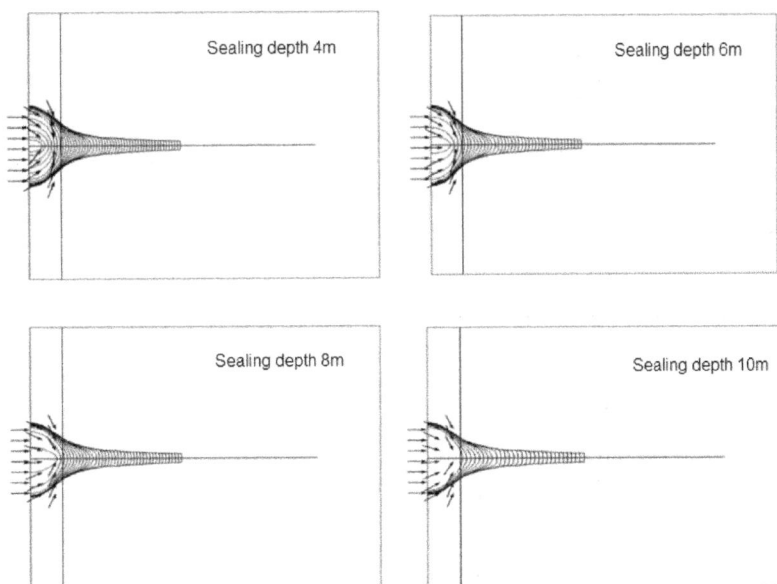

FIG 6 – Borehole leakage line drawing at different hole sealing depth

Analysis of the influence of drainage system suction pressure on borehole air leakage

The borehole air leakage and the leakage flow pattern of the borehole with sealing depth of 4m and drainage system suction pressure of 15kPa, 25kPa and 30kPa are shown in the figure below.

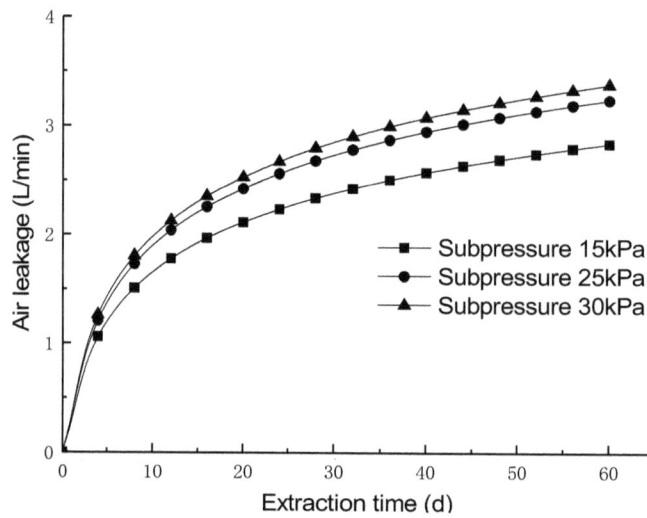

FIG 7 – Air leakage of different negative pressure

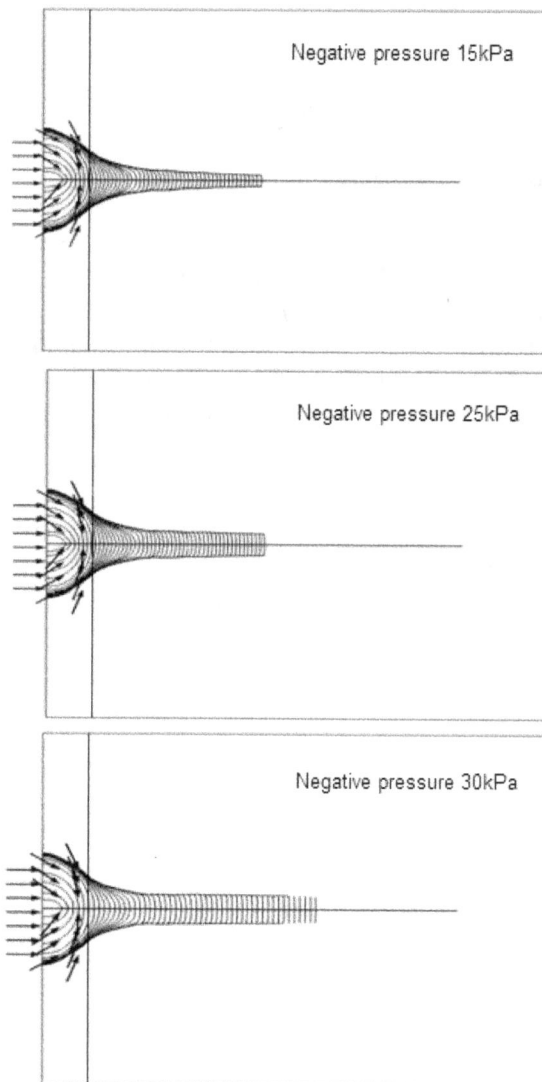

FIG 8 – The leakage flow chart of different negative pressure borehole

It can be seen from Fig.7 that the borehole air leakage increases to some extent with the increase of drainage system suction pressure, but the increase is small, the negative pressure is increased by 10kPa, while the air leakage is increased by about 0.5L/min. As can be seen from Fig. 8, as the drainage system suction pressure rises, the area of the leakage airflow line at the roadway does not expand or decrease, and the range of the leakage airflow around the borehole extends along the radial direction of the borehole to some extent, but the increase of the influence range is very slight, indicating that the increase in drainage system suction pressure has little effect on the amount and range of borehole air leakage.

CONCLUSION

1. As the extraction time increases, the influence range of borehole leakage gradually expands, and the increase rate is increasingly smaller.

2. With the increase of hole sealing depth, the borehole air leakage is gradually reduced, and the leakage range around the borehole is gradually reduced. When the hole sealing depth is close to the depth of fragmentation circle of surrounding rock of the roadway, borehole air leakage range narrowing is obviously slowed down.

3. The borehole air leakage increases to some extent with the increase of drainage system suction pressure, but the influence range is small. As the drainage system suction pressure increases, the range of the air leakage area at the roadway is basically unchanged, while the range of the leakage airflow line around the borehole extends to the periphery to some extent.

REFERENCES

HU SY, LIU HW, 2016. Leakage Mechanism of Coal Seam Gas Drainage Borehole and Its Application Rsearch Progress. Safety in Coal Mines. 47(5):170-173.

HU SY, 2014. Seepage Characteristics around Gas Drainage Borehole and Blockage Mechanism of Powder. Xuzhou:China university of mining and technology.

LI ZJ, CHEN Y, LIN WQ, 2014. Study on forming mechanism and stability of three-phase foam with cement ash. Journal of Safety Science and Technology. (11):54-59.

SUN YN, LU WY, YANG K et al, 2015. Study on disposal technology of air leakage around borehole by three-pouch closure device. Journal of Safety Science and Technology. (3):67-72.

Si, GY, 2019. Parametric Analysis of Slotting Operation Induced Failure Zones to Stimulate Low Permeability Coal Seams. Rock Mechanics Rock Engineering. 52(1):163-182.

Ti ZY, ZHANG F, PAN J, 2018. Permeability enhancement of deep hole pre-splitting blasting in the low permeability coal seam of the Nanting coal mine. Plos One. 13(6).

WANG ZF, ZHOU Y, SUN YN, WANG YL, 2015. Novel gas extraction borehole grouting sealing method and sealing mechanism. Journal Of China Coal Society. 40 (3):588-595.

WANG XW, LI H, 2015. Research on Gas Extraction Leakage Mechanism and New Technology of Plug. Coal Technology, 34(3):121-123.

Wang G, Xu H, Wu MM, Wang Yue, Wang Rui, Zhang XQ, 2018. Porosity model and air leakage flow field simulation of goaf based on DEM-CFD. Arablan Journal Of Geosciences, 148:1-17.

XIA TQ, 2015.Multi-physics Coupling Mechanism of Co-existence Hazards for Coal Spontaneous Combustion and Gas. Xuzhou: China university of mining and technology.

ZHANG XL, 2013. Experimental study on gas leakage mechanism and hole sealing technology of gas extraction borehole in this coal seam. Jiaozuo: Mining engineering of henan university of technology.

Zhang CL, 2018. Dynamic Evolution of Coal Reservoir Parameters in CBM Extraction by Parallel Boreholes Along Coal Seam. Transport In Porous Media, 124（2）:325-343.

Zhao D, 2018. Study on gas seepage from coal seams in the distance between boreholes for gas extraction. Journal Of Loss Prevention In The Process Industries, 54:266-272.

ZHANG C, LIN BQ, ZHOU Y et al, 2013. Strong-weak-strong borehole pressurized sealing technology for horizontal gas drainage borehole in mining seam. Journal of Mining & Safety Engineering, 30(6):935-939.

Ventilation air methane: a simulation of an optimised process of abatement with power and cooling

F J Nadaraju [1], A R Maddocks [2], J Zanganeh [3] and B Moghtaderi [4]

1. MAusIMM, PhD Candidate, The University of Newcastle (Australia), Callaghan, NSW, 2308.
   Email: francis.nadaraju@uon.edu.au
2. Research Associate, The University of Newcastle (Australia), Callaghan, NSW, 2308.
   Email: andrew.maddocks@newcastle.edu.au
3. Research Associate, The University of Newcastle (Australia), Callaghan, NSW, 2308.
   Email: jafar.zanganeh@newcastle.edu.au
4. Professor, The University of Newcastle (Australia), Callaghan, NSW, 2308.
   Email: behdad.moghtaderi@newcastle.edu.au

ABSTRACT

Ventilation air methane (VAM) is ultra-low concentration methane emitted from an underground coal mine. This is a consequence of the high ventilation air volumes (up to 600 m^3/s) that are circulated through the mine to ensure that the methane remains below a safe operating concentration, typically less than 1 vol. % CH_4. Currently, all underground coal mines in Australia vent VAM into the atmosphere, contributing to Australia's national greenhouse gas inventory. In 2016, the Australian Government reported that fugitive emissions of methane from underground coal mines was approximately 19.0 million tonnes (CO_2-equivalent) which was about 4.0 % of Australia's national greenhouse gas emissions. The abatement of VAM therefore becomes important.

The abatement of VAM in a fluidised-bed reactor where the process heat was recovered to produce power in a gas turbine and cooling via absorption chillers has been discussed in earlier work (Nadaraju *et al.*, 2018). The process was simulated using Aspen$^+$ process simulation package to determine the minimum methane concentration to operate the plant and concurrently maintain the oxidation of methane. However, a condensing recuperator and low temperature gas expansion was considered. This work describes an optimised process of heat recovery from the fluidised-bed VAM abatement reactor and the production of power and cooling. For a ventilation flow rate of 20 m^3/s (equivalent to a single abatement unit), the minimum methane concentration for a direct gas turbine was 0.45 vol. % at a reactor temperature of 630 °C and reactor pressure of 1.5 bar. An indirect gas turbine process operated with a minimum methane concentration was 0.4 vol. % at a reactor temperature of 630 °C, compressor outlet pressure of 4.0 bar and turbine flow rate of 2.2 kg/s.

The abatement process is modular and would require 15 units for the ventilation flow of 300 m^3/s. At 0.45 vol. % a total of 5 700 kW_R of cooling would be produced by the direct gas turbine and direct-fired chiller while 5 775 kW_R would be produced by the direct gas turbine and indirect-fired absorption chiller. The indirect gas turbine and indirect-fired absorption chiller would produce 5 100 kW_R of cooling. The cooling produced would offset the cooling requirement for a gassy coal mine with a high geothermal gradient. Furthermore, this translates into an operating cost saving of approximately A $ 0.5 M for the direct gas turbine processes and A $ 0.45 M for the indirect gas turbine process.

INTRODUCTION

Coal has been reported to be Australia's most valuable export in 2018 (Latimer, 2018) and valued at A $ 67.0 billion by the Department of Industry, Innovation and Science (Latimer, 2018). Furthermore, the mining and utilisation of coal has a positive impact on the Australian economy through employment; the supply of electricity and income earned for Australia through exports (Karp, 2018). However, fugitive emissions from underground coal mines in Australia amounted to 19.0 million tonnes of CO_2-equivalent in 2016 (Commonwealth of Australia, Department of the Environment and Energy, 2016). This was about 75 % of fugitive emissions of methane from coal mining or 4.0 % of Australia's national greenhouse gas emissions (Commonwealth of Australia, Department of the Environment and Energy, 2016). Abating emissions from underground coal mines has the potential for significant reductions in greenhouse gas emissions. Additionally, greenhouse gas emissions and climate change concerns are now becoming important for future coal mine approvals (Gloucester Resources Limited v Minister for Planning [2019] NSWLEC 7, 2019).

Fugitive emissions from underground coal mines such as ventilation air methane (VAM) are typically below 1.0 vol. % due to legislative requirements (New South Wales: (Work Health and Safety (Mines) Act 2013, 2013); Queensland: (Coal Mining Safety and Health Act 1999, 1999). The combination of high ventilation air flow rates of up to 600 m^3/s (Tremain et al., 2017) and low methane concentration results in significant quantities of methane which are emitted to the atmosphere (Commonwealth of Australia, Department of the Environment and Energy, 2016) . The abatement of methane in the ventilation air from underground coal mines is therefore a priority for the mining industry, governments and other international organisations engaged in the mitigation of the greenhouse gas (GHG) emissions (Commonwealth of Australia, The Treasury, 2011) .

Through the employment of VAM abatement technology, the methane content of ventilation air is converted to carbon dioxide; reducing the global warming potential (GWP) of the ventilation air (GWP of methane is 25 times higher than that of carbon dioxide). This conversion reduces the overall VAM greenhouse gas emissions on a carbon dioxide equivalent basis by 89%.The heat produced by the combustion of methane can also be used to produce power as well as cooling (Su et al., 2005).

However, the concentration of VAM below the flammability limit of methane (Karacan et al., 2011), does not facilitate the adoption of traditional combustion techniques such as flares, gas engines and turbines. Therefore, abatement technologies have been developed to achieve the oxidation of the low-concentration methane through high temperature thermal and catalytic processes. Of these technologies, the thermal flow-reversal reactor (TFRR) is commercially available and has been installed at mine sites (Q. Li et al., 2015). Other emerging technologies include catalytic flow-reversal reactors (CFRRs) (Z. Li et al., 2017); fluidised-bed reactors (Tremain et al., 2017); chemical looping combustion (Zhang et al., 2014) and porous burners (Wood et al., 2009). The emerging VAM abatement technologies range between laboratory-scale to pilot-plant scale.

The process heat from the oxidation of methane in the ventilation air can be used to sustain the abatement process (Salomons et al., 2003). In this case, the minimum methane concentration is defined where the combustion of methane produces sufficient heat to maintain the conversion of methane. In many studies (Karakurt et al., 2011) the electrical power requirements of the plant are not considered which has a negative impact on the operating cost and the levelised cost of abatement.

Extending power to VAM abatement plants may be impractical or cost prohibitive for some mines, particular over the life of the mine where ventilation shafts are reconfigured or new ventilation shafts are required as the mining area progresses. Although the use of the process heat to maintain the combustion of methane and meet the electrical power requirements of the plant would increase the minimum methane concentration, this arrangement has the benefit of operating the VAM abatement plant without an external power supply. Moreover, for a gassy underground coal mine with a geothermal gradient, the use of absorption chillers would provide operating cost savings through the reduction in electrical power (Broodryk et al., 2015) compared to traditional electrically-driven vapour compression refrigeration.

The recovery of the process heat has been studied in TFRRs. The average operating temperature of the TFRR is 1 000 °C (Gosiewski et al., 2015). The minimum methane concentration in a demonstration TFRR was shown to be 0.2 vol. % for autothermal conditions (Gosiewski et al., 2015) while stable heat recovery was noted at 0.42 vol. % (Gosiewski et al., 2015). At a mine-site in China, a TFRR produced electricity at a minimum methane concentration of 0.6 vol. % (Q. Li et al., 2015) whereas for methane abatement only, the minimum methane concentration was 0.25 vol. % (Q. Li et al., 2015). The high operating temperature of the TFRR would incur high operating costs and have safety concerns.

The replacement of part of the fixed-bed in a TFRR with catalyst results in a CFRR (Su & Agnew, 2006) with an operating temperature up to 670 °C (Z. Li et al., 2017). High conversions of methane can be achieved in CFRRs through the use of expensive catalysts (such as platinum and palladium) that require elaborate catalyst preparation methods (Kucharczyk et al., 2017). Also, dust in the ventilation air could deactivate the catalyst, negatively influencing the performance of the CFRR (Su & Agnew, 2006). In two pilot-scale CFRR studies, Wang et al. (2014) reported a minimum methane concentration of 0.4 vol. % when superheated steam was produced while Z. Li *et al.* (2017) noted a minimum methane concentration of 0.4 vol. % at stable heat recovery. Z. Li *et al.* (2017) also reported a minimum methane concentration of 0.2 vol. % at autothermal operation.

The fluidised-bed reactor has been studied by Tremain *et al.* (2017) for VAM abatement. An operating temperature range between 630 and 740 °C was reported in the Tremain *et al.* (2017) study as well as the ability of the fluidised-bed reactor to handle the variability in methane concentration and methane flow rate better than fixed-bed reactors. However, the fluidised-bed reactor has a higher pressure drop when compared to fixed-bed reactors (Su *et al.*, 2005) leading to a higher operating cost from increased consumption of electricity. However, it is necessary to quantify the minimum methane concentration in fluidised-bed VAM abatement plants in order to assess their technical feasibility in comparison to other emerging VAM abatement technologies.

In this study the recovery of heat from the flue gases of a fluidised-bed ventilation air methane unit was modelled using Aspen[+] to determine the minimum methane concentration for self-sustaining operation. The aim of the work was to identify the minimum methane concentration to achieve VAM abatement and to convert the process heat into power in a gas turbine to operate the plant. Cooling was produced in an absorption chiller where residual process heat was available. The study is an optimisation of work by Nadaraju *et al.* (2018) where the process heat was utilised in a gas turbine to produce power for operating the plant. Nadaraju *et al.* (2018) considered a condensing recuperator which would increase the plant cost through an increased area for heat exchange as well as low temperature gas expansion which resulted in a high minimum methane concentration.

METHODOLOGY

The recovery of the process heat from a fluidised bed ventilation air methane unit was studied in Aspen[+]. Five options were identified (TABLE 1) where the heat was converted into power via a gas turbine and cooling via an absorption chiller. Schematics of each process are presented in FIG 1. Option 1 and Option 4 produced power only via a direct and indirect gas turbine respectively. Options 2, 3 and 5 produced both power and cooling. It was not possible to produce both power and cooling via an indirect gas turbine and direct-fired absorption chiller since the turbine exhaust temperature exceeded the operating limits of the chiller (see Section 3.2.4). For all options 20 m³/s of ventilation air at 30 °C was processed. This represents the flow rate of one fluidised-bed reactor module.

In Option 1 (FIG 1 (a)) the ventilation air was initially compressed before being preheated in the recuperator and supplied to the fluidised-bed reactor. The high-temperature flue gases downstream of the cyclone and electrostatic precipitator (maximum operating temperature of up to 850 °C (Parker, 2003)) were supplied to the gas turbine. In Options 2 and 3, the flue gas downstream of the recuperator is cooled to provide a heat source for the absorption chiller. A heat exchanger in Option 3 was required to produce hot water for supply to the indirect-fired chiller.

For an indirect gas turbine the heat from the flue gases was transferred to a compressed air stream via a heat exchanger (FIG 1 (d)). The turbine exhaust stream was cooled to provide the heat source for the absorption chiller (FIG 1(e)). The flue gases downstream of the recuperator was fixed at 110 °C to prevent the water in the flue gas stream (produced from the methane combustion reaction) from condensing. In a related study, Nadaraju *et al.* (2018) maximised the utilisation of the process heat by minimising the flue gas outlet temperature. In that case the recuperator would effectively be a condenser, increasing the cost of the heat exchanger due to an increased surface area.

TABLE 1 – Summary of VAM abatement processes

Option	Power production	Cooling production
1	Direct gas turbine	None
2	Direct gas turbine	Direct-fired absorption chiller
3	Direct gas turbine	Indirect-fired absorption chiller
4	Indirect gas turbine	None
5	Indirect gas turbine	Indirect-fired absorption chiller

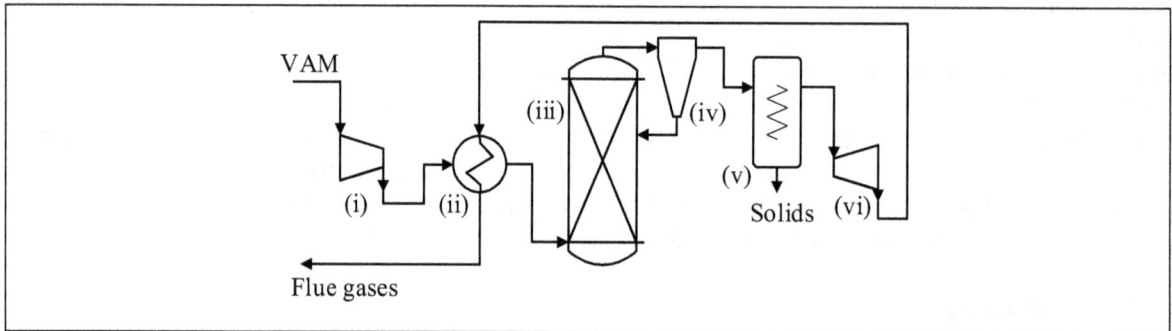

(a) Option 1: Direct gas turbine process

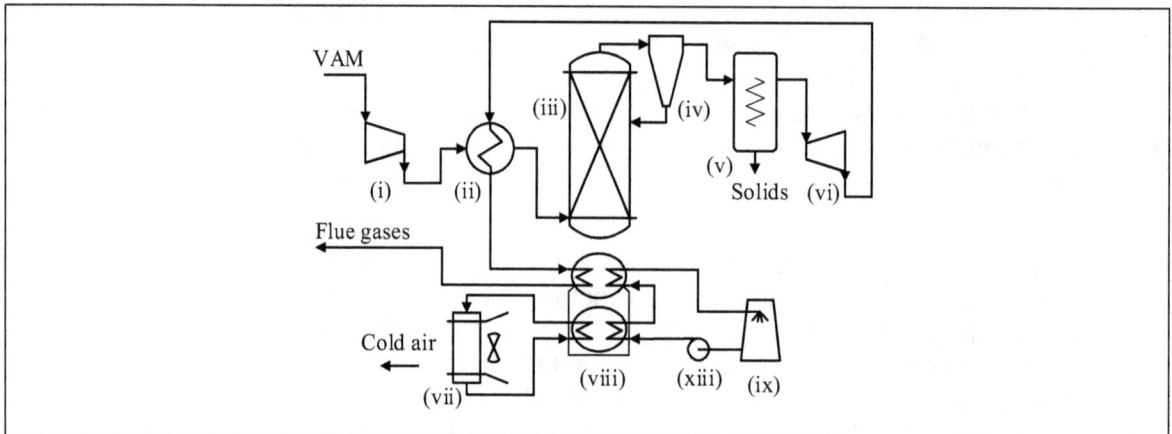

(b) Option 2: Direct gas turbine and direct-fired absorption chiller process

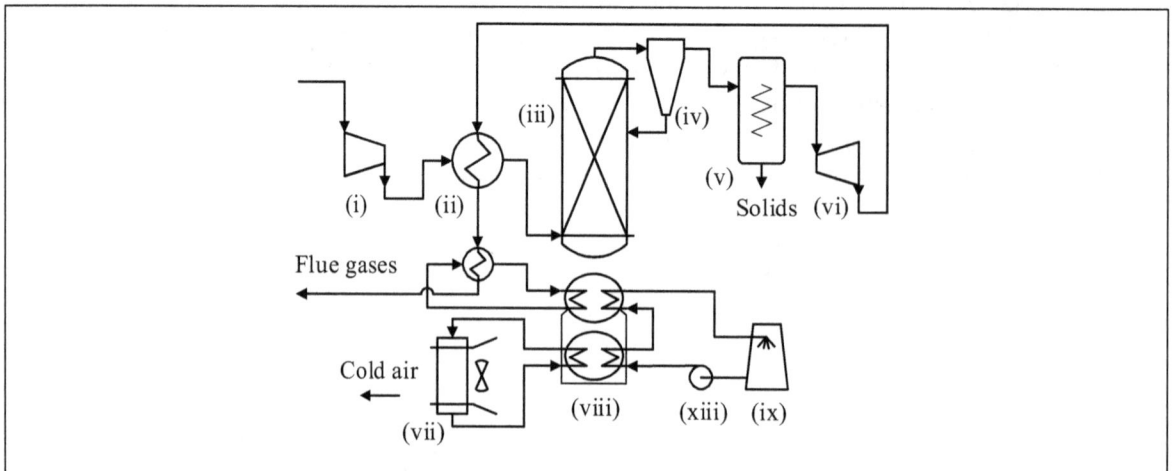

(c) Option 3: Direct gas turbine and indirect-fired absorption chiller process

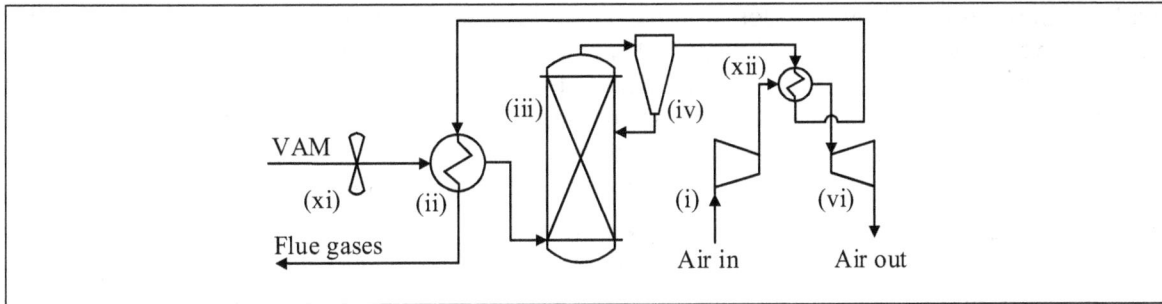

(d) Option 4: Indirect gas turbine process

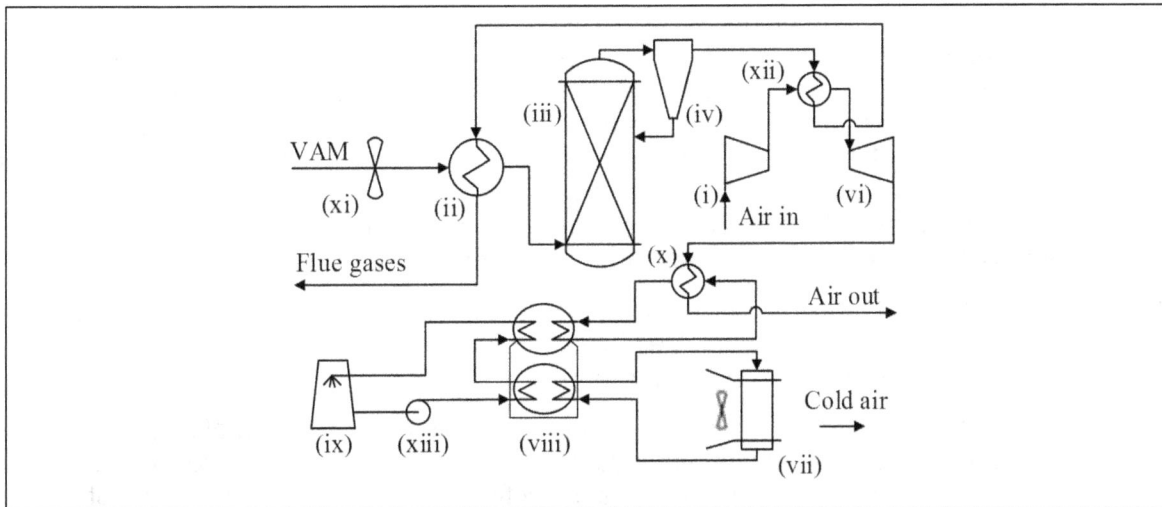

(e) Option 5: Indirect gas turbine and indirect-fired absorption chiller process

FIG 1 – VAM abatement processes producing power and cooling showing a (i) compressor, (ii) recuperator, (iii) fluidised bed reactor, (iv) cyclone, (v) electrostatic precipitator, (vi) gas turbine, (vii) air cooler, (viii) absorption chiller, (ix) cooling tower, (x) water heat exchanger, (xi) fan, (xii) air heat exchanger and (xiii) pump.

The inlet methane concentration of the ventilation air; fluidised-bed reactor temperature and compressor outlet pressure were studied.

The methane concentration was varied between 0.1 and 1.0 vol. %.

The fluidised-bed reactor temperature range was between 500 and 800 °C based on related work where iron-oxide (Zhang *et al.*, 2014) and stone dust (Shah *et al.*, 2015) were recommended for the abatement of low concentration methane. The type of reactor was not considered important in this work; rather, the temperature of the flue gases was of primary concern and assumed to be the same temperature as the bed. In a pilot-scale study of a fluidised-bed VAM abatement reactor, Tremain *et al.* (2017) identified a difference in temperature between the fluidised bed and section above the bed.

The fluidised-bed reactor was modelled in Aspen⁺ using the "RGibbs" reactor block. This reactor block models single phase chemical equilibrium through the minimisation of the Gibbs free energy. A 99 % conversion of methane (Tremain *et al.*, 2017) was considered in the simulation where a bypass stream was used.

To ensure that the methane was oxidised within the fluidised-bed and not upstream of the reactor, the inlet temperature of the VAM was restricted to a maximum temperature 600 °C, which is below the auto-ignition temperature of lean methane-air mixtures (Robinson & Smith, 1984).

The pressure of the air was varied between 1.5 and 4.0 bar. The property method selected for the simulation was the Peng-Robinson equation-of-state. This property method can adequately model the behaviour of non-polar and weakly polar vapour mixtures at high pressures (Pratt, 2001). The largest pressure drop in the system was the fluidised-bed and was calculated at 8.7 kPa using a correlation recommended by Rhodes (1998). This was based on a fluidised bed of 20 tonnes of ilmenite having an average particle size of 161 µm. The bed voidage was 0.4 while the solids density was 4,000 kg/m^3 (the ventilation air density was 0.4 kg/m^3). In addition, the gas distribution system accounted for 30 % of the fluidised-bed pressure drop (based on experience of the research group in operating fluidised beds). A conservative estimate of 15 kPa for the system pressure drop was used based on pressure losses across the fluidised-bed, heat exchangers, piping and pipe fittings.

Heat losses of 49 kW were assumed for Options 1, 2 and 3 while 36 kW of heat losses were assumed for Options 4 and 5. The dimensions of a full-scale fluidised-bed reactor (20 m^3/s) was scaled-up based on a 1.0 m^3/s pilot-scale reactor (Tremain et al., 2017). Heat losses were calculated for the full-scale reactor, cyclone and electrostatic precipitator using correlations for natural and forced convection published in the literature (Incropera & DeWitt, 1996). The reactor dimensions of length 7.0 m, width 4.3 m and height of 6.3 m was determined from the gas-to-solids ratio of 0.5 m^3/s per tonne. Based on an average wind speed of 7.5 km/h(Commonwealth of Australia, Bureau of Meteorology, 2018) and air temperature of 25.2 °C (Commonwealth of Australia, Bureau of Meteorology, 2018) measured at a weather station in Singleton, NSW; heat losses were calculated for free and forced convection where the sides of the reactor and electrostatic precipitator were assumed to be rectangular plates and the cyclone was assumed to be a cylinder. The average value of free and forced convection for the reactor, electrostatic precipitator and cyclone was 49 kW. For the reactor and cyclone, the average value of free and forced convection was 36 kW.

All heat exchangers were modelled using a "Heatx" heat exchanger block. This block calculates the heat exchanger duty and heat exchanger area by specifying a hot and cold stream. The compressor and gas turbine efficiencies were 87 % (Razak, 2007).

A water/lithium bromide absorption chiller was selected for this work. Srikhirin et al. (2001) reported that the alternative, a water/ammonia absorption chiller, was more suited to commercial and industrial refrigeration. Although ammonia poses a low flammability risk, it is an irritant of the eyes, skin and mucous membranes (Calm, 2011). Within a South African mining context, mine cooling is an established technology where ammonia chillers are located a minimum of 200 m away from the closest air intake to the underground workings (SABS-Standards-Division, 2014). It is likely that the VAM abatement plant would be installed adjacent to the ventilation shaft and although an equivalent Australian standard for the location of ammonia refrigeration plants is not available (Van den Berg, 2017); the use of a water/ammonia absorption chiller was not selected.

The temperature of the heat source supplied to a single-effect lithium-bromide absorption chiller ranges between 80 and 120 °C while a double-effect absorption chiller would operate with a heat source temperature between 120 and 170 °C (Deng et al., 2011).

The absorption chiller was simulated in Aspen$^+$ on the basis of a study by Somers et al. (2011). The condenser, absorber and evaporator heat exchangers were modelled as 'Heater' blocks (where simple heat exchanger energy balance calculations are simulated). The generator was modelled as a 'Flash2' block to allow for the separation of water from the lithium bromide solution. The operating temperature of the generator was 15 °C lower than the heat source temperature based on supplier experience (Ercan, 2017). The range of concentration of lithium bromide in absorption chillers is typically between 50 and 65 wt. % (Herold et al., 1996). To prevent crystallisation at the investigated conditions, the strong solution concentration leaving the generator was fixed at 65 wt. % of lithium bromide (ASHRAE, 2009). A weak solution concentration of lithium bromide (50 wt. %) was selected to ensure that the maximum amount of refrigerant would be circulated in the cycle. The solution heat exchanger effectiveness was 64 % (Somers et al., 2011).

An operating pressure of 0.7 kPa was selected for the absorber and evaporator to prevent ice formation on the evaporator tubes. This value is 15 % above the minimum operating pressure of 0.6 kPa (the corresponding evaporating temperature is -0.4 °C).

The temperature of chilled water leaving the evaporator was set at 5.0 °C based on the reported temperature range for absorption chillers of between 5.0 and 10.0 °C (Wu & Wang, 2006). An air cooler return water temperature of 18.0°C was selected since Brake (2002) used this value for design of the largest bulk air cooler in Australia.

For the indirect-fired absorption chiller in Options 3 and 5, hot water at 95.0 °C was produced in the water heat exchanger (component (x), FIG 1). Heat losses of 5.0 °C was assumed in the water distribution system resulting in water at 90.0 °C being supplied to the generator.

Cooling water was supplied to the absorber and condenser from cooling towers. Temperatures of the cooling water were based on typical cooling tower parameters and specified as inputs in the simulation. The inlet temperature of the water was 29.0 °C based on a 5.0 °C cooling tower approach temperature and an average ambient wet-bulb temperature of 24.0 °C (Belle & Biffi, 2013).

A summary of the assumptions for power generation and cooling production are given in TABLE 2.

TABLE 2 – Summary of assumptions and modelling parameters

Parameter	Value
Ventilation air flow rate	20 m³/s
Inlet methane concentration	0.1-1.0 vol. %
Reactor temperature	500-800 °C
Methane conversion	99 %
Maximum reactor feed inlet temperature	600 °C
System pressure drop	15 kPa
Heat loss	49 kW (direct gas turbine)
	36 kW (indirect gas turbine)
Compressor / turbine efficiency	87 %
Compressor outlet pressure	1.5-4.0 bar
Generator supply temperature	Hot water: 90°C
	Flue gas: 80-120 °C (single-effect); 120-170 °C (double-effect)
Strong solution concentration	65 wt. % lithium bromide
Weak solution concentration	50 wt. % lithium bromide
Absorber operating pressure	0.7 kPa
Chilled water temperature	5.0 °C
Air cooler return water temperature	18.0 °C

DISCUSSION

The influence of reactor temperature, compressor outlet pressure and inlet methane concentration on the minimum methane concentration, power and cooling produced is discussed. The water requirements of each process are also discussed.

Direct gas turbine VAM abatement process

Minimum methane concentration

The minimum methane concentration as a function of reactor temperature at different compressor outlet pressures for Option 1 is presented in FIG 2. The minimum methane concentration increased with both reactor temperature and pressure. A step change in the minimum methane concentration was noted above a reactor temperature of 700 °C and was attributed to the restriction of the reactor inlet temperature to 600 °C to prevent oxidation of methane upstream of the reactor. At a given pressure, the reactor feed temperature increased with reactor temperature and reached 600 °C. Above a temperature of 600 °C, additional heat was required to increase the reactor inlet temperature from 600 °C to the reactor temperature and manifested in a higher minimum methane concentration. At higher pressures, the heat produced by compression had a larger influence on the minimum methane concentration and manifested in a less prominent step change in the minimum methane concentration at higher pressures.

The minimum methane concentration also increased with compressor outlet pressure at a given reactor temperature (FIG 2). The compressor power requirement increased at higher pressures. To meet the higher compressor power requirement, the gas turbine produced more power through the supply of more heat (determined by the inlet methane concentration) resulting in an increase in the minimum methane concentration. The maximum methane concentration of 1.0 vol. % limited the maximum pressure to 3.4 bar.

The minimum methane concentration for Option 1 was 0.41 vol. % at a reactor temperature of 500 °C and compressor outlet pressure of 1.5 bar. Tremain *et al.* (2017) showed in a pilot-scale study that VAM was oxidised in a fluidised-bed reactor at a reactor temperature of 630 °C. Considering the experimental work by Tremain *et al.* (2017) would result in a minimum methane concentration of 0.45 vol. % at a reactor temperature of 630 °C.

The minimum methane concentration reported by Nadaraju *et al.* (2018) for the direct gas turbine process was 0.21 vol. % at a compressor pressure of 1.5 bar and reactor temperature of 500 °C. At these conditions, the flue gas temperature leaving the recuperator was 75 °C and the flue gases would require further treatment to process the condensed water vapour (from the methane combustion reaction). Furthermore at a reactor temperature of 500 °C, combustion of VAM will not occur in the fluidised-bed. (Nadaraju *et al.* (2018) did not consider the experimental work by Tremain *et al.* (2017) where combustion occurred at 630 °C).

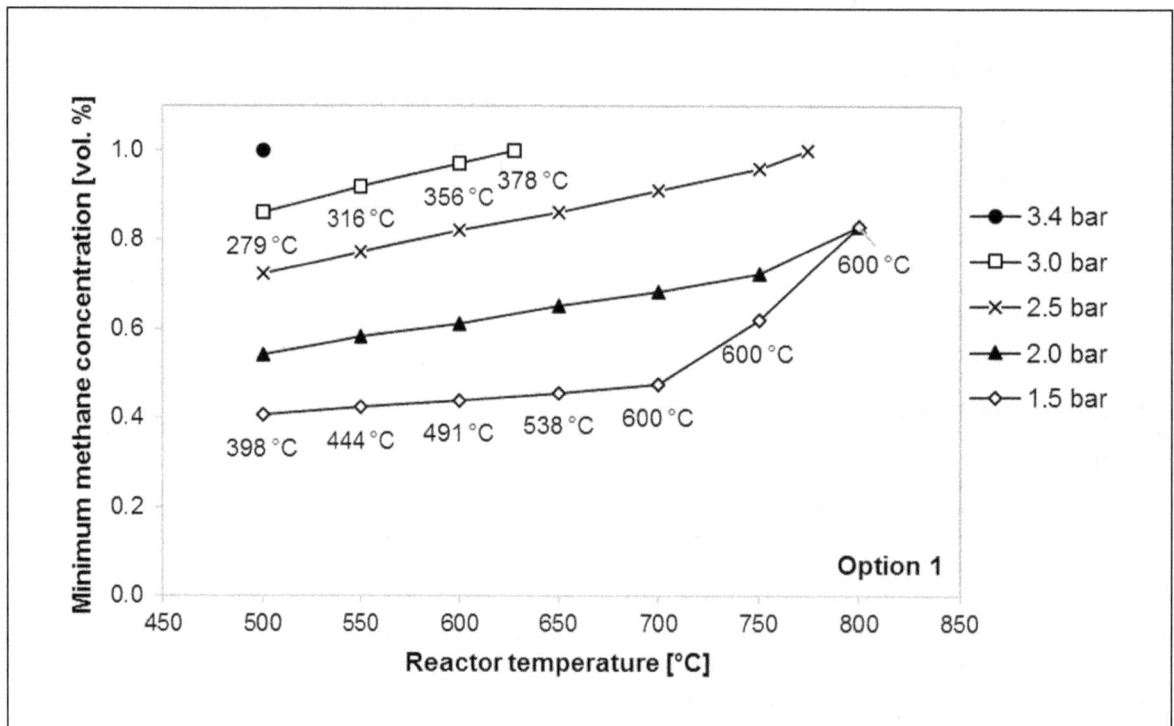

FIG 2 – Minimum methane concentration for the direct gas turbine process as a function of reactor temperature at different compressor outlet pressures (reactor feed temperature for pressures of 1.5 and 3.0 bar is shown)

Net power produced

The net power produced was the sum of the input power of the compressor, input power of the electrostatic precipitator and the output power of the gas turbine. The net power produced as a function of reactor temperature for Option 1 is shown in FIG 3.

The net power output increased with both reactor temperature and compressor outlet pressure. At a given pressure, higher reactor temperatures resulted in more power being produced in the gas turbine due to more heat that was available at higher temperatures and a fixed turbine flow rate

(20 m³/s or 22.8 kg/s). While at a given reactor temperature, more work was done by the turbine when expansion occurred from a higher pressure.

At a reactor temperature of 630 °C, minimum methane concentration of 0.45 vol. % and compressor outlet pressure of 1.5 bar, the net power produced was 960 kW. The excess net power could be fed back into the power grid which would result in a saving in power cost.

Nadaraju *et al.* (2018) reported the net power produced at an inlet methane concentration of 0.5 and 1.0 vol. % and not at the minimum methane concentration; therefore a comparison of the direct gas turbine process with the current work cannot be made.

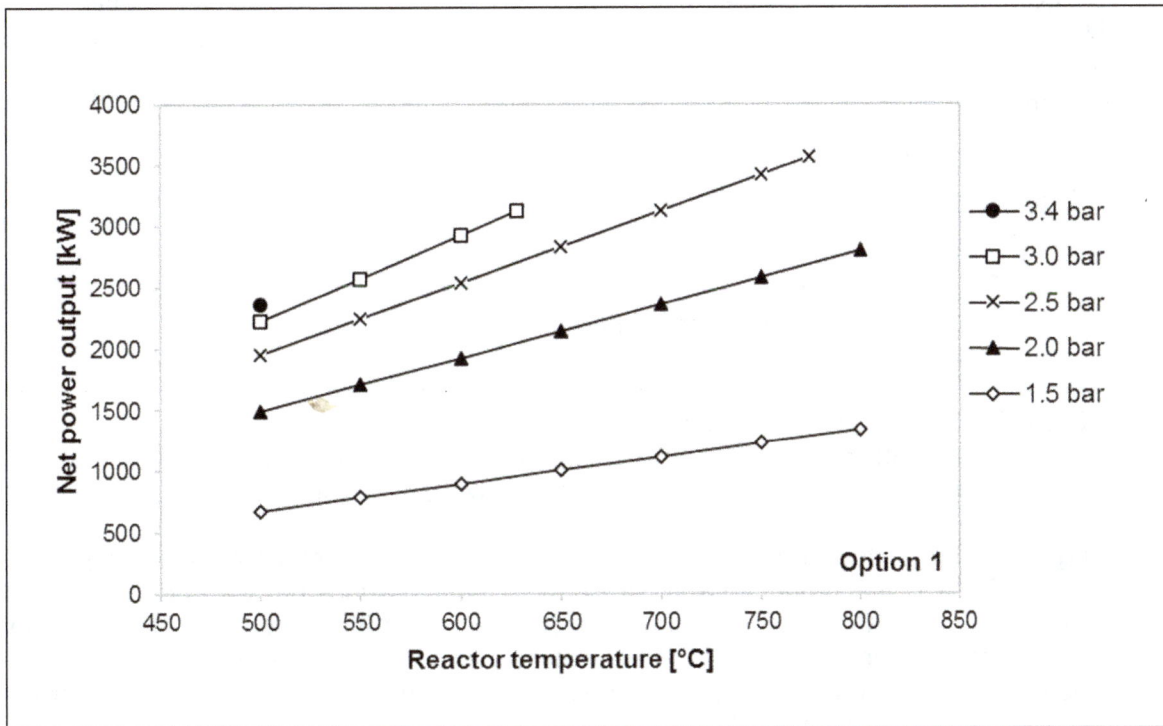

FIG 3 – Net power output for the direct gas turbine process as a function of reactor temperature at different steam pressures corresponding to the minimum methane concentration

Cooling production

The flue gases downstream of the recuperator were passed through the generator section of the absorption chiller to produce cooling either directly (Option 2, FIG 4) or indirectly via a heat exchanger (Option 3, FIG 5).

The cooling produced in Option 2 via a direct-fired absorption chiller is shown in FIG 4. At flue gas supply temperatures above 120 °C, a double-effect chiller was required and is shown in dashed-lines in FIG 4. The cooling produced at a given pressure remained fairly constant with reactor temperature and increased above a reactor temperature of 700 °C. The flue gas temperature determined the amount of cooling that was produced and was related to the minimum methane concentration (FIG 2). The flue gas temperature remained fairly constant below 700 °C and the increase in cooling corresponded with the step-change in minimum methane concentration above 700 °C resulting in more heat being available when the reactor feed temperature was 600 °C. The cooling produced at a given reactor temperature increased with compressor outlet pressure. At higher pressures, the turbine exhaust temperature increased, leading to a higher flue gas temperature downstream of the recuperator resulting in more cooling being produced. Cooling was not produced at 3.4 bar since the flue gas generator supply temperature was 195 °C and exceeded the upper limit of 170 °C.

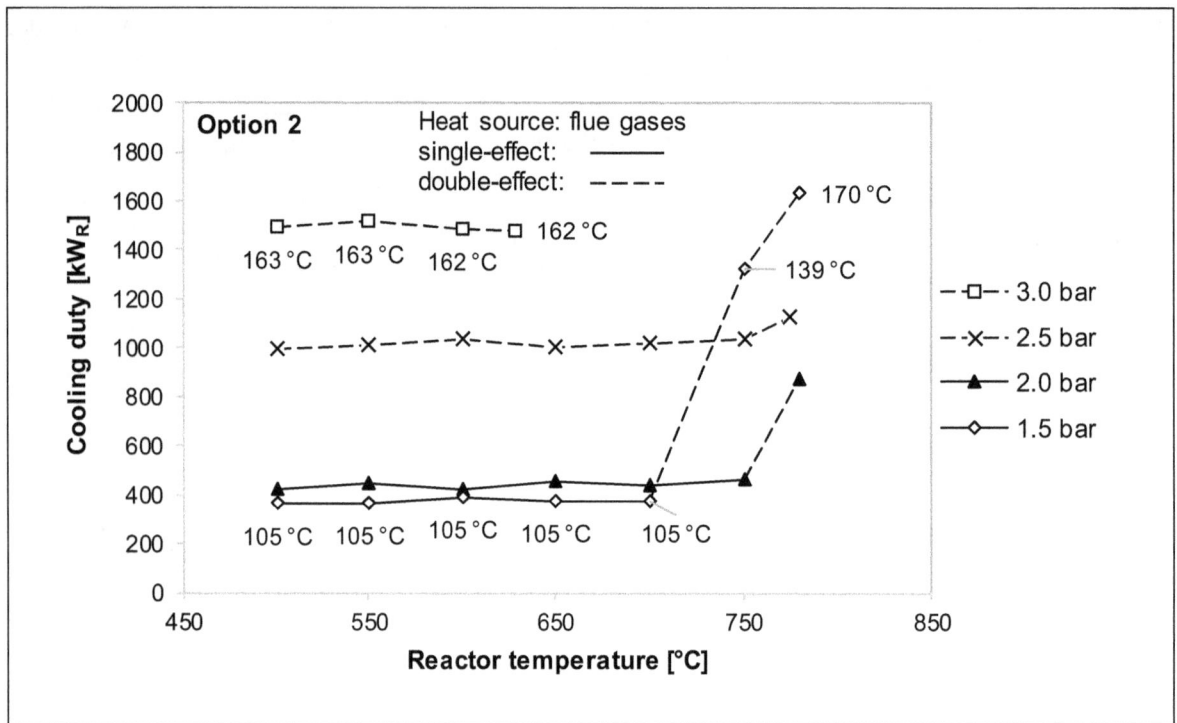

FIG 4 – Cooling produced by the direct gas turbine and direct-fired absorption chiller process as a function of reactor temperature at different steam pressures (flue gas generator supply temperatures are shown for 1.5 and 3.0 bar)

The cooling produced by the direct gas turbine and indirect-fired absorption chiller (Option 3) is shown in FIG 5. Cooling was produced at all operating conditions since hot water was produced by the water heat exchanger and supplied to the generator section at 90 °C. A single-effect chiller was required at all operating conditions since the hot water supplied to the generator section was below 120 °C. The cooling produced as a function of reactor temperature and compressor outlet pressure follows a similar trend when compared to the direct-fired absorption chiller (FIG 4).

At compressor outlet pressures of 1.5 and 2.0 bar, the direct-fired and indirect-fired produced a similar amount of cooling. At 2.5 and 3.5 bar, the direct-fired chiller produced more cooling since the generator supply temperature was higher in the direct-fired chiller.

At the minimum methane of 0.45 vol. %, the direct-fired absorption chiller (Option 2) produced 380 kW$_R$ while the indirect-fired absorption chiller (Option 3) produced 385 kW$_R$ of cooling. The higher temperature heat source (flue gas at 105 °C in Option 2 compared to hot water at 90 °C in Option 3) resulted in less cooling being produced due to a higher operating pressure in the generator section. The higher operating pressure resulted in less refrigerant that boiled in the generator leading to less cooling being produced.

From a safety perspective the indirect-fired absorption chiller operating with a heat source of 90 °C would be preferred over a direct-fired chiller operating with a heat source up to 170 °C. The additional heat exchanger required for the indirect-fired unit to transfer the heat from the steam to the water, would increase the capital and operating costs.

Considering a ventilation air flow of 300 m³/s, fifteen VAM abatement modules would be required. The direct gas turbine and direct-fired absorption chiller process would produce 5 700 kW$_R$ while the indirect-fired absorption chiller process would produce 5 775 kW$_R$. This would offset the cooling requirements at a gassy coal mine and translate into an annual operating cost saving of about A $ 0.5 M for both processes (assuming a power cost of A $ 100 MWh; coefficient of performance of 5.0 and cooling requirements for six months of the year). The provision for electrically-driven vapour compression chillers would be required to ensure that the cooling for the mine is independent of the variable methane concentration and variable methane flow rate. In this case mine cooling will still be available during periods where the methane concentration is below the minimum value.

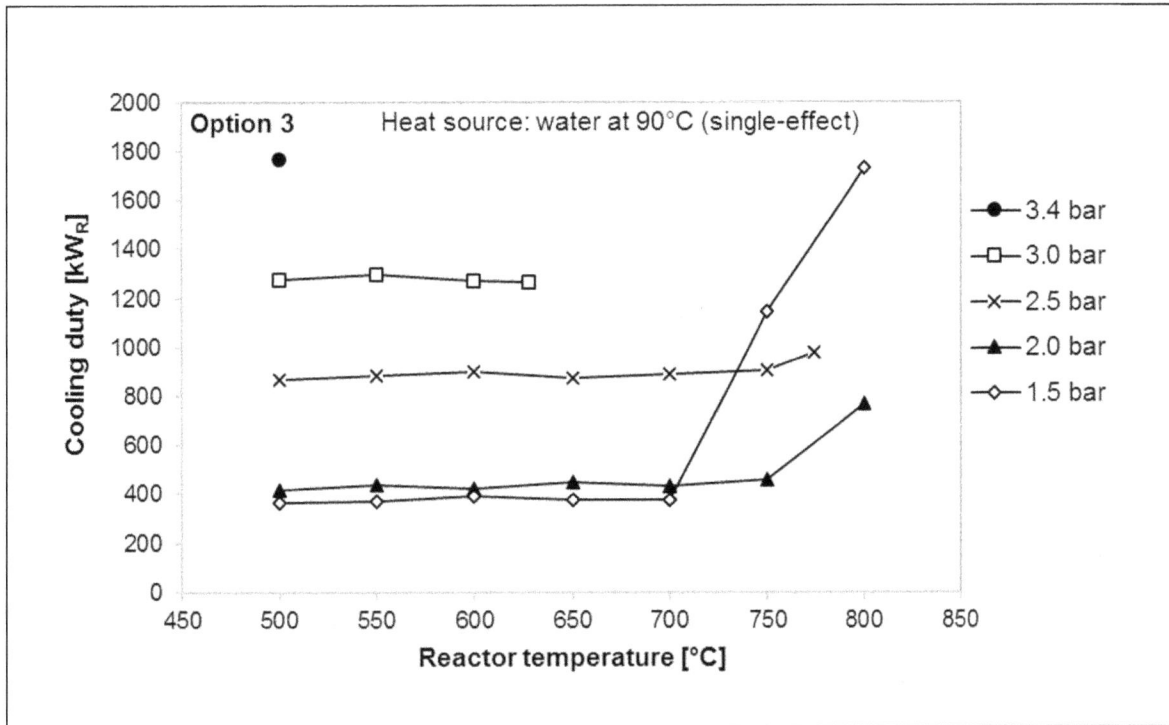

FIG 5 – Cooling produced by the direct gas turbine and indirect-fired absorption chiller process as a function of reactor temperature at different steam pressures

Indirect gas turbine VAM abatement process

Minimum methane concentration

The minimum methane concentration as a function of reactor temperature at different compressor outlet pressures for Option 4 is presented in FIG 6. The minimum methane concentration at a given pressure remained fairly constant with reactor temperature, up to 700 °C. Above 700 °C a step change in the minimum methane concentration was noted when the reactor temperature reached the restricted value of 600 °C similar to Option 1.

At a given pressure, the constant minimum methane concentration below a reactor temperature of 700 °C, was related to a decreasing turbine flow rate (FIG 9). This resulted in the same amount of heat that was transferred to the turbine and the same amount of work that was done by the turbine.

For a given reactor temperature, the minimum methane concentration decreased with pressure. At low pressures the minimum methane concentration was dependent on a high inlet methane concentration in the ventilation air and low compressor heat. At higher pressures the heat generated by compression had a higher contribution to the heat transferred to the turbine compared to the inlet methane concentration which lowered the minimum methane concentration at higher pressures.

The minimum methane concentration was 0.40 vol. % at a reactor temperature of 630 °C and compressor outlet pressure of 4.0 bar.

The minimum methane concentration reported by Nadaraju et al. (2018) for the indirect gas turbine process was 0.49 vol. % at a compressor pressure of 3.5 bar; reactor temperature of 700 °C and turbine flow rate of 7.1 kg/s. At these conditions, the flue gas temperature leaving the recuperator was 55 °C and the flue gases would also require further treatment to process the condensed water vapour (from the methane combustion reaction). The optimised indirect gas turbine process was shown to have a turbine flow rate of 2.2 kg/s (FIG 9) and would result in a smaller turbine compared to the study by Nadaraju et al. (2018).

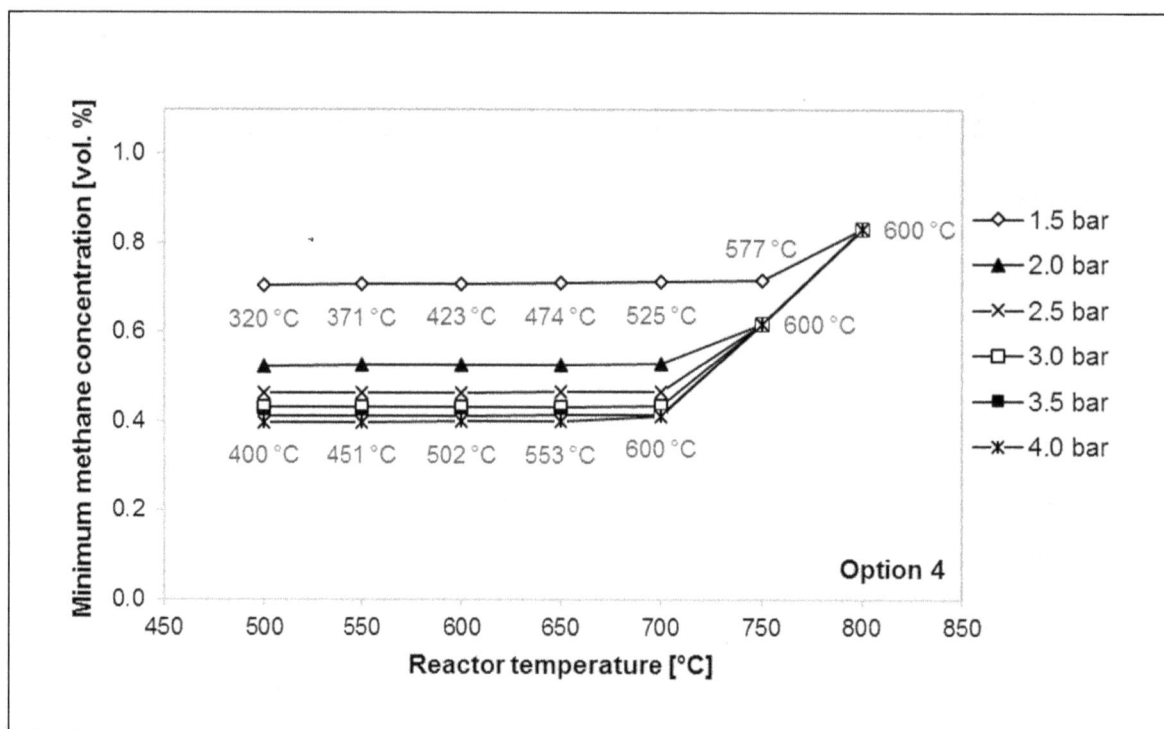

FIG 6 – Minimum methane concentration for the indirect gas turbine process as a function of reactor temperature at different compressor outlet pressures; the reactor feed temperature for pressures of 1.5 and 4.0 bar is also shown

Net power produced

The net power produced was the sum of the input power of the ventilation fan, input power of the compressor and the output power of the turbine. FIG 7 shows the net power produced at a reactor temperature of 630 °C as a function of inlet methane concentration. The net power output increased with both methane concentration and compressor outlet pressure. More heat was transferred to the turbine at higher methane concentrations for a given pressure and more work was done by the turbine when expansion takes place at higher pressures. The minimum methane concentration (FIG 7, horizontal axis) decreased with compressor outlet pressure since the turbine output work and compressor heat was higher at higher pressures which lowered the contribution of heat from the inlet methane concentration.

At a reactor temperature of 630 °C the minimum methane concentration (zero net power output) decreased from 0.71 vol. % at 1.5 bar to 0.40 vol. % at 4.0 bar.

The net power output of zero for the indirect gas turbine process was also reported by Nadaraju *et al.* (2018).

The net power output as a function of reactor temperature, corresponding to the minimum methane concentration, is shown in FIG 8. Below a reactor temperature of 650 °C zero net power was produced at all compressor outlet pressures. Above a reactor temperature of 650 °C the net power produced was greater than zero i.e. the output power of the turbine was greater than the sum of the input power of the ventilation air fan and compressor. Above 650 °C, the excess net power increased with both reactor temperature and compressor outlet pressure.

The net power output was influenced by the inlet methane concentration. Below a reactor temperature of 650 °C the reactor feed temperature remained below the restricted value of 600 °C resulting in the output power of the turbine balancing the input power of the fan and compressor. Above a reactor temperature of 650 °C more heat was available in the flue gases due to the higher minimum methane concentrations that were required to raise the reactor feed from 600 °C to the reactor temperature. The higher inlet methane concentration contributed to more heat being available in the flue gas stream that was transferred to the air heat exchanger. This resulted in the turbine producing excess power when compared to the input power of the fan and compressor.

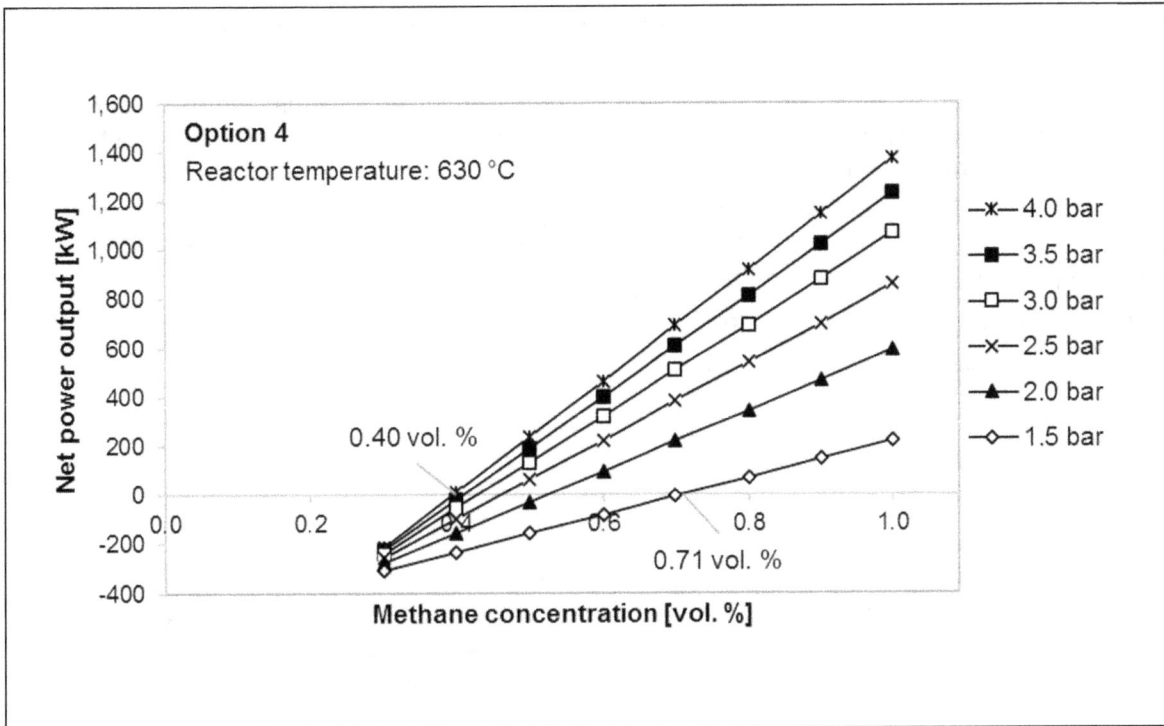

FIG 7 – Net power produced by the indirect gas turbine process at a reactor temperature of 630 °C at different compressor outlet pressures (the minimum methane concentration 1.5 and 4.0 bar is also shown)

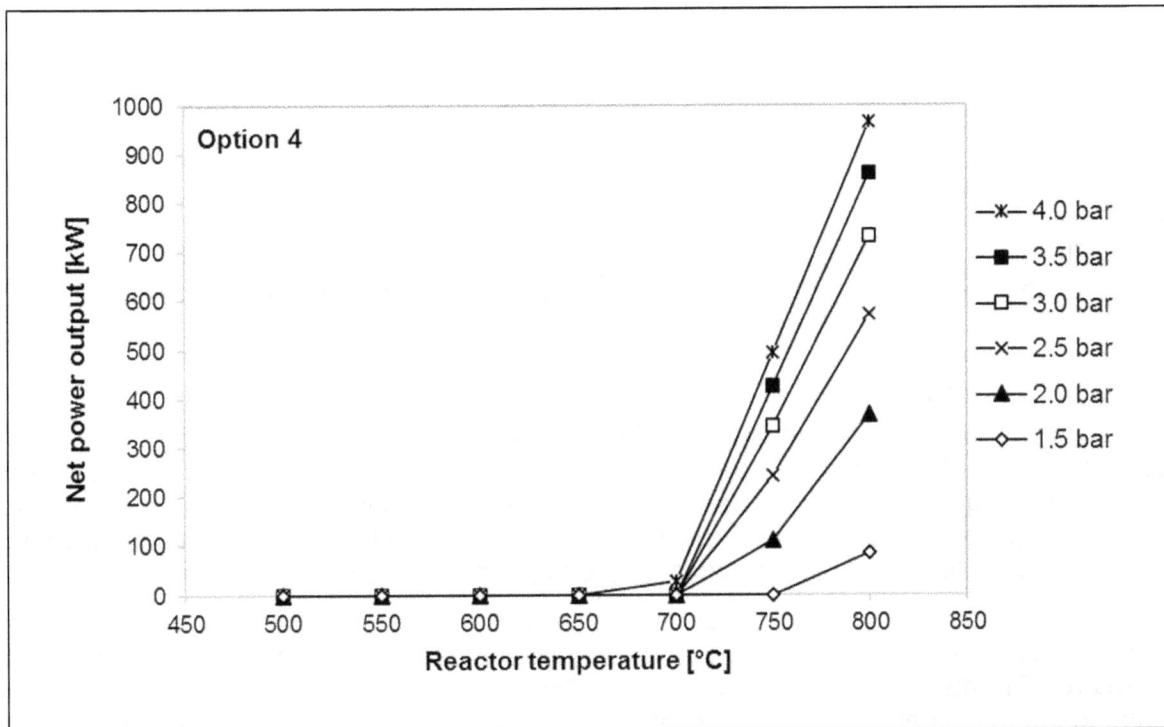

FIG 8 – Net power output for the indirect gas turbine process as a function of reactor temperature at different compressor outlet pressures (at the minimum methane concentration)

Turbine flow rate

The turbine working fluid mass flow rate as a function of steam pressure is shown in FIG 9. The turbine flow rate was an additional variable when compared to the direct gas turbine (Option 1). The flow rate decreased with both compressor outlet pressure and reactor temperature for reactor temperatures below 700 °C. Above 700 °C the turbine flow rate increased with both compressor outlet pressure and reactor temperature.

The shape of the curves was determined by the combination of minimum methane concentration (FIG 6) and heat transferred to the turbine. Below a reactor temperature of 700 °C, the turbine flow rate decreased to maintain the same amount of heat that was transferred to the turbine. The step-change in minimum methane concentrations above a reactor temperature of 700 °C resulted in more heat being transferred to the turbine with more work being done by the turbine and a corresponding increase in the turbine flow rate was noted.

The turbine flow rate was 2.2 kg/s at the minimum methane concentration of 0.40 vol. %.

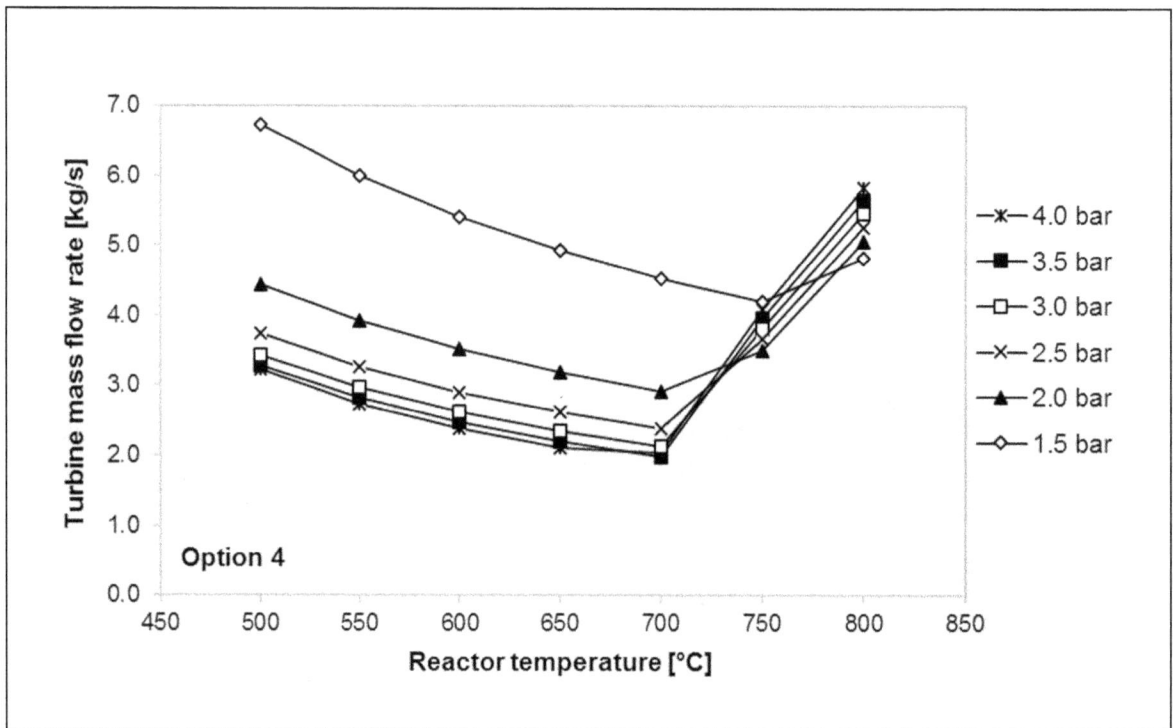

FIG 9 – Turbine flow rate for indirect gas turbine process as a function of reactor temperature at different compressor outlet pressures (at the minimum methane concentration)

Cooling production

The cooling produced by the indirect gas turbine and indirect-fired absorption chiller (Option 5) is presented in FIG 10 . Cooling was produced at all reactor temperatures and compressor outlet pressures since hot water at 90 °C was supplied to the absorption chiller. It was not possible to produce cooling via a direct-fired absorption chiller since the turbine exhaust temperature ranged between 220 and 670 °C and was greater than the upper operating limit of 170 °C.

The cooling produced by the indirect-fired absorption chiller increased with reactor temperature for a given pressure. This trend was related to the minimum methane concentration and turbine outlet temperature. The methane concentration (and turbine flow rate) determined the amount of heat supplied to the turbine and the turbine outlet temperature increased with reactor temperature. The step change in minimum methane concentration (FIG 6) resulted in more heat being transferred to the turbine which increased the turbine exhaust temperature.

At a given reactor temperature, a decrease in cooling produced with compressor outlet pressure is noted. At higher pressures, both the minimum methane concentration (FIG 6) and turbine outlet temperature decreased which resulted in less heat being available.

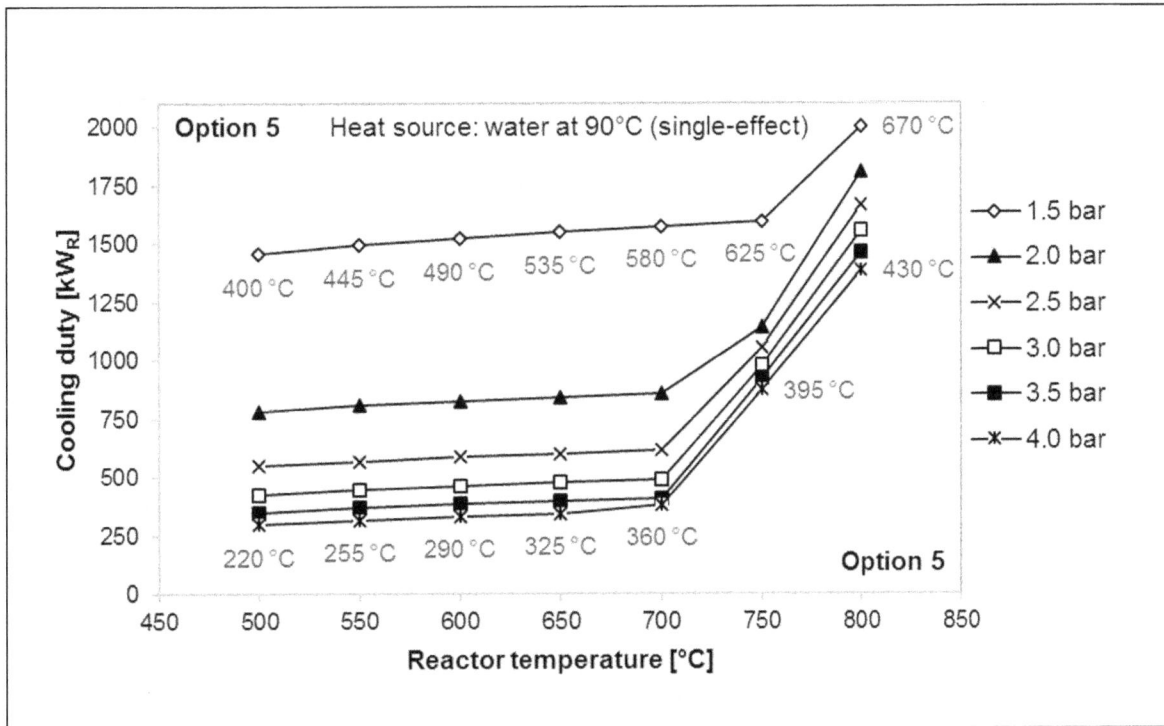

FIG 10 – Cooling produced by the indirect gas turbine and indirect-fired absorption chiller process as a function of reactor temperature at different steam pressures (turbine exhaust temperatures for 1.5 and 4.0 bar are shown)

At the minimum methane concentration of 0.40 vol. %, 340 kW_R of cooling was produced. To process a ventilation air flow of 300 m^3/s, fifteen modules would produce 5 100 kW_R and result in an operating cost saving of A $ 0.45 M per year.

The gas turbine processes discussed by Nadaraju *et al.* (2018) excluded any configurations of the production of cooling by the flue gases or turbine exhaust gases.

Water requirements

The water requirements for each option are summarised in TABLE 3. The direct and indirect gas turbine processes do not require water when producing power only. When power and cooling are produced the direct gas turbine and direct-fired chiller process required 32.1 L/s per module compared to 38.5 L/s per module when an indirect-fired chiller was used. The indirect gas turbine process and indirect-fired absorption chiller required 33.7 L/s per module.

For a ventilation flow of 300 m^3/s, the direct gas turbine and direct-fired chiller process required 41.6 ML/day of water while the indirect gas turbine and indirect-fired chiller water requirement was 43.7 ML/day. The direct gas turbine and indirect-fired chiller required 49.9 ML/day of water.

CONCLUSION

A ventilation air methane abatement process that converts the process heat from the flue gases of a fluidised-bed reactor into power via a gas turbine was simulated using Aspen[+]. The power produced by the gas turbine was used to operate the plant equipment (fan, compressor and electrostatic precipitator). For the direct gas turbine process, the minimum methane concentration of a single module that can process 20 m^3/s of ventilation air was 0.45 vol. % at a reactor temperature of 630 °C and compressor outlet pressure of 1.5 bar. The flue gases were a heat source in direct-fired and indirect-fired absorption chillers where 380 kW_R and 385 kW_R were produced, respectively. An indirect gas turbine process was also studied and the minimum methane concentration was 0.4 vol. % at a reactor temperature of 630 °C, compressor outlet pressure of 4.0 bar and turbine flow rate of 2.2 kg/s. The water required to operate each module was 41.6 ML/day for the direct gas turbine and direct-fired chiller; 49.9 ML/day for the direct gas turbine and indirect-fired chiller and 43.7 ML/day for the indirect gas turbine and indirect-fired chiller.

TABLE 3 – Summary of water requirements

	Option 2 Direct gas turbine + direct-fired cooling	Option 3 Direct gas turbine + indirect-fired cooling	Option 5 Indirect gas turbine + indirect-fired cooling
Hot water supply to generator	–	6.2 L/s	5.6 L/s
Chilled water flow	7.1 L/s	7.1 L/s	6.2 L/s
Condenser / absorber cooling water flow	25.0 L/s	25.0 L/s	21.9 L/s
Total water flow (per module)	32.1 L/s	38.3 L/s	33.7 L/s
Total water flow (300 m³/s, fifteen modules)	481.5 L/s	574.5 L/s	505.5 L/s
	41.6 ML/day	49.6 ML/day	43.7 ML/day

For a gassy coal mine with a ventilation flow of 300 m^3/s, the abatement plant would comprise fifteen modules. The direct turbine and direct-fired chillers would produce a total of 5 700 kW_R of cooling while the direct gas turbine and indirect-fired chillers would produce 5 775 kW_R. The indirect gas turbine and indirect-fired chillers would produce 5 100 kW_R of cooling. The cooling would offset the cooling requirement of 7 000 kW_R of a typical Australian gassy coal mine (Belle & Biffi, 2013). The cooling produced by the abatement plant translates into an annual operating cost saving of about A $ 0.5 M for the direct gas turbine processes or A $ 0.45 M for the indirect gas turbine process (at a power cost of A $ 100 MWh). Electrically-driven vapour compression chillers would also be required to ensure that the mine cooling is not affected by the variable methane concentration in the ventilation air. Standby refrigeration may also be required during periods where the methane concentration is below the minimum value.

ACKNOWLEDGEMENTS

The authors acknowledge the financial support received from The University of Newcastle, Australia and the Priority Research Centre for Frontier Energy Technologies & Utilisation for the work presented in this paper.

REFERENCES

ASHRAE. (2009). *2009 ASHRAE Handbook - Fundamentals (SI Edition)*. United States: American Society of Heating, Refrigerating and Air-Conditioning Engineers, Inc.

Belle, B., & Biffi, M. (2013). *Cooling pathways for deep Australian longwall coal mines of the future*. Paper presented at the The Australian Mine Ventilation Conference, Adelaide, Australia.

Brake, D. (2002). *Design of the world's largest bulk air cooler for the Enterprise mine in northern Australia*. Paper presented at the 9th US mine ventilation symposium, Kingston, Canada.

Broodryk, A., DeVries, J., Kyselica, P., & McLean, K. (2015). *The Design, Installation and Commissioning of an Absorption Refrigeration System at Gwalia Gold Mine in Western Australia*. Paper presented at the The Australian Mine Ventilation Conference, Sydney, Australia.

Calm, J. M. (2011). *Refrigerants for Deep Mines*. Paper presented at the 23rd International Congress of Refrigeration (ICR2011), Prague, Czech Republic.

Coal Mining Safety and Health Act 1999, Coal Mining Safety and Health Regulation 2017 (Queensland Government (legislation) 1999).

Commonwealth-of-Australia, & Bureau-of-Meteorology. (2018). Climate statistics for Australian locations (Singleton STP). Retrieved 4 April 2018, from http://www.bom.gov.au/climate/averages/tables/cw_061397_All.shtml

Commonwealth-of-Australia, & Department-of-the-Environment-and-Energy. (2016). Australian Greenhouse Emissions Information System. Retrieved 6 June 2018, from http://ageis.climatechange.gov.au/

Commonwealth-of-Australia, & The-Treasury. (2011). *Strong growth, low pollution*. Retrieved from Australia: http://carbonpricemodelling.treasury.gov.au/content/report.asp

Deng, J., Wang, R. Z., & Han, G. Y. (2011). A review of thermally activated cooling technologies for combined cooling, heating and power systems. *Institute of Refrigeration and Cryogenics, Shanghai Jio Tong University, 37*, 172-203.

Ercan, S. (2017). [Personal communication: absorption chiller operating temperature].

Gloucester Resources Limited v Minister for Planning [2019] NSWLEC 7 (Land and Environment Court, New South Wales 2019).

Gosiewski, K., Pawlaczyk, A., & Jaschik, M. (2015). Energy recovery from ventilation air methane via reverse-flow reactors. *Energy, 92*, 13-23.

Herold, K. E., Radermacher, R., & Klein, S. (1996). *Absorption chillers and heat pumps*. Florida, United States: CRC Press.

Incropera, F. P., & DeWitt, D. P. (1996). *Fundamentals of Heat and Mass Transfer* (Fourth ed.): John Wiley & Sons.

Karacan, C. Ö., Ruiz, F. A., Cotè, M., & Phipps, S. (2011). Coal mine methane: A review of capture and utilization practices with benefits to mining safety and to greenhouse gas reduction. *International Journal of Coal Geology, 86*.

Karakurt, I., Aydin, G., & Aydiner, K. (2011). Mine ventilation air methane as a sustainable energy source. *Renewable and Sustainable Energy Reviews, 15*(2), 1042-1049.

Karp, P. (2018). Australian government backs coal in defiance of IPCC climate warning Retrieved from Australian government backs coal in defiance of IPCC climate warning website: https://www.theguardian.com/australia-news/2018/oct/09/australian-government-backs-coal-defiance-ipcc-climate-warning?CMP=share_btn_link

Kucharczyk, B., Stasińska, B., & Nawrat, S. (2017). Studies on work of a prototype installation with two types of catalytic bed in the reactor for oxidation of methane from mine ventilation air. *Fuel Processing Technology, 166*, 8-16.

Latimer, C. (2018). Coal is Australia's most valuable export in 2018. Retrieved from https://www.smh.com.au/business/the-economy/coal-is-australia-s-most-valuable-export-in-2018-20181220-p50nd4.html

Li, Q., Lin, B., Yuan, D., & Chen, G. (2015). Demonstration and its validation for ventilation air methane (VAM) thermal oxidation and energy recovery project. *Applied Thermal Engineering, 90*, 78-85.

Li, Z., Wu, Z., Qin, Z., Zhu, H., Wu, J., Wang, R., . . . Wang, J. (2017). Demonstration of mitigation and utilization of ventilation air methane in a pilot scale catalytic reverse flow reactor. *Fuel Processing Technology, 160*, 102-108.

Nadaraju, F. J., Maddocks, A. R., Zanganeh, J., & Moghtaderi, B. (2018). Thermodynamic assessment of heat recovery from a fluidized-bed ventilation air methane abatement unit. *Energy Fuels, 32*(4), 4579-4585.

Parker, K. (2003). *Electrical Operation of Electrostatic Precipitators*. London, United Kingdom: Institution of Engineering and Technology.

Pratt, R. M. (2001). Thermodynamic properties involving derivatives: using the Peng-Robinson Equation of State. *Chemical Engineering Education, 1*, 112-115.

Razak, A. M. Y. (2007). *Industrial Gas Turbines - Performance and Operability*: Woodhead Publishing.

Rhodes, M. (1998). *Introduction to Particle Technology*. England, United Kingdom: Wiley.

Robinson, C., & Smith, D. B. (1984). The auto-ignition temperature of methane. *Journal of Hazardous Materials, 8*(3), 199-203.

SABS-Standards-Division. (2014). Refrigerating systems, including plants associated with air-conditioning systems. In (Vol. SANS 10147:2014). South Africa: SABS.

Salomons, S., Hayes, R. E., Poirier, M., & Sapoundjiev, H. (2003). Flow reversal reactor for the catalytic combustion of lean methane mixtures. *Catalysis Today, 83*, 59-69.

Shah, K., Moghtaderi, B., Doroodchi, E., & Sandford, J. (2015). A feasibility study on a novel stone dust looping process for abatement of ventilation air methane. *Fuel Processing Technology, 140*, 285-296.

Somers, C., Mortazavi, A., Hwang, Y., Radermacher, R., Rodgers, P., & Al-Hashimi, S. (2011). Modelling water/lithium bromide absorption chillers in ASPEN Plus. *Applied Energy, 88*(11), 4197-4205.

Srikhirin, P., Aphornratana, S., & Chungpaibulpatana, S. (2001). A review of absorption refrigeration technologies. *Renewable and Sustainable Energy Reviews, 5*, 343-372.

Su, S., & Agnew, J. (2006). Catalytic combustion of coal mine ventilation air methane. *Fuel, 85*(9), 1201-1210.

Su, S., Beath, A., Guo, H., & Mallett, C. (2005). An assessment of mine methane mitigation and utilisation technologies. *Progress in Energy and Combustion Science, 31*, 123-170.

Tremain, P., Maddocks, A. R., & Moghtaderi, B. (2017). *A pilot-scale study on the oxidation of ventilation air methane (VAM) using ilmenite*. Paper presented at the 11th Asia-Pacific Conference on Combustion, The University of Sydney, NSW Australia.

Van den Berg, L. (2017). [Personal communication: location of ammonia plants in Australia].

Wang, S., Gao, D., & Wang, S. (2014). Steady and Transient Characteristics of Catalytic Flow Reverse Reactor Integrated with Central Heat Exchanger. *Industrial & Engineering Chemistry Research, 53*, 12644–12654.

Wood, S., Fletcher, D. F., Joseph, S. D., Dawson, A., & Harris, A. T. (2009). Design and Evaluation of a Porous Burner for the Mitigation of Anthropogenic Methane Emissions. *Environmental Science and Technology, 43*(24), 9329-9334.

Work Health and Safety (Mines) Act 2013, Work Health and Safety (Mines) Regulation 2014 (New South Wales Government (legislation) 2013).

Wu, D. W., & Wang, R. Z. (2006). Combined cooling, heating and power: A review. *Institute of Refrigeration and Cryogenics, Shanghai Jio Tong University*, 459-495. doi:10.1016/j.pecs.2006.02.001

Zhang, Y., Doroodchi, E., & Moghtaderi, B. (2014). Chemical looping combustion of ultra low concentration of methane with Fe2O3/Al2O3 and CuO/SiO2. *Applied Energy, 113*, 1916-1923.

Coal Mine Ventilation

Ventilation control system multiphase implementation approach driven by safety and production

E Acuña[1], C Visage[2], H Mohle[3] and G Durandt[4]

1. Ventilation Specialist, BBE Consulting Canada, Canada, enrique.acuna@bbegroup.ca
2. Ventilation Specialist, BBE Automation, South Africa, cvisagie@bbe.co.za
3. Principal Engineer, BBE Consulting Australasia, Australia, hmohle@bbegroup.com.au
4. Principal Engineer, BBE Consulting Australasia, Australia, gdurandt@bbegroup.com.au

ABSTRACT

Ventilation Controls Systems (VCS), also commonly referred as Ventilation-on-Demand (VOD) systems, have historically been offered to the mining industry with the initial promise of reducing electrical energy consumption. Their promotion was also further refined to permit greater production opportunity and improving an energy per tonne metric. Such ventilation management systems have been suggested, or attempted, as step-change implementations at mine sites. However, the visualisation or control to provide a fully utilised asset-tagging-based or environmentally driven system may be absent. Operationally, throughout life of mine, using energy savings alone as the VCS/VOD driver is a challenge to sustain as it can compete against typical production incentives. Where site-specific KPIs include EBITDA, they are more heavily influenced by higher production output than energy savings. A step-change implementation, in practice, is difficult to achieve due to the complexity of the logistics, the learning curve, and the maintenance required. Today, industry is beginning to shift away from the VCS/VOD energy-savings driver focus, diversifying it into three pillars: safety, production, and energy savings. Also, there is a trend away from step-change implementations, with industry preferring low-risk, multi-phase, and multi-year projects. Herein, a multiphase approach is suggested, driven by these the three pillars, to maximise production versus the perceived goal of maximising energy savings. It is expected that this approach aligns better with existing mine management methodologies and KPIs, with the potential to result in more realistic operational levels of implementation.

Key words: VCS, VOD, health and safety, production, energy savings

INTRODUCTION

The concept of introducing Ventilation Control Systems (VCS), widely mentioned as Ventilation On Demand (VOD), has become part of the common language when referring to the optimised management and operation of ventilation systems. Generally, the target is to make the best use of the installed ventilation systems to:

1. maintain or improve the health and safety of the work force;

2. provide flexibility and enhance the development and production plans; and

3. generate energy savings.

These three items can also be referred as the three pillars of VCS/VOD and are provided in order of relevance. The health and safety of workers is an underground mine's licence to operate. From an optimisation perspective, the first pillar – health and safety – secures a feasible solution. The second pillar, ventilation flexibility, is used to achieve or exceed the development and production plan to reduce the unit cost ($/tonne) and bring the highest return on investment. Finally, the unit cost can be further reduced through energy savings. However, it is expected that lowering the risk, or any enhancement of the production plan, will have significantly larger returns as compared to energy savings only.

Another important aspect of the potential in energy savings cited for such systems is that they can be used to pay for the implementation of the system. Nevertheless, the benefits of a system that can deliver VCS/VOD are far beyond ventilation itself. They provide "eyes" into the mine from surface at any point in time, opening the door for other technology related implementations. These can be significantly more attractive from a business perspective but are not in the scope of this study.

In evaluating VCS/VOD, the proper definition of the investment, the benefits, the pay back, and the return over investment must be calculated for each scenario or phase of implementation. This informs the decision on which level of control is the appropriate. Should maintenance of the systems not be properly understood, considered, or implemented over time for all its components, the system will underperform or eventually fail. This has been the fate of many historic ventilation control trials, for example the timed operation of auxiliary fans rarely lasts. Properly resourcing the maintenance of a VCS/VOD system, like any system, is key to its operation.

VCS/VOD system justifications have primary relied on energy savings benefits, which can be attractive from a business perspective, but are hard to sustain from an operational perspective. This is especially the case if they are not actually measured. When the incentives are aligned with safety and production, then the argument can be made that complementing the energy savings with the first two pillars is an overall better selling strategy and that it should get significantly more buy-in from the operational end. In this instance, it is expected that it will prove to work better over time.

From an operator and worker perspective, it is attractive to hear about how a system will safeguard and provide the means of meeting or exceeding the production target, with respect to ventilation. News that the company will save energy and increase revenue without front-line impact is not well received. This is relevant particularly in production-incentivised environments. Considering that the front-line personnel will be the most affected stakeholder during the process of installation and adoption of ventilation control systems, their buy-in is critical for a successful implementation.

Optimisation can provide several tools to make decisions based on a certain driver. A simple first approach could be a greedy algorithm, which is the one that seeks the easier next step with the highest return, this is commonly referred to as getting the low hanging fruit. A simple approach is to make a ranking of the steps considered, or in this case the levels of control, based on a KPI. This could be the benefit over investment (estimated value of benefit divided by estimated value of investment and the associated effort or timeline required for implementation). By this method, the highest-ranking step will be the one reporting the most benefit at the least cost. The application as employed here, is to use such an approach from the starting point of no installation, to then decide the levels of control to implement. This would be opposed to jumping directly to ventilation on demand driven by tagging (VOD level 4). The overall objective is to generate a methodology to decide what is the best way of implementing a VCS, considering multiple phases, to find the solution that better fits each site.

This paper introduces the levels of controls that have been describe in previous works and then provides the mayor tasks associated with their implementation. A greedy multiple phase approach is then presented, including the possible variants that could be introduced. This considers the different levels of control and the infrastructure in place at the main and auxiliary fan systems. This is followed by a discussion that explores the application of the proposed ranking approach and the robustness of the decision plus a sensitivity analysis. This leads to a conclusion as to if and when to introduce tagging based control.

BACKGROUND

When referring to any control system, in this case for ventilation, a few initial concepts are required. For example, start with the view of "what cannot be measured cannot be controlled". It is also fundamental to understand what the issues are, how accurately a parameter can be measured and how frequently it can be polled with the technology available.

For ventilation the required measurements revolve around, airflow (as volume), differential pressure, barometric pressure, gas concentrations, ambient temperature, relative humidity, dust and diesel particulate levels among others. Their relevance can vary widely depending on mine-site condition needs. Currently, there is a variety of instrumentation manufacturers that have sensors available on the market as standalone units or packaged multi parameter 'air quality' stations. However, for mineral and diesel particulate, the sensor choices are limited with some still in the test phase for underground mining. Many of the particulate monitors for permanent installation may be based upon an indirect measurement.

After understanding what can be measured and its associated limitations, the next step begins to look at the control strategies that can be used and how change would be triggered. However, at the first stage of sensor deployment the central command could be used solely to see if pre-defined events could be confidently seen in a timely manner. A very particular condition could present itself in which the central command could potentially see what is happening across different locations of the mine. This can be very valuable even without yet having the necessary capacity to initiate manual or automated (centrally and remotely) actions. For control, this step would require the ventilation department to generate and update a list of acceptable measurement ranges, actions associated to either a single input or a combination of inputs from the sensors, and the trigger for defined actions. A dedicated control room operator could be required to follow specified procedures at certain times of the day, for the full shift or part thereof, to make sure the sensor values are monitored, and actions are taken accordingly. The resulting system would be classed manual, as opposed to remote control, however it could be argued that it is still represents a rudimentary level of control. This is referred as Level 0 Visualisation (Acuña, 2018).

In practical terms, progressing from manual field measurements with manual control performed at regular intervals, maybe weekly at best, by the ventilation department or a contractor, to an online measurement, is a significant step transformation. This change presents a major shift in terms of what can be done considering the three pillars for VCS (health and safety, production and energy savings).

Five control strategies (Allen & Tran-Valade, 2016) have previously been presented and conceptually explained with the following reference definitions:

- Level 1 - Manual Control;
- Level 2 - Time of Day Scheduling;
- Level 3 - Event Based;
- Level 4 – Tagging; and,
- Level 5 – Environmental.

More than describing each strategy separately, the focus of this study is to give guidance on how they could be coupled together for deployment, and how-to multi-phase their implementation. The aim being a better return over investment for such an implementation, while specifically considering the learning curve required for both the operation and the maintenance of the system.

The key is to understand the amount of software programming and hardware required for each level of control, the resources (personnel, training, equipment, supplies, cage time, and logistics) that will be needed for the system's deployment and maintenance.

In terms of resources allocated to the project, the following needs to be considered:

- Configuration of software (including any higher-Level implementation or new feature)

- Maintenance of software/IT components (over time)

- Underground deployment team (for a higher-Level implementation)

- Underground maintenance team (over time)

Control software is commercially available (Acuña et al, 2016), it has been developed based on the needs of the mine sites where it has been implemented. When deciding on a software to use, or an in-house development, it is recommended to have a clear understanding of the features required for each control strategy. These may be general, common to most mines, and site specific. It is not uncommon that the software will need additional development to accommodate a mine's specific requirements (Acuña & Allen, 2017). Any implementation package should include focussed training for all the users and the architecture team that will oversee maintaining all the software components needed (servers, clients, databases, hard drives, and communication network).

In terms of resourcing the deployment underground, the key is to keep the ventilation control system aligned with the mine needs; namely, needing a good understanding of the initial deployment requirements, plus the ongoing deployment demands as the mine develops.

Additionally, a second functionality team will most likely be required to be dedicated to the maintenance of the installed sensor, communication and mechanical controller hardware underground. Certain sensors, specifically for certain gasses, contain chemical cells that need periodic replacement, others may need total replacement upon failure. Maintaining connectivity to the local or central decision base can be a challenge, however wireless options continue to be developed.

The overall timely availability and performance of a VCS will depend on these two teams.

Giving consideration to the appropriate learning curve for all the key players is critically important, including:

- Surface control room operators
- Ventilation department
- Workers, supervision, and management
- Deployment, commissioning, and maintenance team

The capacities and expectations of the Ventilation Control System (VCS) or Ventilation On Demand (VOD) have to be clearly stated for all to understand what to do in case of malfunction.

Control Room Operators (CROs) must be provided with very specific rules in terms of what they can do or not do to accommodate airflow requests for development and production. They should also be equipped with guideline documents and operational procedures provided by the ventilation department for regular operation and to support blast clearing based on the controls available. These will have to be updated as higher Levels get deployed. CROs are usually the main controller of the settings of the VCS, hence their need for training and follow up on all the features that they need to work effectively. Troubleshooting between the ventilation department, control room, supervision, and workers should also be expected. Resourcing a ventilation department for all the additional tasks that a VCS implementation requires might be overlooked initially; this has the potential to jeopardize the implementation and its use over time.

Workers underground are the front-line users of the system, they have to understand that the system is expected to run according to its capacities most of the time, but not necessarily 100% of the time. Procedures must be developed, and the underground personnel trained to check the operationality of the VCS before entering their workplace, how to detect if the system is not working properly, and corrective actions if required. Initially, setting worker expectations of benefits in the right range according to the control strategies is key to a successful implementation. Their timely reaction, in case of an issue, to minimise any possible impact to the development and production plan is also critical.

Worker education needs to be focussed. For example, they understand that the company benefits from shutting down or changing the speed of fans to save energy, however if the training sessions are put together and delivered accordingly, they will also quickly appreciate that having less fans operating, or running at a lower duty, can help to lower working face temperatures as opposed to have all fans working. They should also understand that their health and safety benefits from a monitored environment, plus the rationalised use of the airflow can give additional flexibility to achieve development and production plans, and in some cases exceed it.

This is particularly important in an incentive driven environment. If workers understand that a VCS implementation is meant to be a win-win, and they see the expected returns then they will be in alignment to make it a success. This requires an important effort over time in terms of training sessions as the VCS is implemented and new Levels of control are introduced. Just like the workers, supervision and management must similarly learn and understand the capacities of the VCS and how to use it properly.

The deployment, commissioning, and maintenance personnel should consider automation, instrumentation, and IT specialist capability to deal with the different pieces of the VCS: control actuators, fan starters, sensors, communications networks, servers, and clients among others. Considering that an important piece of the technology available could be relatively new, possibly a prototype and still improving (Flores & Acuña, 2016), most of the training and learning to troubleshoot relies on supplier support and knowledge gathered through experience at the mine site.

MULTIPLE PHASES SUGGESTED IMPLEMENTATION

Estimating the level of effort and cost associated to each control strategy should be done on a site-specific basis and could significantly vary depending on each location. For that reason, a qualitative estimate will be used in this section to calculate an order-of-magnitude level of cost and effort required to include the main drivers to estimate implementation.

In terms of the expected benefits of the three pillars, the argument is similar, nevertheless, some estimates are available in the literature for energy savings (Acuña & Allen, 2017) and will be used as a base to estimate benefits in percentage (%).

The greedy approach, described earlier, will then use the ratio of the benefit divided by the effort to prioritize the levels of control to be implemented in terms of which one will be the best choice for each step of the multiple phased implementation.

Estimate effort required

In order to estimate the level of effort required to implement every control strategy, the main elements needed to deploy and maintain the hardware and software will be considered for each Level of implementation and based on that a relative effort qualitative value is determined.

In any of the control strategies suggested a control room is required to remotely (or centrally) visualise and control the system, however the deployment of the control room is only needed once and then needs to be maintained and upgraded accordingly. Since this effort is required for all strategies, it will not be used to evaluate the specific effort of each.

The deployment of standalone sensors or sensor packages is usually the first step of any implementation. The most common sensors to deploy are airflow; temperature (dry bulb); relative humidity; and carbon monoxide (CO). Airflow sensors are needed to monitor air volumes and confirm compliance with the ventilation design basis, i.e. an air volume per diesel engine BHP regulation, or just to validate the airflow distribution of the mine is as expected. Temperature and relative humidity sensors coupled with a barometric pressure sensor (which can be built into the same sensor package) enables the calculation of wet bulb temperature and used to assess heat stress conditions, or predict their expectation based on a level of work effort, in a mine. Finally, CO sensors are used to assist re-entry after gas clearance (blasting). They can also be indicative of the ventilation system's performance for diluting diesel fumes and potentially DPM levels. Other sensors can be deployed as part of the mix, but that would be a mine site specific definition.

Sensor packages require rack-mounting (sometimes excavation is required). Communication and power cable must be run from the nearest electrical station, and then data transmission routed somehow to surface and to the control room. This deployment step is labour intensive initially, and its maintenance needs to be considered according to the manufacturer recommendations for both the software and hardware components both on surface and underground (U/G).

Visualisation alone, or coupled field manual control, is probably not a practical or sustainable control strategy for an active mine. However, it is a realistic first step towards central control and helps build the business case for other control levels that may be considered. Also, it enables the early identification and implementation of out-of-range alarms for the installed sensors.

Incrementally, for Manual Control the capability of executing actions becomes available. The range of actions normally includes turning on and off fans, opening and closing of doors and regulators, or setting a fixed opening %. The devices to be controlled, need to be powered and communications established to the central control. However, in some cases, specific local control modules in the field with one example being that required for doors or regulators.

Time-of-Day Scheduling, as a step up from Manual Control, only needs further programming and commissioning efforts if the communications to the fans and controls are already in place.

Event Based control will enable the implementation of actions based on available inputs that are not already implemented in any of the Levels of control. This control strategy again only needs programming and commissioning efforts if the communications to the sensors, fans, and control infrastructure are already in place. Again, this is a case by case and site-specific definition.

A vehicle/person tagging based control strategy, is a step-change as compared to the previous strategies. This is due to the level of effort required, on many fronts, that have to work simultaneously to generate dynamic control (with moving inputs) as opposed to the previous static control.

For personnel tracking, tags have to be deployed in all cap-lamps or equivalent for comprehensive monitoring of the work force's movement. Such systems require the means to be tested before going U/G; this requires a personnel tag validation station or a means to check that the system is recognising the tag. Additionally, tags for equipment have to be deployed on all U/G mobile equipment, which in itself can be a daunting task due to the dynamic nature of most equipment's location. Equipment tags will also require a validation station U/G or a visual indication of them working properly. Caution has to be exercised with tags as some were not designed to work U/G and will most likely require an enclosure to withstand an adverse underground environment. Additionally, the tag verification for equipment and cap lamp has to be part of the pre-operational check with a defined frequency.

To profit from the tag deployment, software-defined capture/ventilation zones have to be established with appropriate Wi-Fi (or other data portals) deployed to determine if the tags (personnel and equipment), enter, are present in or depart the zones. Each tagging solution is usually provided with a specific software that is used to monitor tag location, this may also require the capacity to check that vehicles and personnel locations are reported correctly without ghosting. All the previous components mentioned (tags, wi-fi or portals, and tag software) have to work together simultaneously with the VCS to provide control. They must also be monitored and maintained as required or defined.

Fail-safe procedures and training are required for the work force to recognise a failure and understand how to react both on surface and U/G., this may require fall-back or fail safe specified settings of the various controls in regard to an air volume or regulator setting Depending on how the system is implemented, two fail-safes may be required: first a way to inform if an area or ventilation zone is safe to access and second, a way to alert if an area is unsafe and evacuation is required.

Environmental control, as a standalone control strategy, might not be suitable depending on local jurisdictional regulations for airflow volumes and might be suitable only in the determination of fail-safe arrangements. However, if applicable, the incremental requirement is only on the software side, provided that the required communications and granularity of sensors are in place. Table 1 presents an attempt to list the effort components required per level of control.

TABLE 1 – List of components to be considered per level of control (control strategies)

Control strategy Level 0 through 5	Programming	Communications and power	Sensors
Visualisation	SCADA/HMI/Out of range alarms	Comms + power to sensors	Airflow, temp, CO, others
Manual Control	SCADA/HMI/VCS SOFTWARE	Comms to fans and controls (doors/regulators)	-
Time of Day Scheduling	VCS SOFTWARE	-	-
Event Based	VCS SOFTWARE	-	-
Tagging	VCS SOFTWARE/ TAG SOFTWARE	Comms + power to wi-fi or portals	Tags (personnel and equipment)
Environmental	VCS SOFTWARE	-	-

Table 1 was assessed and turned into qualitative efforts and presented in Table 2. This is only an attempt to determine the value for the effort required, it identifies the mayor components that will drive the amount of effort required on a site by site base. Consequently, the value presented is only indicative of the quantity of components that will be required as a minimum to be provide a Level of control in terms of deployment but also in terms of maintenance after deployment.

It is expected that the cost is proportional to the effort required, however, from a practical perspective the cost is still important. In a working, or new, mine where everybody is competing for resources including hoist time, logistics support, U/G equipment and personnel availability, the complexity of the task and the impact on the development or production plan have the potential to become the main driver to estimate the overall cost of implementing a ventilation control system.

TABLE 2 – Qualitative effort estimate per level of control (control strategies)

Control strategy Level 0 through 5	Programming	Communications and power	Sensors	Total
Visualisation	0.5	1	1	2.5
Manual Control	1	2	-	3
Time of Day Scheduling	0.5	-	-	0.5
Event Based	0.5	-	-	0.5
Tagging	2	2	3	7
Environmental	1	-	-	1

As presented in Table 2, the most demanding control strategies are: Tagging, Manual Control and Visualisation in that order, with Tagging being comparable to the sum of efforts required for the other three.

Estimated benefits according to control strategy

The benefits of each Level of implementation are also site-specific and must be estimated considering the development and production plan targets. From the three pillars perspective (health and safety, production, and energy savings), achieving health and safety is an enabling condition and as such no value, as it is considered the minimum requirement for the plan to work.

Production flexibility, or a potential enhancement, are hard to measure, and from a practical perspective enabling the mine to consistently achieve the development and production plan by minimising ventilation constraints is a significant contribution. Ultimately, the most measurable indicator is the energy usage (electrical and heating fuels as applicable) and the reduction or savings that could be achieved depending on the control strategy implemented. This assumes that a baseline can be estimated in each case, which is a non-trivial task over time.

From a business case perspective considering a positive Net Present Value (NPV) within a reasonable time horizon as the metric to drive the decision making. If the energy savings can cover the cost of implementing the ventilation control system, then the decision is sound. However, deciding the level of implementation might still be a function of the effort, the resources available, and the technology acceptance by the work force.

The estimates of energy savings achievable per level of control as have been presented by Acuña and Allen, 2017 will be used as an indicator to develop the estimate presented in this study. Table 3 presents the estimated savings per Level of control. Note, Visualisation was not considered as a standalone control strategy, consequently in this estimation Visualisation and Manual Control were coupled together.

TABLE 3 – Percentage energy saving estimates per level of control (control strategies)

Control strategy Level 0/1 through to 2-5	Energy savings (%)
Visualisation + Manual Control	25
Time of Day Scheduling	12
Event Based	0
Tagging	25
Environmental	0
Total	62

As presented in Table 3, Manual Control and Tagging are the control strategies reporting the greatest savings followed by Time of day Scheduling. Tagging can be split in two categories: tagging driving main fan control (@15%) and tagging driving auxiliary fan operation (@10%).

Greedy approach and results

A greedy or low-hanging-fruit approach is a blind selection process that chooses the next step based upon the direction of focussing on the highest return. From an optimisation perspective, this is a simple strategy that has the potential to deliver good results. However, the constraints should be understood and complied with.

For this study the ratio between benefit and effort (deployment + maintenance) is considered. The highest benefit has the greatest chance of being chosen and vice versa, the strategy requiring the highest effort has the least chances to be chosen.

Table 4 presents the estimated ratios for each control strategy. Based on these, the strategies to be implemented are Time of Day Scheduling, Visualisation and Manual Control and Tagging, in that order. However, Time of Day Scheduling can only be implemented after Visualisation and Manual Control according to the efforts and benefits presented in Table 4. This would suggest coupling the first three control strategies as a first phase of implementation and Tagging as a second phase of implementation.

Event Based and Environmental implementation would need additional work to define its benefit before deciding in which phase to implement it. The same argument could be proposed for the phase 0 Visualisation.

TABLE 4 – Benefit divided by Effort ratio per level of control (control strategies)

Control strategy Level 0/1 through to 2-5	Effort (E)	Benefit (B)	Ratio B/E
Visualisation + Manual Control	5.5	25	4.5
Time of Day Scheduling	0.5	12	24
Event Based	0.5	0	0
Tagging	7	25	3.6
Environmental	1	0	0

DISCUSSION OF PROPOSED APPROACH

From the results of the previous section, two main outcomes can be drawn for the multiple phase approach considering 3 steps of implementation:

- Control strategies Level 0/1 to 2, then 5 and then 4 depending on legislation, or

- Control strategies Level 0/1 to 2, then 4 and 5

It becomes apparent that merging the implementation of the first three control strategies could be beneficial because the location and nature of the work required is similar or at least groupable.

The main reason for the split in multiple phases introduction, is the effort and resources required to implement the complete ventilation control system for each control strategy. This needs to be considered in conjunction with the limited resources available underground. To spread it over time, over multiple horizons (mining levels) instead of concentrating on a single horizon, is a more practical approach considering the development and production plan is competing for U/G resources simultaneously. Additionally, the learning curve and training required is significant for both deployment and maintenance. Each horizon's implementation will probably take in the range of a year to implement or more.

If a sensitivity analysis would be conducted for both the benefits and efforts required to determine the impact in the final solution, then most of the analysis would revolve around the level of effort required to implement control Levels 0/1 to 2, followed by tagging, and their relative value. From an implementation perspective, respecting an effort argument, Level 4 cannot be implemented before the other pieces needed for Levels 0/1 and 2 are in place.

Additionally, as presented by Flores and Acuña, 2016, the level of effort required to implement all the components required, by tagging considering wi-fi or portals, for a dynamic system is considerably larger than the requirements to manually control a static system. The case can be made that the solution can vary based on the estimated benefit value of each level of control, but alternatives could be limited by regulations.

Variants of implementation

The features available within some of the software packages allow for flexibility in terms of how the savings are achieved. This in turn enables some variations of the implementation strategy that will impact the level of effort and benefit that can be obtained.

For example, when considering fan control, a common split is between main and auxiliary fan control and savings. To stop, start or change the speed of an auxiliary fan normally only involves the headings supplied by individual fans. For the main system fans, a change in their operational set point can affect the complete mine distribution, or at least the airflow on certain levels. A decision has to be made in terms of changing the volumes available to the mine or keeping them stable and just re-assigning the volume as needed or scheduled. A few variants will come from controlling the secondary ventilation alone, primary ventilation distribution alone, or the combination of both. The primary flows should not change if the secondary ventilation and the distribution remain unchanged. However, for a practical and successful implementation, in any of these cases the driver for design and decision making should be worker safety and production flexibility as opposed to just energy savings.

Main fan implementation requires additional effort in maintenance due to the need to vary RPM or other fan flow control options. But then they benefit from allowing unchanged volume due to seasonal and daily temperature on surface (natural ventilation pressure (NVP) influence to fixed RPM as compared to constant and known airflow volume through the mine with a variable RPM to compensate for NVP).

Main-fan-based Tagging savings can be partially recovered from Time of Day Scheduling if consistent airflow requirement patterns are identified for days through seasons for the mine, or if this is jointly implemented with other technologies like short interval control. In such case Time of Day Scheduling can recover significant savings that could otherwise be assigned to Tagging, but fail-safes, operational procedures, and training will also be required.

Alternatively, if the Environmental control strategy is implemented before Tagging, some of the savings indicated for Tagging can be attributed to Environmental.

To properly assess the split of the savings per Level of control, a site-specific study has to be performed based on the order of implementing the control strategies, the practicalities of each operation, the operational philosophy, culture, and the existing regulations.

Safety and production suggested implementation

Based on the solution provided by the greedy approach, the practicalities of U/G implementations, as related to the resources available, and using only safety and production as drivers, the suggested implementation program is as follows.

The three ordered main phases of implementation should be:

- Phase 1 – it should consider control Levels 0 to 2, and maybe two or three sub phases to deploy each level of control. The prime target should be the auxiliary fans, doors, and regulators. Main fans could also be targeted, with their control limited to being either manual and/or scheduled.

- Phase 2 - environmental could be introduced to operate only between shift change, during blast clearance, to reduce and/or give consistency to the re-entry time after blasting. For certain operations, this could be applied during the regular shift but with the necessary fail-safes, controls, operational procedures, and training in place.

- Phase 3 – the implementation of a Tagging control strategy. The variants of each implementation will have to be assessed on a site-by-site basis and driven by a business case.

CONCLUSIONS

A qualitative effort estimate, coupled with a previously determined energy savings evaluation, and applying a greedy approach was used to determine the most suitable multiphase implementation of a ventilation control system. Greater consideration was given to the three decision pillars: health and safety, production, and energy savings, as opposed to just energy savings. A discussion of the results was proposed, it took into account changes in the estimated level of effort or benefits. Further, the additional variants of implementing a ventilation control system were considered for the primary and secondary fan operation. Finally, a suggestion for three phases of implementation is proposed if only safety and production flexibility, or enhancement, were considered. Interestingly, the Tagging control strategy becomes one of the last phases of implementation. It is also recommended that it become subject to a business case evaluation to be performed only once the other levels of control are already implemented.

REFERENCES

Acuña, E.I. Desafíos presentes y futuros en la implementación de un sistema de control de la ventilación. 1er International Mine Ventilation Symposium, Santiago, Chile, November 2018.

Acuña, E.I., and Allen, C. Ventilation Control System implementation and energy consumption reduction at Totten Mine with Level 4 'Tagging' and future plans. 1st International Conference on Underground Mining Technology, Sudbury, Canada, October 2017.

Acuña, E.I., Alvarez, R.A., and Hurtado, J.P. Updated Ventilation On Demand review: implementation and savings achieved. Proceedings of the 1st International Conference of Underground Mining, Santiago, Chile, October 2016.

Flores, O. and Acuña, E.I. Improving monitoring and control hardware cost at Totten Mine. Proceedings of the CIM MEMO Conference, Sudbury, Canada, October 2016.

Tran-Valade, T & Allen. Ventilation-On-Demand key consideration for the business case. Proceedings of the Toronto Canadian Institute of Mining (CIM) Conference, Toronto, Canada, May 2013.

Implementation of energy efficiency and ventilation on demand at the Aguas Teñidas mine

J de Miguel Fuentevilla[1] and R Castro Pérez[2]

1. Ventilation engineer, MATSA, Mina de Aguas Teñidas, Huelva, Spain,
   Email: javier.fuentevilla@matsamining.com
2. Ventilation superintendent, MATSA, Mina de Aguas Teñidas, Huelva, Spain,
   Email: ruben.castro@matsamining.com

ABSTRACT

The Aguas Teñidas mine, hereinafter referred to as the ATE mine, is a copper and polymetallic (zinc, lead) underground mine located in the south of Spain, and uses primary exhaust ventilation.

The physical limitations of the network due to the location of the Aguas Teñidas mine, the limitations of electric power availability from the grid and the constant increase in the price of electricity were some of the reasons that accelerated the search for energy efficiency solutions in primary and auxiliary ventilation.

Several energy saving strategies were examined. These ranged from controlling the fans by the worker in the stope, to having a remote manual control from the control room. At the same time, automatic control of the surface primary fans is being done during the shift changes, and work is being done on the implementation of a ventilation on-demand system according to work modes manually from the control room.

Other energy efficiency opportunities that have been adopted include the replacement of underground secondary ventilation duct by an alternative that has a lower friction factor, using vent duct fittings with lower shock losses and lower leakage at couplings, as well as the replacement of a twin surface horizontal primary fan with a combined motor rating of 1,4 MW by a single vertical axial with a motor rating of 0,71 MW.

"Matsa, an energetic island"The Aguas Teñidas mine is located near the town of Almonaster la Real, in the province of Huelva, Spain. It is a mine of massive hydrothermal sulphides, with the presence of Cu, Zn and Pb. The mine produces 2 Mt per year of polymetallic ore and copper. Along with the ATE mine, MATSA has two other nearby mines, Sotiel and Magdalena, in operation. The three mines together produce 4.4 Mt of ore per year. The final product is a copper concentrate and a polymetallic concentrate.

The method of mining is sublevel stoping, with some orebodies using drift and fill. The ore is transferred by diesel LHD to ore passes and then both ore and waste is trucked to surface via dedicated haulage ramps.

The mine has five intakes: VR1 shaft, haulage ramp, service ramp and shafts VR8 and VRC1. There are three primary exhausts: VR2 of 3,1 m, VR3 of 4,5 m and VR7 of 4,1 m, allowing semi-independent primary ventilation districts for the various production zones.

The primary fans of the mine extract up to 600 m³/s, through the following surface exhaust fans:

- VR3: two (twin) horizontal fans in parallel of 0,7 MW each (total 1,4 MW).

- VR2 fan: single horizontal fan of 710 kW.

- VR7 fans: two (twin) horizontal fans in parallel of 500 kW each (total 1,0 MW).

All the fans are equipped with VFD and with control panels of different protection parameters, all connected and controlled by the SCADA. The VR3 will be replaced by vertical fans at the end of 2019, to reduce the installed and absorbed power.

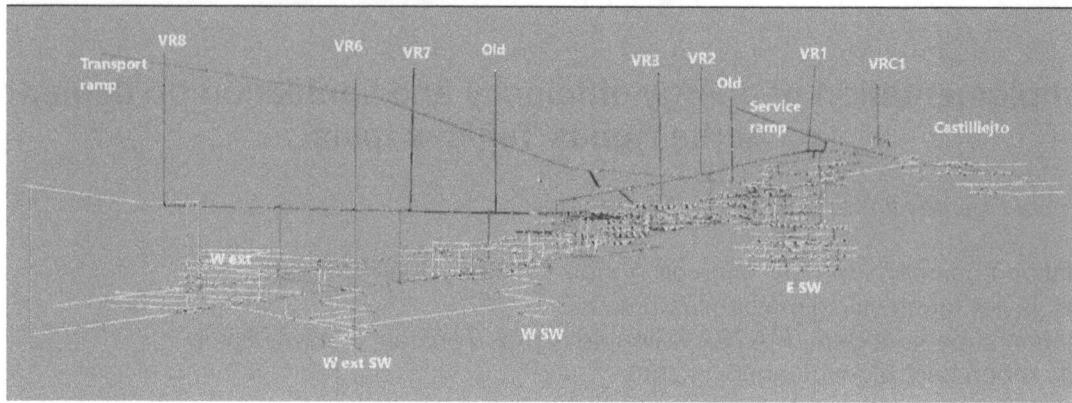

FIG 1 - View E - W of the ATE mine

JUSTIFICATION OF SEARCH FOR ENERGY EFFICIENCY

Matsa is located in what is known as an "energy island" because we depend on an electric power substation that is supplied by the county and that can only supply a fixed quantity of MWh. The power consumption of the overall mine site, including the concentrator is very close to the available incoming power limit (36 MW) from the utility company. In addition, the contract with the electrical supplier provides for a continuous increase in the price of electricity. For both these reasons, it became imperative that the operation identified ways to save both average and peak demand power.

Thus, in the northern hemisphere summer of 2017, we began to connect the auxiliary and main fans in our network to our energy management system and began to design concepts of energy saving in ventilation while seeking to maintain or improve the environmental and flow conditions in our mines.

SITUATION OF THE VENTILATION IN AGUAS TEÑIDAS BEFORE THE INSTALLATION OF THE VENTILATION ON DEMAND SYSTEM

At Aguas Teñidas mine, the auxiliary fans were controlled manually and locally by the workers, leaving the auxiliary fans running continuously, even during the shift change and causing delays in the re-entry of the stopes.

None of the fans were connected to our data network, so there was no remote control of the fans, nor the recording of the power consumption or hours of operation, the latter being vital for correct preventive maintenance.

On the other hand, the surface primary fans operated continuously at their maximum capacity without considering the necessary duty point according to the work that should be done or the time of mining in a ventilation district, and without having remote telemetry of the readings of their control parameters (vibrations, bearing temperatures, fan pressures, etc.).

OPTIMIZATION PLAN

This project was born as a result of our site electrical power limitations, but with the intention to simultaneously improve our key ventilation KPI (kWh/ $m^{3)}$ of air circulated through our primary and secondary ventilation).

Primary ventilation

In 2018, the Ventilation Department developed an action plan that aimed to optimize MATSA's Ventilation system, following the development of the overall site's strategic plan:

Optimization of our KPI kW/m³ used in our primary ventilation through:

- Remote monitoring and control of primary and secondary fans from surface control room.

- Surface fans changed from twin horizontal to single vertical.

- Installation of new automatic dampers as underground regulators.

Before implementing our VOD system, measurements were taken to set up the initial KPIs and in late 2017-early 2018, our primary KPIs (kW-hr/m³) circulated was:

Scenery primary fans in 2018 (Aguas teñidas).	Q(m3/s)	kWh
VR3 (2 horizontal fans)	236	1122
VR2	125	320
VR7	190	632
Total consumption primary fans kWh		2074
Q total m3/s	m3 total	551
kpi´s kWh/m3	kWh/m3	3.76

Scenery primary fans in 2019 (Aguas teñidas)	Q(m3/s)	kWh
VR3 VERTICAL(new axial fan)	240	590
VR8 (new axial fan)	210	589
VR2 (new fan according requirements)	35	53
VRC1 CASTILLEJITOS (new fan)	125	275
Total consumption primary fans	kWh	1507
Q total m3/s	m3/s	610
kpi´s kWh/m3	kWh/m3	2.47

FIG 2 - Comparative between primary energy consumption and KPIs kWh/m³ 2018/2019

VRC1 and VR8 are currently both being installed and will become operational later this year.

During 2019, we will begin production in new mining areas, which will require the installation of new surface fans, VRC1 and VR8, increasing our flow(Q) from 551 m³/s to 610 m³/s. An absorbed power reduction of 37% is expected only with the changes made on primary ventilation for 2019. By the end of 2019 all of our fans will be vertical, in addition to the elimination of horizontal fans VR7 will become an intake shaft. It is expected that our energy efficiency will be improve from KPI´s 3.76 kWh/m³/ to 2.47 kWh/m³, a 34.3% improvement, saving, **4035 MWh-year** respect 2018,((24h - 4,5h per each change shift) x 365 days x 567 kWh).

On the other hand, the primary ventilation control by districts from the control room during the shifts will further improve this KPI as follows:

Testing carried out with VR2 and VR7 fans have shown that lowering the speed (rpm) is a fact, so the potential of power consumption save to 2019 during shift changes can achieve a reduction of 50% in districts where there is no blasting and which can be considered as independent circuits.

In addition the installation of automatic regulators will allow further energy savings, by allowing us to lower the RPM of the 4 available fans at ATE, as it is considered that there will always be at least two districts where fan speed can be lowered. This is expected to be a conservative estimate since there are many shifts where there is no blasting, thus an expected average of 1449 kW can be saved per day (fig.3), noting that the ATE operation works 3 shifts with 1.5 h during each shift change, 4.5 h-day, resulting in a total saving per year of 724 MW

E1-Estimation after probes of the power consumption .(during shift changes), reducing rpm of the primary fans in 2019

	kWh	kW- day	MW-year
VR3 reduction of the consumption 50 %	295	3969	1449
VR8 reduction of the power consumption 50%	425		
VR2 50 %	27		
VRC1 CASTILLEJITOS 50%.	135		
E1-Total save power consumption MW-year considering only 50 % of our fans with this strategy			724 (MW-year)

FIG 3 - Total energy savings per day and per year due to reduce power consumption of the primary fans during shift changes.

Additionally, the installation of louvers by levels will allow reallocating our ventilation in two different levels of work or aperture: one in operations and other in main access. We have started this year to install these automated regulators in the VR2 area. Additionally, this year we expect to install new louvers in 17 levels in the VR8, VR3, VR7 and VR2 areas. It is estimated that will reduce approximately 20% of the energy consumption of the primary ventilation in 2019. This estimate is based on our measured performance during trials in the VR2 area, which will detail later on in this paper. Basically automatic louvers will be controlled from the control room and will allow us to have 2 working modes for the levels where they are installed, so they will extract more or less air flow, as we have divided the mine by districts. These louvers will allow us to reduce the air flow needed for each level depending on the work done, so the total requirement of the main fans will be reduced to an oversized but much more optimal reality. Fig. 4 summarises these savings.

Strategy 3, instalation of louvers per levels			
	kWh	kW- day	MW-year
VR3 VERTICAL	118	2,832	1,034
VR8	118	2,827	1,032
VR2	42.4	1,018	371
VRC1 CASTILLEJITOS	55	1,320	482
Energy save: Reduction of 20% of the energy consumption (estimated)			2,919(MW-year)

FIG 4 - Total energy savings per day and per year due to reduced fan speed make possible by the louvers strategy.

Installation of the automatic regulators is expected to result in an annual energy savings 2,919MW-year. However, with the current primary ventilation system operating in semi-parallel there are a series of technical constraints that could negatively affect this outcome. We are delimiting the constraints with our ventilation consultant. Thus, our planned energy consumption in 2020 is approximately 31% lower than in 2018. This implies an improvement of 46% of our KPI (kWh/m^3) of circulated air.

Secondary ventilation

We describe below our action plan and the expected power savings in auxiliary ventilation, based on the following:

- Manual switch-on and switch-off of fans in working areas and during changes shifts.

- Reduction of the fan motor absorbed power, as a result of the new underground secondary ventilation ducts

Between 2017 and 2018, the auxiliary ventilation was not well design and sized. It was after 2018 when we started to quantify and measure the real needs. However, the actual quantitative and qualitative improvement was achieved with the new communications network within the mine, which allowed us through SCADA to monitor and control the status of the secondary fans. Nonetheless, the robustness of the communications network has taken some time to rectify to the point where communication is no longer being lost with the measuring and control air devices.

Manual switch-on and switch-off of fans (2018)

FIG5 below shows the real consumptions of the auxiliary ventilation by type as of January 2019. We have taken as a reference this table (FIG-5) to absorb the deviations during the year. Thus, the scenario showed is an average of 2018 where it includes the total of fans with communication at 70% (availability). Out of this percentage, the total equipment with functioning communication or shutdown, in addition equipment without communication, all shown real absorbed power installed, so we quantify our real consumptions and be able to quantify the different resistance of each ventilated level. (This absorbed power measurement is key to determine and quantify the resistance with the new vent ducts).

E1-2-*,** Manual on/off control of the auxiliary fans from room control and ***70%** **(availability) during 2018

Fans model	kW	55	220	110	22	30	75	45	37
Nº fans per model	uds	27	3	0	5	3	5	2	3
Air flow delivery (Q)	m3/s	23	21	16	8	13	17	15	14
Subtotal m3/s (Q)	m3/s	618	104	16	40	13	85	44	41
Power consumption	kWh	35	65	75	13	18	44	30	24
Sub total power consumption		950	194	0	65	54	218	60	71
Fans in mode "off" (considering power consumption)	kWh	246	0	0	13	18	87	30	24
Fans wthout comunication(considering fans working) potencia sin control/dia	kWh	282	129	0	0	18	44	30	24
M02		35							
L58		35							
L17									24
L06					13				
L60		35							
L80									
L53						13.05			
L97-98			65						
L99-100			65						
L78,79,87		106							
L35							44		
L51									24
L54								29	

		h	kWh-day	MWh-year
Total of power install	**kW**	**2,921**		
Total power consumption	**kWh**	**1,613**	**34,683**	**13,754**
Average power consumption fans on-line (turn on)	**kWh**	**668**	**18,545**	**6,769**
Average power consumption fans on-line (turn off)	**kWh**	**418**	**13,068**	**4,770**
Average power consumption fans without comunication	**kWh**	**526**	**6,070**	**2,216**

FIG 5 – Power consumption strategy manual on/off control of the auxiliary fans.

The FIG 6 below shows the source used in this detailed study, which corresponds to production in a point-of-time used as a reference from Scada.

Nº	Tag Ventilador	Tag Cuadro	Potencia KW	Nivel	Labor	Control Manual	Auto Parada	Auto Arranque	Grupo	Horas	Nº	Tag Ventilador	Tag Cuadro	Potencia KW
1	L83 y L84	CL84	220	540	BASE IVR23	ON OFF	ACTIVO	OFF	0	5770	26	L53	CL53	22
2	M02	CL2	55	1020 CAST	AIVR 1020 510	ON OFF	ACTIVO	OFF	0	2075	27	L29	CL29	30
3	L48	CL3	55	750 S	AL 750	ON OFF	ACTIVO	OFF	0	1104	28	L65	CL65	55
4	L03	CL4	55	600 EXTO	RP EXTO	ON OFF	ACTIVO	OFF	0	6954	29	L78	CL78	55
5	L8	CL8	75	720 W	AIVR 720 9930	ON OFF	ACTIVO	OFF	0	6954	30	L43	CL51	75
6	L50	CL50	37	660 W	LT 660 W	ON OFF	ACTIVO	OFF	0	3992	31	L62	CL80	55
7	L82	CL48	55	700 E	AIVR 690 980	ON OFF	ACTIVO	OFF	0	9346	32	L89	CL89	55
8	L76	CL9	55	1000	RPK 950	ON OFF	ACTIVO	OFF	0	5562	33	L81	CL81	55
9	L24	CL30	45	660 W	AIVR 660 934 (IVR12)	ON OFF	ACTIVO	OFF	0	5822	34	L79	CL79	55
10	L42	CL42	75	750 E	RPE 750	ON OFF	ACTIVO	OFF	0	6576	35			
11	L101	CL101	55	580 EXTO	RP EXTO	ON OFF	ACTIVO	OFF	0	256	36	L35	CL35	75
12	L11	CL20	55	700E	AIVR 690 980	ON OFF	ACTIVO	OFF	0	6450	37	L87	CL87	55
13	L58	CL58	55	540 EXTO	RSB 540 EXTO	ON OFF	ACTIVO	OFF	0	0	38	L86	CL86	55
14	L17	CL17	30	720 S	RP 750 - 720	ON OFF	ACTIVO	OFF	0	0	39	L63	CL63	55
15	L06	CL102	22	840	GL 840 770	ON OFF	ACTIVO	OFF	0	1931	40	L33	CL10	22
16	L44	CL44	30	670 E	LT 670 E	ON OFF	ACTIVO	OFF	0	6998	41	L97 y L98	CL90	220
17	L59	CL59	55	670 E	AIVR 670 1090 (IVR 21)	ON OFF	ACTIVO	OFF	0	***	42	L99 y L100	CL97	220
18	L60	CL60	55	630 E	AIVR 630 1090 (IVR21)	ON OFF	ACTIVO	OFF	0	0	43	B01	CB01	55
19	L61	CL61	55	750 S	RP SUR (N750)	ON OFF	ACTIVO	OFF	0	7137	44	L28	CL18	75
20	L64	CL64	55	720 E	AIVR 720 (IVR21)	ON OFF	ACTIVO	OFF	0	1014	45	L51	CL52	37
21	L59	CL66	55	670 E	LT 670 E	ON OFF	ACTIVO	OFF	0	4158	46	L39	CL39	22
22	L21	CL12	55	885	AL 900	ON OFF	ACTIVO	OFF	0	6582	47	L64	CL54	45
23	L14	CL15	37	750 E	RPE 750 E	ON OFF	ACTIVO	OFF	0	6793	48	L31		55
24	L69	CL69	55	630 W	AP 630 9880 (PIQ.31)	ON OFF	ACTIVO	OFF	0	7686	49			
25	L80	CL5	55	520 EXTO	RSB 520 EXTO	ON OFF	ACTIVO	OFF	0	0	50	L19	CL19	22

FIG 6 - Status of each of the installed fans, communication status (on/off-line, or turn on/off), and the fans on emergency mode for critical locations shown in red).

ONLINE MARCHA REMOTO   ONLINE PARADO LOCAL
ONLINE MARCHA LOCAL   ONLINE EMERGENCIA
ONLINE PARADO REMOTO   OFFLINE

FIG 7 - Map key for the different fans status is as follows:

The definition of each vent status, stated from top down and left to right:

Fan with communication (on-line) and "on" and turned on from the control room. (Green color).

Fan with communication (on-line) and "on" and turned on from the mine.(blue color)

Fan with communication (on-line) and "off" turned off from the control room.(black color)

Fan with communication (on-line) and "on" and turned off from the mine.(grey color).

Fan with communication (on-line) and "on" monitored from control room and with alarm on (critical locations).(red color).

Fan without communication (off-line) which we don't know if they are turned "on" or "off".(orange color)

Thus, out of the total installed power of 2,921 kWh., the real consumption is 1,613 kWh.

In summary, the average in 2018:

Total power installed   2,921 KWh.

Total power consumption of all fans without strategy   1,613 KWh.

Results strategy manual on/off control of the auxiliary fans from room control in 2018

Real power consumption of fans online in mode"on"   668 KWh.

Real power consumption saved of fans in mode "off"   418 KWh.

Power consumption estimated of fans without communication   526 KWh.

With the above data, the power savings achieved in auxiliary ventilation during 2018, with control from the mine has set a monthly average of 247 MWh/month, which represents a 2,975 MWh/year.

Manual switch-on and switch-off of fans during shift changes (2018)

As shown below in the SCADA image (Figure8.), within our action plan, during the shifts we turn off all of the auxiliary ventilation in the areas that will not have work. As explained in primary ventilation, we have the mine allocated by districts ,so we could achieve three times a day during 1.5h (crew change time), reduce the power consumption up to a 44% during those hours, the savings is of 830 MWh/year.

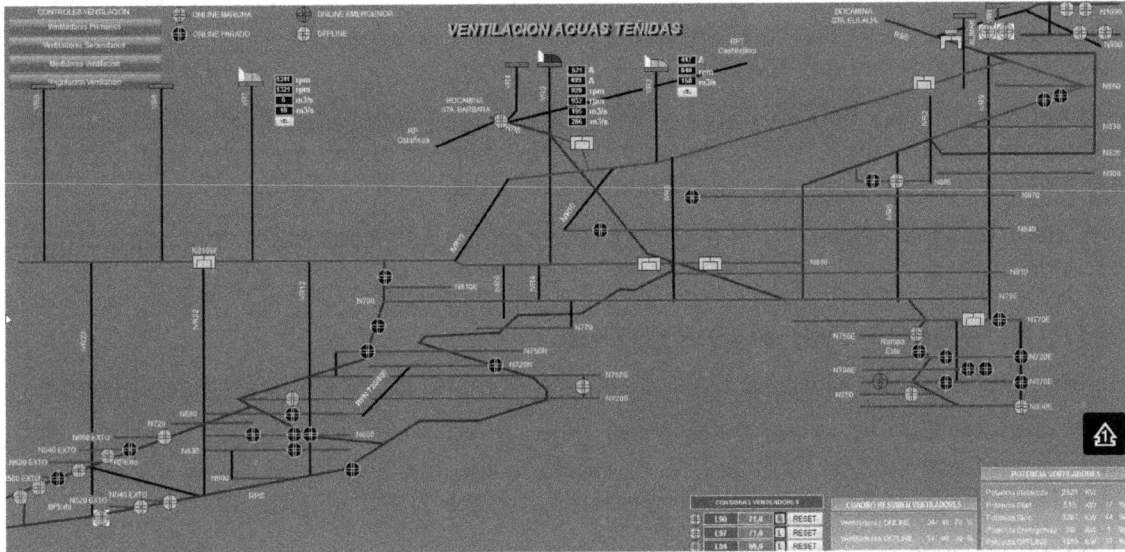

FIG 8 - ATE, Scada image showing the status work of the fans during shift change

Thus our plan for the end of 2019, is to have 100% of our ventilation on-line, and the total power absorbed of 1.643KW/h will be reduce to a 56% during change shifts, representing 1,187 MWh/year.

Manual switch-on and switch-off of fans (2019/2020).

The data of the auxiliary ventilation in 2019 is showing better results. In relation to communication, it has gotten better achieving 92% of the ventilation controlled. We believe that by the end of 2019, we will have 100% of the ventilation controlled from the control room thanks to the new procedures that have been established and that will be discussed further below. Based on the above, for end of 2019/2020 we estimate the following results:

E1-2-, room control with 100% fans on-line 2019			kWh-day	MWh-year
Total of power install	kW	2921		
Power consumption of the fans	kWh	1613	31454	11481
Average power consumption fans on-line (turn on)	kWh	990	19305	7046
Average power consumption fans on-line (turn off)	kWh	623	12149	4434
Average power consumption fans without comunication	kWh	0		

FIG 9 - Estimation of auxiliary ventilation consumption in 2019/2020 (Setting per plan for 2019).

Total power installed	2,921 kWh.
Total power consumption of all fans without strategy	1,613 kWh.
Results strategy manual on/off control of the auxiliary fans from room control in 2019 with 100%(availability)	
Real power consumption of fans online in mode"on"	990 kWh.
Real power consumption saved of fans online in mode "off"	623 kWh.
Power consumption estimated of fans without communication	0 kWh.

Thus the first year, with 100% of auxiliary ventilation online and with this settings and considering 19.5 operating hours., the saving in each shifts, we can achieve an annual savings of 4,400MWh-year.

MOVING TOWARDS THE VOD SYSTEM APPROACH

Before implementing ventilation on demand, several factors had to be considered, such as mining methods, the division of primary ventilation into ventilation districts, types of ventilation circuits, equipment used and personnel, contaminants present in the operation, available network in mine and the work standards used in MATSA.

On the other hand, it was observed that although the weekly mine plan is complied with, the order of activities varied over time, thus ruling out the possibility of using software that could program auxiliary and / or primary ventilation according to a schedule.

Besides, there was the availability of a control room with possibilities of manual and / or automatic ventilation control.

CONSTRUCTION OF THE VOD SYSTEM

Our remote control system of fans and ventilation on demand has the following components:

- Fan starters: they have been used star triangle drive mainly, although we have began to implement soft starter. This device connects to the fan and the network via PLC. Through the PLC, they can be controlled remotely through the TCP / IP protocol.

- Communications network: is the network that allows us to connect the SCADA to the PLCs, and also directly to the sensors (AQMS, see below). Our network is formed by optical fiber and the connection is ETHERNET, with MODBUS TCP / IP communication protocol.

- PLC: connects with the fan starter and also the communications network. It can be integrated with other devices for specific VOD tasks.

- SCADA: connects with devices connected to the network, allowing the programming of complex operation algorithms. It enables us to see the status of the device (regulators, AQMS sensors and fans), control the device manually or automatically, view and store databases of the outputs of the devices.

- Air quality monitoring stations (AQMS): we have acquired MAESTRO AQMS, and these are directly connected to our network. Its application serves to improve productivity and security. There are 02 types:

- VIGILANTE: this station measures the concentration of gases (CO, O2, SO2 and NO2), air speed, relative air humidity, barometric pressure, dust and dry temperature of the wet bulb.

- AIRSCOUTS: these are sensors that measure the air speed by means of ultrasound. In addition, the dry air temperature is measured.

- Variable frequency drives: they allow to vary the speed of the motor of the fans, according to the demand of air and we use them in all the primary fans, and in the auxiliary fans used for ramp development and in some booster fans.

- Automated louvers: to provide for automation of the regulators, we have replaced the old style of drop board wooden regulators. The ability to remotely control the regulators is an essential component to carry out the ventilation system on demand.

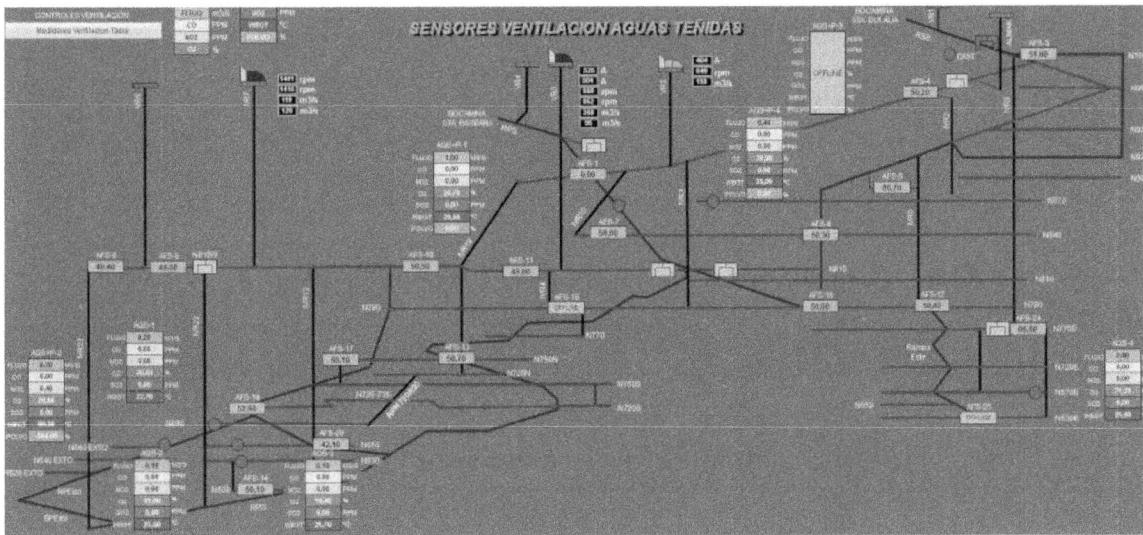

FIG 10 - ATE mine diagram showing the AQMS

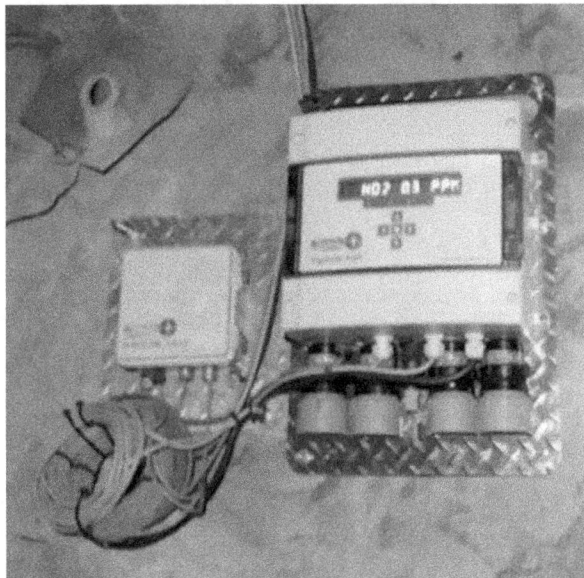

FIG 11 - AQMS VIGILANTE of MAESTRO

FIG 12 - Image of the ZITRON stainless steel louver installed at L630 W

LEVELS OF VENTILATION ON DEMAND

Manual control of auxiliary ventilation

To control the auxiliary ventilation from the control room, the following strategy is being used:

- Manual on/off control of the auxiliary fans.

- Auto shut off of all secondary fans at all shift changes, and control room operator then remotely turns on those fans that are needed to remove blasting gases or are needed due to the requirement for personnel to be in a particular area.

- Only the control room operator can control the fans, except for malfunctions or faults in the network.

- Elimination of the local (at the fan starter) manual control selector. Fans cannot be operated in local mode anymore.

In August 2017, manual control of the auxiliary ventilation was started from the mine control room of Aguas Teñidas.

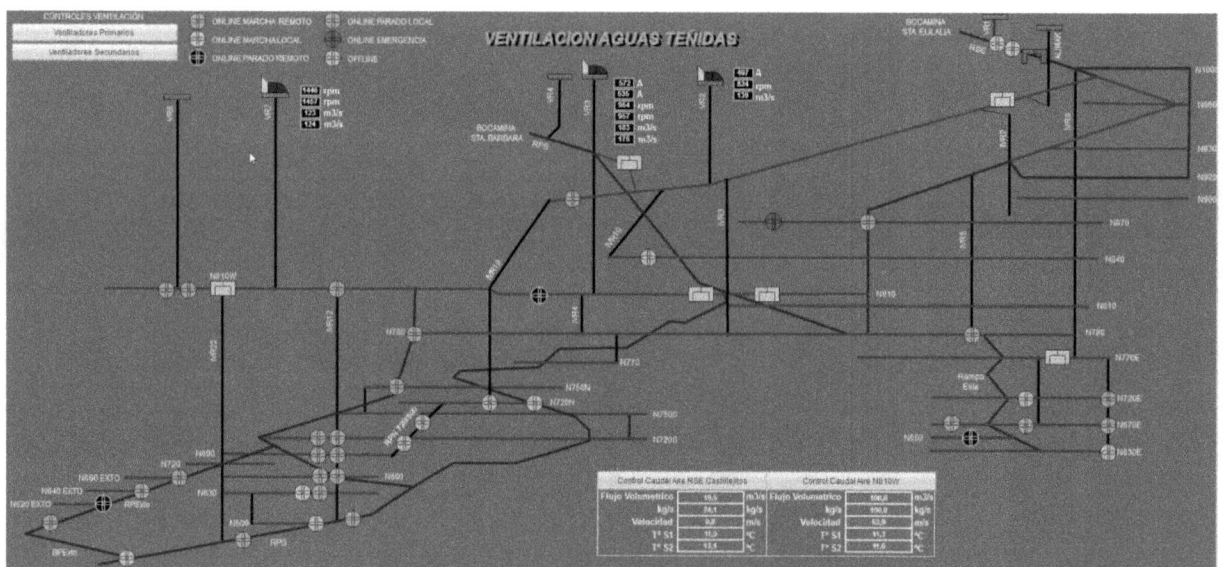

FIG 13 - Diagram of ATE mine with auxiliary and main fans

Energy impact

An improvement in productivity and safety in the environment was observed, as the on-coming shift could enter faster and safer to the stopes with the blast gases evacuated, once the appropriate fan(s) were turned on after the blasting, without exposing the workers to gases.

Figure 14 shows that since this system was implemented in August 2017, the total number of auxiliary fans in the mine has remained approximately equal (red line), however as we have converted more and more of these fans to the VOD system (green line), the total electrical power consumption has trended downward (black line).

The average monthly electricity savings since the auxiliary fans were controlled from the control room is 247 MWh / month.

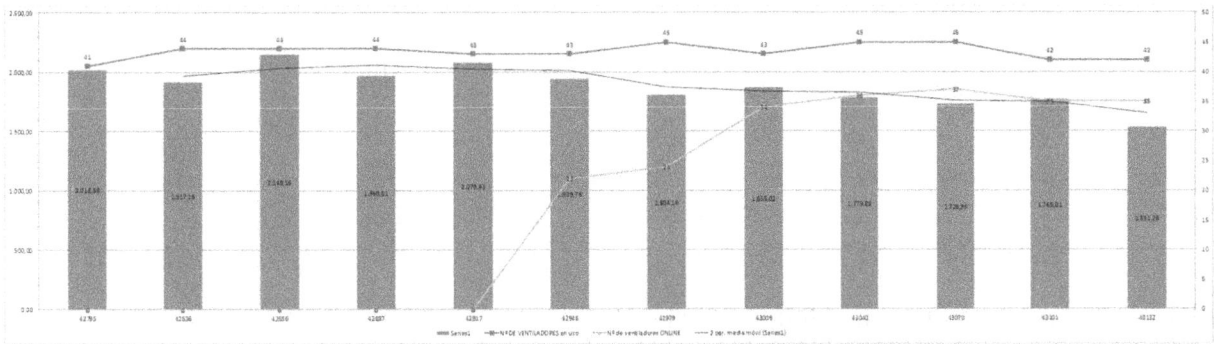

FIG 14 - Graph of the annual electricity consumption of the mine

Figure 15 shows the decrease in the ATE mine peak power demand (descending green line). At the same time a step change has formed in the shift changes as it is seen in the red horizontal line.

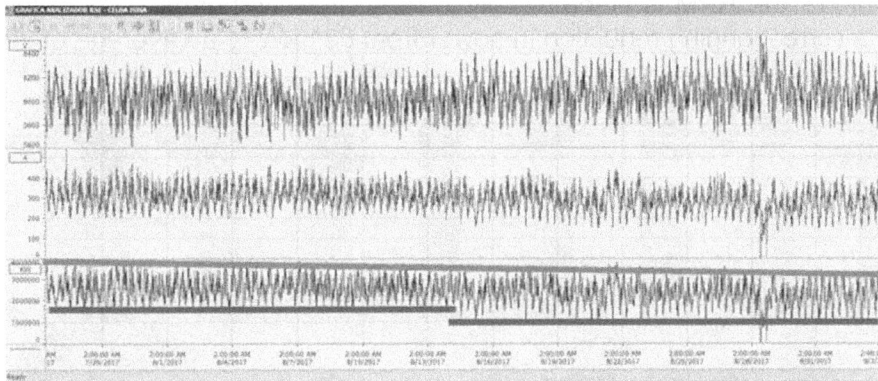

FIG 15 - Graph of the effects of VOD on the electric consumption of mine

Auto scheduling and manual control of the main fans and manual control

In addition to converting the auxiliary fans to a VOD system, two of the primary fans on surface (VR2 and VR7) now automatically reduce and then increase their RPM over the shift changes, and in case of blasting in this area, the RPM of the fan is manually raised from the control room.

VR2 fan

This fan ventilates the central ATE zone.

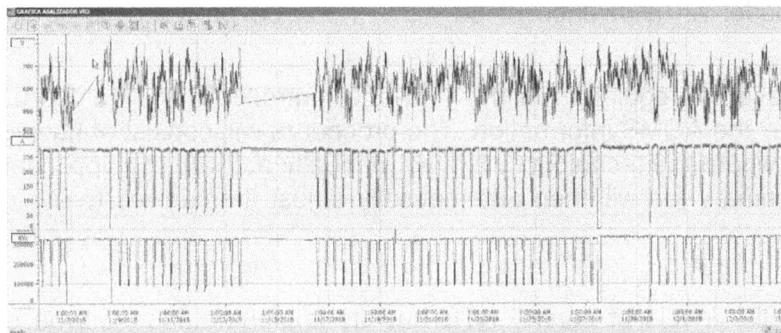

FIG 16 - Monthly consumption of the VR2 fan for November 2018, showing the decrease in rpm and kW during shift change

The VR2 fan is reduced to 50% RPM at shift change, the minimum that can be used before the fan goes into stall and/or the minimum required for motor cooling), reducing the power consumption at this time from 320 kW (84%) to 92 kW (50%) [Figure 17]

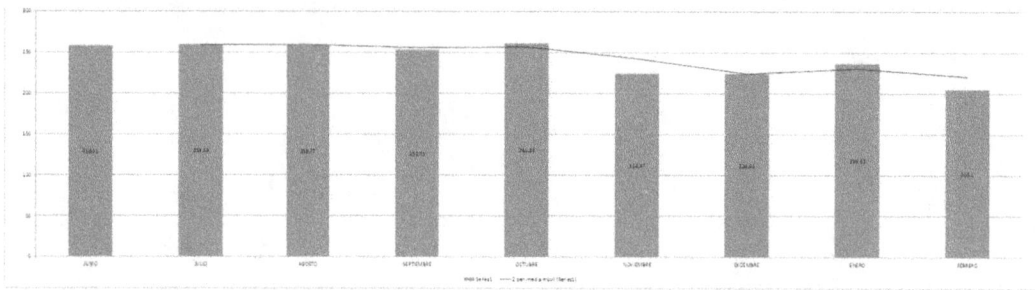

FIG 17 - Power consumption of the VR2 fan over the past 9 months where there is a reduction of MWh by the auto lowering of rpm in the shift changes

VR7 fan

This fan ventilates the area of ATE West and EXTO.

At each shift change, the RPM is automatically lowered from its normal 87% of full speed to 500 rpm (60%) with a corresponding reduction in power from 631 kW to an average of 220 kW.

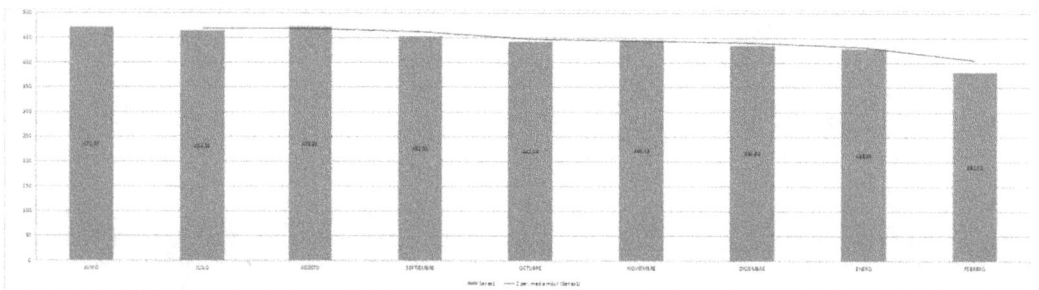

FIG 18 - Power consumption of the VR7 fan over the past 9 months where there is a the reduction of MWh by the auto lowering of rpm in the shift changes

Ventilation on-demand based on manual control of auxiliary fans according to working modes

In this phase of the project, airflow to each level within each primary ventilation district (VR2, VR7 and VR3) will be based on one of two 'work modes':

- Access mode: an allowance of 15 m3 / s will be provided for drilling, inspections, explosive loading and any level without activity.

- Production mode: an allowance of 30 m3 / s will be provided for shotcreting and scoop operations.

It is essential to have air regulators in the network controlled from the SCADA and which can measure and display the AQMS information. The procedure will consist of having a daily mine plan in the control room where the technician will then manually activate the applicable work mode, and the regulator and primary fan will then automatically adjust themselves to achieve that duty point on the level.

District VR2

This fan ventilates the levels of central ATE and its implementation will be carried out on two levels.

ANALYSIS OF OPPORTUNITIES AND WEAKNESSES

Weaknesses

Our biggest challenge is to ensure we have robust network with 100% availability, despite the continuous changes in the mine and the movement of fans. Prioritizing the maintenance of the network is key to the stability of the VOD system.

Louvers and ventilation doors are devices that require continuous maintenance since a breakdown in one of these devices will force us to stop VOD on that level until they are repaired.

Air monitoring stations require continuous maintenance and calibration to show reliable information.

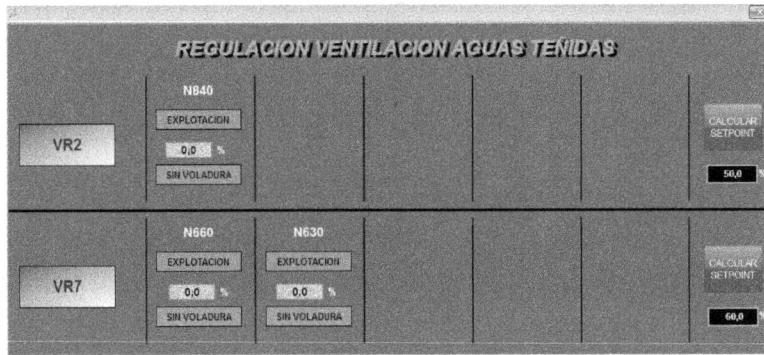

FIG 19 - Menu for manual control of ventilation levels according to ventilation work modes

Etapa	Descripción de la etapa	Q total	rpm	Consigna (%)	absorvida (kV	Apertura (m2)	Apertura (%)
3	MODO ACCESO N840	100	690	69,0%	183,9	3,70	59,7%
4	MODO PRODUCCIÓN N840	122	840	84,0%	331,8	6,20	100,0%

FIG 20 - Table work modes for the central ATE L840

Opportunities

Automatic response of the system according to the gases. This is something we are actively considering.

Systematic installation of variable frequency drives in all the auxiliary, according to its economic payback.

Software implementation study to complement the SCADA.

Ventilation on demand according to the tagging of machinery.

SUBSTITUTION OF THE PREVIOUS VENT DUCTS BY MORE EFFICIENT DUCT SYSTEMS

Other actions towards energy efficiency, has included the replacement of the duct system used so far, by ones (AC INDUSTRIES) with better mechanical properties, lighter, with fittings with geometries with lower shock losses and a lower friction factor to the old ones.

To demonstrate the performance of the replaced duct system, it has been done an example where both vent ducts are compared for the same configuration:

A case study has been undertaken to compare predicted performances of an auxiliary ventilation system connecting with a typical sewn type duct with standard duct fittings and the performance of a ducting system using the same fan but with TURBO-DUKT and streamline design (low aerodynamic resistance) duct fittings such as AC Industries' Rhino-Fittings product series.

Case 1, use 1400 mm typical sewn type duct, the fan (CC1400 2 x 110kW) with no high flow junctions or elbows, and only a T-piece in the split. For the take-offs along the drives only standard T-pieces are used. Standard eyelet coupling is used. A friction factor, K, of 0.00283 Ns2/m4 and a leakage coefficient, LC of 0.522 m3/s/km

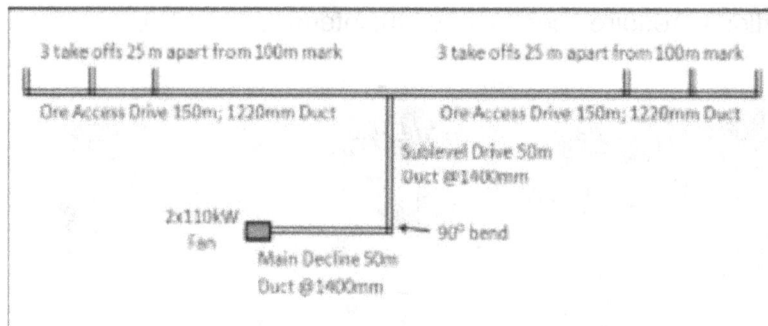

FIG 21 - Plan view of the comparative illustration of TURBO-DUKT and Typical sewn type duct

Case 2 - use 1400 TURBO-DUKT with high flow design Lobster-Bak and Rams Horn pieces in the split. In the take-offs radius branch pieces instead of T pieces are used. For the sewn duct K value is 0.00317 Ns2/m4 and LC value is 0.522.

The ratio of radius to diameter is 4:1. Therefore, shock factor for the Lobster-Bak/Ram Horns is calculated at 0.17 and shock factor for the bending of duct at 90° or T-pieces is assumed at about 1.12.

The following table shows the predicted performance of two ventilation ducting systems over 250m:

System	Friction factor, Ns^2/m^4	Leakage factor, mm2/m2	Shock factors	FTP, Pa	Z	C	B	A	D	E	F	Total delivered	Fan elec power, kW	€/yr
A. Ordinary	0.00283	239	1.12	3491	42.9	10.1	4.8	4.1	10.1	4.8	4.1	38.0	198	138758
B. "Turbodukt"	0.00317	239	0.17	2082	47.3	11.6	5.9	4.4	11.6	5.9	4.4	43.8 (extra 5.8 m³/s)	147	103018 (savings €35,740)
C. "Turbodukt"	0.00317	48	0.17	2146	47.1	12.6	5.9	4.4	12.6	5.9	4.4	45.8	152	106521

FIG 22 - Table of results of the comparative of TURBO-DUKT and Typical sewn type duct

REPLACEMENT OF TWO PARALLEL HORIZONTAL FANS BY A VERTICAL FAN IN THE RETURN AIR RAISE, VR3

A further opportunity for saving peak demand and energy consumption was identified in one of the existing surface twin horizontal fans. The VR3 raise is an exhaust shaft, in which two fans are installed on the surface working in parallel with an installed power of 1,4 MW. An investigation was completed to assess the potential to reduce the power requirement of these fans, while maintaining the existing flow of 250 m3 / s, at both the current and future resistances to which the fan will be subject. The alternative option examined was to replace these two fans by a vertical fan of with a single 710 kW motor, making it possible to reduce the operating pressure of the fan due to elimination of the substantial above-collar losses.

With this change, the amount of power installed and the electricity consumed will be reduced, given the power limitations of the energy "island" of MATSA.

Features and fan curves of the current VR3 fans

The average measurement, made of Collar TP was 1419 Pa and 240 m3 / s, well below the theoretical 3000 required. These fans are high pressure.

Total Air flow (2 Parallel Fan)		250	m³/s
Total Air flow (Single Fan)		125	
Air density		1.2	kg/m³

Velocity pressure at:	(Drawing ref.)		
Exhaust shaft	1	148	Pa
Adaptor cone shaft-bend	2	405	Pa
Bend connection to shaft	3	405	Pa
Plenum with 2 outlets	4	333	Pa
Inlet Cone - Contraction	5	458	Pa
Fan ring area	6	1802	Pa
Fan evase outlet	7	114	Pa

Pressure drop Summary:		
Shaft collar TOTAL pressure	3000	Pa
90 degree plenum with 2 outlets	473	Pa
Inlet Cone - Contraction	121	Pa
Fan evase outlet	108	Pa
Discharge velocity pressure	114	Pa
TOTAL pressure drop	**3816**	**Pa**

Duty Point Summary		
Air Flow	**125**	**m³/s**
Total Pressure	**3816**	**Pa**
Fan efficiency @ duty point	**80**	**%**
Shaft Power	**596**	**kW**
Motor Speed	**1000**	**r.p.m.**

Accesories reference

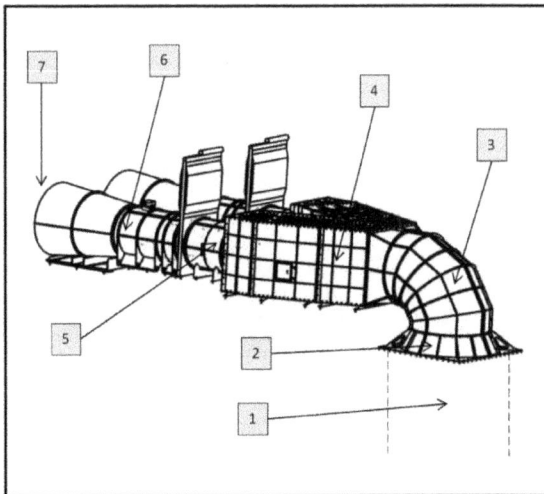

FIG 23 - Fan features VR3 ZITRON

On the other hand, pressure losses through the accessories on the fan collar are 816 Pa.

Fan curves of the vertical fan design VR3

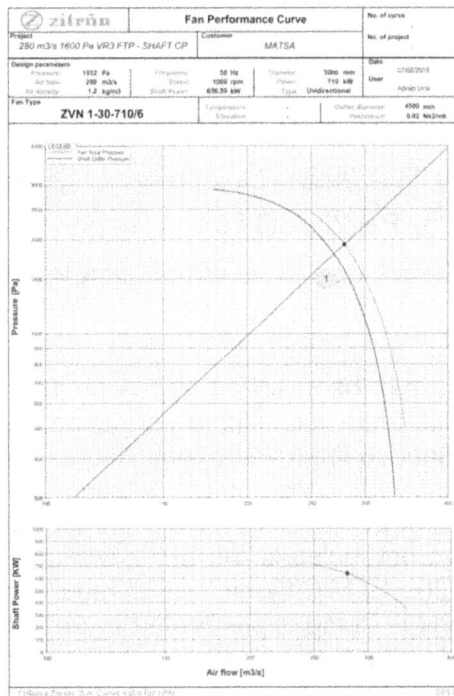

FIG 24 - Proposed single vertical fan at 280 m3/s

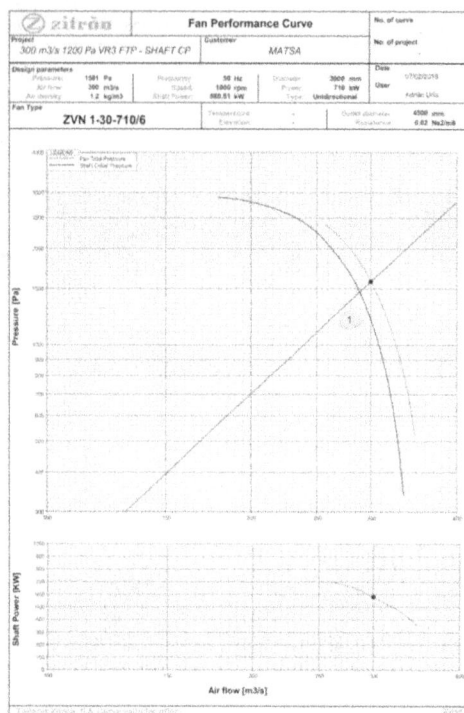

FIG 25 - Proposed single vertical fan at 300 m3/s

These curves shows two scenarios that coincide with the current work pressures and those projected above 2025. The duty points give us 2 scenarios.

- Scenario 1: Design total pressure up to 1581 Pa, 300 m3/s and fan power of 710 Kw.

- Scenario 2: duty point of 1932 Pa, 280 m3 / s and fan power of 710 Kw.

As we can see this 2 designs will allow us to deliver more flow with much less consumption.

CONCLUSIONS

The ventilation strategy in Matsa has been an ambitious project whose pilars have built on the technical excellence training of its team (continuing education), according to all the parameters (that which can be measured cannot be improved), consultant support and a step-by-step strategy. Thus Matsa, taking into account real consumption totaling 3.641 kWh in 2018 and 31.895 MWh/year, the total implantation of our strategy in the primary and auxiliary ventilation, we will save 11.747 MWh/year (we have not included the potential savings of changing the vent ducts) obtaining a real consumption of 20.149 MWh/year, a total reduction of 37% of energy savings just for "Aguas Teñidas". This strategy is being implemented in the other two mines, Magdalena and Sotiel, in different levels of development, this paper has focused on Aguas Teñidas because it is the most complicated mine, thus the implementation of this strategy in the other two mines will be much easier. It is important to note that the total investment presents a payback of 1.5 years.

REFERENCES

R. Dave Brokering1*, D.M. Loring1, C.J. Rutter, Practical Implementation of VOD at the Henderson Mine. 2017,11Climax Molybdenum Company, Henderson Operations, USA

Enrique Acuña*, Cheryl Allen, Totten Mine Ventilation Control System update: implementation and savings achieved with Level 1 "User Control" and future plans, 2017, Totten Mine, Vale Canada Limited, Canada

Risk analysis of ventilation system design regulations for mines operating in Kazakhstan

S Sabanov[1], Y Tussupbekov[2], A Karzhau[3] and B Aldamzharov[4]

1. Associate Professor, School of Mining and Geosciences, Nazarbayev University, Astana, 010000, Kazakhstan. Email: sergei.sabanov@nu.edu.kz
2. Research Assistant, School of Mining and Geosciences, Nazarbayev University, Astana, 010000, Kazakhstan. Email: yerbol.tussupbekov@nu.edu.kz
3. Research Assistant, School of Mining and Geosciences, Nazarbayev University, Astana, 010000, Kazakhstan. Email: abusaadi.karzhau@nu.edu.kz
4. Research Assistant, School of Mining and Geosciences, Nazarbayev University, Astana, 010000, Kazakhstan. Email: Bekbol.Aldamzharov@nu.edu.kz

ABSTRACT

Underground mine ventilation system design in Kazakhstan is based on standards and regulations approved in early nineties of last century, which were adopted from Soviet era normative. Those normative utilise calculation methodologies based on requirements for outmoded equipment. Main problem is that old outdated local regulations use overestimated airflow quantity to dilute diesel emissions and dust, and a complicated ventilation network. The aim of this study is to analyse financial risks associated with use of the outdated local regulations compare to more modern world regulations.

To solve outlined above issues, a case study has been conducted at a metal mine operating in Kazakhstan. The mine needs additional investments for optimisation of its ventilation systems, and therefore the financial risks associated with such modernisation should be estimated. Ventilation systems at the zinc-lead mine under the old outdated local regulations and under more modern world regulations have been designed. Risk analysis considers problems related with overstated expenses for the ventilation system using the outdated local regulations.

Utilising the modern regulation requirements for dilution of diesel toxic gases, the air quantity can be decreased on 30%. Stochastic modeling of Internal Rate of Return ('IRR') for both regulations ventilation systems design have been produced with help of Monte Carlo simulation. Financial model use electricity pricing as a major variability for the simulation. As result, the probability distribution for the outdated regulation ventilation system design demonstrates the IRR has the range of 10-18% within the corresponding defined value of 90%, and for the modern regulations – 16-24%, correspondingly.

Performed risk analysis recommends implementing the modern regulations for Kazakhstan's mines the ventilation system design that will benefit for saving energy costs. Results of this study will help to analyse ventilation efficiency in agreement with underground mine safety regulations and might assist Kazakhstan change the outdated regulations for the better.

Keywords: mining regulations, ventilation design, risk analysis

INTRODUCTION

Costs for mine ventilation typically increase when mines expand to deeper levels. Some underground mines in Kazakhstan still use continuously operating fans at their maximum capacity. This does not help to save energy and costs, which in many cases are very significant. However, mine safety is related to risks associated with deficit of fresh air for mine gases dilution and their removal from the mine workings (Sabanov, 2018). Underground mine ventilation system design in Kazakhstan is based on standards and regulations approved in early nineties of last century, which were adopted from Soviet era normative. Those normative are not up-to-date and utilise standards for outmoded equipment. Main problem is that old outdated local regulations use overestimated airflow quantity to dilute diesel emissions and dust, and a complicated ventilation network for radical ventilation. The aim of this study is to analyse financial risks associated with use of the outdated local regulations compare to more modern world regulations.

The studied mine uses sublevel stoping with backfill that produces ore utilising conventional drill and blast method. This type of mines are characterized by a complex network systems which require a large quantity of airflow, especially in downcast levels. The mine was restarted with a production of 2.25 million ton per annum (Mtpa) of ore. The mine's output accounts for most of overall zinc and copper concentrates. The mine operates with trackless mining equipment on a 3 x 8 hour shifts per day, seven days per week roster. There are mid-shift lunch breaks to allow blasting to occur in the ore headings to maintain production efficiency. The mine uses a boundary force-exhaust ventilation system (Figure 1). The amount of intake air required to ventilate mining operations is 352 m³/s. Fresh air comes through the shafts 'Malv', Shaft #3 and 'Vent' and exhausted via the shaft "Vozdvyh". Forced ventilation uses an axial fan 'VOD-40', arranged on the surface of 'Malv' shaft and 'VOD-30 2M' installed on the surface of "Ventilation" shaft.

There are overall 4 main fans installed on the surface:
1. 'VOD-40' force fan supplies 130 m³/s of air and located at the shaft 'Malv'. The shaft supplies air to Level 18;
2. 'VOD-30' force fan supplies 113 m³/s of air and located at the shaft 'Vent'. The shaft Supplies air to Level 13 and after that using ventilation raises air is delivered to below levels;
3. 'VOD-30 2M' force fan supplies 109 m³/s of air and located at Shaft #3. It delivers air from surface to Level 5, then through ventilation raises the air reaches Level 7 and Level 9;
4. 'VCD-47.5' exhaust fan is exhausting 224 m³/s of air and located at the shaft 'Vozdvyh'. There are collective ventilation cross-cuts at 9, 12, 13 and 14 horizons, which are currently directly connected to the exhaust shaft, the deep horizons are exhausted by ventilation raises.

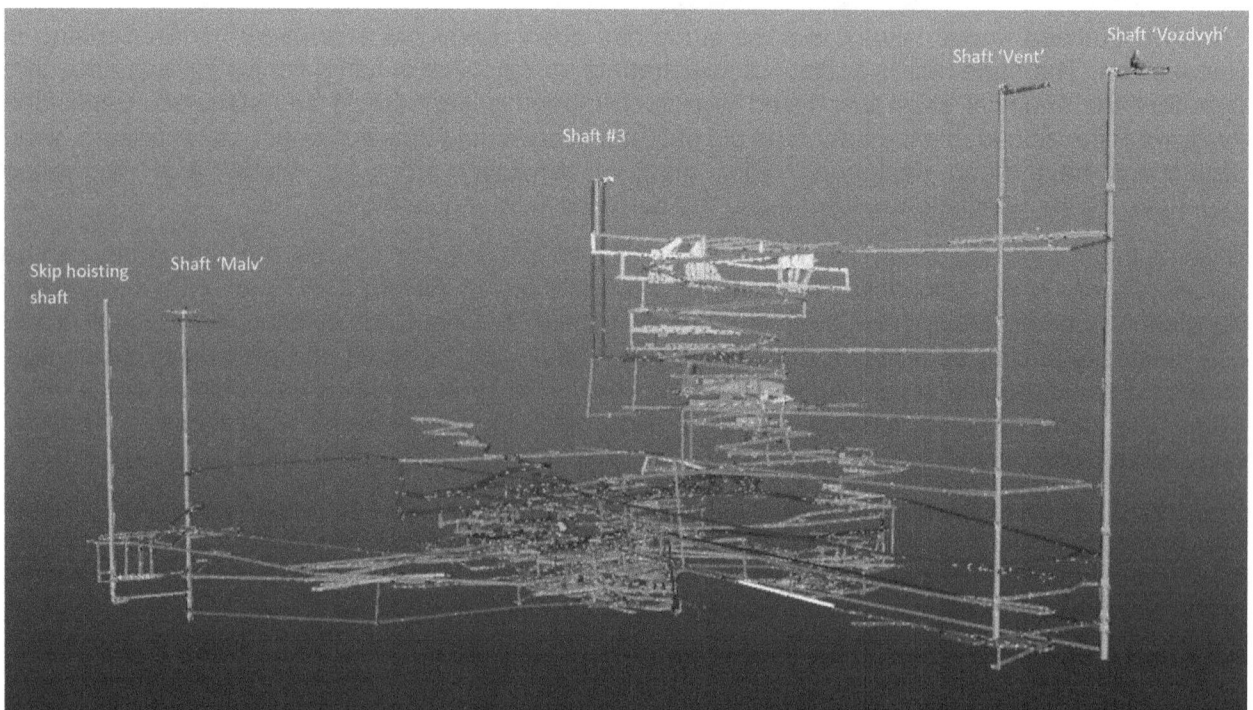

FIG 1 - Mine Ventilation Network Layout

Auxiliary ventilation required to force air into blind headings using axial flow fans and flexible ventilation ducting. Fans located in the main access deliver air from the fresh side of the primary ventilation circuit without recirculating the blast fumes.

The mine ventilation network modeling has been undertaken to be compliant with Kazakhstan's mining regulations requirements for underground mine ventilation (Kazakhstan mining regulations, 2004). Minimum airflow velocity at working faces should be 0.5 m/s and development headings - 0.25 m/s, correspondingly. Table 1 shows Kazakhstan mine ventilation requirements used to model the mine ventilation.

TABLE 1 – Kazakhstan Mine Ventilation Requirements

Maximum air velocities	Requirements
Cleaning and preparation drives (footwall drive)	4 m/s
Cross cuts, vent, main haulage ways and main decline	8 m/s
Other places	6 m/s
Crossings and main vent shafts	10 m/s
Shafts for lifting persons	8 m/s
Shafts for lifting cargo	12 m/s
Shafts for lifting persons in emergency	15 m/s
Vent shafts and vertical development without ladders	No restrictions

The airflow requirements for the mine have been calculated based on the underground equipment and the blast ventilation requirements. Accepted by Kazakhstan´s mining regulations, the method of determining ventilation requirements considers that the removal of diesel fumes is based on a diesel dilution rate of 5m³/min/hp of diesel engine horsepower, which is equal of 0.112 m³/s per kW. The quantity of air required for the purposes of respiration of personnel is about 6 m³/min of air for each person. Mine dust limit is 2 mg/m³. Threshold Limit Value ('TLV') for NOx – 0.0001%, CO – 0.0016%, NO₂ -0.0025%, SO₂ - 0.00035%, H₂S -0.00066% (Kazakhstan mining regulations, 2004).

Ventilation systems for the zinc-lead mine under the old outdated local regulations and more modern world regulations have been designed for this study.

In this study Risk Analysis considers problems related with overstated expenses for ventilation system at use of the outdated local regulations. Risk analysis approach considers ventilation system design, event risks estimations and financial estimations.

Ventilation system design is comprised of ventilation network optimisation processes taking into consideration financial risks related with mining development capital expenses and ventilation operational expenses.

Event risks estimates the number of potential risks occurred in case of use outdated regulations and the total impact from these events. Events can be analysed based on information received from the mine's historical data.

Financial estimation uses stochastic cash flow models comparing IRR for both regulations ventilation systems design produced with help of Monte Carlo simulations. Stochastic modelling provides ranges of IRR outcomes with confidence limits as well as mean values in the specified case. Decision-makers can view not only the mean outcome value but also the range of possible outcome range values. Monte Carlo simulation utilized the randomly select values from the probabilistic distribution. Palisade @Risk7.6 plug-in was used to construct and analyse stochastic models in Microsoft Excel spreadsheets.

GENERAL THEORY

Optimisation

El-Nagdy and Shoaib (2015) showed that in normal ventilation situation, the target is safe, economic and feasible. Safety and feasibility are usually reflected from the required airflow quantity, the lower and upper limit of airflow quantity and control variables and the controllability of branches. Thus, the main objectives are to minimise the power consumption, and the overall cost of ventilation. The overall air power of a ventilation network comprises the power consumed to overcome branch resistances and regulator resistances. The objective function can then be defined as follows (Nyaaba, Frimpong, El-Nagdy 2014):

$$z = \sum_{j=1}^{B} (r_j q_j^2 |q_j| + s_j |q_j|) \qquad (1)$$

Where z, rj, qj and sj are the total power consumed by the network in Watts, the j is branch resistance factor in Ns^2/m^8; air quantity in m^3/s, regulator pressure loss in Pa.

'During emergencies, the economic factor is minor and the difficulty of control facility installation is relatively important. So that, the optimisation during mine crisis period mainly aims to make the number of control facilities as less as possible and the control quantity as lower as possible, which is convenient for temporary control measures. Because the control facility number is the calculated number of branches, whose control variables are not zero. Then it is an object optimisation problem about integer programming, which is difficult to get the result. In order to simplify the solving process, it is necessary to combine control variables, control facility number into one objective function, and make the optimum scheme, which is a kind of compromise between them. However, it is not easy to find an idea objective function' (Wu and Li, 1993). Depending upon these hypotheses the main objective function in ventilation model will be a compromise to include economic and safety factors (El-Nagdy and Shoaib, 2015).

Air quantity required to dilute diesel emissions

Halim (2017) found out that in underground mines where diesel-powered equipment is used, it is common to estimate ventilation requirement according to engine power. The airflow quantity for an underground mine is usually based on the engine power of diesel vehicles used in the mine, multiplied by unit airflow requirement, such as 0.05 to 0.06 m^3/s per kW used in Australia or 0.047 to 0.092 m^3/s per kW used in Canada. These unit airflow requirements are stated in local mining Occupational Health & Safety (OH&S) regulations.

'The review of the history of ventilation requirements in the past regulations found that the current values are incorrect due to the differences of TLV-TWA between countries and longer working hour in many of today's mines compared to that when the values were derived. OH&S regulators in these countries should review these values and derive new ones based on an engine testing program. Adjusted safe concentration limits should be used in this testing program when working hour is not 8 hours per day. DPM and heat must be included in this testing program' (Halim, 2017).

The ventilation requirement based on gases is what the mine must provide for the operation of the diesel engine in underground (Haney, 2012 and McGinn, 2007). Some examples of MSHA's certified ventilation requirements for modern diesel engines from different manufacturers used in underground mine demonstrate 0.03 to 0.05 m^3/s per kW (Halim, 2017).

Underground mines in Kazakhstan adopted the value of 0.112 m^3/s per kW, which were derived from temporary methodological guidelines in early nineties last century. Those Soviet Union normative that use calculations on old-fashioned diesel engines from sixties last century. This value is higher comparing to the values used around the world.

Risk analysis method

Monte Carlo simulation presents risk analysis by building models of possible outcomes. When creating such models, any factor that is characterized by uncertainty is replaced by a range of values - a probability distribution. Then multiple calculations of the results are performed, each time using a different set of random values of the probability functions. Monte Carlo simulation allows to obtain the distribution of the values of possible consequences. A Monte Carlo simulation is a broad term for computational algorithms that generates random numbers (realisations) given a specific density function (Pyrcz and Deutsch, 2014). The Monte Carlo simulation of random numbers from an arbitrary probability or cumulative distribution function can be obtained by Deutsch and Journel.

Wang, Xiong and Xie (2018) described Monte Carlo method as a statistical test method, which be used to solve the mathematical physics problem, that is difficult to determine the formula, through statistical sampling theory. For a statistic with uncertainty probability density function, Monte Carlo method is easier to solve the synthetic uncertainty which approaches the real solution (Ni, 2015). Monte Carlo method evaluates measurement uncertainty by a numerical method to achieve distributed propagation (Wang, Xiong and Xie, 2018). 'Figure 2 depicts the independent input distributions transmission, and the probability density function of output quantity Y is obtained through three probability density functions of input quantities X1, X2, X3 with the model propagation. The three probability density functions of input quantities X1, X2, X3 in Figure 2 are normal distributions, triangular distributions and normal distributions respectively, which are independent, and the probability density function of output quantity Y is an asymmetry distribution. If the probability density function of output quantity Y is asymmetry like in Figure 2, then the measurement uncertainty evaluation. Monte Carlo method carries out the discrete sampling of the input quantity probability density function, and then calculates the discrete sampling of output quantity Y by the measurement model. Next, the best estimate, the standard uncertainty, and the inclusion interval of the output quantity Y are calculated by the discrete distribution value of the output quantity, which are improved as the number of sample M of the input X increases' (Wang, Xiong and Xie, 2018).

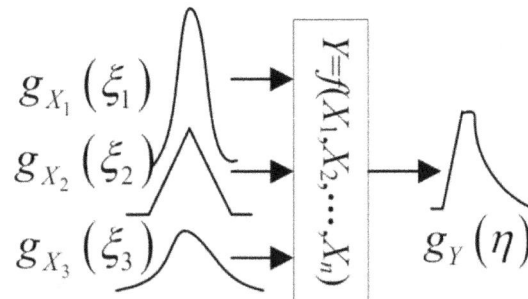

FIG 2 - Independent Input Distributions Transmission (Wang, Xiong and Xie, 2018)

The Bernoulli distribution is a discrete distribution having two possible outcomes labelled by n=0 and n=1 in which n=1 ("success") and n=0 ("failure"). It therefore has probability density function (Evans et al. 2000):

$$P(n) = \begin{cases} 1-p \ for \ n-0 \\ p \ \ for \ n = 1 \end{cases} \tag{2}$$

Which can be written:

$$P(n) = p^n \ (1-p)^{1-n} \tag{3}$$

The PERT distribution, P (a, m, b) is a special case of the four parameter beta distribution whereby: 1) the parameters **a** and **b** are the maximum and minimum bounds of the distribution; 2) the mode, m is explicitly defined; and 3) the mean and variance obey strict definitions (Clark, 1962):

Mean, $\qquad\qquad \mu = E[X] = (a+4m+b)/6 \tag{4}$

Variance, $\qquad\qquad Var \ (X) = (b-a)^2 \ / \ 36 \tag{5}$

To hold true for the PERT distribution, the standard beta parameters α and β, are derived from P (a, m, b) by:

$A= ((\mu-a) \ (2m-a-b)) \ / \ ((m-\mu) \ (b-a)) \tag{6}$

$B= (a \ (b- \mu)) \ / \ (\mu-a) \tag{7}$

For the symmetric case, the standard beta parameters α and β must satisfy this condition (Raymond 2013):

If m= (b+a)/2, than α = 3, β = 3 $\tag{8}$

The proposed methodology estimates propagation of uncertainty sources through an operational risk framework that considers impacts on the existing mine ventilation system and optimisation benefits based on calculations of Net Present Value (NPV) and IRR. The propagation is based on Monte Carlo simulations with variations in key aggregated processes and variables rather than varying all the variables individually (Zhou, Q. and Arnbjerg-Nielsen, K. 2018). The methodology can contribute to decision making on acceptance of global practices for mine ventilation regulations considering potential economical profit associated with utilization of modern regulations in Kazakhstan.

RESULTS AND DISCUSSIONS

Risk Analysis

Main problem with adopted in Kazakhstan outdated regulations for mine ventilation is that the current ventilation system utilise approximately double the airflow quantity required to dilute diesel toxic gases. Dust intensity calculations based on old-fashioned mining equipment are overestimated. Fire safety for mine ventilation system design requires splitting a mine by sections. Such design typically involve additional shafts, fans, infrastructure to supply fresh air to each section of a mine in case of sealing. Thus underground mine ventilation system in Kazakhstan is very complicated and result in overestimated electric power consumption.

Risk analysis takes into consideration problems related with overstated ventilation system expenses. Proposed risk analysis approach considers ventilation system design, event risks estimations and financial estimations.

Ventilation system design

The principal objectives of the ventilation system design are to remove the diesel toxic gases from mechanised mobile equipment and remove blasting fumes from the workings to provide a reasonable re-entry period. Based on the latest mine ventilation system surveys, a computerised model was developed for simulation processes aimed to provide with sufficient amount of fresh air to working faces. Calculated airflow requirements take into account the mining fleet. Calculations used the entire fleet with their modelled availability (85%) and utilization (85%). Ventilation simulation were produced by using 'Ventsim 4.8' software. To support mining equipment and leakages the original circuit airflow calculated by use of the outdated regulations made 352 m^3/s. Network efficiency was 64% and fans power consumption equalled 486 kW.Utilising the modern regulations requirements for dilution of diesel toxic gases 0.06 m^3/s per kW, it is possible to save about 195 kW of input power electrical for the fans. With these modern regulations, the mine will require only 210 m^3/s rather than currently supplied 352 m^3/s, which calculated based on the outdated regulations required to have 0.112 m^3/s per kW.

Event risks estimation

There are seven problems have been studied at this underground mine. High consumption of electricity is mostly associated with the powerful fans that are aimed to dilute diesel toxic emissions. Leakages are proportionally related to the overestimated airflow quantity. Complicated ventilation network related with the outdated regulations design that requires additional maintenance for infrastructure. Mine ventilation regulators and doors sophisticated and high maintenance. Mine production scheduling is poorly planned and needs to be adjusted according to a simplified ventilation network. Problems with fans assume non-efficient use of operational fan settings. Management of the mine ventilation system cannot efficiently operate with outdated standards and should be improved.

Event and operational risk estimation comprises a fit comparison for the number of risks occurring and financial impact from these events with the use of the current ventilation system. Seven issues or events with various probabilities of occurrence per year were considered. The events are listed below:

- High consumption of electricity;
- Leakages;
- Complicated ventilation network;
- Regulators and doors;

- Production scheduling;

- Main fans; and

- Mine ventilation management.

These events have historically happened in the studied mine, and therefore probabilities of their occurrence were estimated on the received statistical data. Probability of event occurrence used a Bernoulli distribution. The historical financial impact of these events in the studied mine was analysed by minimum, most likely and maximum expense values, using a Pert distribution.

Figure 3 demonstrates the probability of occurrence for these events within one year. As a result, from three to seven events can occur with a probability of 90% (Figure 3).

Thus according to the input data three events per year can occur with a probability of 11%, four events - with a probability of 27%, five events - with a probability of 34%, six events - with a probability of 21%, and seven events – with a probability of 5% (Figure 3).

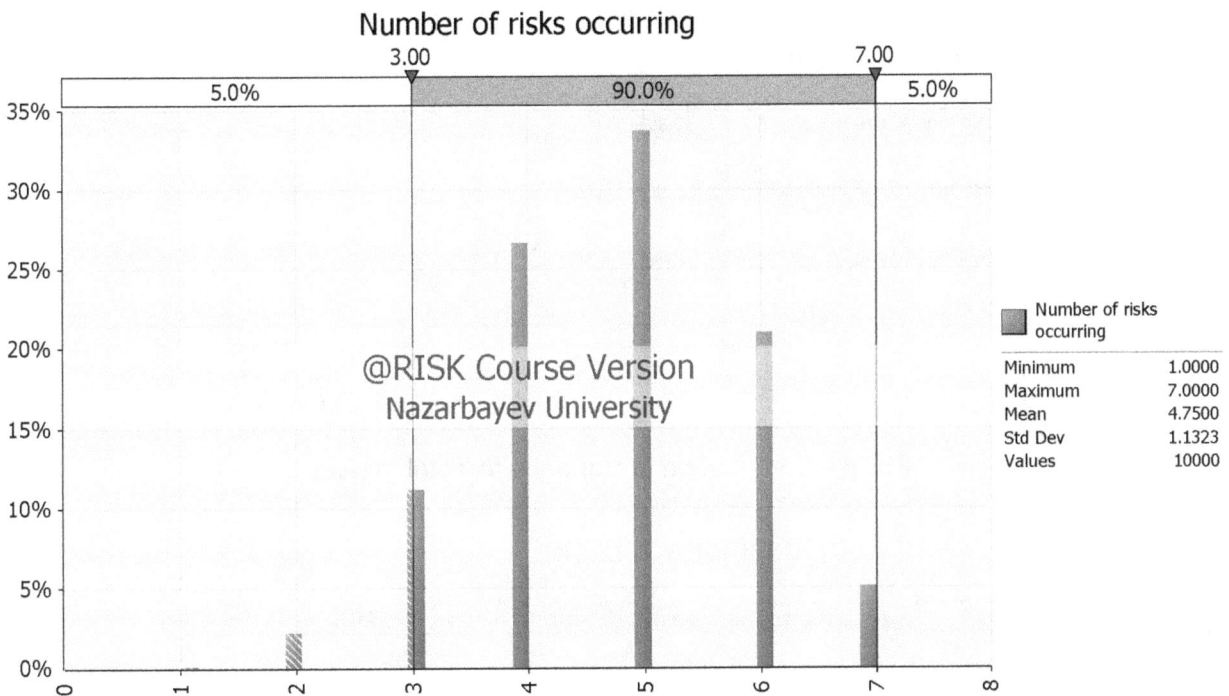

FIG 3 - Fit Comparison for number of risk occurring

Figure 4 present events' Contribution to Variance to Total Impact. The highest Contribution to Variance 41% was associated with 'High consumption of electricity' and the smallest about 1% - with 'Mine ventilation management' (Figure 4).

Total Impact in terms of financial losses for the occurring, was estimated with help of Monte Carlo simulation, and lies in the range of 78K–198K US$ in the corresponding defined value of 90% (Figure 5).

Financial Estimation

Stochastic modeling of IRR for both regulations ventilation systems design was produced with the help of Monte Carlo simulation (Figures 6 and 7). Financial model use electricity pricing as a major variability for the simulation.

Y-axis of the histograms is relative frequency presented in percentage and demonstrates number of times that an event occurs during Monte Carlo simulation experimental trials, divided by the total number of trials conducted. After 10 000 simulated iterations Monte Carlo simulation produced representation of possible outcomes for IRR, considering the major variability of electricity pricing fluctuation.

As result probability distribution for the outdated regulation ventilation system design demonstrates IRR in a range of 10-18% in the corresponding defined value of 90% (Figure 6), and for the modern regulations – 16-24%, correspondingly (Figure 7). Moreover, there is only 5% chance that IRR will be less than 10% for the outdated regulations design and 15% for modern regulations design, correspondingly. This can be observed from the histogram in Figures 6 and 7 that the distribution's values lie below the corresponding defined value of 90%.

Ventilation system designed on bases of the modern regulations demonstrates mean value of IRR on 6% higher that the ventilation system designed on bases of the outdated regulations.

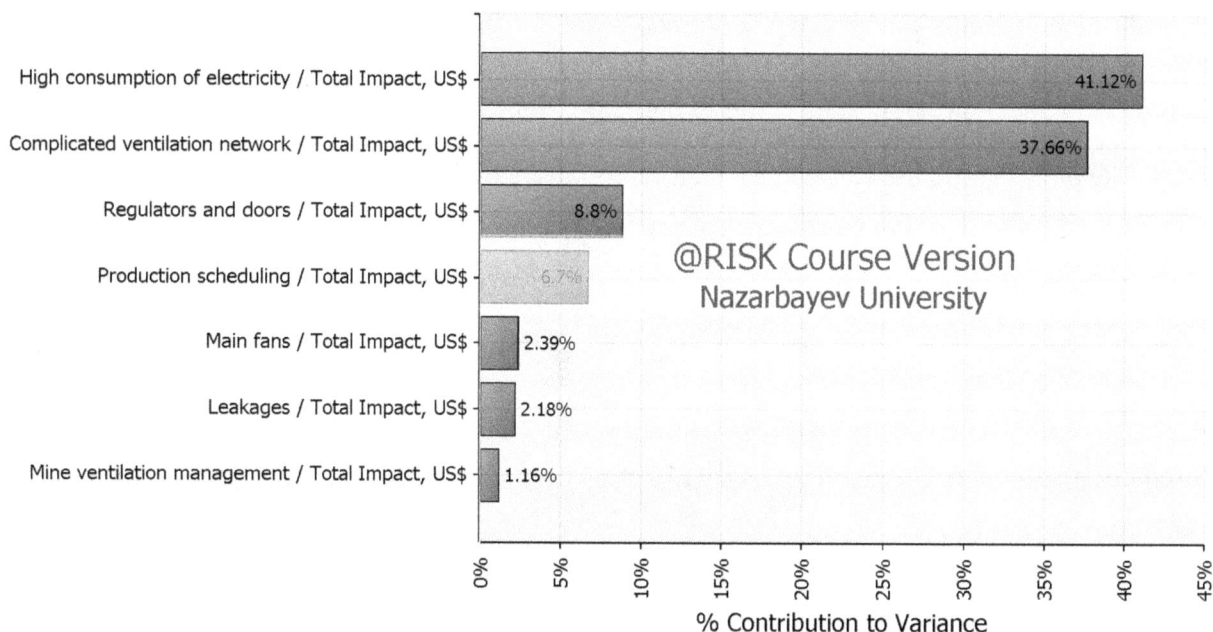

FIG 4 - Contribution of Variance to Total Impact

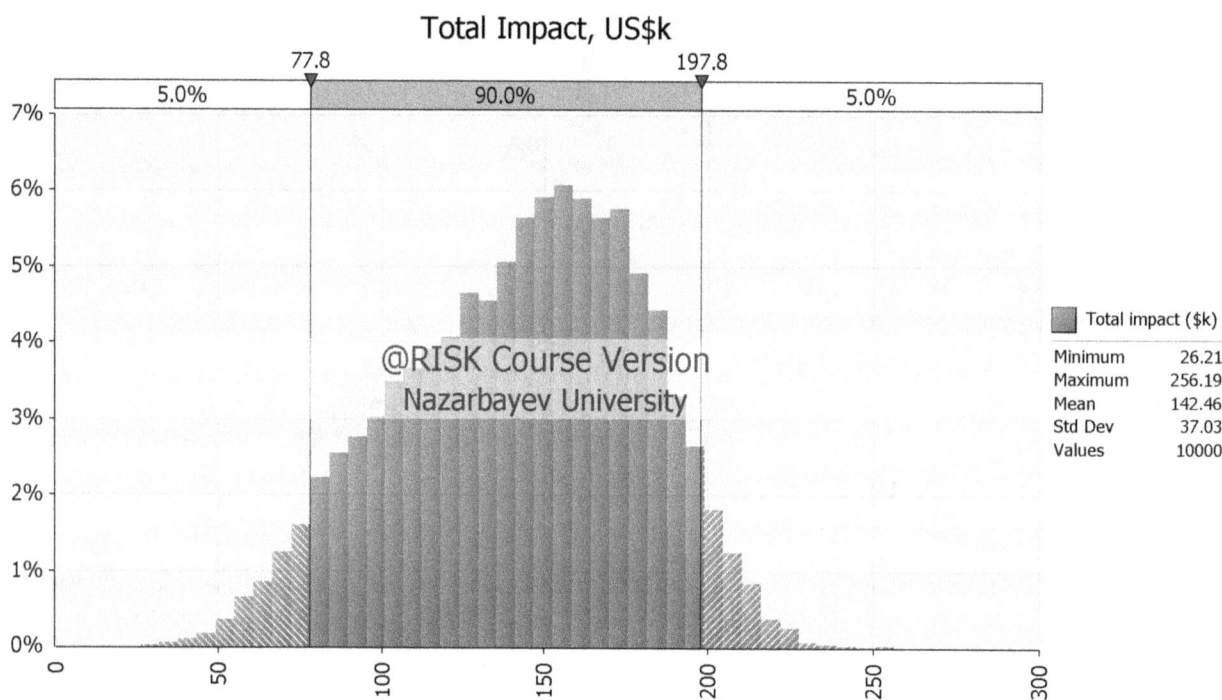

FIG 5 – Total Impact

IRR at Outdated Regulations

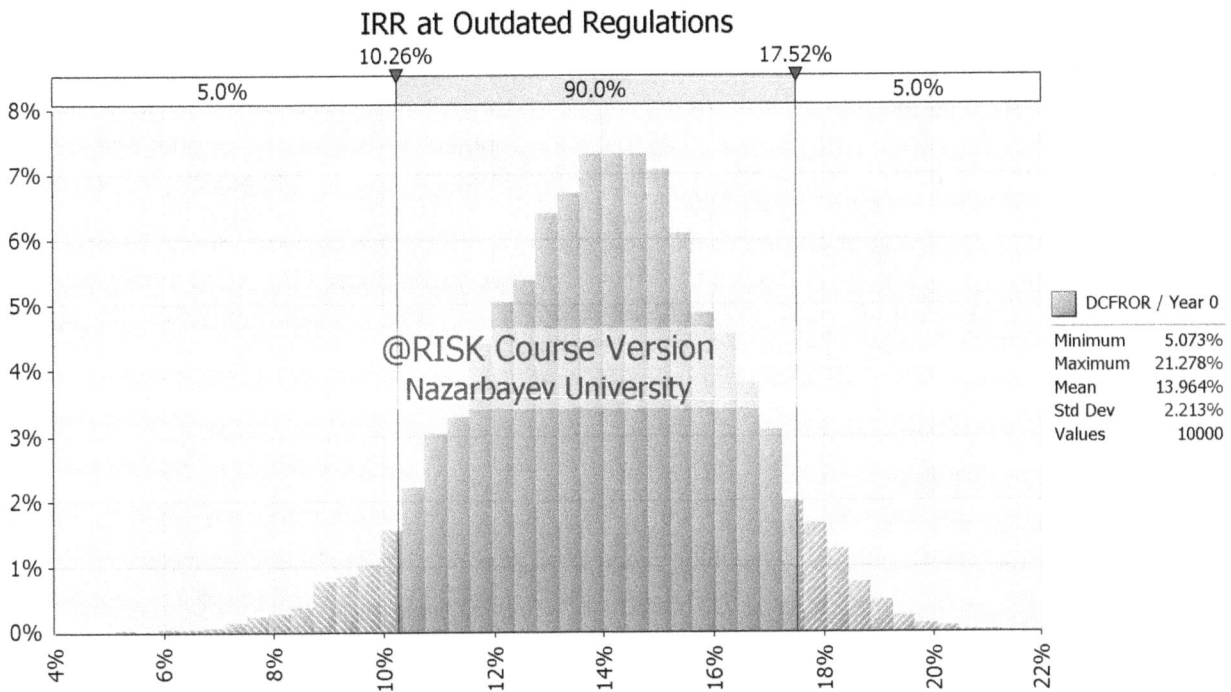

FIG 6 - IRR distribution for the outdated regulations ventilation design.

IRR at Modern Regulations

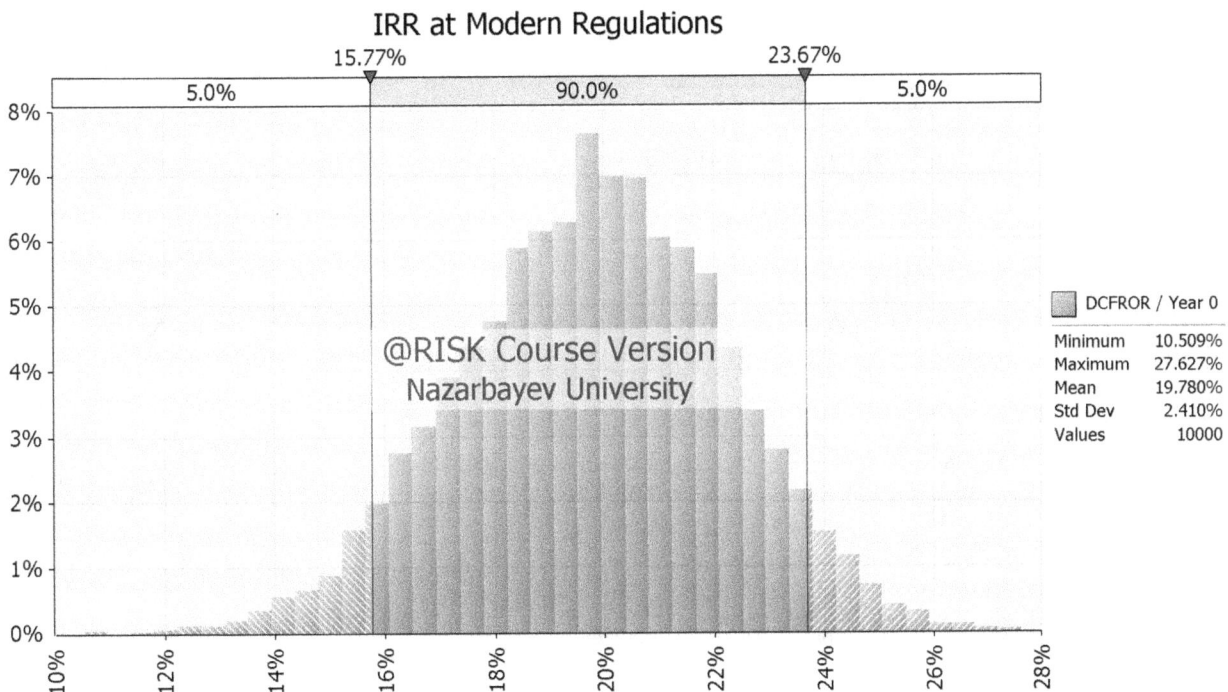

FIG 7 - IRR distribution for the modern regulations ventilation design

CONCLUSIONS

Risk analysis of ventilation design approaches for the operating zinc-lead mine located on Kazakhstan has been conducted. Ventilation systems for the zinc-lead mine under the old outdated local regulations and more modern world regulations have been designed. Risk analysis considers problems related with overstated expenses for ventilation system using the outdated local regulations. Fit Comparison for number of risks occurring at use of the outdated regulations for the ventilation system design has been produced with demonstration of Total Impact in terms of financial losses for the happened events. In total seven events that might occur within one year of the mine operation have been analysed. For these events Total Impact in terms of financial losses estimated in the range of 78K–198K US$ at 90% of confidence level. Utilising the modern regulations requirement for calculation of air quantity to dilute diesel toxic gases 0.06 m^3/s per kW the received savings made 195 kW of input power electrical for the fans. Under this, the mine will use 210 m^3/s instead of currently supplied 352 m^3/s that was based on the outdated regulations required to have 0.112 m^3/s per kW. Stochastic modeling of IRR for both regulations ventilation systems design have been produced with help of Monte Carlo simulations. The financial model use electricity pricing as a major variability for the simulation. As result probability distribution for the outdated regulation ventilation system design demonstrates IRR in a range of 10-18% in the corresponding defined value of 90%, and for the modern regulations – 16-24%, correspondingly. Performed risk analysis recommends implementing the modern regulations for Kazakhstan's mines the ventilation system design that will benefit for saving energy costs. Results of this study will help to analyse ventilation efficiency in agreement with underground mine safety regulations and will assist Kazakhstan change the outdated regulations for the better.

ACKNOWLEDGMENTS

This study supported by Nazarbayev University Grant Program for Research Grant (#090118FD5337) 'Risk Analysis Methodology for Automated Mine Ventilation Systems'.

REFERENCES

El-Nagdy, K., A. M. Shoaib, A. (2015). Alternate solutions for mine ventilation network to keep a preassigned fixed quantity in a working place. Int J Coal Sci Technol (2015) 2(4):269–278.

Halim, A. (2017). Ventilation requirements for diesel equipment in underground mines – Are we using the correct values? North American Mine Ventilation Symposium NAMVS 2017.

Haney, R.A., 2012. Ventilation requirements for modern diesel engines, in Proceedings of 14th US/North American Mine Ventilation Symposium, Salt Lake City, Utah, USA, (eds: F. Calizaya and M.Nelson), pp. 249-256.

Kazakhstan mining regulations requirements for underground mine ventilation (2004). Retrieved from http://egov.kz/cms/ru/law/list/P090002207_

McGinn, S., 2007. Controlling diesel emissions in underground mining within an evolving regulatory structure in Canada and the United States of America, paper presented to Queensland Mining Industry Health and Safety Conference 2007, Townsville, Queensland, Australia, 7 August.

Ni, Y., (2015). Practical evaluation of uncertainty in measurement (5th edition) (Beijing: China zhijian publishing house, Standards press of China) pp 35-54.

Nyaaba, W., S. Frimpong, S., El-Nagdy, K. A. (2014). Optimisation of mine ventilation networks using the Lagrangia algorithm for equality constraints. In: 12th international conference on mining, petroleum and metallurgical engineering MPM12, Suez University, 20–22 Oct 2014.

Pyrcz, M. J., Deutsch, C. V. (2014). Geostatistical reservoir modeling. USA: OUP.

Sabanov S. (2019) Financial Risk Analysis of Optimized Ventilation System in the Gold Mine. In: Widzyk-Capehart E., Hekmat A., Singhal R. (eds) Proceedings of the 27th International Symposium on Mine Planning and Equipment Selection - MPES 2018. Springer, Cham.

Wang, X., Xiong, J., and Xie, J. (2018). Evaluation of Measurement Uncertainty Based on Monte Carlo Method. MATEC Web of Conferences 206, 04004 (2018), ICCEMS 2018.

Wu, ZL., Li, HS. (1993). Simulation of mine ventilation under the influence of mine fires. In: Proceedings of the US mine ventilation symposium, University of Utah, pp. 359–363, June 21–23.

Evans, M.; Hastings, N.; and Peacock, B. (2000) 'Bernoulli Distribution.' Ch. 4 in Statistical Distributions, 3rd ed. New York: Wiley, pp. 31-33, 2000

Clark CE (1962). The PERT model for the distribution of an activity. Operations Research 10, pp. 405-406

Zhou, Q., Arnbjerg-Nielsen, K. (2018). Uncertainty Assessment of Climate Change Adaptation Options Using an Economic Pluvial Flood Risk Framework. Water 2018, 10, 1877.

Coal Seam Gas, including Gas Drainage/Storage and Utilisation

Study on microstructure differences of coal samples before and after loading

J Liu[1*], J Gao[2], G Jia[3] and M Yang[4]

1. Associate Professor, School of Safety Science and Engineering, Henan Polytechnic University, Jiaozuo, China, 454000. Email: liujiajia@hpu.edu.cn
2. Professor, School of Safety Science and Engineering, Henan Polytechnic University, Jiaozuo, China, 454000. Email: gao@hpu.edu.cn
3. Master, School of Safety Science and Engineering, Henan Polytechnic University, Jiaozuo Henan, China, 454000. Email: 849768108@qq.com
4. Associate Professor, The collaborative Innovation center of coal safety production of Henan Province, Jiaozuo, China, 454000. Email: yming@163.com

ABSTRACT

To study the difference of pore structure morphology and permeability before and after loading of different coal samples under certain loading conditions. The pore size of coal samples before loading was tested by mercury intrusion method. According to the results of mercury intrusion test adsorption pores (<100 nm) accounted for more than 60% of the total pores. Scanning electron microscopy (SEM), transmission electron microscopy (TEM) and nuclear magnetic resonance (NMR) techniques were used to compare and analyze the fracture morphology and permeability of different coal samples before and after loading. The experimental results revealed that the compact structure of the coal sample was destroyed after loading, a large number of cracks were formed, and the original pores were further connected, and a large number of shear cracks were obviously observed. NMR measurements revealed that the total area of the T_2 spectrum for parallel bedding of saturated coal sample before loading was 12692 (dimensionless) and that it increased to 46735, after loading. The vertical bedding before loading was 12295 and that it increased to 37623, after loading. The permeability of the coal body was significantly increased, and the same loading mode, the permeability of the parallel bedding after loading was larger than that of the vertical bedding.

KEYWORDS: Loaded coal sample, Microstructures, NMR, SEM, Differences

INTRODUCTION

The pore space of coal (pore size distribution and pore structure of coal) has significant influences on the storage and transportation of gas in coal. On the one hand, it affects the flow of gas in coal, on the other hand, it provides an excellent place for the adsorption of gas in coal (Li Z et al; Clarkson and Bustin et al; Nie B et al; Li J et al; Sun M et al; Xu C et al; Yao Y et al.). Based on its favorable conditions for the storage and transportation of coalbed methane, pores in coals are divided into adsorption-pores including micropores (<10nm) and mesopores (10–100nm), and seepage-pores (>100nm). (IUPAC; Liu D et al; Sang S et al; Yao Y et al). Micropores constitute the adsorption space of gas in coal, and the mesopores and seepage holes provide a good channel for gas migration in coal.

Many methods for studying pores and fissures, such as: mercury intrusion, scanning electron microscopy(SEM), transmission electron microscopy(TEM), small angle neutron scattering (SANS), nuclear magnetic resonance(NMR), etc. (Lin B et al; Li Z et al; Kang Z et al; Zhang J et al; Song X et al; Yao Y et al; Song L et al; Zhou S et al; Ouyang et al), among which the mercury injection is the most commonly used method (Song B et al; Meng Q et al; Xu H et al; Cai Y et al.)

Many scholars have studied the pore structure of unloaded raw coal samples and observed that the pore space distribution of different coal samples was different (Sun L et al). In recent years, with the application of some techniques to change the permeability of coal reservoirs (Zhang W et al; Xue W al; Hao F et al). Researchers have studied the pores of loaded coal samples, which could be divided into single-axis loading, pseudo-triaxial and true triaxial, and the stress path was divided into three-way isobaric and unequal pressure and multiple stress paths (Liu Y et al; Liu Q et al; Huang F et al;

Gao B et al). The coal seam permeability was improved by different loading methods (Guo J et al; Pan R et al), thereby improving gas drainage efficiency, reducing gas risk hazards and maximizing the utilization of coalbed methane. However, different layered coal samples do have anisotropic characteristics (Lu F et al; Lu F et al).

In general, for unloaded coal samples, parallel layered fissures had good communication conditions and coal sample penetration. It was better than vertical bedding coal samples (Lin B et al; Jiang T et al; Pan R et al; Liu Y et al).

However, the change of pore morphology and permeability of different coal samples after loading, and the quantitative characterization of coal sample morphology and permeability after loading have not been studied. Based on this, the author considered the morphological and permeability changes of different coals before and after loading under a certain loading path, and quantified these changes by NMR techniques. On the one hand, to extract gas at the site in the future, the drilling arrangement should indeed provide a theoretical reference in consideration of coal seam bedding, and on the other hand, verify the effect of the permeability change of the coal sample after loading.

The engineering background of this paper is based on the study of cracks in the mining face, which provides a good channel for coalbed methane migration and plays an important role in ensuring normal, safe coal mine operation.

PORE DISTRIBUTION AND BASIC PHYSICAL PARAMETERS OF COAL

Pore distribution of coal

The pore volume at each stage is an important parameter that reflects the pore structure of coal. Currently, mercury intrusion methods are commonly used to determine the ratio of pores in coal. The basic principle of mercury intrusion is as follows. According to the Laplace formula, mercury with a contact angle greater than 90° cannot enter the micropores of coal under any pressure conditions, and the applied pressure can overcome the resistance caused by the surface tension of mercury. In this way, a function of the pressure required to fill a certain pore and the pore size can be established. The pressure required to force mercury into a given pore is consistent with the Laplace formula (Du J et al):

$$P_C = -\frac{2\sigma \cos\theta}{R}$$

Where R is the pore radius; σ is the surface tension, often taken as $480*10\text{-}5\text{N/cm}^3$; θ is the contact angle of the mercury with the sample to be tested, adopting 140°; P_C is the intrusion pressure of mercury: 105MPa.

We simplify the preceding formula to obtain the following formula:

$$r = \frac{7500}{P}$$

Mercury intrusion test procedure:

(2)A coal sample was taken from the 29031 working face of a mine in Shanxi Yuwu Mine. First, the collected fresh large coal samples were prepared into 50*100mm columns along parallel bedding and vertical bedding. The remaining small coal sample was crushed with a small iron hammer to a particle size of 3~6mm, to remove mineral impurities in the sample and avoid the influence of artificial cracks, thus ensuring experimental data accuracy. Prior to the test, the coal sample was dried in an incubator at 70~80 °C for at least 12h. Then, the sample was placed in a dilatometer, vacuumed (<6.67Pa) and then tested (Equipment model: Auto Pore 9505 IV pressure).The Automatic Mercury Porosimeter is shown in Fig.1. The test results are shown in Fig.2 and Tab.1.

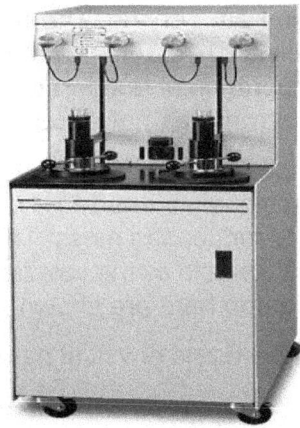

FIG 1 –Auto Pore 9505 IV

TABLE 1 – Pore volume parameter test results

Pore classification	Microporous	Small pore	Mesopore	Macropore	Total mercury intake mL/g	Mercury withdrawal mL/g	Porosity %
	<10nm	10~100nm	100~1000nm	>1000nm	0.0809	0.063	9.1819
	71.9%	27.8%	0.276%	0.046%			

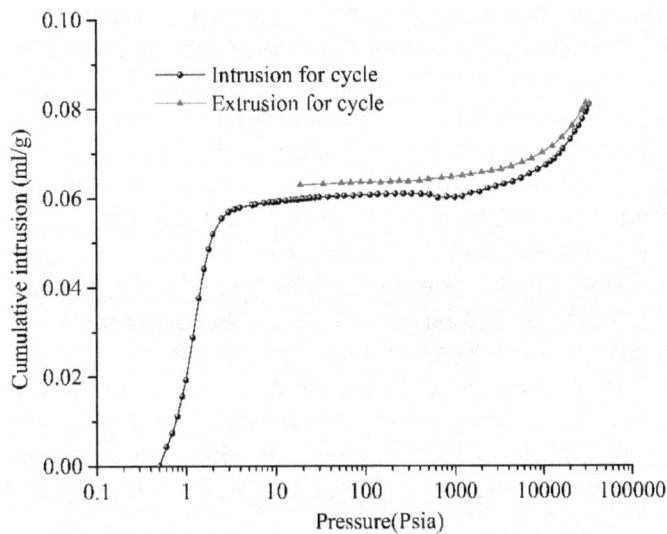

FIG 2 –Experimental curves of cumulative intrusion (ml/g) with pressure (Psia)

As shown in Fig.2, the mercury intrusion curve and the mercury removal curve do not coincide, and the former has a concave hysteresis loop, indicating that the coal sample contains a number of open and semi-closed pores. Using the mercury intrusion method, the pore distribution characteristics of more than 5.5nm in each coal sample could be obtained, whereby the Hodot decimal classification is adopted. As shown in Table.1, in the Yuwu coal sample, micropores and transition pores account for a large proportion, and the proportion of mesopores is the least. Transition pores and micropores predominate, accounting for more than 60% (Ren J et al).

COAL SAMPLE BASIC PARAMETERS AND INDUSTRIAL ANALYSIS

The R_0 of the coal is 2.19 (bituminous coal) (Liu S et al). Industrial analysis of coal samples refers to GB/T212-2008. First, the moisture level was determined using the air-drying method. A total of 1 ± 0.1g coal powder with a particle size of less than 0.02mm was placed in a blast drying oven and then dried under airflow for 1h to a constant temperature. The moisture of the coal sample was determined by the difference in mass before and after drying. The ash content was determined using the slow ashing method. Here, approximately 1±0.1g coal powder with a particle size less than 0.2mm was placed in a muffle furnace. The experimental temperature was controlled so as to burn the sample for 1h at 815±10° until the quality error does not exceed 0.001g after two consecutive burnings.

Volatile matter was determined as follows. A total of 1±0.1g coal powder with a particle size of 0.2mm or less was placed in a crucible previously burned to a constant weight at 900 °C and then heated in the air for 7 minutes at 900±10° to reduce the mass. The mass of the coal sample was subtracted from the moisture content of the coal sample as the volatile matter of the sample. The analysis results for the specific coal sample investigated here are shown in Table.2.

TABLE 2 – Basic physical parameters of coal

Moisture/%	Ash/%	Volatile/%	R0
0.83	10.02	18.19	2.19

As shown in Tab.2, the moisture content of the coal sample is less than 50%, indicating that the coal sample has a higher degree of coalification. The lower ash yield indicates better combustion characteristics. The content of coal-derived base volatiles is less than 30%, which further demonstrates the high degree of coalification of the coal samples from Yuwu, Shanxi.

LOW FIELD NMR EXPERIMENTAL TEST

In recent years, NMR technology has been favored by scholars at home and abroad for its rapid, nondestructive and visual technical advantages, and it has been established as a new method for studying the pore characteristics of coal bodies. The technology is based on the principle that hydrogen nuclei will be aligned under the action of an external magnetic field. The method can be used to measure the relaxation characteristics of hydrogen-containing nuclear fluid (water) in the pores of coal rock and to obtain the nuclear magnetic resonance intensity and the transverse relaxation time T_2. The relationship curve is obtained to determine the pore fracture distribution characteristics of a coal sample. According to the principle of NMR technology, the T_2 spectrum reflects the amount and distribution of pores in the coal sample. The area of the closed pattern enclosed by the T_2 spectrum and the X-axis, referred to as the T_2 spectrum area, represents the volume of pores in the coal sample. The pattern change of the spectrum reflects the distribution of the pore size in the coal sample. The larger that T_2 is, the larger the pore size. The smaller that the pore size is and the closer that the T_2 spectrum is to the left, the larger the proportion of tiny pores in the coal sample. Using low field NMR methods a fine quantitative characterization of microstructures in coal can be achieved (Ren H et al; Yao Y et al; Xie S et al Yao Y et al; Lu F et al.); after the coal sample is saturated and centrifuged, the core is tested and analyzed by NMR technology to obtain the nuclear magnetic porosity, effective porosity and direct permeability of the coal sample.

$$\frac{1}{T_2} = \frac{S}{V} \rho = F_s \frac{\rho}{r} = CT_2$$

Where ρ is the lateral relaxation rate of coal, which is a parameter for characterizing rock properties; $\frac{S}{r}$ is the specific surface area of coal; F_s is the pore shape factor (spherical pores, F_s =3; tubular pores, F_s =2; plate shape); r is the pore radius; C is the conversion factor.

As indicated by the preceding formula, there are different types of pore relaxation time, and the smaller that T_2 is, the smaller the pore radius of the pores.

NMR EXPERIMENT AND RESULTS ANALYSIS

The prepared natural cylindrical coal samples were weighed and their data recorded separately. Using a nuclear magnetic resonance instrument (MesoMR23-060H-I; Fig. 3(a)), the measurement parameters were set as follows: SW was 100, NS was 2, and TW was 1500, (1) First, the T_2 spectrum of the natural state coal sample was measured. (2) Then, the coal sample was placed in a constant temperature drying oven (Fig. 3(b)), and the temperature was set to 60 °C for drying for 10 h. The natural cooling weight was recorded, and the sample was dried again under this condition for 2 hours until the weight error of the coal sample was within 0.5%. (3) The thoroughly dried coal sample was naturally cooled to room temperature, and the T_2 spectrum of the dried coal sample was measured again. (4) The complete coal sample was removed and placed in a vacuum water-saturated device (Fig. 3(c)) and was maintained for 10 hours, the surface of the dry portion was removed, the sample was weighed, and the result was recorded. After 2 hours of water saturation, the weight error was less than 0.5% before and after the coal sample, and the water was gone. The T_2 spectrum of the coal sample was measured again.

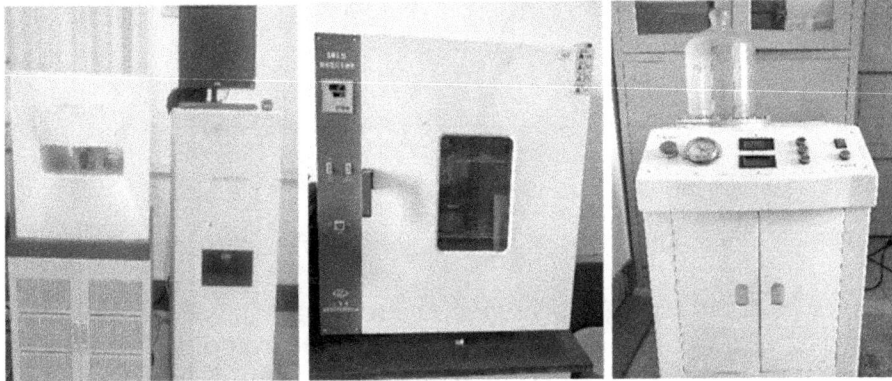

MesoMR23-060H-I (b) Drying oven (c) Vacuum water-saturated device
FIG 3– Experiment apparatus

The T_2 spectra of three different coal samples were tested as follows :

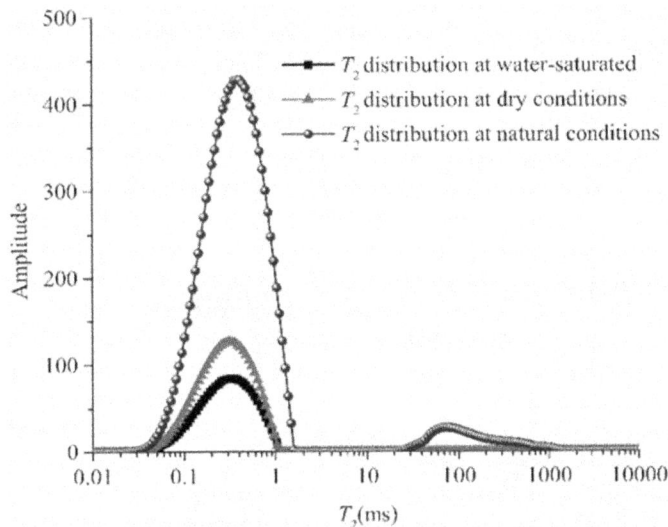

FIG 4– T_2 distribution of parallel bedding at different water conditions

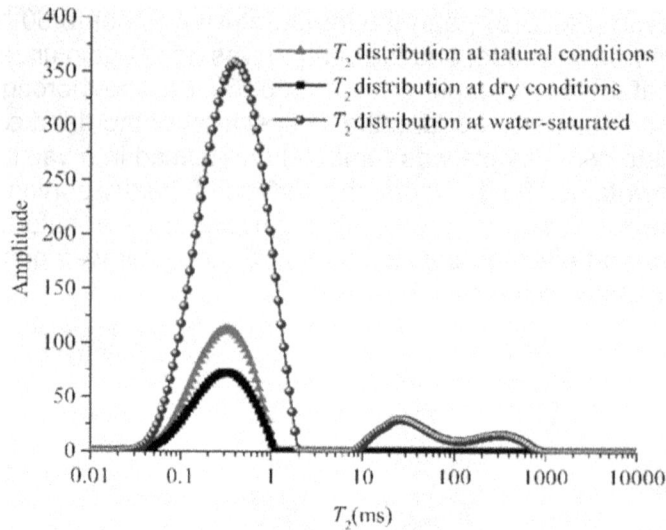

FIG 5– T2 distribution of vertical bedding at different water conditions

As clearly shown in Fig. 4 and Fig. 5, the T2 spectral areas of the coal samples are different under different water conditions. The order is as follows: T2 spectral area of the coal sample after drying (2040 dimensionless), <T2 spectral area of coal sample under natural water condition (3080),

<T2 spectral area of saturated coal samples (12692). The T2 spectrum of the parallel bedding under different water conditions is larger than that of the vertical bedding. The T2 spectrum of the vertical coalbed of saturated water samples displays a three-peak shape, the peak of the T2 spectrum is 12295, and the parallel bedding exhibits a bimodal morphology. The peak area of the T2 spectrum is 12692, and the coal sample exhibits anisotropic characteristics in the bedding. The area of the T2 spectrum decreases from natural to dry, indicating that the moisture in the coal evaporates continuously from the micropores under dry conditions and is lost along the pore channels. The area of the T2 spectrum increased after the dry coal sample was saturated, indicating that the water entered the coal pores and the mesopores from the coal fissure channel and then entered the micropores. These experimental results indicate that the connectivity between the pores in the coal is good as a whole.

COAL SAMPLE LOADING AND NMR TEST ANALYSIS

The loading experimental equipment adopts the true triaxial fluid-solid coupling test system. The saturated coal sample was dried again under the previous drying conditions.

Applied stress path: First, triaxial stress was gradually applied according to the three-direction isostatic state until the predetermined stress axial pressure and the confining pressure are 2MPa. While maintaining the confining pressure, loading was performed at a speed of 0.1MPa/step, and the axial pressure was applied stepwise until coal sample destruction.

The coal sample after the fracturing was again placed in a vacuum-saturated device for 12h, and the T2 spectrum of the coal sample after the fracturing was tested by NMR. The NMR measurement parameters were unchanged. The test results are shown in Fig. 6 and Fig. 7.

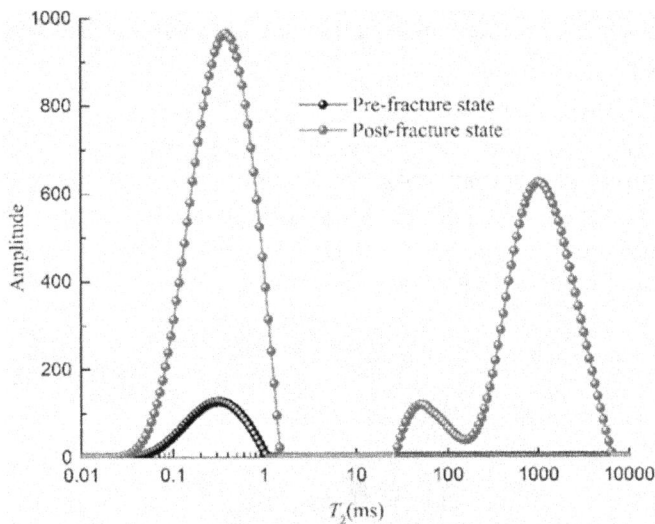

FIG 6 – T_2 distribution of parallel layers at different water conditions

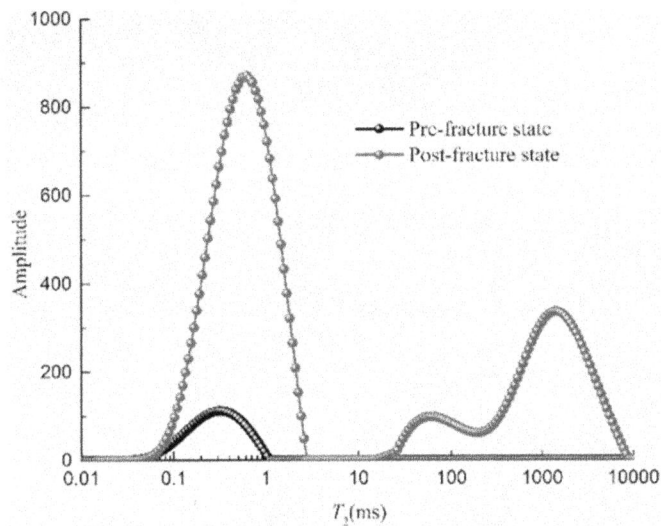

FIG 7 – T_2 distribution of vertical layers at different water conditions

The T_2 spectrum of the coal sample was measured by NMR. As shown in Fig. 6 and Fig. 7, the T_2 spectrum area of the coal sample is significantly increased after loading, and the parallel layer T_2 spectrum area is 46735, which is 9112 more than that of the vertical layering. The first peak area of the microscopic pores in the parallel bedding accounted for 57.6% of the total area, and the vertical bedding was 66.4%. The third peak area for characterizing macropores and fissures in parallel bedding accounts for 38.2% of the total area and 27.5% for the vertical. The difference in T_2 spectral area of coal samples with the same loading conditions is primarily due to the difference in bedding. Under the loading conditions, the pore and fracture area of the coal sample increased significantly, indicating that the gas migration channel of the coal sample was unblocked.

When comparing the T_2 spectra before and after coal sample fracturing, it can be noted the T_2 spectrum area of the parallel coal seam sample after fracturing was 2.68 times that before fracturing. In addition, the T_2 spectrum area of the vertical bedding coal sample was 2.33 times that before fracture. After fracturing, the effective pores of the coal sample increased. Therefore, under the saturated water condition, more water entered the pore-fracture in the coal through the effective pores. This outcome indicates that the loading mode under this experimental condition effectively increases the effective porosity of the coal sample.

The first peak areas of the parallel and vertical coal samples after loading were 2.6 and 2.13 times, respectively, the second peak areas were 5.7 times and 8.6 times, respectively, and the third peak area was approximately 300 times.

COAL BODY BEFORE AND AFTER LOADING: SCANNING ELECTRON MICROSCOPE TEST

The CamScan MX2600 thermal field emission scanning electron microscope was used to test the microscale before and after the coal sample was loaded. For the discreteness of the scanning results, a cubic coal sample with a complete appearance of a small size of approximately 0.5cm^3 should be used, and a relatively flat section should be selected as the observation surface. The CamScan MX2600 thermal field emission scanning electron microscopy results are shown in Fig. 8. The specific SEM test results are shown in Fig. 9.

FIG 8 – CamScan MX2600 type thermal field emission scanning electron microscopy

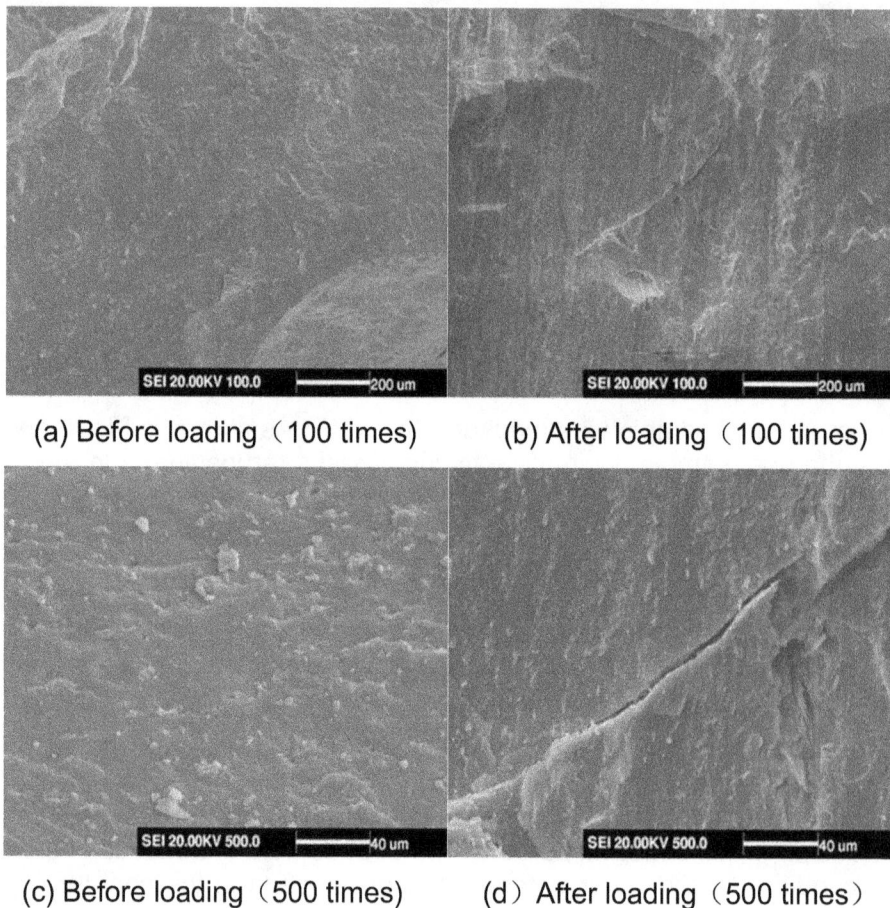

(a) Before loading（100 times）   (b) After loading（100 times）

(c) Before loading（500 times）   (d）After loading（500 times）

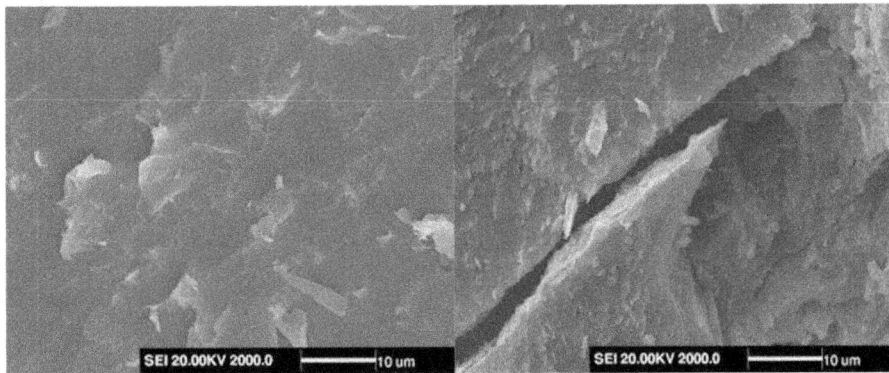

（e）Before loading（2000 times)(f) After loading（2000 times）

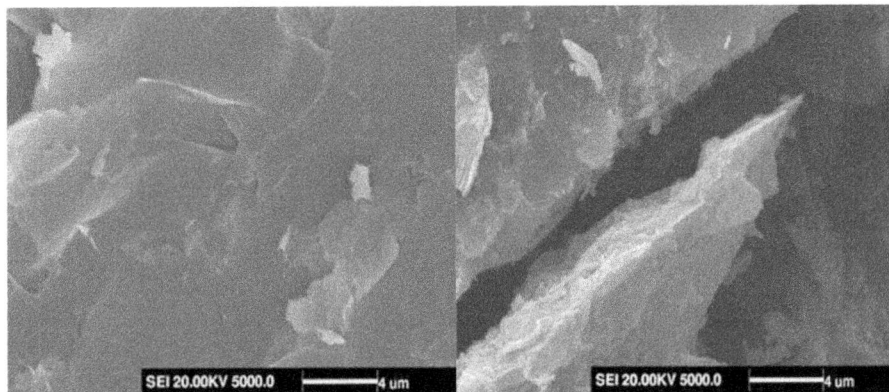

(g) Before loading（5000 times) (h) After loading（5000 times

FIG 9 – Scanning electron micrograph of coal body before and after loading

As shown in Fig.9, the structure of the coal body is dense, and the surface is rough. There are many hollow holes of different sizes and irregular shapes. However, there is no obvious crack in the coal body before loading, which is not conducive to coal body gas seepage. After the coal body was loaded, its dense structure was destroyed, a large number of fissures formed, and the original pores were further connected. A large number of shear fissures were observed, which greatly increased the permeability of the coal body.

COAL BEFORE AND AFTER LOADING: TRANSMISSION ELECTRON MICROSCOPY

To grasp the changes in the internal microstructure of the coal before and after loading and the penetration and expansion characteristics of the crack, a high-resolution transmission electron microscope (Made in Japan (model JEM-2100) with magnification 2~800,000 times high resolution transmission electron microscope Japanese model JEM-2100, as shown in Fig. 10) was used to test the morphology and microstructure of the coal before and after loading. From the microscopic angle, the influence of the internal microstructure on the seepage characteristics was further analyzed.

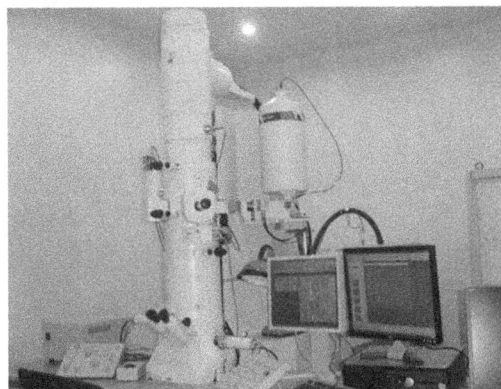

FIG 10 – JEM-2100 type transmission electron microscope

High-resolution transmission electron microscopy was used to examine the internal microstructure and morphology of the coal body before loading and after loading. As shown in Fig.10 and 11, the microstructure of the coal sample before loading was smooth, as was the outer edge of the coal particles. The end face was also relatively complete. By analyzing the original coal body before loading, it was determined that the microstructure in the original coal body was not conducive to the circulation of gas in the coal body.

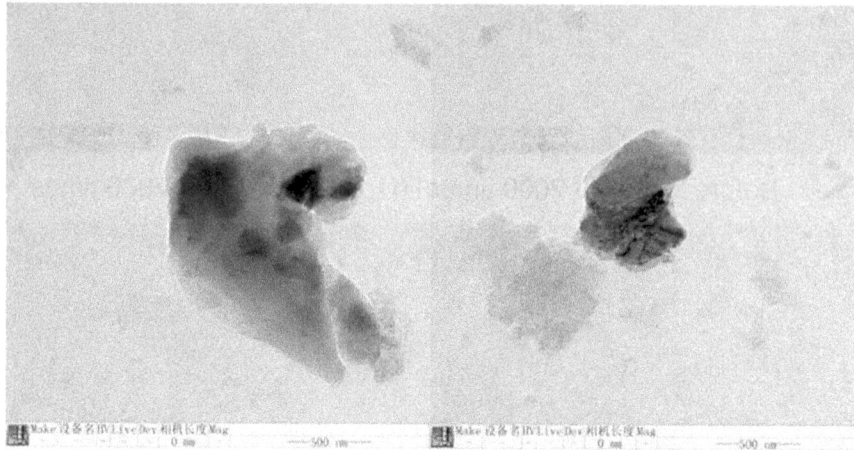

a) 500nm

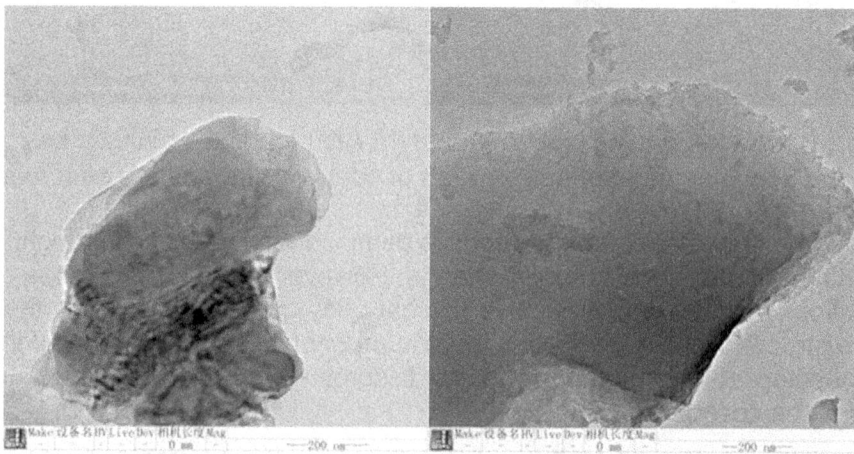

b) 200nm

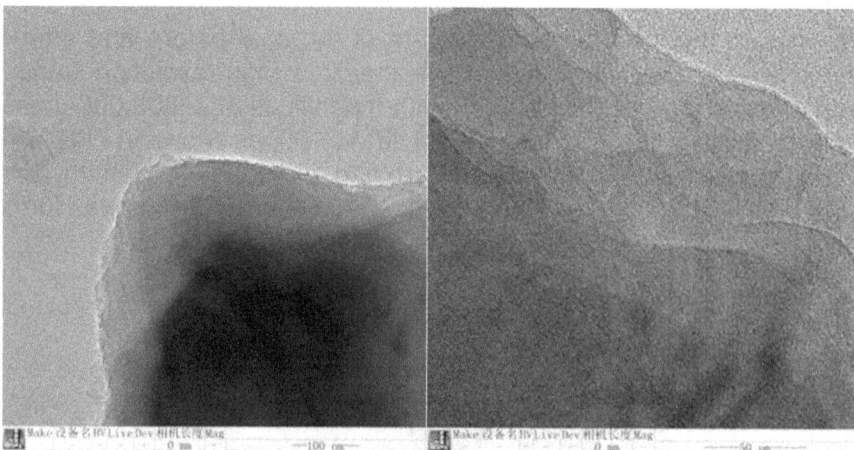

c) 100nm

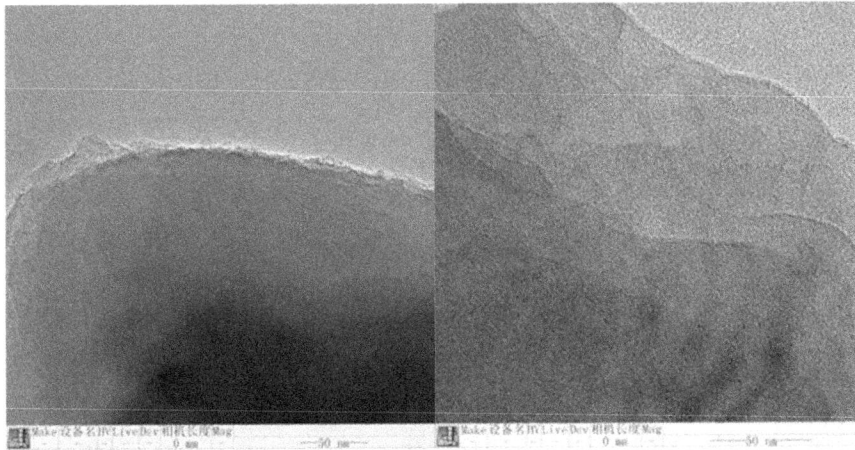

d) 50nm

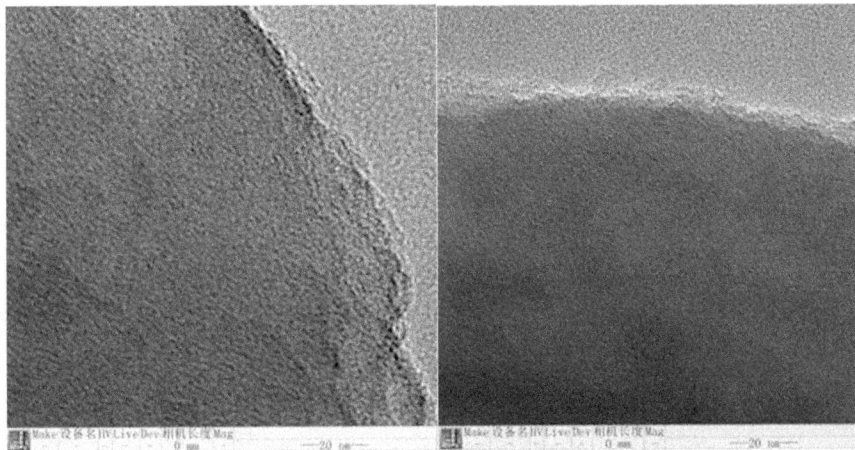

e) 20nm

FIG 11 – Microstructure and morphology of coal body before loading

High-resolution transmission electron microscopy of the internal structure of the coal before loading, the microstructure, and the morphology are shown in Fig.11. Analysis of Fig.11 shows that the microstructure of coal before loading is a relatively smooth surface, there are no obvious pores or cracks, the shape is relatively regular, and the outer edge of the coal particle surface is relatively complete; analysis of raw coal before loading shows that the microstructure of the original coal is not conducive to the flow of gas in the coal.

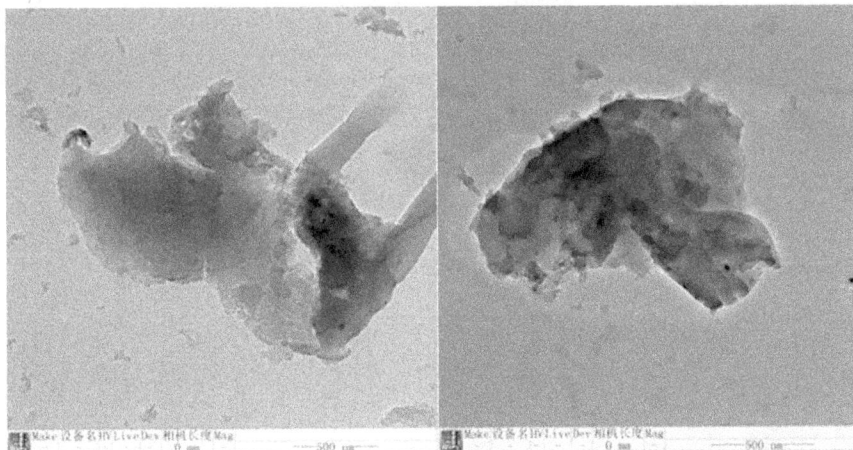

(a) 500nm

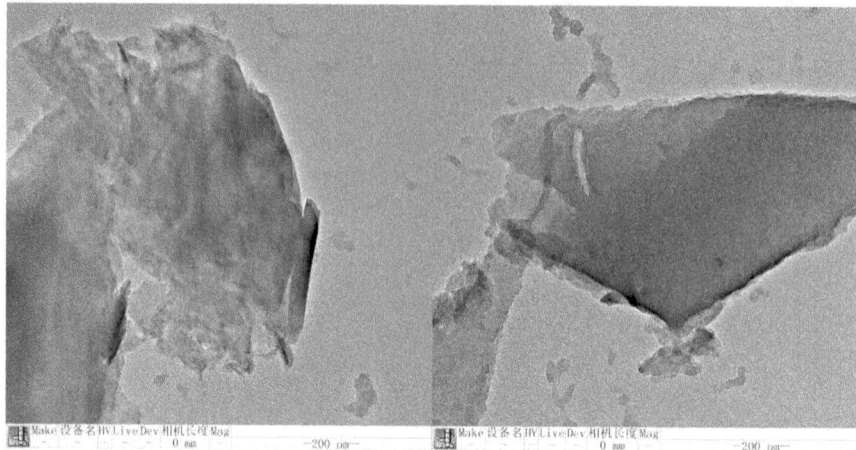
(b) 200nm

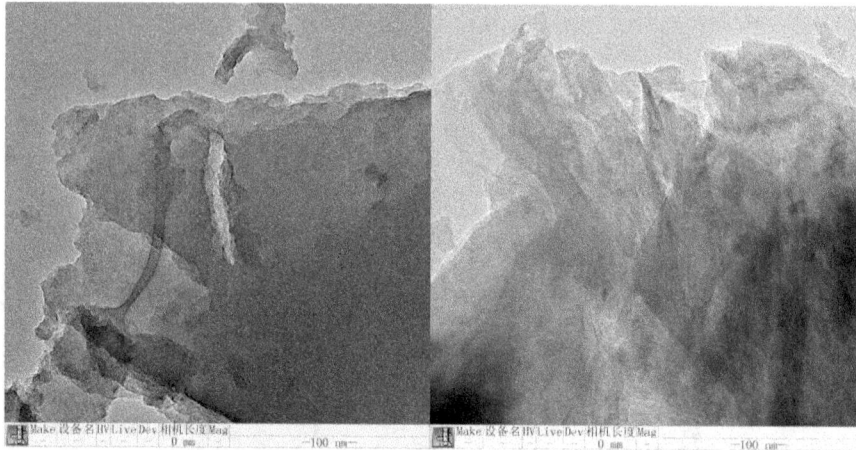
(c) 100nm

(d) 50nm

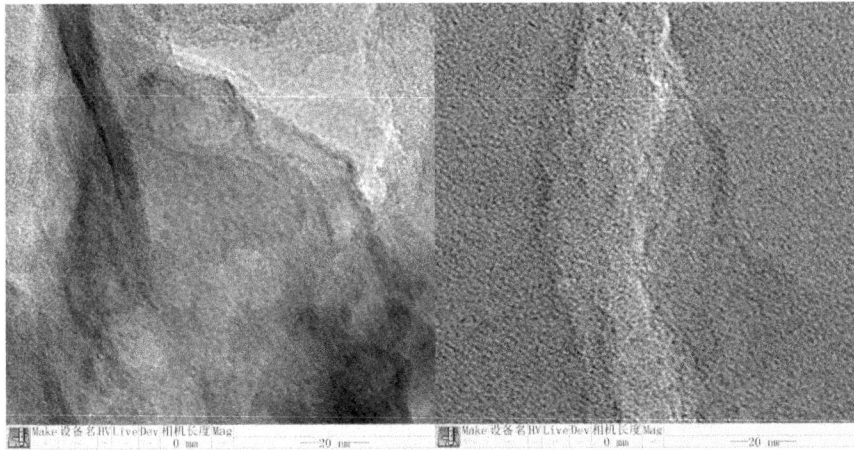

(e) 20nm

FIG 12 – Microstructure and morphology of coal after loading

The internal microstructure of the coal after loading is shown in Fig.12. As shown in the figure, after the coal body is damaged by the load, obvious pores and cracks appear, and the local area collapses and forms a crushing zone. The outer edge of the coal particles is irregular and uneven, the surface of the coal body is weakened, and the coal powder area is clear. Traces of gas flow can be observed.

Further analysis of the microstructure at 50nm and 20nm (Fig. 12(d) (e)) reveals a main fracture of the coal body. A large number of secondary fractures appear around the main fracture. The main fracture and the secondary fracture both develop in the same direction. The fractures are relatively developed and exhibit obvious fractal characteristics. The tip of the fracture also exhibits divergent characteristics. These fractures are the origin of microcracks and pore development. In addition, they represent passages for gas seepage, diffusion channels and enrichment zones. Through the high-resolution projection electron microscopy analysis of the internal microstructure of the coal before and after loading, it can be observed that the number of pores and cracks of the coal body after the load is damaged is obviously increased, which is beneficial to the expansion, development and penetration of the crack and finally the coal body. The seepage characteristics are significantly increased.

CONCLUSIONS

The T_2 spectral area of coal samples before and after loading measured by NMR method under different loading coal samples under different loading paths revealed that the T_2 spectral area of the parallel layer before loading was 500 (dimensionless) than that of vertical bedding. After loading, the parallel layered T_2 spectral area was 9000 more than the vertical bedding. It indicated that there was indeed a difference in the bedding of the coal sample, and the permeability of the parallel bedding after loading was better.

A microscale inspection of the coal body before and after loading was performed by scanning electron microscopy. Before the coal body was loaded, the surface was relatively flat, there were many hollow holes of different sizes, the shape is irregular and the coal body was ignorant After the coal body was loaded, its dense structure was destroyed, a large number of cracks formed, and the original pores were further penetrated, which further increased the permeability of the coal body.

High-resolution transmission electron microscopy was used to examine the internal microstructure of coal before and after loading. There were no obvious pores and fissures before the coal body was loaded. The outer edge of the coal particles was also complete. After the coal was loaded, obvious pores and fissures appeared. In addition, a weakened coal powder area appeared on the surface of the coal body, the local area collapsed and formed a fracture zone, and gas flow traces could be clearly observed. The increase in cracks is conducive to the expansion and penetration of cracks and improves the permeability of coal.

ACKNOWLEDGEMENTS

This work was supported by National Key Research and Development Program of China (2018YFC0808100), Funded by National Natural Science Foundation of China (51604101; 51704099; 51734007), This project was also supported by the Henan Postdoctoral Foundation (001801016), Key R&D and Promotion Special Projects of Henan Province (192102310511), the Doctoral Fund of Henan Polytechnic University (No.B2018-59) and the State Key Laboratory Cultivation Base for Gas Geology and Gas Control (Henan Polytechnic University) (WS2017B06).

REFERENCES

C.R. Clarkson, R.M. Bustin.1999. The effect of pore structure and gas pressure upon the transport properties of coal: a laboratory and modeling study.1. Isotherms and pore volume distributions. Fuel. (78):1333-1344.

Cai, Y D., Liu, D M., Yao, Y B.et al.2011. Fractal Characteristics of coal pores based on classic geometry and thermodynamics models. Acta Geological Sinica-English Edition. 85(5):1150- 1162.

Du, J W. 2015. Study on pore of the low permeability coal seam in Lu'an Mining Area. Henan Polytechnic University.

Fan, C J., Li, S., Lan, T W., Li, M S.et al.2017. Influences of different factors on enhancing methane drainage by hydraulic fracturing. China Safety Science Journal. 12:97-102.

Gao, B B.2010. Expriment study on evolution of cracks and permeability of mining coal. Beijing Jiaotong University.

Guo, J J.2017. Study on crack evolution and permeability change mechanism of medium and high rank coal in the process of bearing stress. Southwest Petroleum University Doctoral Dissertation.

Hao, F C., Sun, L J., Zuo, W Q.et al.2016. Hydraulic flushing aperture variation and anti-blocking technology considering rheological property. Journal of China Coal Society.6:1434-1440.

Huang, F., Li, T Y., Gao, X Y., Yang, X.et al. 2019. Study on the macro-micro failure mechanism of granite and its geometry effect under the different conditions of confining pressure by discrete element. Journal of China Coal Society. 3:924-933.

IUPAC (International Union of Pure and Applied Chemistry), 1994. Physical Chemistry Division Commission on Characterization of Porous Solids. Recommendations for the characterization of porous solids (Technical report). Pure and Applied Chemistry. 66(8):1739-1758.

Jiang, T T., Zhang, J H., Huang, G. et al 2017. Study on the permeability characteristics of coal rock at different bedding directions. Science Technology and Engineering.17:206-211.

Kang, Z Q., Li, X., Li, W., Zhao, J.et al 2018. Experimental investigation of methane adsorption/desorption behavior in coals with different coal-body structure and its revelation. Journal of China Coal Society. 5:1400-1407.

Li, Z W.,2015. Study on microstructure characteristics of low rank coal and its control mechanism on gas adsorption/desorption. China University of Mining and Technology (Beijing).

Li, Z W., Lin, B Q Guo, Z Y., Gao, B.et al 2013. Characteristics of pore size distribution and its impact on gas adsorption. Journal of China University of Mining& Technology. 42(6):1047-1053.

Liu, D M., Huang, W H., Tang, D Z., Tang, S H.et al.2006. Research on the pore-fractures system properties of coalbed methane reservoirs and recovery in Huainan and Huaibei coal fields. Journal of China Coal Society. 2:163-168.

Lin, B Q., Wu, H J., Zhu, C J., Lu, Z G.et al.2010. Structural characteristics and fractal laws research in coal and rock ultrafine pore. Journal of Hunan University of Science & Technology (Natural Science Edition) 3:15-18+28.

Li, J Q, Liu, D M., Yao, Y B.et al. 2013. Physical characterization of the pore-fracture system in coals, Northeastern China. Energy exploration & exploitation. 31(2):267-285.

Liu, Q Q.2015. Damage and permeability evolution mechanism of dual-porosity coal under multiple stress paths and its application. China University of Mining and Technology (Beijing).

Liu, Y B.2009. Study on the Deformation and damage rules of gas-filled coal base on meso- mechanical experiments. Chongqing University.

Lu, F C., Gao, J L., Zhang, Y G., Liu., J J.et al.2018. Anisotropic ultrasonic characteristics of coal pores and fractures under uniaxial loading. Progress in Geophysics.6:2555-2562.

Lu, F C., Zhang, Y G., Jiang, L H.et al.2018. Anisotropic characteristics of nuclear magnetic resonance of pores and fractures in coal under uniaxial loading. Coal Geology & Exploration.1:66-72.

Lin, B Q.1988. Discussion on permeability of gas containing coal. China Mining Institute.12:15-20+65.

Liu, Y B., Li, M H., Yin, G Z.et al 2018. Permeability evolution of anthracite coal considering true triaxial stress conditions and structural anisotropy. Journal of Gas Science and Engineering 52:492-506.

Liu, S P.2017. Simulation Research on Fluid Process Continuity associated with CO2-ECBM of High- Rank Coal Reservoir in Southern Qinshui Basin. China University of Mining (Beijing).

Meng, Q R., Zhao, Y S., Hu, Y Q., Feng, Z C et al.2011. Experimental study on pore structure and pore shape of coking coal. Journal of China Coal Society. 3:487-490.

Nie, B S., Lun, J Y., Wang, K D., Shen, J S., Ju, Y W., et al. 2018. Characteristics of nanometer pore structure coal reservoir. Earth Science. 43(5):1755-1762.

Ouyang, Zhong Q., Liu, D M., Cai, Y D et al.2016.Fractal Analysis on Heterogeneity of Pore- Fractures in Middle-High Rank Coals with NMR. Energy & Fuels. 30(7):5449-5458.

Pan, R K., Cheng, Y P., Yuan, L.et al.2014. Effect of bedding structural diversity of coal on permeability evolution and gas disasters control with coal mining. Natural Hazarads.73(2):531- 546

Ren, J G.2016. Gas diffusion law and its controlling mechanism study on medium and high rank tectonically deformed coalsin central and southern parts of north china plate. Henan Polytechnic University.

Ren, H.K., Wang, A.M., Li, C.F., Cao, D.Y.et al 2017. Study on porosity characteristics of low-rank coal reservoirs based on nuclear magnetic resonance technology. Journal of Natural Gas Science and Engineering. 45(4):143-148.

Sun, M D., Yu, B S., Hu, Q H.et al. 2017. Pore connectivity and tracer migration of typical shales in south China. Fuel. 203:32-46.

Sang, S X., Zhu, Y M., Zhang, S Y., Zhang, J., Tang, J X et al, 2005. Solid-gas interaction mechanism of coal-adsorbed gas(I)-coal pore structure and solid-gas interaction. Natural Gas Industry.1:13-15+205.

Song, X X., Tang, Y G., Li, W., Zeng, F G.et al.2014. Pore structure intectonically deformed coals by small angle X-ray scattering, Journal of China Coal Society. (4):719-724.

Song, L., Ning, Z F., Sun, Y D., Ding, G Y.et al.2017. Pore structure characterization of tight oil reservoirs by a combined mercury method. Petroleum Geology & Experiment.5:700-705.

Song, B Y., Song, D Y., Li, C H., Yuan, L.et al.2017. Influence of magmatic intrusion on the coal pore on the basis of mercury intrusion porosimetry. Coal Geology & Exploration.3:7-12.

Sun, L.2013.Soft and Hard Coal Adsorption-desorption Law of Different Coal Rank and It`s Application. China University of Mining and Technology, Beijing.

Xu, C W., Lv, Y F., Tan, B D.et al.2017. Micro/Nano-Pore Characterization of Coalbed Methane Reservoir in Jixi Basin, Northeast China. Journal of Nanoscience and nanotechnology .17(9):6268-6275.

Xu, H., Zhang, S., Leng, X., Tang, D., Wang M.et al. 2005. Analysis of pore system model and physical property of coal reservoir in the Qinshui Basin. Chinese Science Bulletin. (50):52-58

Xue, W. 2018.Study on pressure relief mechanism of hydraulic flushing in coal seams and its application. China University of Mining and Technology, Beijing.

Xie, S.B., Yao, Y.B., Chen, J.Y., Yao, W., 2015. Research of micro-pore structure in coal reservoir using low-field NMR. Journal of China Coal Society. 40(1):170-176.

Yao, Y B., Liu, D M., Tang, D Z.2009. Preliminary evaluation of the coalbed methane production potential and its geological controls in the Weibei Coalfield, Southeastern Ordos Basin, China. International Journal of Coal Geology. 78(1):1-15.

Yao, Y., Liu, D W. Huang, D. Tang, Tang, S.et al. 2006. Research on the pore-fractures system properties of coalbed methane reservoirs and recovery in Huainan and Huaibei coal-fields Journal of China Coal Society.31 (2):163-168.

Yao, Y B., Liu, D M., Cai, Y D.et al.2010. Advanced characterization of pores and fractures in coals by nuclear magnetic resonance and X-ray computed tomography. Science China-earth Sciences 53(6):854-862

Yao, Y.B., Liu, D.M., 2016. Petrophysics and fluid properties characterizations of coalbed methane reservoir by using NMR relaxation time analysis. Coal Science and Technology 44(6):14-22.

Zhang, J Z., Li, X Q., Zou, X Y. et al. 2019. Characterization of the Full-Sized Pore Structure of Coal- Bearing Shales and Its Effect on Shale Gas Content Energy & Fuels. 33(3):1969-1982.

Zhou, S D., Liu, D M., Cai, Y D.et al.2016. Fractal characterization of pore-fracture in low-rank coals using a low-field NMR relaxation method.Fuel. 181:218-226.

Zhang, W. 2015. The Application and Parameter Optimization of the CBM Horizontal Well Section Hydraulic Fracturing in Hebi Mine Area. China University of Mining and Technology, Beijing.

Study on the heat removal efficiency of thermal probe on coal stockpile under different arrangement and ambient air velocity

W Gaoming[1], M Li[2], W Hu[3], L Yanfei[4] and W Weifeng[5]

1. Doctor, College of Safety Science and Engineering, Xi'an University of Science and Technology, Xi'an Shaanxi Province, PR China, 710054. Email: wgm20180326@163.com
2. Student, College of Safety Science and Engineering, Xi'an University of Science and Technology, Xi'an Shaanxi Province, PR China, 710054. Email: mal@xust.edu.cn
3. Professor, College of Safety Science and Engineering, Xi'an University of Science and Technology, Xi'an Shaanxi Province, PR China, 710054. Email: 2240142043@qq.com
4. Student, College of Safety Science and Engineering, Xi'an University of Science and Technology, Xi'an Shaanxi Province, PR China, 710054. Email: 1054894337@qq.com
5. Professor, College of Safety Science and Engineering, Xi'an University of Science and Technology, Xi'an Shaanxi Province, PR China, 710054. Email: 251044098@qq.com

ABSTRACT

In recent decades, the spontaneous combustion of large-scale coal stockpile was frequent, seriously causing the waste of resources and environmental pollution. The ambient airflow was one of the factors that induced it occurred. In this paper, based on the principle of heat removal by thermal probe, we conducted a simulated experiment to study the heat removal effect, such as the temperature difference and heat removal ratio, heat removal radius, and cumulative heat removal volume, in a coal stockpile as its internal temperature increased from 100 to 150 °C under different experimental conditions. The results showed that the scale of temperature drop at designated point in the coal stockpile were 46.3%, 10.8%, and 32.8% that under the above experimental conditions, respectively. A comparison study found that as the insertion angle of thermal probe in the coal stockpile was 60 degree, the heat removal efficiency of thermal probe was significantly affected by outside circumstance, and is highly sensitive to airflow. For instance, as the thermal probe insertion scale was 1:4, outside air velocity was 2 m/s, and inserted angle of thermal probe was 60 degree, the average radius of heat removal by thermal probe was 5.79 m (maximum heat removal radius), and the overall heat removal efficiency of thermal probe was enhanced by 79% and 93% that in a constant temperature circumstance (100 and 150 °C), respectively. Therefore, it can be concluded that the suitable arrangement of thermal probe in the coal stockpile was advantageous to heat removal. This paper provided a new technical approach and theoretical basis for the prevention and control of coal spontaneous combustion for large-scale coal stockpile under different conditions.

Keywords: Coal stockpile; Thermal probe; Heat removal efficiency; Arrangement

INTRODUCTION

Coal fire disasters almost spread in every countries and regions all over the world. Coal spontaneous combustion not only induced the burning of a great deal of coal resources, but also the release of toxic and harmful gases that seriously affect the ecological environment [1-3]. In China, statistics indicated that coal fires account for about 0.1% to 0.22% of global carbon emissions every year [4]. Ground coal stockpile was regarded as an important place for coal transportation and temporary storage. To effectively prevent high temperature agglomeration caused by contact oxidation with air in storage environment, and then spontaneous combustion of coal, is one of the important works at present in the disaster prevention [5-7]. The main causes of large-scale fire accidents in coal stockpile were as follows: There was countless air leakage channels in coal stockpile with spontaneous combustion tendency, or its surface contacted with air for a long time, which resulted the heat accumulation sharply by the slow oxidation induced the temperature rapidly raised, and eventually caused to spontaneous combustion of coal stockpile [8,9]. Krishnaswamy S. et al [10] concluded that as the air leakage rate of loose coal in coal stockpile is greater than 1.2 m3/min or less than 0.06 m³/min, the coal stockpile was not easy to spontaneous combustion. Based on the large number of experimental investigations, Brooks K et al [11] concluded that as the air leakage rate of the coal stockpile was between 0.4 and 0.8 m^3/min, the coal stockpile was prone to spontaneous combustion. Therefore, spontaneous combustion in the coal stockpile could happened while the air leakage intensity was in a certain range. The internal temperature of coal stockpile was another important factor leading to coal spontaneous combustion, which was mainly a result of a dynamic process of coal oxidation and heat accumulation [12-14]. Therefore, heat removal from coal oxidation was an effective way to prevent spontaneous combustion of coal stockpile.

Thermal probe, as a tool with strong heat transfer capacity, great heat transfer uniform property, single heat transfer safety and economy, is widely used in heat transfer and removal of high temperature heat source [15,16]. The principle of using thermal probe removal technology to prevent coal spontaneous combustion is to remove heat from deep heat source in the regenerative area of coal spontaneous combustion, thus to drop the temperature, accelerate heat loss in the high temperature area and destroy the regenerative environment in the coal stockpile, so as to realize the purpose of preventing coal spontaneous combustion [17-19]. However, in the process of using heat removal technology to prevent spontaneous combustion in the coal stockpile, relatively little research has been conducted on the heat transfer efficiency of thermal probe under different working conditions. Therefore, a self-developed experiment was carried out to analyze the heat removal effect of thermal probe under different arrangement and air conditions. The main purpose of this paper was to provide a theoretical basis for the heat removal technology of thermal probe to prevent spontaneous combustion of coal stockpile.

EXPERIMENTAL AND METHODOLOGY

Experimental systems

The experimental system is shown in Figure 1, which includes temperature control system, monitor system, coal heat and air supply system. Temperature control system consists of temperature control box, plate heater, temperature patrol instrument, thermocouple and monitoring software. The air supply system consists of a fan and mechanical anemometer, the components and functions of the system are shown in Table 1.

TABLE 1 - Experimental system parameters of the devices.

Experimental system	Device	Specification	Property
Temperature monitor system	Temperature monitor	XSLC09	Monitoring temperature
	Thermal resistance	Pt100	Collect temperature data
	Infrared thermal imager	DM60-S	Surface heat detection
	Computer	M400	Data processing
Air supply system	Wind turbine	YWF4E-500	Provide airflow
	Anemometer	CFJ25	Measuring air velocity
Temperature control system	Temperature control box	ZWK/WCK	Temperature control
	Heater	/	Heating of coal-box
Heating system	Thermal probe	/	Heat transfer and cooling
	Sil pad	/	Heat insulation
	Coal-box	/	Coal storage

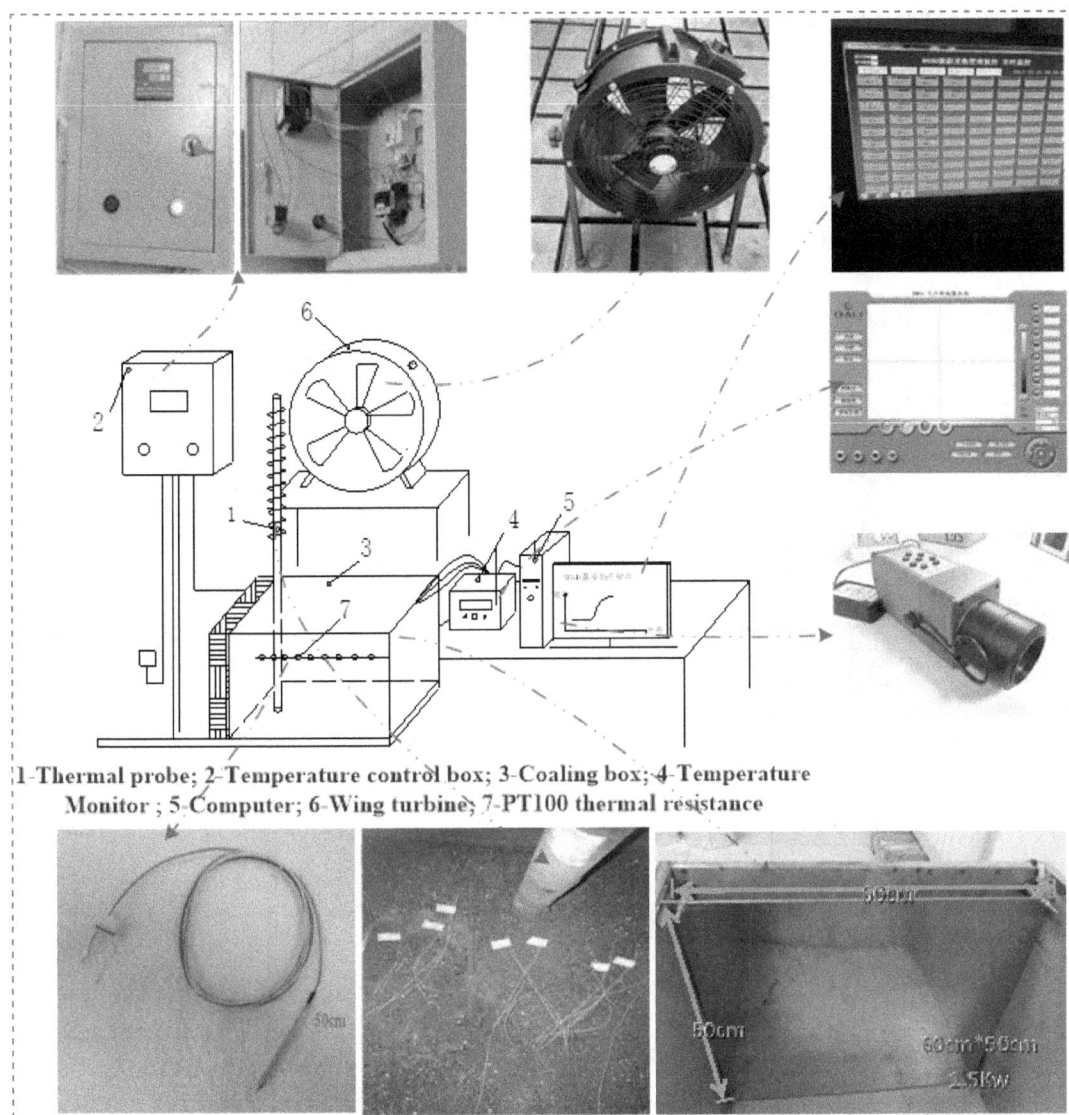

FIG. 1 - Experimental systems of heat removal technology by thermal probe.

Carbon steel-hydrothermal probe with fins was used in this experiment. The working temperature ranged from 30 to 250 °C, it was made by carbon steel with a total length of 1500 mm and a diameter of 45 mm, which had good reflective and radiative properties with an emissivity of 0.95. The size of the coal heating box was 700 *600 *500 mm, the total volume of coal box was 0.24 m³, and the maximum temperature resistance was 800 °C. The plate heater was 600 *500 mm in size and 5 mm in thickness, which can single heat the coal samples in coal heating box, which has the characteristics of fast and uniform heating the coal samples. According to the requirements of the experiment, a plate heater with the power of 1.5 kW and a surface load of 4 W/cm² was selected in this experiment. Glass wool with low density and low thermal conductivity was filled between the heater and the out layer of the coal box as insulation material, which could play the role of heat preservation and insulation.

MATERIAL AND METHOD

Experimental coal was collected from a large-scale coal stockpile on the ground in Changcheng No. 1 Coal Mine of Inner Mongolia New Mining Group. The coal was mainly bright and vitrinite, black, brittle and fragile, with a small amount of dark and silk charcoal. The coal seam was easy to spontaneous combustion with high calorific value. The industrial analysis and true density test results of coal samples are shown in Table 2.

Coal samples No	Proximate and ultimate				
	M,ad/%	A,d/%	V,daf/%	FC,ad/%	Φ /(g/cm³)
Great wall mining area	3.39	10.49	17.68	68.44	1.51

Note: Mad is the moisture content; Aad is the ash yield; Vdaf is the volatile matter; FCad is the fixed carbon content; Φ is the true density.

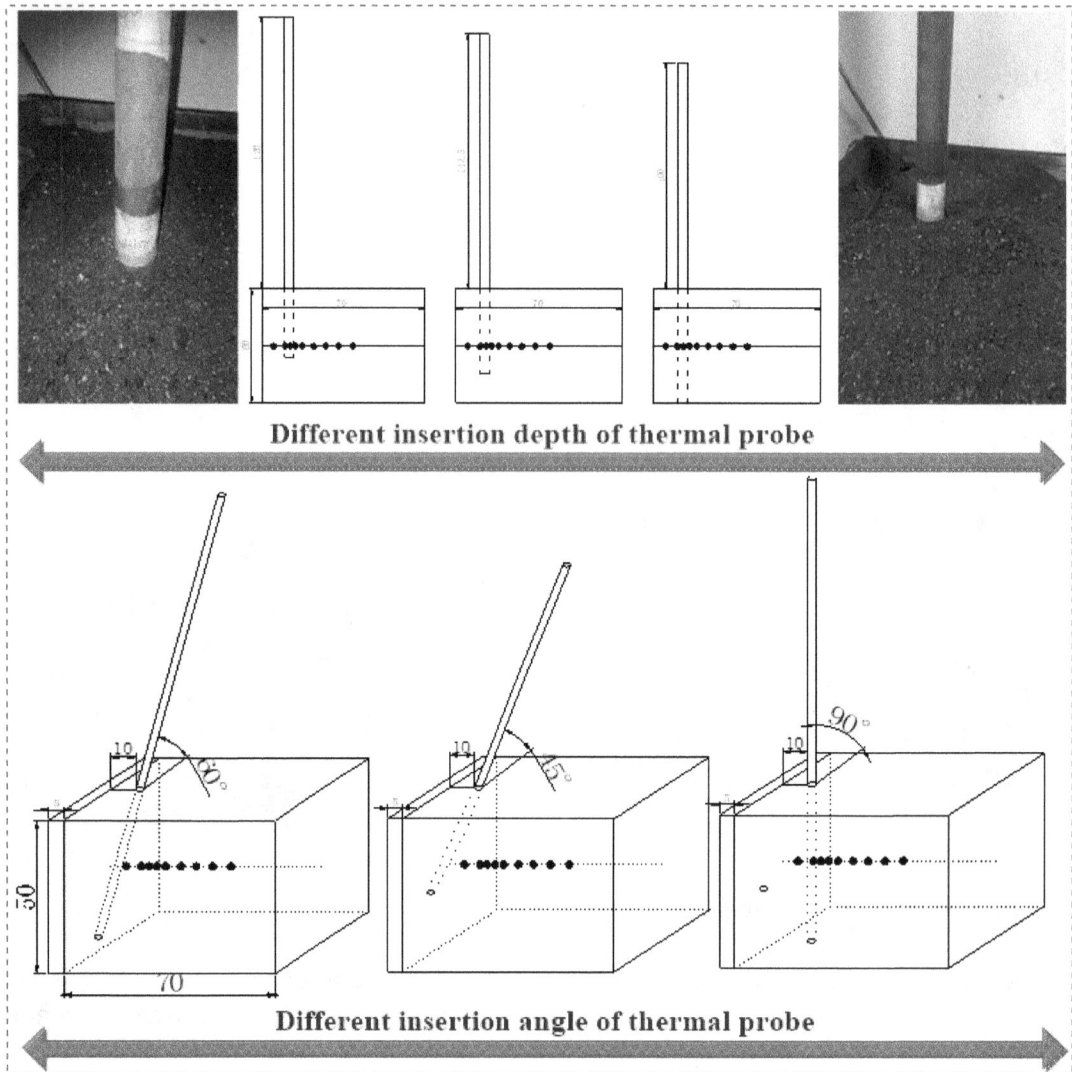

FIG. 2 - Schematic diagram of different arrangement of thermal probe.

Because of the heat removal effect of thermal probe was greatly influenced by outside air velocity, the insertion scale and angle of the condensation section were the main factors. Therefore, in the process of experiment design, the control variable experiments were carried out on the heating temperature of the coal stockpile and arrangement of thermal probe. Firstly, the "H-T" experiment was carried out on the insertion scale (H) and heating temperature (T) of thermal probe. The steps were as follows:

(1) Coal box was heated by plate heater, the temperature of plate heater kept a stable circumstance in the coal box and then came to a next stage. When the temperature varied was lesser than 0.1 °C, the temperature reached a stable status.

(2) When the thermal probe was inserted into a fixed position, the first step was to test the heating temperature at 100 °C, and the insertion ratio of the thermal probe was 1:4. The insertion scale is shown in Figure 2, recording the tested data as the temperature reached equilibrium.

(3) When the temperature measurement finished, the thermal probe was taken out and the heating process of coal box was suspended. After the coal temperature reached balanced once again, 1:3 and 1:2 experimental operations were carried out.

(4) After the completion of the experiment at 100 °C, as the temperature was stabilized, the coal stockpile was then heated to 150 °C. Then, then repeat steps 1 to 4 to collect and analyse the data of the internal temperature circumstance in the coal stockpile.

Next, the "θ-υ" experiment of thermal probe with insertion angle (θ) and air velocity (υ) in the coal stockpile was carried out. It is different with the "H-T" experiment, the heating temperature in the coal stockpile was 150 °C, the insertion ratio was 1:2, the inclined angle plane was parallel to the thermal probe heating plate, and the insertion angles of the thermal probe were 45° 60° and 90°, respectively. The experimental procedures were consistent with the "H-T" experiment (1), and the other experimental steps were as follows:

(5) When the temperature of each measuring points inside the coal stockpile was stable, the experimental coal stockpile was stabilized at 150 °C. Inserting the thermal probe and providing 1.5 m/s and 2 m/s experimental air velocity, respectively. The insertion method was shown in Figure 2.

(6) When the experimental temperature and air velocity were consistent with the stated value, stopping air supply and recording the temperature data of each measure point, collecting the heat transfer data of thermal probe at the same time. Finally, analysed heat removal efficiency of thermal probe under different experimental conditions.

RESULT AND DISCUSSION

Because of different storage environment of the coal stockpile, the distribution of the temperature circumstance was different. Therefore, the experiment studied coal stockpile by controlling heating rate of coal at different measure points and insertion ratios, the results are shown in Table 3. Eight sets of comparative tests were carried out to analyse the influence of thermal probe on the distribution of temperature circumstance inside the coal under different experimental conditions. The temperature variation curves of each measure point in the coal heating process was shown in Figure 3. From the analysis of heating time, it took about 80 hr for the coal stockpile to heat up to stability, the effect of the thermal probe on different measure points would have a "delay effect" with the increased of the distance of the measure point that based on the same heating rate. According to the analysis of temperature curves, the delay time of each measure point would decrease about one hour after the temperature of coal stockpile reached stable. Compared with the temperature field distribution before inserting the thermal probe, the temperature of each measure point inside the coal stockpile have a large drop, which indicated that the thermal probe had an obvious effect on cooling the temperature in the coal stockpile, and the closer distance between thermal probe and measurement point, the greater cooling range, the better the heat transfer effect of the thermal probe. When the heating rate of coal increased, while the temperature of the heating plate was 150 °C, the variation law of the temperature field of the coal stock was consistent with the previous experimental results. Meanwhile, before the thermal probe inserted into the coal stockpile, the temperature of the 1# measuring point reached 94.1 °C higher than the coal spontaneous combustion critical temperature value, while the coal heat up rapidly until the coal was spontaneously ignited [20], after inserting the thermal probe into coal stockpile could be effectively reduced to 64.4 °C, restraining the oxidative heat release rate of coal and the risk of spontaneous combustion of coal storage [21].

TABLE 3 - Experimental design of thermal probe arrangement parameters.

Lab conditions	Heat/°C	Insertion Scale	Inclined angle/ °
1	100	1 : 2	90
2	100	1 : 3	90
3	100	1 : 4	90
4	150	1 : 2	90
5	150	1 : 3	90
6	150	1 : 4	90
7	150	1 : 2	60
8	150	1 : 2	45

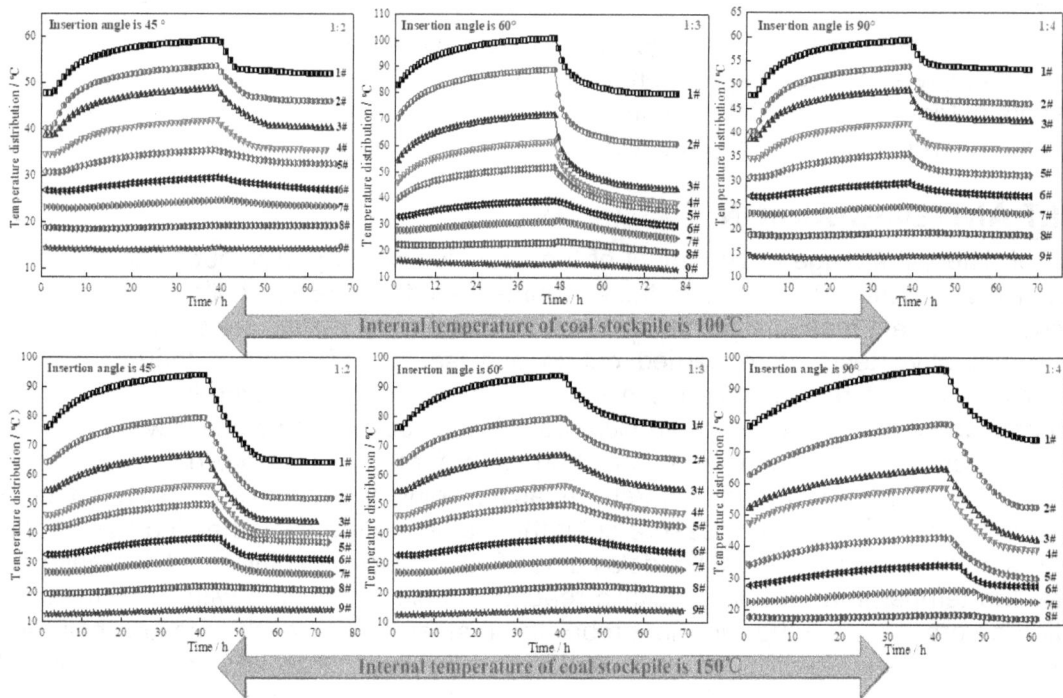

FIG. 3 - Temperature distribution of each measurement point in the coal stockpile under different arrangement and temperature circumstance.

HEAT REMOVAL EFFECT OF THERMAL PROBE IN COAL STOCKPILE WITH NO AIR

Heat removal effect of thermal probe under different insertion scale

The cooling range and efficiency with different insertion scales are shown in Figure 4, Figure 4(a) showed that the internal temperature circumstance for the coal, under a heating temperature of 100 °C, the heat removal effect was the best as the thermal probe insertion scale was 1:4, and the worst as the thermal probe insertion scale was 1:2, that is in the effective cooling section of the thermal probe, the heat removal efficiency was proportional to the insertion scale. It can be seen from Fig 4(b) that the distance between measurement points and thermal probe was larger, the heat removal effect gradually decreased. On the contrary, the heat removal efficiency was higher. Moreover, when the circumstance temperature in the coal stockpile reached 150 °C, the effect of thermal probe was more obvious as the insertion scale was 1:2, whereas the worst effect performed as the insertion scale was 1:4, which indicated that the heat removal efficiency of thermal probe with different insertion scale was closely related to the temperature environment inside the coal stockpile.

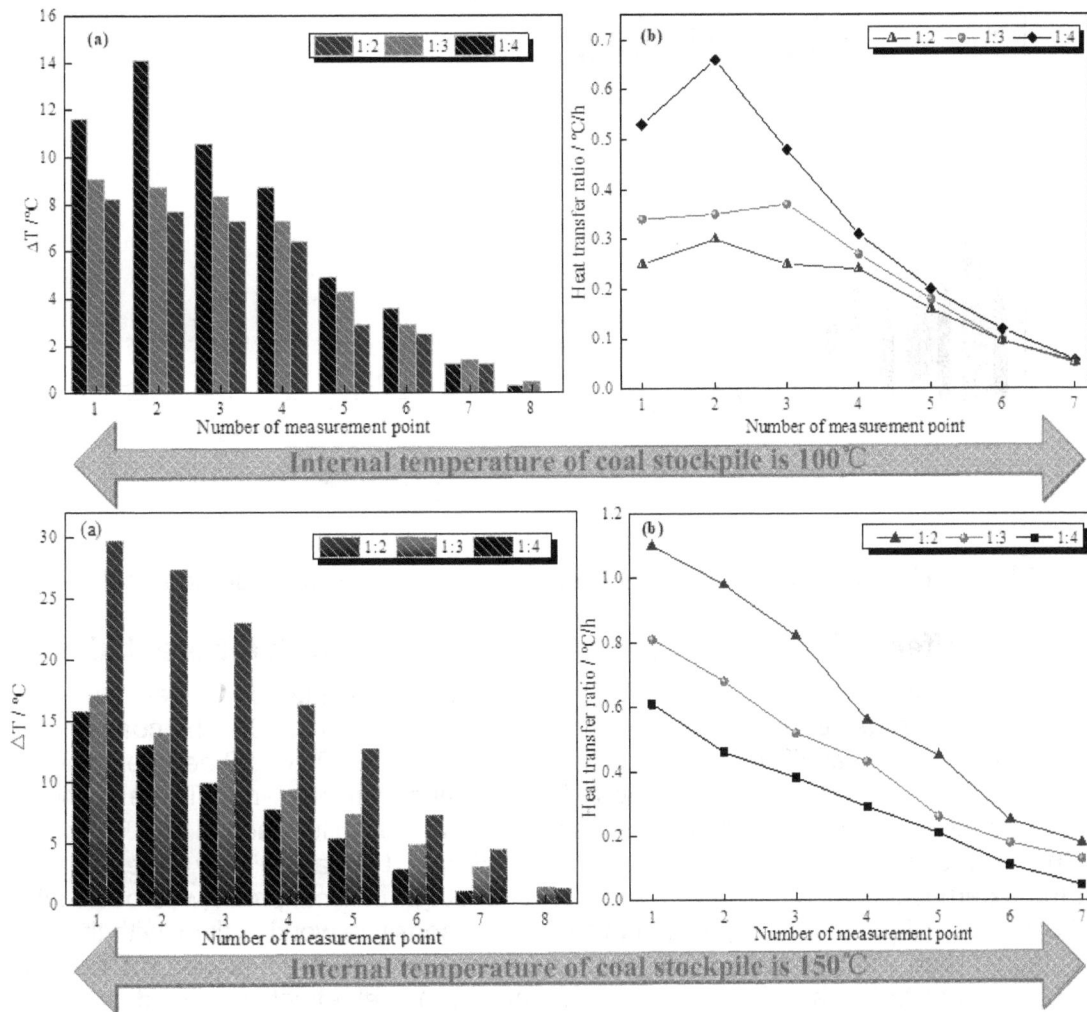

FIG. 4 - Heat removal effect of thermal probe under different
temperature circumstance and insertion scale.

Heat removal effect of thermal probe under different insertion angle

In order to compare the cooling range and rate of thermal probe in the coal stockpile at 45, 60 and 90 degrees of insertion angle, within no air velocity and 150 °C temperature circumstance in coal stockpile, the results were shown in Figure 5. It can be seen from Figure 5(a) and (b) that as the thermal probe insertion angle was 60°, the internal temperature dropped rate and unit time decrease rate were the highest in coal stockpile, then the effect the insertion angle was 45 degree was lower than 60 degree and the lowest at 90 degree. It was showed that the proper insertion angle of thermal probe had an important influence on the heat removal efficiency in the coal stockpile. Moreover, the temperature difference and heat removal efficiency were related to the distance distribution of the measurement points, the farther the distance was, the worse the heat removal effect of the thermal probe was, which illustrated that it was very important for heat removal efficiency to keep reasonable distance of thermal probe in the coal stockpile.

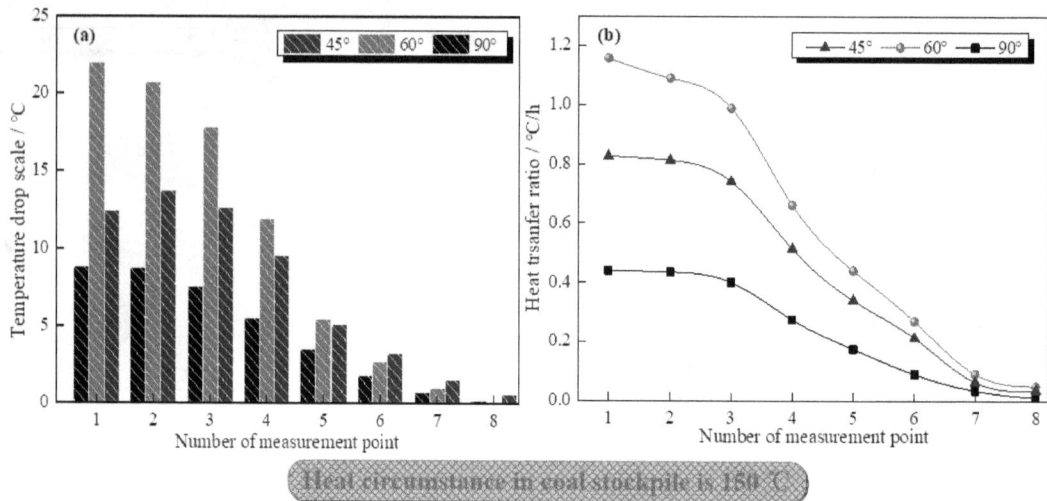

FIG. - 5 Heat removal effect of thermal probe under different insertion angle.

Heat removal effect of thermal probe in coal stockpile with different air velocity.

To compare and analyze the heat transfer efficiency of thermal probe under different air velocity and insertion angle in the outside temperature environment of 100 and 150 °C in the coal stockpile, the experiments of heat removal efficiency of thermal probe under 45, 60 and 90 degree insertion angle and 0, 1.5 and 2 m/s air velocity were designed, the results were shown in Table 4 and Fig 6. It showed that as the temperature environment inside the coal stockpile reached 100 °C, the insertion angle of the thermal probe increased from 45 to 60 and 90, and the air velocity varied from 0, 1.5 to 2 m/s in the external circumstance, the heat removal effect of thermal probe was obviously higher than that of the non-air environment, as a high air velocity sensitive working fluid, with the increased of the contacted area with the airflow, the heat and mass transfer rate between the coal stockpile and outside circumstance accelerated, then the heat removal effect was evidently improved. Moreover, when the environment temperature of coal stockpile increased to150 °C, the heat removal effect of the thermal probe was better than that of 100 °C.

TABLE 4 - Comparison of heat removal effect in coal stockpile under
different insertion angle and air velocity.

Insertion angle /°	90°		60°		45°	
Air velocity / m/s	C_{max} /°C	CR / °C/h	C_{max}/°C	CR / °C/h	C_{max}/°C	CR / °C/h
2	16	0.368	27	0.66	22.3	0.64
1.5	10.4	0.329	24.7	0.58	16	0.59
0	9.3	0.269	22	0.558	10.8	0.47

Note: C_{max} is the maximum cooling temperature; CR is the average cooling ratio of thermal probe.

From the comparative analysis of Fig 5, it can be seen that as the insertion angle of thermal probe was 60 degrees and the air velocity was 2 m/s, with the increased of air speed, the heat removal temperature difference and efficiency of the thermal probe to each measurement point in the coal stockpile, which were obviously higher than that of the insertion angle of 45 and 90 degree, respectively. From Table 4, it could be seen that the heat removal efficiency of thermal probe in the coal stockpile was the highest at an inclination of 60° that under the condition of no air, the heat removal efficiency of 60 degree was higher than that of 90 and 45 by 107% and 11.6%. It increased 76% under 1.5 m/s and 79% and 3% under 2 m/s air velocity. Therefore, a certain inclination angle can help to transfer accumulative heat, compared with the vertical insertion of thermal probe into coal stockpile, the heat removal efficiency was highest as the angle between the condensation section of thermal probe and the air flow was 60 degree. Because of the air velocity enhanced the heat dissipation effect on the cooling section of the thermal probe, which speeds up the circulation of heat removal working medium inside the thermal probe. Meanwhile, there was no inclination angle to reduce the speed limitation and the inner limit of thermal probe, the equivalent thermal resistance of the thermal probe increased. When the air velocity was 1.5 m/s, the heat removal of thermal probe has no obvious increase. As the air velocity increased, the thermal resistance of the heat-condensing section of thermal probe decreased, the air-cooling effect was enhanced. The heat transfer medium could be more efficiently liquefied and returned to the evaporation section of thermal probe to reduce the increase of the internal resistance of the heat removal by the carried limitation. The effective phase absorption of working liquid in the evaporation section could reduce the evaporation section of thermal probe. Moreover, as the temperature reduces the influence of the flow obstruction of the heat transfer working gas, and further reduces the internal resistance of thermal probe. Therefore, when the air velocity reached 2 m/s, the thermal probe had a significant enhancement on heat removal effect in the coal stockpile. However, it can be concluded from Fig 6, with the distance between measurement points increased, the heat removal temperature difference and the efficiency in the coal stockpile shown an obvious decreasing trend, thus it was necessary to analyse the heat removal radius of thermal probe in the coal stockpile.

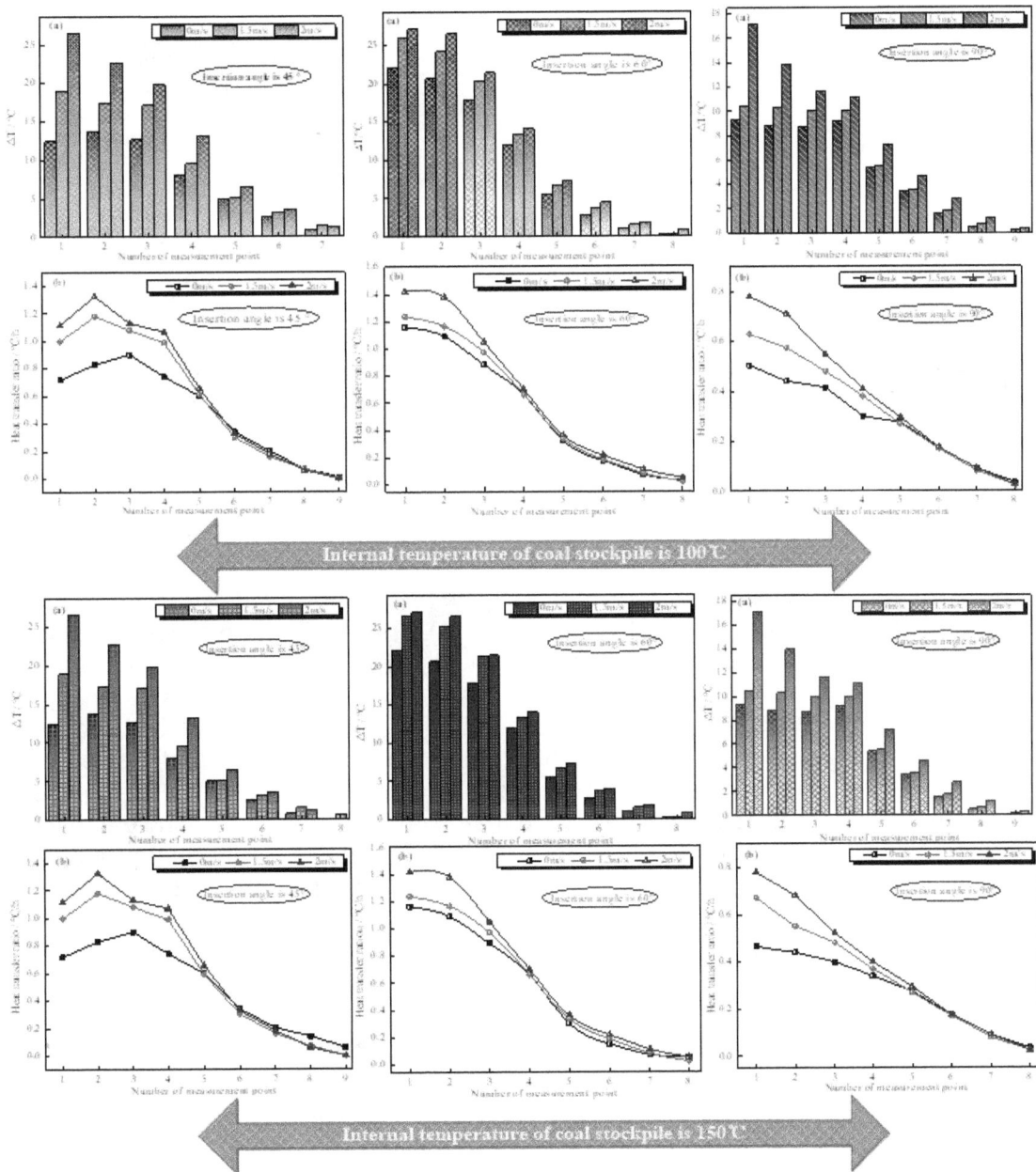

FIG 6 - Heat removal effect of thermal probe under different air velocity and insertion angle.

HEAT REMOVAL RADIUS AND VOLUME OF THERMAL PROBE IN COAL STOCKPILE

Heat removal radius of thermal probe under different conditions

The heat removal radius can be described by the biggest cooling areas as the thermal probe insertion into the coal stockpile at a high temperature circumstance. The efficiency of thermal probe for coal stockpile was different under the different experimental conditions, it can be calculated by Eq (1):

$$R = \frac{\lambda(t_1 - t_2)}{l}$$

(1)

where: t_1 and t_2 were the temperature on both sides of coal stockpile, λ is the average thermal conductivity of coal at corresponding temperature, l is the thickness of coal stockpile.

TABLE 5 - Heat removal radius of thermal probe under different arrangement without wind.

Heating temperature	Arrangement	Radius/ cm
100°C	1 : 2	45.1
	1 : 3	47
	1 : 4	47.68
150°C	1 : 2	43
	1 : 3	45.9
	1 : 4	46.5
	45°	45.9
	60°	44.7
	90°	41.3
	0 m/s	44.7
	1.5 m/s	47.6
	2.0 m/s	49.2

According to the arrangement of the thermal probes, the heat removal radius of thermal probe in the coal stockpile at 100 and 150 °C was calculated by Eq (1). The results were shown in Table 5. When the temperature of outside circumstance in the coal stockpile was constant, the heat removal radius of the thermal probe was proportional to the insertion scale, which explained the obviously results of the heat removal effect with the increase of the insertion scale of thermal probe condensation section. In order to analyse the constant temperature of the coal stockpile and the variation of the effective heat removal radius as the thermal probes were arranged differently, the temperature of the coal stockpile was set to 150 °C, it was found that as the insertion scale, angle and external air velocity were 1:4, 60 degree and 2.0 m/s, respectively. The radius of thermal probe in the coal stockpile was largest and the heat removal effect was more significantly. Most interestingly, when the outside circumstance in the coal stockpile raised from 100 to 150 °C, the heat removal radius was reduced, this phenomenon can be explained by the inefficient heat removal power of the experimental thermal probe.

Heat removal volume of thermal probe under different conditions.

Different arrangements of thermal probe in the coal stockpile was anther important parameters to measure its own heat removal effect. Therefore, it was significant importance to calculate the cumulative heat removal volume under different working conditions, it can be calculated by Eq (2):

$$Q = \frac{T_c - T_a}{R_h + R_c} t$$

(2)

where: Tc is the average temperature inside the coal stockpile, °C; Ta is the average temperature in experimental environment, °C; Rh is the average heat removal resistance between coal and the thermal probe evaporation section, °C/W; RC is the thermal probe condensation section and average thermal resistance of heat dissipation, °C/W; t is the time from the beginning of the "thermal probe-coal stack" system to reach equilibrium.

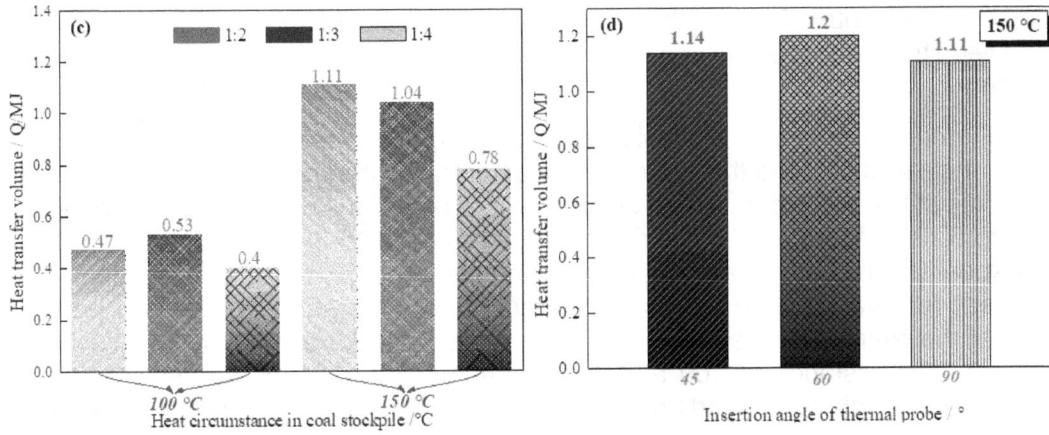

FIG. 7 - Heat removal volume of thermal probe under different insertion scale and angle with no air.

Especially, the calculation process of each parameter under different working conditions will not be described again, this paper only analysed the cumulative heat removal volume with different temperature conditions in the coal stockpile. The results were shown in Fig 7 and Fig 8, it can be seen that as the ambient temperature in coal stockpile was constant, the cumulative heat removal volume of thermal probe in the coal stockpile was proportional to the insertion scale. When the temperature environment in the coal stockpile rose to 150 °C, the cumulative heat removal volume of thermal probe was nearly doubled increased. Furthermore, from the comparison between Figure 7 and Figure 8 shows that as the insertion angle of thermal probe was 60°, and the external ambient air velocity was 2 m/s, the cumulative heat removal volume of thermal probe in the coal stockpile was significantly higher than that the effect under other working conditions. Similarly, when the temperature in the coal stockpile reached at 150 °C, the cumulative heat removal volume under the set working conditions still increases exponentially.

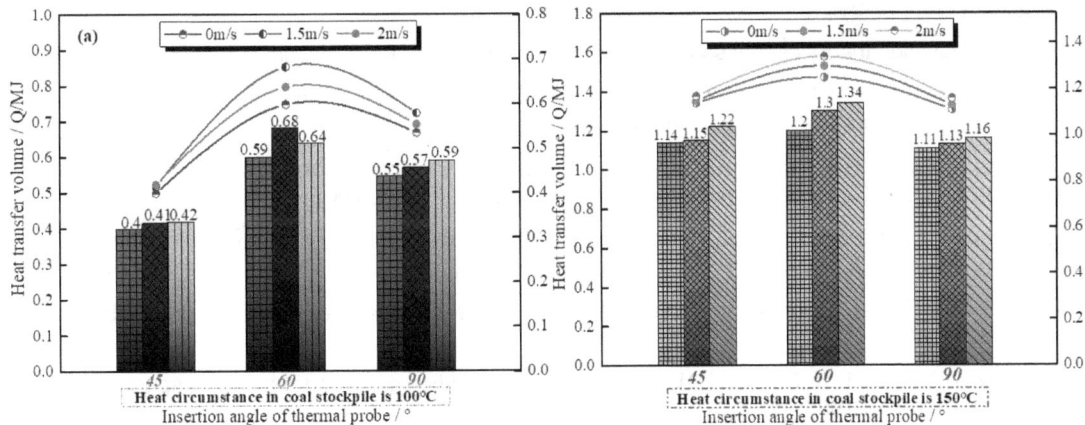

FIG. 8 - Heat removal volume of thermal probe under different air velocity and circumstance in the coal stockpile.

CONCLUSIONS

In this paper, we conducted a simulation experiment to research the heat removal by thermal probe in coal stockpile as the internal temperature were 100 and 150 °C under different experimental conditions. The experimental results were shown as follows:

(1) From the analysis of temperature curves, it can be seen that as the temperature of coal stockpile was stable, the heating time was 80 hrs, and the cooling delay time of each measurement point was about 1hr. Compared with the temperature distribution before inserting the thermal probe, the temperature of each measuring point in coal stockpile decreased considerably, it indicated that the thermal probe had obvious effect on the heat removal in coal stockpile. Meanwhile, the closer the distance between the thermal probe and the measurement points, the greater the heat removal effect of thermal probe.

(2) When the temperature in coal stockpile was constant, the insertion scale of thermal probe was 1:4, insertion angle was 60°, and the air velocity of external environment was 2 m/s, the heat removal effect and radius of thermal probe in coal stockpile increased significantly. Such as the scale of drop in temperature of designated point in the coal storage heap are 46.3%, 10.8%, and 32.8% that under the above experimental conditions, respectively, and the average radius of heat removal is 5.79 m (maximum heat removal radius), the axial difference in temperature is 1.2 °C. Furthermore, we concluded that the overall heat removal efficiency of thermal probe increased by 79% and 93% in the temperature circumstance of 100 and 150 °C, respectively, which further proved that it was helpful to select reasonable arranged parameters for the prevention and control of coal fire hazards in coal stockpile. This paper provided a new technical approach and theoretical basis for the prevention and control of coal spontaneous combustion for large-scale coal storage heap that under a particular airflow condition.

ACKNOWLEDGMENT

This research was financially supported by the following funds: National Key Research and Development Plan (2018YFC0808104).

REFERENCE

1. Justyna Ciesielczuk. Coal Mining and Combustion in the Coal Waste Dumps of Poland. Coal and Peat Fires: A Global Perspective 2012, 464−473.

2. Kuenzer C, Zhang J, Tetzlaff A, et al. Uncontrolled coal fires and their environmental impacts: Investigating two arid mining regions in north-central China. Applied Geography 2007; 27: 42-62.

3. Stracher, G.B., Taylor, T.P. Coal fires burning out of control around the world: thermodynamic recipe for environmental catastrophe. In: Stracher, G.B. (Ed.), Coal fires burning around the world, a global catastrophe. Int. J. Coal Geol 2004; 59: 7-17.

4. Song Z, Kuenzer C. Coal fires in China over the last decade: A comprehensive review. International Journal of Coal Geolog 2014; 133: 72-99.

5. Xu Jingcai. Determination theory of coal spontaneous combustion zone. Beijing: Coal Industry Press 2001.

6. Xiao Y, Wang Z P, Li M A, et al. Research on correspondence relationship between coal spontaneous combustion index gas and feature temperature. Coal Science & Technology 2008.

7. Brooks K, Bradshaw S, Glasser D. Spontaneous combustion of coal stockpiles-an unusual chemical reaction engineering problem. Chemical Engineering Science 1988; 43: 2139-2145.

8. Fierro V, Miranda J L, Romero C, et al. Prevention of spontaneous combustion in coal stockpiles. Fuel Processing Technology 1999; 59: 23-34.

9. Zhang R X, Xie H P. Experimental study of the propensity of coal stockpiles to spontaneous combustion. Journal of China Coal Society 2001.

10. Krishnaswamy S, Agarwal P K, Gunn R D. Low-temperature oxidation of coal. 3. Modelling spontaneous combustion in coal stockpiles. Fuel 1996; 75: 353-362.

11. Brooks K, Svanas N, Glasser D. Evaluating the risk of spontaneous combustion in coal stockpiles. Fuel 1988; 67: 651-656.

12. Wang Q, Song G, Sun J. Spontaneous Combustion Prediction of Coal by C80 and ARC Techniques. Energy & Fuels 2009; 23: 4871-4876.

13. Qi X, Wang D, Xue H, et al. Oxidation and Self-Reaction of Carboxyl Groups During Coal Spontaneous Combustion. Spectroscopy Letters 2015; 48: 173-178.

14. Wen H, Yu Z, Deng J, et al. Spontaneous ignition characteristics of coal in a large-scale furnace: An experimental and numerical investigation. Applied Thermal Engineering 2017; 114: 583-592.

15. Wullschleger S D, Childs K W, King A W, et al. A model of heat transfer in sapwood and implications for sap flux density measurements using thermal dissipation probes. Tree Physiology 2011; 31: 669-679.

16. Ma L, Li B, Deng J, Li Z, Zhang Y. Deep heat transfer technology using thermal probe in high temperature region of coal storage pile (gangue hill) spontaneous combustion. Science and Technology Review 2014; 32: 76-80.

17. Senthil Kumar R, Vaidyanathan S, Sivaraman B. Effect of copper nanofluid in aqueous solution of long chain alcohols in the performance of heat pipes. Heat and Mass Transfer 2015; 51: 181-193.

18. Deng J, Xiao Y, Li Q, et al. Experimental studies of spontaneous combustion and anaerobic cooling of coal. Fuel 2015; 157: 261-269.

19. Kole M, Dey T K. Thermal performance of screen mesh wick heat pipes using water-based copper nanofluids. Applied Thermal Engineering 2013; 50: 763-770.

20. Salinger AG, Aris R, Derby J J. Modelling of Spontaneous Ignition of Coal Stockpiles. AICh E Journal 1994; 40: 991-1004.

21. Senthilkumar R, Vaidyanathan S, Sivaraman B. Effect of copper nano fluid in aqueous solution of long chain alcohols in the performance of heat pipes. Heat and Mass Transfer 2015; 51: 181-193.

Detection and Control of Spontaneous Combustion, and Diesel Emissions Control and Measurement

Investigation of combustion properties and burning behaviour of Australian Hunter Valley coal dust

M J A Al-Zuraiji[1], J Zanganeh[2] and B Moghtaderi[3]

1. Research Associate, Discipline of Chemical Engineering, University of Newcastle, Callaghan, NSW 2308, Email: c3176251@uon.edu.au
2. Project Manager (VAMSP), Discipline of Chemical Engineering, University of Newcastle, Callaghan, NSW 2308, Email: Jafar.Zanganeh@newcastle.edu.au
3. Professor, Discipline of Chemical Engineering, University of Newcastle, Callaghan, NSW 2308, Email: Behdad.Moghtaderi@newcastle.edu.au

ABSTRACT

The risk of coal dust fire and explosion is a major safety concern in underground coal mines. A better understanding of the coal dust thermal properties which drive the fire and explosion phenomena would greatly assist in adopting an appropriate and effective protection approach to avoid such fires and explosions. The aim of this work is to address the thermal and explosion characteristics of coal dust, such as minimum ignition energy (MIE), minimum ignition temperature (MIT) and minimum explosive concentration (MEC). Each of these parameters are considered to be the minimum requirement to initiate the combustion process. The heat release rate and burning behaviour of the coal dust samples are also examined in detail. The experimental work was conducted on two typical coal dust samples (Samples A represent the run of mine coal dust and B represent a fine coal dust collected before the washing process) collected from an underground coal mine in the Hunter Valley region of Australia. The thermal and explosive characteristics of the coal dusts were examined by employing a variety of relevant apparatus such as cone calorimeter, hot plate furnace, Hartman glass tube and 20 L explosion chamber.

The results of this study indicated a significant discrepancy between the fire and explosion characteristics of these two samples. The differences in fire and explosion properties examined are related to the thermo-physical and chemical properties of the coal dust samples. The dust layer MIT of Sample A was 270 °C, which is lower than the dust layer MIT for Sample B by about 110 °C. Moreover, Sample A required only 251 mJ to ignite, whereas Sample B required 740 mJ. Both Sample A and Sample B exploded using 1 kJ of ignition energy, however, the maximum pressure rise was achieved at 450 g.m^{-3}. The heat release rate (HRR) experiments for coal dust samples, exposed to 50 kWm^{-2} heat flux, showed that the peak heat release rate for Sample A was 93.1 kWm^{-2} and 83.1 kWm^{-2} for Sample B. The oxidation reaction consumed 63.2 g and 76.0 g of oxygen during the combustion of samples A and B, respectively.

INTRODUCTION

Accidental fires and explosions caused by combustible dusts are known to occur in many chemical plant and coal mining industries. According to NFPA 654 (2005) the combustible dusts are those particles which have less than a 420 μm diameter and are able to cause an explosion or fire under certain conditions in the presence of an oxidiser. To take proper action to eliminate the risk of fire and explosion, a better understanding of the coal dust fire and explosion characteristics is essential. In a typical coal mine the coal dust particles usually are present in different forms such as dust layers and dust clouds. Therefore, accidental fire and explosion caused by either of these coal dust forms can, jeopardise lives and damage properties.

The deposit of coal dust on hot surfaces represents a potential source of ignition and fire (Palmer and Tonkin, 1957). Self-heating and an exothermic reaction can also contribute to coal dust fire initiation by increasing the coal dust bed temperature to the auto ignition temperature of the coal dust. The influence of the dust layer depth, environmental conditions and coal properties on the minimum ignition temperature (MIT) of coal dust have been investigated by (Ajrash et al., 2016a; Bowes and S. E.Townshend, 1962; El-Sayed and Abdel-Latif, 2000; Litton, 1992). These researchers found that as the layer thickness is increased by 1.5 cm, the MIT of the dust layer is reduced by about 30 °C, and that the MIT of the dust layer is reduced by about 75 °C as the coal dust size fraction increases from the range of 0-74 μm to 74-125 μm.

The risk of explosion due to the exposure of a coal dust cloud to an ignition source is another potential major safety risk for the mining industry. The pressure wave and turbulence generated from the explosion of a coal dust cloud can lift the deposited coal dust off the surfaces, thereby increase the fire and explosion severity. The outcomes of past studies show that the particle size, the physio-chemical composition of coal dust and the ignition energy are the main contributors to coal dust cloud explosion severity (Cashdollar, 2000; Proust et al., 2007; Torrent et al., 1988; Yuan et al., 2012). Cashdollar (2000) indicated that the minimum explosion concentration of coal dust could be increased three fold as the particle diameters increase from 20 μm to 2000 μm, for coal dust particles of 50 μm mean diameter. The amount of heat release from the combustion of coal dust layer can assist in measuring the potential energy of the coal dust layer as an ignition source. The literature shows that a number of studies have used cone calorimeter to determine the amount of heat release from wool, cotton and polyethylene, however, not much studies have been carried out to examine the heat release from the coal dust combustion (Holbrow et al., 2000; Wachowicz, 2008). Knowing the contribution percentage of each of these variables would assist in determining the explosion characteristics, such as the pressure rise and deflagration index.

The ignition energy which is usually manifested as thermal or electrical energy is the key element in coal dust fires and explosions. Without the ignition energy the fire and explosion would not occur, even with the right quantity of fuel and oxidiser agent. The experimental and theoretical studies carried out in the past, Ajrash (2016b), have indicated that the probability of fire and explosion is directly proportional to the ignition energy. Furthermore, increasing the ignition energy causes the pressure rise to increase accordingly. The minimum ignition energy (MIE) for a dust cloud usually is measured via the Hartmann glass tube apparatus. This apparatus uses an electrical spark with different energy levels to ignite the dust cloud. The previous researches by (Addai et al., 2016a, 2016b; Janes et al., 2008; Olsen et al., 2015) have shown that the high volatile matter content in the coal dust significantly reduces the minimum ignition energy (MIE) of the coal dust. In a study conducted by Torrent (1988) the MIE of coal dust was reduced by about 15 times due to an increase in the volatile matter content of about 4%. Moreover, the lower MIE is related to the stoichiometric concentration of the dust cloud.

METHODOLOGY AND TECHNIQUE

The particle size distribution of the coal dust samples were measured before and after the sieving process. The sieving was carried out in a standard sieve with the mesh size of 125 μm. Table 1 presents the coal dust samples sieving results as a function of mean diameter.

Table 1: D10, D50 and D90 of coal dust samples before and after sieving

	Coal Sample A (Run of mine)		Coal Sample B (Fresh fine coal dust)	
	Before sieving	After sieving	Before sieving	After sieving
$D_{10}(\mu m)$	12.74	2.32	73.30	8.52
$D_{50}(\mu m)$	138.94	25.38	201.76	68.87
$D_{90}(\mu m)$	425.67	107.96	460.74	178.17

D_{10}, or particle size such that 10% of the distribution is less than this size.

D_{50}, particle size such that 50% of the distribution is less than this size. This could also be considered as the median diameter of coal dust particles.

D_{90}, or particle size such that 90% of the distribution is less than this size.

Table 2 shows the ultimate and proximate analysis results for Sample A and Sample B (sieved samples).

Table 2: Proximate analysis of Sample A and Sample B dusts

				Dust Samples	
	Parameter	Unit	Method	Sample A	Sample B
Proximate analysis	Moisture	%	ISO 11722	1.9	1.6
	Ash	%	ISO 1171	16.7	12.1
	Volatile matter	%	ISO 562	34.4	37.5
	Fixed Carbon	%	ISO 17246	47.0	48.8

MIT of coal dust layer

The experimental procedure to determine the MIT of the coal dust layers was carried out in accordance with ASTM E2021 (ASTM E2021, 2013). The experimental assessment was conducted on a hot surface plate furnace in controlled environmental conditions. The apparatus consist of a 200 mm diameter heating plate, controlling block, data logger, a series of thermocouples, coal dust ring holders and a computer as seen in Figure 1.

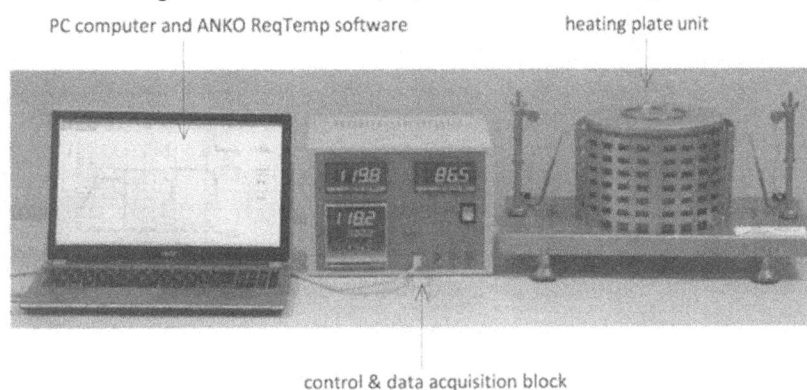

Figure 1: MIT of dust layer apparatus.

Explosion chamber

The explosion properties of coal dust Samples A and B were determined using an explosion chamber apparatus. The apparatus consists of a 20 L spherical explosion chamber, coal dust dispersion chamber, trigger and control panel and a computer to run the program, (Figure 2). The experimental work was carried out in accordance with ASTM E1226 (ASTM Standard, 2014).

MIE of coal dust cloud

A Hartmann glass tube apparatus was employed to measure the MIE of the coal dust cloud. The system consists of a glass tube (230 mm height), dust disperser, data logger and electrical ignitors (Figure 3). The experimental work was carried out according to ASTM E-2019 standard.

Heat release rate and burning behaviour

The apparatus commonly used to measure the rate of heat release is known as cone calorimeter (Figure 4). This fire testing device works based on the principle of oxygen consumption during combustion. Apart from the heat release rate (HRR), this device provides the time lapse to ignition, gas analyses and other parameters related to the burning properties of the tested material. The experimental work on both coal dust samples was carried out in accordance with ASTM E1354-17 standard (American and Standard, 2004).

RESULTS AND DISCUSSION

MIT of coal dust layer

Based on the ASTM E2021 standard, the ignition of the coal dust layer on the hot plate is considered to have occurred once glowing and surface cracking signs appear (ASTM E2021, 2013). However, the minimum ignition temperature (MIT) is also achieved once the sample temperature rises 20 °C above the plate temperature for particle sizes below 212 μm (Ajrash et al., 2016b).

Figure 2: 20 L explosion apparatus.

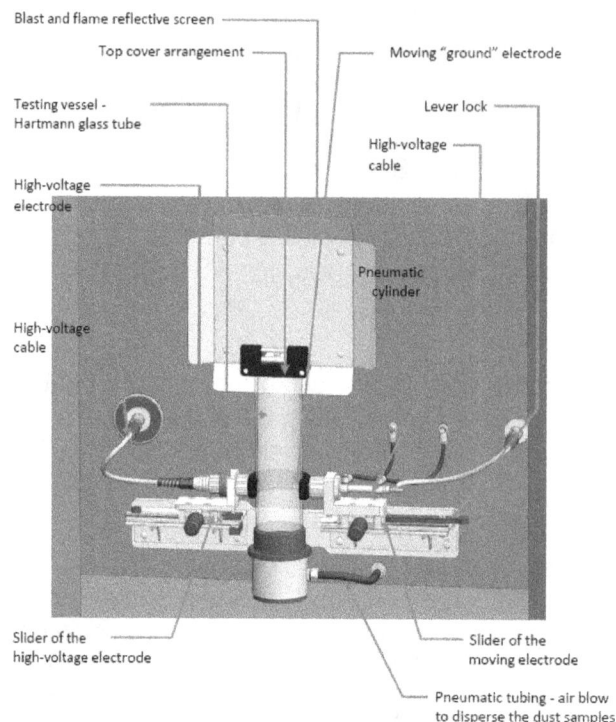

Figure 3: Hartmann glass tube apparatus.

Figure 4: Cone calorimeter.

The experimental results for MIT of coal dust samples A and B indicate that the particle size has a significant impact on the MIT. The MIT for Sample A with average coal dust particle size of 25.38 μm was about 270 °C, while the MIT for Sample B, with mean coal particle size of 68.87 μm, was 380 °C. Moreover, the particle size was observed to have a significant impact on the temperature-time trend and ignition behaviour. Figure 5 shows the temperature-time profile of Sample A for ignition and no-ignition tests. In no-ignition tests, the sample temperature did not reach the plate temperature. The temperature started to increase after the loading of the sample and reached 245 °C after 415 seconds. The temperature then gradually reduced and stabilised at 228 °C. The rise in temperature can be attributed to the self-heating of the particles and the partial combustion process. However, the delivered thermal energy and the generated heat from the partial combustion process was not sufficient to achieve ignition. By increasing the plate temperature to 270 °C, Sample A successfully ignited. The sample temperature profile shows two peaks referring to the ignition occurrence. In the first peak, the sample temperature reached just 20 °C over the plate temperature, which is considered as ignition according to IEC 61241(IEC 61241-2-1, 1994). The sample temperature in the second peak reached 115 °C over the plate temperature. The time and the length of the second ignition, indicated by the temperature time profile, are mostly attributed to the combustion of the coal dust that is positioned further away from the hot plate, where a longer time is required for the coal dust particles to retain the heat until it reaches the MIT.

Figure 6 shows the temperature profile for ignited coal Sample B. The temperature profile includes one peak of ignition only at which the sample temperature reaches to about 20 °C over the plate temperature. The ignition of both A and B coal samples was accompanied by smouldering and cracking on the surface of the samples (Figure 7) which is in agreement with the standards.

Coal dust explosion properties

Coal dust explosion tests for the two coal dust samples (i.e. Sample A and Sample B) was carried out in a 20 L explosion vessel.

To determine the explosive properties of coal dust samples (i.e. A and B) five coal dust concentrations of 300, 450, 600, 750 and 1000 g.m^{-3} were examined. The ignition energy source was a 1 kJ chemical ignitor. The maximum pressure rise was achieved at coal dust concertation in the range of 450 g.m^{-3} - 600 g.m^{-3} coal dust concentration for both Samples A and B.

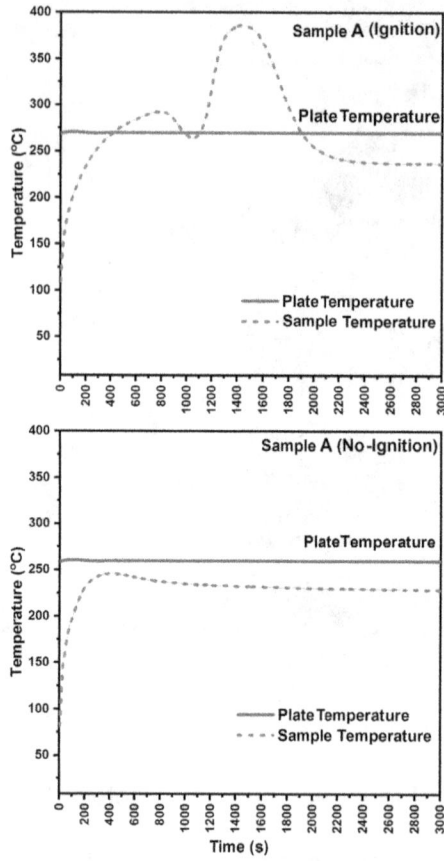

Figure 5: Temperature profile for ignited and non-ignited of Sample A.

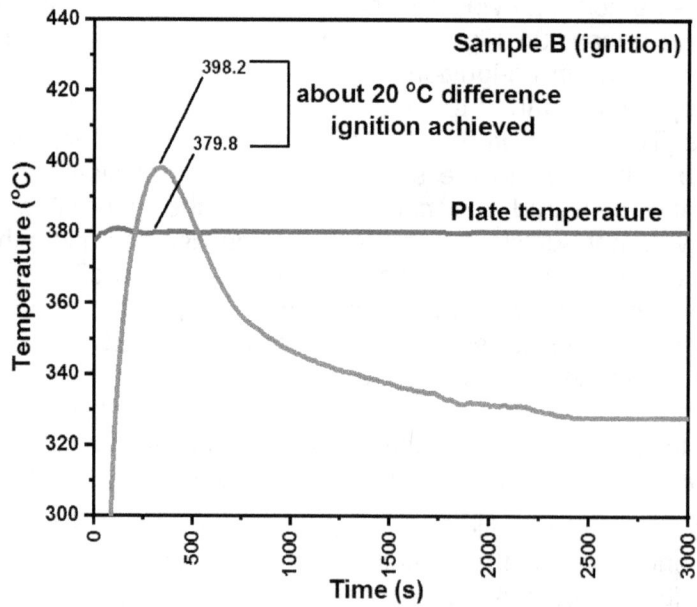

Figure 6: Temperature profile for ignited Sample B.

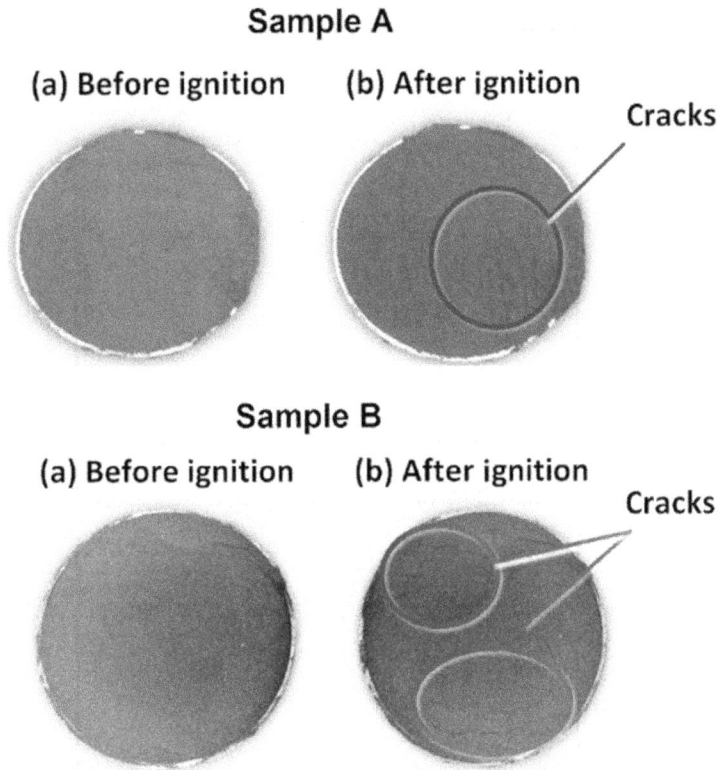

Figure 7: Footage of Samples A and B before and after ignition on hot plate.

The maximum pressure rise of Sample A was 5.91 bar, while the maximum pressure rise of Sample B was 5.55 bar. The pressure rise for Sample A was higher than that of Sample B for all of the examined coal dust concentrations. The pressure rise for both A and B samples started to decline steadily as the coal dust concentration increased from 450 to 1000 $g.m^{-3}$. The reduction in the pressure rise as the coal dust concentrations increased can be attributed to the fact that there was an excess amount of fuel. As a result, the burning rate of the coal dust reduced causing a reduction in the combustion products. These results are in good agreement with the study by Cashdollar (2000) where the maximum pressure rise was at 600 $g.m^{-3}$ then it reduced as the coal dust concentration increased (Figure 8).

The deflagration index (Kst) is defined as the normalised explosion pressure rise and can be mathematically calculated according to Equation 1.

$$K_{st} = \left(\frac{dp}{dt}\right)_{max} . V^{\left(\frac{1}{3}\right)}$$

(1)

where $(dp/dt)_{max}$ is the maximum rate of pressure rise, and V is the volume of the reaction vessel.

Table 3 shows the explosion class classification of fuel according to its deflagration index.

Table 3: Explosion class

Explosion class	K_{st} (bar.m.s^{-1})	Explosion severity
St 0	0	non explosive
St 1	1–200	weak
St 2	201–300	strong
St 3	>300	extreme

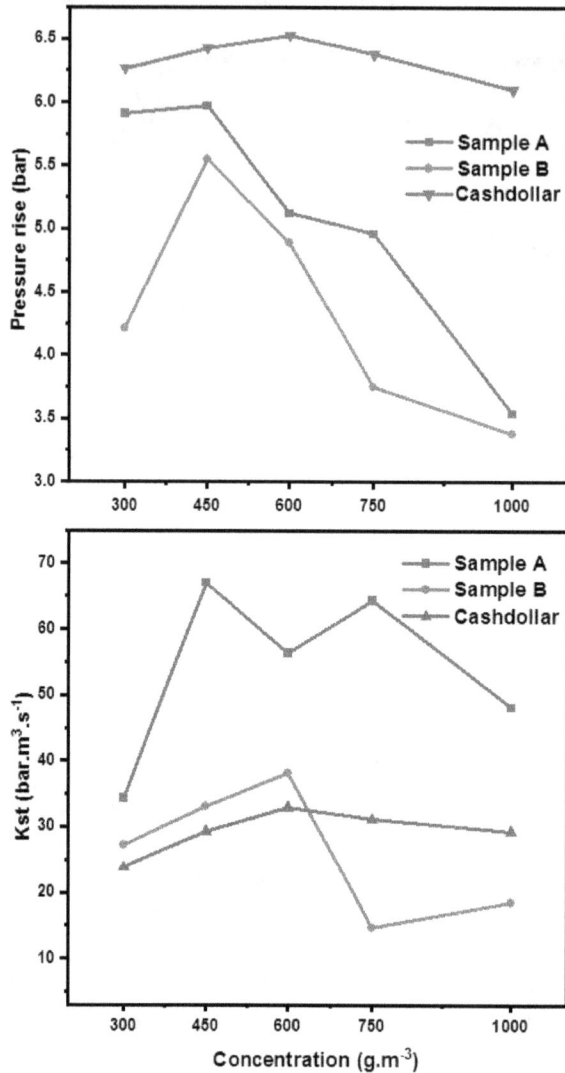

Figure 8: Explosion pressure rise and deflagration index for coal dust
Samples A, B and previous literature (Cashdollar, 2000).

According to the results acquired from the 20 L explosion chamber and the explosion classifications (Table 3), both Samples A and B fall in Class 1 (ST1) explosion category, which is the lowest class of explosion.

MIE of coal dust Samples A and B

Coal dust samples (Sample A and B) at different concentrations were subjected to varying ignition energies to determine the MIE using a Hartmann apparatus. The mass of the weighed samples and the concentrations are presented in Table 4.

Table 4: Sample concentrations corresponding to weight of samples

Mass of weighed sample (mg)	180	360	540	900	1200	1800	2400
Concentration (g.m^{-3})	150	300	450	750	1000	1500	2000

MIE for coal dust Sample A

Figure 9 presents the MIE results for coal dust Sample A. The solid symbols refer to the tests at which the ignition occurred and the hollow symbols refer to the tests where no ignition was observed. Numbers labelled beside each symbol represent the total number of successive repeat tests for each concentration. As per the MIE range description presented earlier, the MIE for the coal dust Sample A was between 100-300 mJ (100 mJ < MIE < 300 mJ).

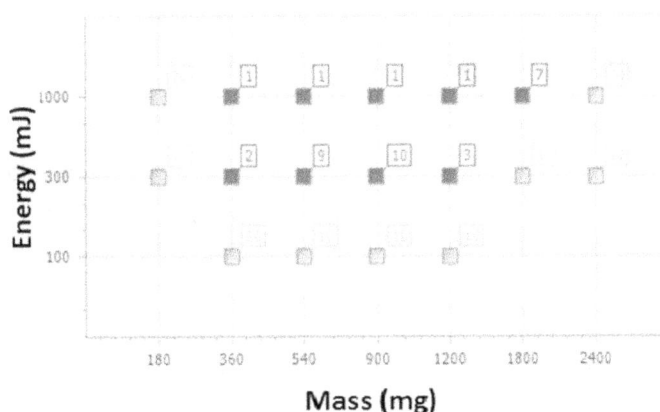

Figure 9: MIE measurement of Sample A (Solid symbols refer to ignition, hollow symbols refer to no-ignition).

A single value for the MIE can be estimated using the probability of ignition as stated below (CEN 2003):

$$\log MIE = \log E_2 - I[E_2].\frac{(\log E_2 - \log E_1)}{(NI+I).[E_2]+1} \tag{2}$$

where I [E_2] is the number of tests with successful ignition at E_2 and (NI + I) [E_2] stands for the total number of tests at the energy level of E_2. The values obtained using the above formula will have a maximum uncertainty of 1 mJ.

Using the calculation suggested by Equation 2, a single value for the MIE of the coal Sample A is estimated to be 251 mJ.

MIE for coal dust Sample B

Figure 10 shows the MIE results for the coal dust Sample B. As observed, the minimum required energy to ignite the coal dust is significantly higher for the coal dust Sample B when compared with Sample A. The MIE for the coal Sample B was 300 mJ < MIE < 1000 mJ which is approximately three times higher than MIE for Sample A.

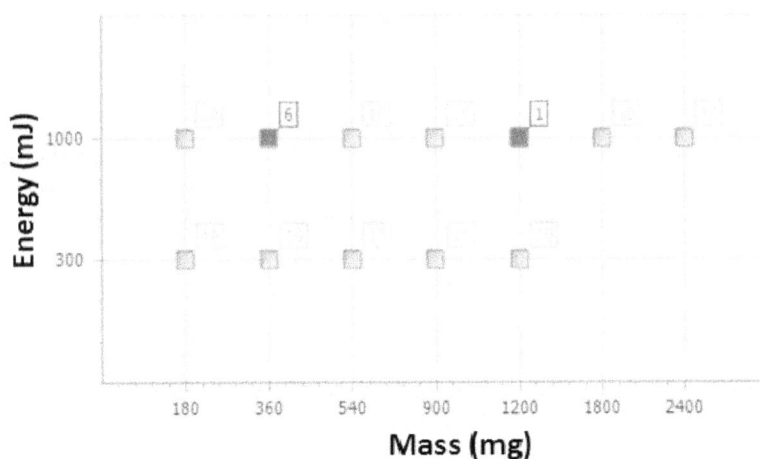

Figure 10: MIE measurement of Sample B (Solid symbols refer to ignition, hollow symbols refer to no-ignition).

Using the calculation suggested by Equation 2, a single value for the MIE of the coal dust Sample B is estimated at about 740 mJ, see Table 5.

Testing for minimum ignition energy gives guidance as to whether ignition by electrostatic discharge from plant, personnel or process conditions is likely to occur in practice. Sample B has a much higher minimum ignition energy (740 mJ) than Sample A (251 mJ) indicating that its dust clouds are less sensitive to electrical spark ignition.

Table 5: Comparison between the MIE of this study and existing relevant literature (Addai et al., 2016b)

	MIE (mJ)	D50 (µm)	Volatile matter (%)
Sample A	251	25.38	34.4
Sample B	740	68.87	37.5
Brown coal (Addai et al.,2016)	18	95	89.5

Figure 11 shows the moment of coal dust ignition for coal dust Sample B captured using a NIKON SLR camera. This test included a coal dust concentration of 1000 $g.m^{-3}$ which was ignited by an ignition energy of 1000 mJ.

Figure 11: Flame propagation for 1000 $g.m^{-3}$ Sample B ignited with 1000 mJ energy.

The flame travelled the full distance of 230 mm of the Hartmann's tube in 401 ms; yielding an upward flame propagation velocity of 0.6 $m.s^{-1}$.

Heat release rate and burning behaviour

The results concerning the heat release rate and burning behaviour of the two coal dust samples indicated that Sample A started burning well before Sample B combustion started. The visible flame appeared after 15 seconds and 49 seconds from applying the radiant heat on Sample A and Sample B, respectively. The visible flame then gradually reduced until it died out after 1575 and 1849 seconds for Sample A and Sample B, respectively. The heat release rate (HRR) profile indicates that Sample A burned intensively at the beginning of the combustion process and reached the highest HRR value at about 50 seconds from the combustion initiation. The HRR curve plateaued until about 1220 seconds, then gradually reduced and died out (Figure 12 (a)). However, for Sample B the HRR peak occurred at about 1841 seconds before the flame died out. This behaviour is in line with the oxygen consumption profile presented in Figure 12 (c), where the higher oxygen consumption improved the combustion efficiency leading to a higher HRR. Figure 12 (b) shows the mass loss associated with Samples A and B during the combustion process.

Figure 12 (d) and (e) show the CO_2 and CO production profiles for both coal dust samples. The results indicate that the maximum amount of CO_2 is produced at the beginning of the burning process. This can be attributed to the partial incomplete combustion in the lower layers of the coal dust. During the combustion process heat penetrates into the lower layers of the bed burning surface. This heat is enough to initiate the combustion of the lower layers, however, it is not sufficient to achieve complete combustion leading to generation of CO_2 and other gases. Once the coal bed layers reached a uniform temperature the CO_2 production was gradually reduced. The combustion process and production of CO and CO_2 is also controlled by the oxygen availability at the combustion zone, particularly at lower layers. The coal particles which received more oxygen burned more efficiently and produced less CO, whereas the shortage of oxygen at the combustion location drove the combustion to proceed inefficiently, leading to more CO production. The coal dust bed compaction also showed a significant impact on the combustion process by letting more oxygen penetrates into the bed, which allowed the volatiles to be liberated and reach the surface quicker. Figure 13 shows a footage of Sample B burning captured using a NIKON SLR camera.

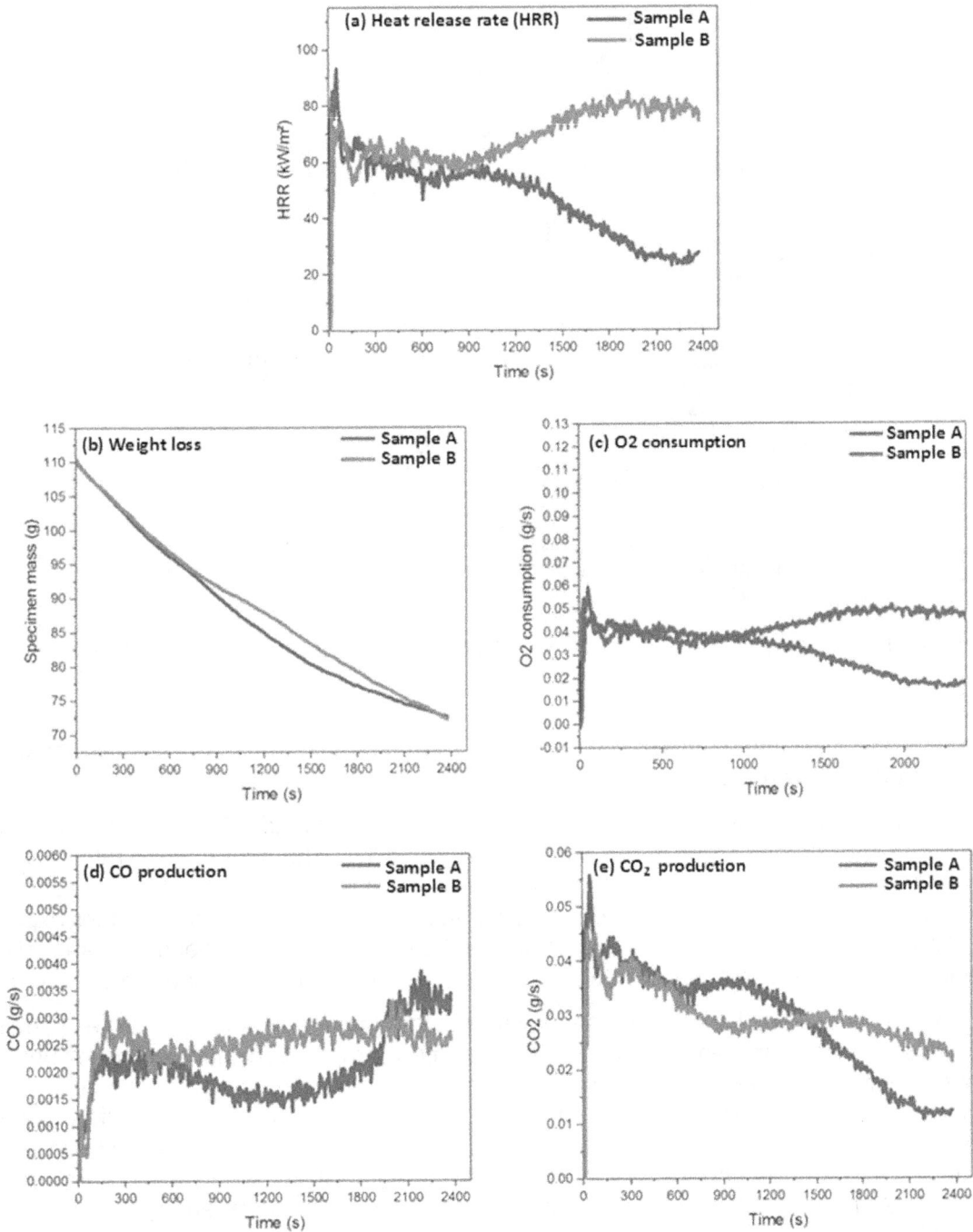

Figure 12: burning parameters of Sample A and Sample B using cone calorimeter.

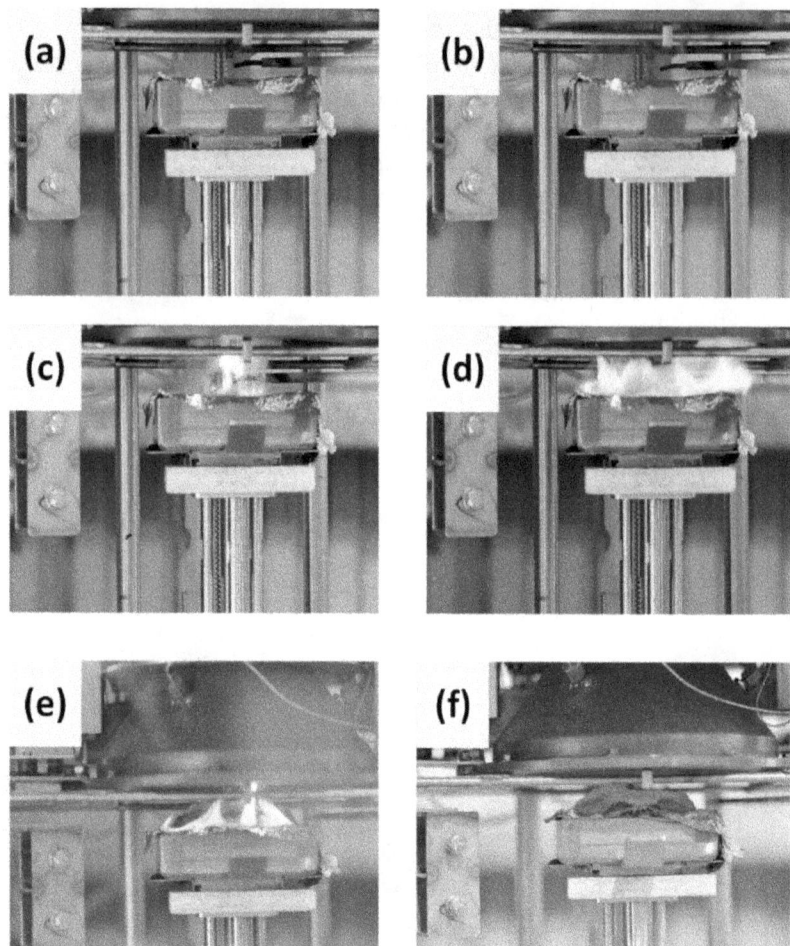

Figure 13: Footage of Sample B burning under 50 kW/m^2 radiant heat exposure.

CONCLUSION

This research examined the combustion properties and burning behaviour of two coal dust samples, collected from a typical mine in the Hunter Valley Australia.

The optimum concentration for explosion of both samples was around 450 g.m^{-3}. The explosion index of both coal dust samples falls in the 'weak explosion' class. The heat release rate of Sample B was greater than that of Sample A, which was determined to be due to the higher percentage of fixed carbon and volatile matter in Sample B. Finally, finer coal particles produce higher explosion severity and require less energy to ignite. However, the heat released from coal dust combustion is more dependent on the percentage of volatile mater and fixed carbon rather than the particle size.

The outcomes of this work have been employed by the mine development professionals to determine the safe installation and operation of a new coal preparation plant at the mining site. Experimental results indicated that the coal dust Samples A and B could be auto ignited when deposited on a hot surface at temperatures of 270 °C and 380 °C, respectively. So the maximum surface temperature of the electrical and mechanical equipment installed in this hazardous area has to be designed to operate below 195 °C for sample A and 305 °C for Sample B according to AS/NZS-61241(2005). In addition, any potential ignition source with spark energy of 251 mJ and 740 mJ, must be eliminated from this area.

ACKNOWLEDGMENT

The authors wish to acknowledge the financial support provided to them by Australian Coal Association and Low Emission Technology (ACALET), the Australian Department of Industry, and the University of Newcastle (Australia).

REFERENCES

Addai, E.K., Gabel, D., Krause, U., 2016a. Experimental investigation on the minimum ignition temperature of hybrid mixtures of dusts and gases or solvents. J. Hazard. Mater. 301, 314–326.

Addai, E.K., Gabel, D., Krause, U., 2016b. Experimental investigations of the minimum ignition energy and the minimum ignition temperature of inert and combustible dust cloud mixtures. J. Hazard. Mater. 307, 302–311.

Ajrash, M.J., Zanganeh, J., Moghtaderi, B., 2016a. Experimental investigation of the minimum auto-ignition temperature (MAIT) of the coal dust layer in a hot and humid environment. Fire Saf. J. 82, 12–22.

Ajrash, M.J., Zanganeh, J., Moghtaderi, B., 2016b. Experimental investigation of the minimum auto-ignition temperature (MAIT) of the coal dust layer in a hot and humid environment. Fire Saf. J. 82, 12–22.

American, A., Standard, N., 2004. Standard test method for heat and visible smoke release rates for materials and products using an oxygen consumption calorimeter 1–22.

AS/NZS-61241, 2005. Australian / New Zealand Standard ™ Electrical apparatus for use in the presence of combustible dust : General requirements.

ASTM E2021, 2013. ASTM E2021: Standard test method for hot-surface ignition temperature of dust layers.

ASTM Standard, 2014. Standard test method for pressure and rate of pressure rise for combustible dusts E 1226. West Conshohocke.

Bowes and S. E.Townshend, 1962. Ignition of combustible dusts on hot surfaces. APPL. PHYS 13.

Cashdollar, K.L., 2000. Overview of dust explosibility characteristics. J. Loss Prev. Process Ind. 13, 183–199.

El-Sayed, Abdel-Latif, A.M., 2000. Smoldering combustion of dust layer on hot surface. J. Loss Prev. Process Ind. 13, 509–517.

Holbrow, P., Hawksworth, S.J., Tyldesley, A., 2000. Thermal radiation from vented dust explosions 13, 467–476.

IEC 61241-2-1, 1994. Methods for determining the minimum ignition temperatures of dust. Geneva, Switzerland,, Switzerland,.

Janes, A., Chaineaux, J., Carson, D., Lore, P.A. Le, 2008. Mike 3 versus Hartmann apparatus: comparison of measured minimum ignition energy (MIE) 152, 32–39.

Litton, C.D., 1992. Ignition of combustible dust layers on a hot surface, in: Fire Safety Science. pp. 187–196.

Olsen, W., Arntzen, B.J., Eckhoff, R.K., 2015. Electrostatic dust explosion hazards e towards a < 1 mJ synchronized-spark generator for determination of MIEs of ignition sensitive transient dust clouds. J. Electrostat. 74, 66–72.

Palmer, K.N., Tonkin, P.S., 1957. The ignition of dust layers on a hot surface. Combust. Flame 1, 14–18.

Proust, C., Accorsi, A., Dupont, L., 2007. Measuring the violence of dust explosions with the "20l sphere" and with the standard "ISO 1m3 vessel." J. Loss Prev. Process Ind. 20, 599–606.

Torrent, J.G., Armada, I.S., Pedreira, R. a., 1988. A correlation between composition and explosibility index for coal dust. Fuel 67, 1629–1632.

Wachowicz, J., 2008. Analysis of underground fires in Polish hard coal mines. J. China Univ. Min. Technol. 18, 332–336.

Yuan, J., Huang, W., Ji, H., Kuai, N., Wu, Y., 2012. Experimental investigation of dust MEC measurement. Powder Technol. 217, 245–251.

Mechanisms for spontaneous combustion events in low intrinsic reactivity coals and carbonaceous shales

J Theiler[1] and B B Beamish[2]

1.(AAusIMM) Senior Mining Engineer, CB3 Mine Services Pty Ltd, Darra Qld 4076.
    Email: j.theiler@cb3minesevices.com
2.(MAusIMM CP (Min)) Managing Director, B3 Mining Services Pty Ltd, Kenmore Qld 4069.
    Email: basil@b3miningservices.com

ABSTRACT

The R_{70} self-heating rate index provides a measure of the intrinsic reactivity to oxygen of organic carbon in coal and shales at temperatures between 40-70°C. It has been routinely used for nearly 40 years by the coal industry to rate spontaneous combustion propensity for hazard assessment. Low propensity, and hence low intrinsic reactivity, is assigned to R_{70} values less than 0.5°C/h. Despite this rating, spontaneous combustion events have occurred at mines that fit this low propensity category. Adiabatic oven incubation testing has shown that under normal mine conditions at ambient temperatures, self-heating leading to thermal runaway is not possible for coals or carbonaceous shales with this low intrinsic reactivity. However, there are two possible mechanisms that can alter this outcome. Firstly, if reactive pyrite is present then the oxidation reaction of the pyrite with water and oxygen can act as an initiator of self-heating. This raises the temperature to the point where the carbon oxidation reaction takes over producing thermal runaway to ignition if sufficient carbon is present. Secondly, if an external heat source comes into contact with the coal or carbonaceous shale it can have the same effect of raising the temperature to the point where the carbon oxidation reaction rate is sufficient to sustain self-heating. This paper presents examples of both these mechanisms.

INTRODUCTION

There are numerous index tests available for assessing the spontaneous combustion propensity of coal (Nelson and Chen, 2007). The Australian coal industry has routinely used the R_{70} self-heating rate test for the past 40 years to rate the intrinsic spontaneous combustion propensity of coal (Beamish, 2005; Humphreys, Rowlands and Cudmore, 1981). More recently the industry has adopted the use of adiabatic Incubation testing to evaluate the likelihood of spontaneous combustion for the conditions existing on the mine site (Beamish and Beamish, 2011, 2012). This has eventuated due to the recognition of deficiencies in the R_{70} test to identify the presence of reactive pyrite (Beamish and Theiler, 2017a). Also a number of recent incidents have occurred in coals with low propensity ratings based on the R_{70} value (Beamish and Theiler, 2017b) primarily due to the coal temperature being artificially lifted appreciably above ambient temperature (Cliff, Brady and Watkinson, 2014). Examples of both these circumstances and the self-heating mechanisms involved, as applied to coal and carbonaceous shales, are explained in this paper.

ADIABATIC TESTING METHODS AND SAMPLES

R_{70} self-heating rate testing

Full details of the test methodology are given in Beamish, Barakat and St George (2000). The coal sample is dried under nitrogen at 110°C before reintroduction of oxygen at 40°C and subsequent tracking of increase in thermal oxidation over time. The test measures the time taken for the dry, crushed coal to increase in temperature from 40 °C to 70 °C due to oxidation under adiabatic conditions. A relative scale developed by Beamish and Beamish (2011) is used to rate the intrinsic spontaneous combustion propensity of the coal. The rating classification has evolved to differentiate between coals from Queensland and New South Wales as well as regions with similar climatic conditions to these two states of Australia.

Incubation testing

This test is designed to replicate coal incubation self-heating behaviour from low ambient temperature. As such, the normal in-mine temperature is used as the starting point for the test. The nature of the test also assumes that in the operational situation there is a critical pile thickness present that minimises any heat dissipation (represented by the adiabatic oven testing environment) and there is a sufficient supply of oxygen present to maintain the oxidation reaction. A larger sample mass and lower oxygen flow rate is used, compared to the R_{70} test method, to produce conditions that more closely match reality. The sample either reaches thermal runaway, or begins to decrease in temperature due to insufficient reactivity to overcome heat loss from moisture liberation/evaporation and/or heat sink effects from non-reactive mineral matter.

Coal and carbonaceous shale samples

Details of the coal and carbonaceous shale samples used in this study are contained in Table 1. Coal A is a high volatile A bituminous coal from the top portion of a seam. It has a high pyrite content as indicated by the pyritic sulphur content of the coal. Coal B is from a medium volatile bituminous high quality coking coal seam with a low pyritic sulphur content. Carbonaceous Shale A is from the immediate roof strata sequence of a high volatile C bituminous coal seam. It has an elevated pyrite content. Carbonaceous Shale B is from a high ash band within a high volatile A bituminous coal seam. All of these samples have R_{70} self-heating rate values less than 0.5, which rates them as having a low intrinsic spontaneous combustion propensity.

TABLE 1 – Coal sample properties.

Sample	Moisture Content (%)	Ash Content (%)	Pyritic Sulphur Content (%)	R_{70} Value (°C/h)	Propensity Rating
Coal A	7.1	16.1	9.00	0.29	Low
Coal B	2.1	11.0	0.30	0.36	Low
Carbonaceous Shale A	8.1	75.3	1.62	0.37	Low
Carbonaceous Shale B	2.0	80.9	0.07	0.15	Low

ADIABATIC SELF-HEATING BEHAVIOUR

Reactive pyrite initiated self-heating

The adiabatic self-heating curves for Coal A are shown in Figure 1a. The low R_{70} value of the coal indicates that the organic carbon present has a low reactivity to oxygen and hence the coal has a low spontaneous combustion propensity rating based on this parameter. However, it should be noted that the R_{70} test is conducted on a dry basis and therefore there is no moisture available for the pyrite oxidation reaction. The results for the Incubation test (Figure 1a) clearly show that the pyrite in this coal is reactive and at the low ambient mine temperature the pyrite initiates self-heating, which then rapidly accelerates as the coal temperature exceeds 40°C. The organic carbon in the coal also begins to oxidise more rapidly as the coal temperature increases and once the temperature exceeds 110°C it becomes the dominant reaction leading to thermal runaway.

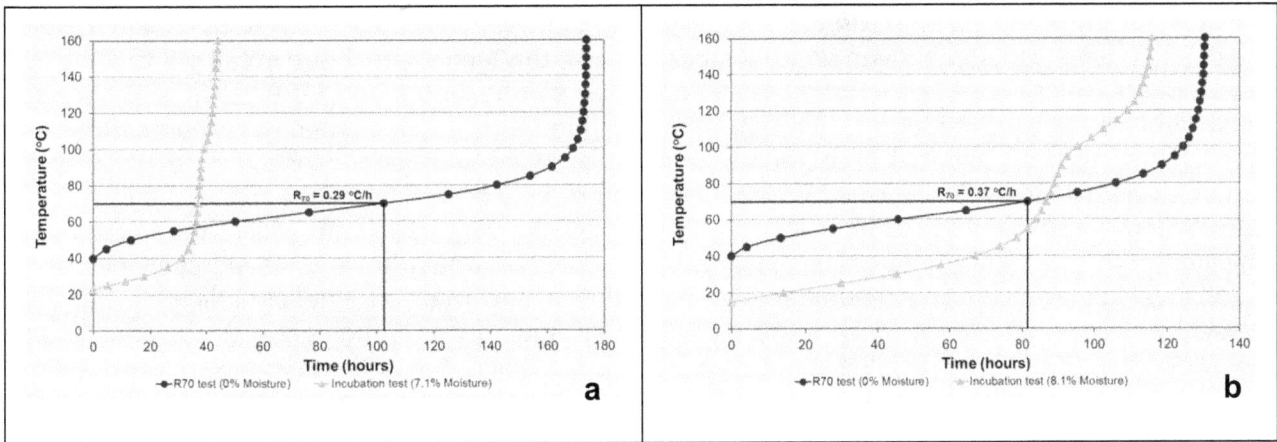

FIG 1 – Adiabatic oven test results for: a) Coal A and b) Carbonaceous Shale A, showing accelerated self-heating from reactive pyrite oxidation when moisture is present (Incubation test) compared to intrinsic coal self-heating with moisture absent (R_{70} test).

The adiabatic self-heating curves for Carbonaceous Shale A are shown in Figure 1b. Again, when moisture is available for pyrite oxidation it can be seen that the pyrite in the sample is reactive. It initiates self-heating from an even lower ambient mine temperature than Coal A. There is also a noticeable slowdown in the self-heating rate as the sample temperature reaches the moisture liberation and evaporation transition shoulder from 95-105°C. The organic carbon oxidation reaction dominates once this temperature is exceeded due to the moisture removal. At the moisture content of 8.1% the carbonaceous shale would not have self-heated without the reactive pyrite being present.

External heat source initiated self-heating

The adiabatic self-heating curves for Coal B are shown in Figure 2a. In a dry state, which does not occur on the mine site, the coal self-heats to thermal runaway. However, when moisture is present the coal initially self-heats from the mine ambient temperature then begins to decrease in temperature due to the heat loss mechanism of moisture liberation and evaporation. The test was terminated after 40 hours as it was evident that thermal runaway was not possible for these initial mine conditions. Carbonaceous Shale B shows a similar trend (Figure 2b), although the combined effect of the high mineral matter content (as indicated by the high ash content of the sample) and the moisture presence at a much lower ambient mine temperature produces a more dramatic heat sink effect. As with Coal B the test was terminated after 40 hours since it was decreasing in temperature.

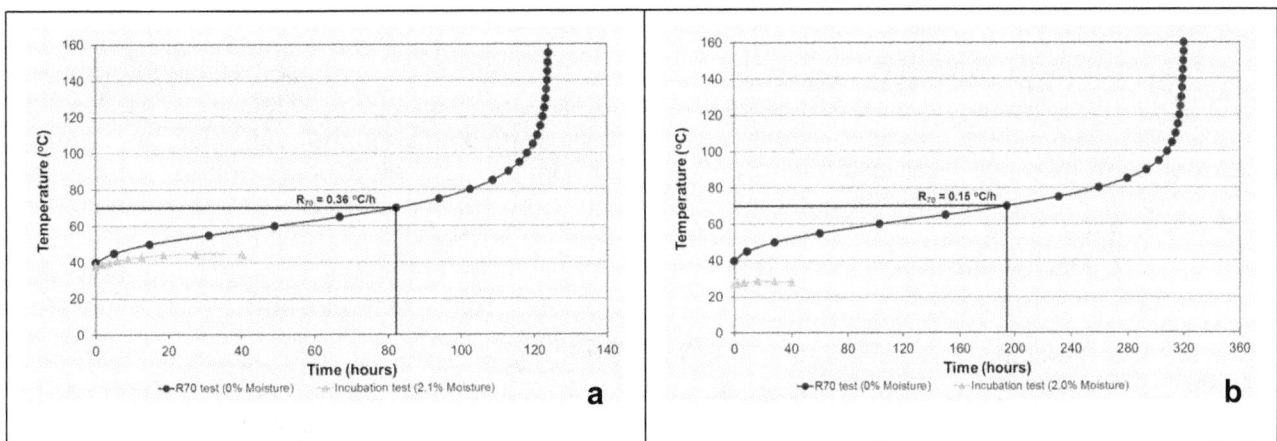

FIG 2 – Adiabatic oven test results for: a) Coal B and b) Carbonaceous Shale B, showing moderated self-heating and no thermal runaway when moisture is present (Incubation test) compared to intrinsic self-heating to thermal runaway with moisture absent (R_{70} test).

To assess the effect of an external heat source on Coal B and Carbonaceous Shale B, the adiabatic oven was used in oven heating mode to raise the temperature of the sample to a higher temperature before turning the oven back to adiabatic self-heating mode. At 97°C Coal B showed sustained self-heating through to thermal runaway (Figure 3a). Consequently, if the coal comes into contact with an external heat source such as a curing compound, for example PUR (Cliff, Beamish and Cuddihy, 2009) or other products that produce an exothermic reaction, the likelihood of creating a spontaneous combustion event is increased. The ease with which the self-heating initiated from this higher temperature suggests that the coal would have achieved sustained self-heating at an even lower temperature. Extrapolation of the reaction kinetics indicates that this could have been as low as 80°C. If the external heat source was in contact with the coal for an extended period of time this value could have been even lower, due to the drying out of the coal by the heat source rather than the heat of oxidation being used up in moisture liberation and evaporation.

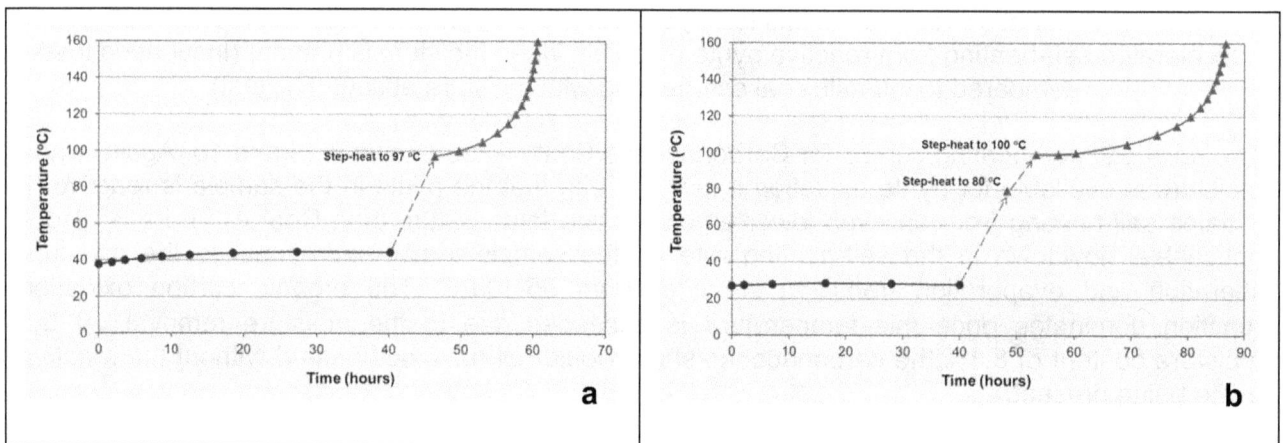

FIG 3 – Adiabatic oven test results for: a) Coal B and b) Carbonaceous Shale B, showing sustained self-heating from an elevated temperature.

Initially, an attempt was made to heat Carbonaceous Shale B to a temperature of 80°C, but this proved insufficient to cause sustained self-heating in the sample as it was unable to reach the set temperature of the oven. However, at 100°C Carbonaceous Shale B showed sustained self-heating through to thermal runaway (Figure 3b). The initial rate of self-heating from this elevated temperature was extremely low indicating that this is most likely the minimum temperature required to overcome the combined heat sink effects of the high mineral matter content and moisture content of the sample.

CONCLUSIONS

Low spontaneous combustion propensity ratings for both coal and carbonaceous shales determined from R_{70} self-heating rate tests do not necessarily rule out the possibility of developing a spontaneous combustion event. There are two possible mechanisms that can alter the balance between heat gain and heat loss during coal and carbonaceous shale oxidation in favour of heat gain. Firstly, if reactive pyrite is present with sufficient moisture it can cause a pyrite oxidation reaction to take place. As such, it can initiate self-heating in low intrinsic reactivity coal measure rocks, even at low ambient temperatures. As the pyrite oxidation continues it creates a mutual self-heating rate acceleration with the carbon oxidation that leads to thermal runaway.

The second mechanism is self-heating that is initiated by an external heat source coming into contact with the coal or carbonaceous shale. Under these circumstances the coal or carbonaceous shale temperature is artificially elevated to a point where sustained self-heating is possible. This is mainly due to the temperature dependence of the carbon oxidation reaction rate, which progressively increases at higher temperatures, and also the external heat source helps to remove moisture.

Both these self-heating initiation mechanisms have major implications for spontaneous combustion likelihood evaluation as part of the overall mine risk assessment process. They need to be considered in the context of each mining situation.

REFERENCES

Beamish, B B, 2005. Comparison of the R_{70} self-heating rate of New Zealand and Australian coals to Suggate rank parameter, *International Journal of Coal Geology*, 64(1-2):139-144.

Beamish, B and Beamish, R, 2012. Benchmarking coal self-heating using a moist adiabatic oven test, in *Proceedings of the 14th US/North American Mine Ventilation Symposium* (eds:F Calizaya and M G Nelson), pp 423-427 (University of Utah, Department of Mining Engineering, Utah, USA).

Beamish, B and Beamish, R, 2011a. Testing and sampling requirements for input to spontaneous combustion risk assessment, in *Proceedings of the Australian Mine Ventilation Conference* (eds: B Beamish and D Chalmers), pp 15-21 (The Australasian Institute of Mining and Metallurgy: Melbourne).

Beamish, B B and Theiler, J, 2017a. Assessing the reactivity of pyrite, in *Proceedings 17th Coal Operators' Conference* (eds: N Aziz and B Kininmonth), pp 391-394, (University of Wollongong).

Beamish, B B and Theiler, J, 2017a. Recognising the deficiencies of current spontaneous combustion propensity index parameters, in *Proceedings of the Australian Mine Ventilation Conference*, pp 113-117 (The Australasian Institute of Mining and Metallurgy: Melbourne).

Beamish, B B, Barakat, M A and St George, J D, 2000. Adiabatic testing procedures for determining the self-heating propensity of coal and sample ageing effects, *Thermochimica Acta*, 362 (1-2):79-87.

Cliff, D, Beamish, B and Cuddihy, P, 2009. Explosions, fires and spontaneous combustion, in *Monograph 12, Australasian Coal Mining Practice – Third Edition*, pp 421-435 (The Australasian Institute of Mining and Metallurgy: Melbourne).

Cliff, D, Brady, D and Watkinson, M, 2014. Developments in the management of spontaneous combustion in Australian underground coal mines, in *Proceedings 14th Coal Operators' Conference*, pp 330-338, (University of Wollongong).

Humphreys, D, Rowlands, D and Cudmore, J F, 1981. Spontaneous combustion of some Queensland coals, in *Proceedings of Ignitions, Explosions and Fires in Coal Mines Symposium*, pp 5-1 - 5-19 (The AusIMM Illawarra Branch).

Nelson, M I and Chen, X D, 2007. Survey of experimental work on the self-heating and spontaneous combustion of coal, *The Geological Society of America Reviews in Engineering Geology*, 18:31-83.

Health and Safety Hazard Management

Tracer gas study of nano diesel particulate matter (nDPM) behaviour in secondary ventilation practices

S Black[1], S Wilkinson[2], L van den Berg[3] and K Manns[4]

1. FAusIMM, Manager Project Development, ChemCentre, Building 500, Corner Townsing Drive and Manning Road, Bentley WA 6102. Email: sblack@chemcentre.wa.gov.au.
2. Consultant, C/O ChemCentre, Building 500, Corner Townsing Drive and Manning Road, Bentley WA 6102. Email: steven.wilkinson@bigpond.com.
3. MAusIMM, Senior Ventilation Engineer, BBE Consulting (Australasia), Suite 6, 89 Winton Road, City of Joondalup, Perth, WA, 6027. Email: lvandenberg@bbegroup.com.au.
4. MAusIMM, Ventilation Consultant, BBE Consulting (Australasia), Suite 6, 89 Winton Road, City of Joondalup, Perth, WA, 6027. Email: kmanns@bbegroup.com.au.

ABSTRACT

The mine ventilation system plays an important role in mitigating human exposure to diesel particulate matter (DPM) emissions, exhaust gases and heat in underground mines. The mine's secondary ventilation systems are critical for diluting contaminants when working at the face of a heading.

It is important to understand the localized flow profiles on a "micro scale" to identify areas of improvement and ensure that the ventilation system is optimized to maximise dilution of exhaust gases. The application of tracer gas technology enables reliable measurements of flow behavior and differentiation between the contribution of various contaminant sources including nano DPM (nDPM).

A tracer gas study was conducted at the underground Sunrise Dam gold mine focusing on mining activities under auxiliary ventilation at the face of a development heading.

The tracer gas sulfur hexafluoride (SF_6) was used to study the transport of nDPM between source and proximate equipment operators to assess the potential real-time exposure of underground workers (e.g. Service crews, Shotcreters, Jumbo drill operators, Bogger operators and Shift Supervisors) and the efficiency of secondary ventilation systems at each worksite over similar periods of time. The key benefit of tracer gas studies is that they enable differentiation between nDPM sources and allow accurate characterisation of air flow behaviour.

This paper describes how the use of tracer gas technology identified anomalies and concentration differences for various activities. The information obtained can then be used to better inform the planning of administrative controls to manage activities around other major diesel activities.

This work is part of a project funded by Department of Mines, Industry Regulation and Safety (DMIRS) and the Minerals Research Institute of Western Australia (MRIWA).

INTRODUCTION

Underground mining in Western Australia is heavily reliant upon diesel-powered equipment. This is likely to remain so for some time to come, especially when expanding the mining envelope to deeper mines. This is largely because cost and production efficiencies offered by diesel-powered equipment are unrivalled by any competing systems (e.g. battery-electric or plug-in electric equipment). Mining in the Eastern Goldfields is progressively getting deeper as shallow orebodies are depleted. The economics of underground mining will progressively become more reliant on the cost and production efficiencies offered by using diesel equipment. Future green fields discoveries are likely to be under deep mineralogical cover in the east of the state and are likely to have dependence upon the cost and production efficiencies offered by using diesel equipment.

The use of diesel-powered equipment has been continuously increasing in underground metal/non-metal mines. This extensive utilization of diesel-powered equipment generates the potential for exposure of underground miners to diesel particulate matter (DPM), including nano diesel particulate matter (nDPM). The majority (by numbers) of the diesel particles are in the nano size range (<100

NM), hence the term nDPM. In 2012, the International Agency for Research on Cancer (IARC) classified diesel engine exhaust as carcinogenic to humans, Group 1 (WHO, 2012).

The health implications of nDPM are known to extend beyond the lungs and the particles are small enough to diffuse throughout the body (Nemmar *et al*, 2010; Oberdörster *et al*, 2002) and even penetrate the blood brain barrier (Heidari Nejad *et al*, 2014).

The management of diesel emissions in underground mines is a significant challenge. Although emissions from plant are increasingly lowered (through OEM or after-market exhaust after-treatment), deeper mines and more stringent guidelines/limits increase management difficulty. There is increasing interest in understanding the importance of nDPM in underground mines and managing such particles effectively. Nanoparticles are generally defined as particles <100 nm, which typically represent the majority of diesel exhaust, in terms of both mass and number.

There are critical gaps of knowledge on the character of nDPM, its behaviour in underground hard rock mine environments.

To ensure the continued availability of using diesel equipment into the future as an option for operations in underground mines as the mining envelop expands with deeper mines, the characterisation of nDPM is critical as it will need good management and controls going forward.

The mine ventilation system plays an important role in mitigating human exposure to diesel particulate matter (DPM) emissions, exhaust gases and heat in underground mines. The mine's secondary ventilation systems are critical for diluting contaminants when working at the face of a heading.

It is important to understand the localized flow profiles on a "micro scale" to identify areas of improvement and ensure that the ventilation system is optimized to maximise dilution of exhaust gases. The application of tracer gas technology enables reliable measurements of flow behavior and differentiation between the contribution of various contaminant sources including nano DPM (nDPM).

Previous studies performed by ChemCentre in collaboration with USA researchers (confidential work not published) showed that the smaller the particle size (< 80 nm) the more likely the particle would behave like a gas. Hence, tracer gas technology is an appropriate tool for studying nanoparticles, such as nDPM.

The ideal tracer gas should be chemically and thermally stable, safe, non-toxic, non-corrosive readily attainable and easily transportable, inexpensive, odorless and naturally occurring in the environment. In addition, it should be easily detected at very low levels and should maintain its stability in the container holding the air sample during transport to a laboratory for analysis. Sulfur hexafluoride (SF_6) meets these requirements. It can be obtained as a liquid under pressure in cylinders, is a gas under ambient conditions, is inet, non-toxic and can readily be detected at the parts per billion (ppb) level using data logging instruments and at parts per trillion (ppt) level when analysed in the laboratory. At these concentrations the gas does not disturb the system to be investigated.

This paper describes a study of diesel exhaust flow behaviour and source contribution in the underground mine at Sunrise Dam Gold Mine (SDGM) using tracer gas technology.

PROCEDURE

On-site tracer gas study on the transport of nDPM between source and proximate equipment operators was performed to assess the real-time exposure of mineworkers (e.g. Service crews, Shotcreters, Jumbo drill operators, Boggers and Supervisory staff) and the efficiency of secondary ventilation systems at each worksite over similar extended periods of time. The key benefit of tracer gas studies is that they enable differentiation between different potential nDPM sources and allow accurate characterisation of air flow behaviour.

The following activities were studied at a development heading (Figure 1) in the underground SDGM;

- Charge-up;
- Bogger;

- Hydro-scaling - spraymec; and
- Shotcreting – Spraymec and Agi truck.

Other activities also studied included: (i) Truck and (ii) Traverse study.

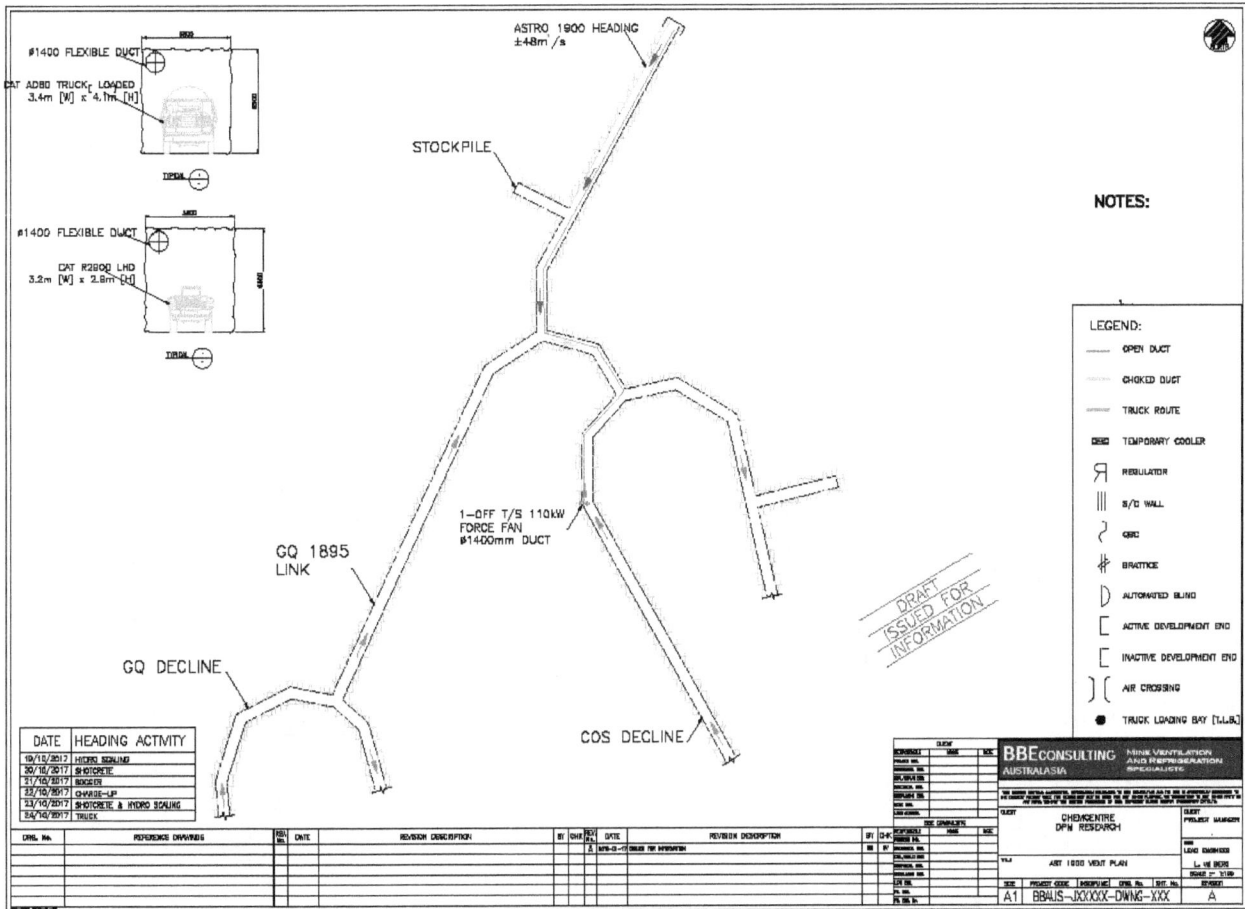

FIG 1 - Schematic of the ventilation air circuits at the studied development heading in SDGM.

The ventilation arrangement in the development heading where the study was undertaken comprised of a typical twin stage 110kW axial flow fans, force ventilation at the development heading via 1400mm duct. The secondary fans were located in the trucking decline. The distance from the fans to the face was approximately 200 m. The fans typically supply about 45m³/s with approximately 35m³/s supplied to the face after leakage. This yields an air speed of 1.4m/s. The duct discharge end was located approximately 25 m from the working face.

A summary of the heading ventilation measurements taken during different activities is given in Table 1.

The tracer gas sulfur hexafluoride (SF₆) was released (from a pressurised cylinder) via Teflon tubing directly into the exhaust stream at the exhaust outlet pipe of the diesel engine under study (Figure 2). The aim was to achieve the best possible mixing of tracer gas with the exhaust as is practically possible and taking in consideration the high temperature of the exhaust within the exhaust pipe. Close mixing of the tracer gas with the exhaust gas would then allow the tracer gas to behave in a similar manner to the exhaust gases and nDPM in the heading. Measurements of the concentrations of SF₆ were taken in real time at various receiving sites using Fourier-Transform Infrared Spectrometers (referred to as MIRAN, M). Thus, the tracer gas was used as a surrogate to characterise and "visualise" the flow profile and the build-up and decay of nDPM in the development heading and other relevant areas in the mine.

TABLE 1 - Ventilation Conditions. Note: The linear velocity was between 0.8 and 1.0 m/s.

Activity at Heading	Size of drive	Rated kW	Required airflow	Measured airflow	Condition by visual inspection
Hydro-scaling	5.5 m x 6.0 m	90 kW	4.5 m³/s	30 m³/s	Very good
Shotcreting	5.5 m x 6.0 m	346 kW [90 +256 kW]	17.3 m³/s	31 m³/s	Good
Charging	5.5 m x 6.0 m	110 kW	5.5 m³/s	29 m³/s	Good
Bogging	5.5 m x 6.0 m	305 kW	15.25 m³/s	28 m³/s	Good

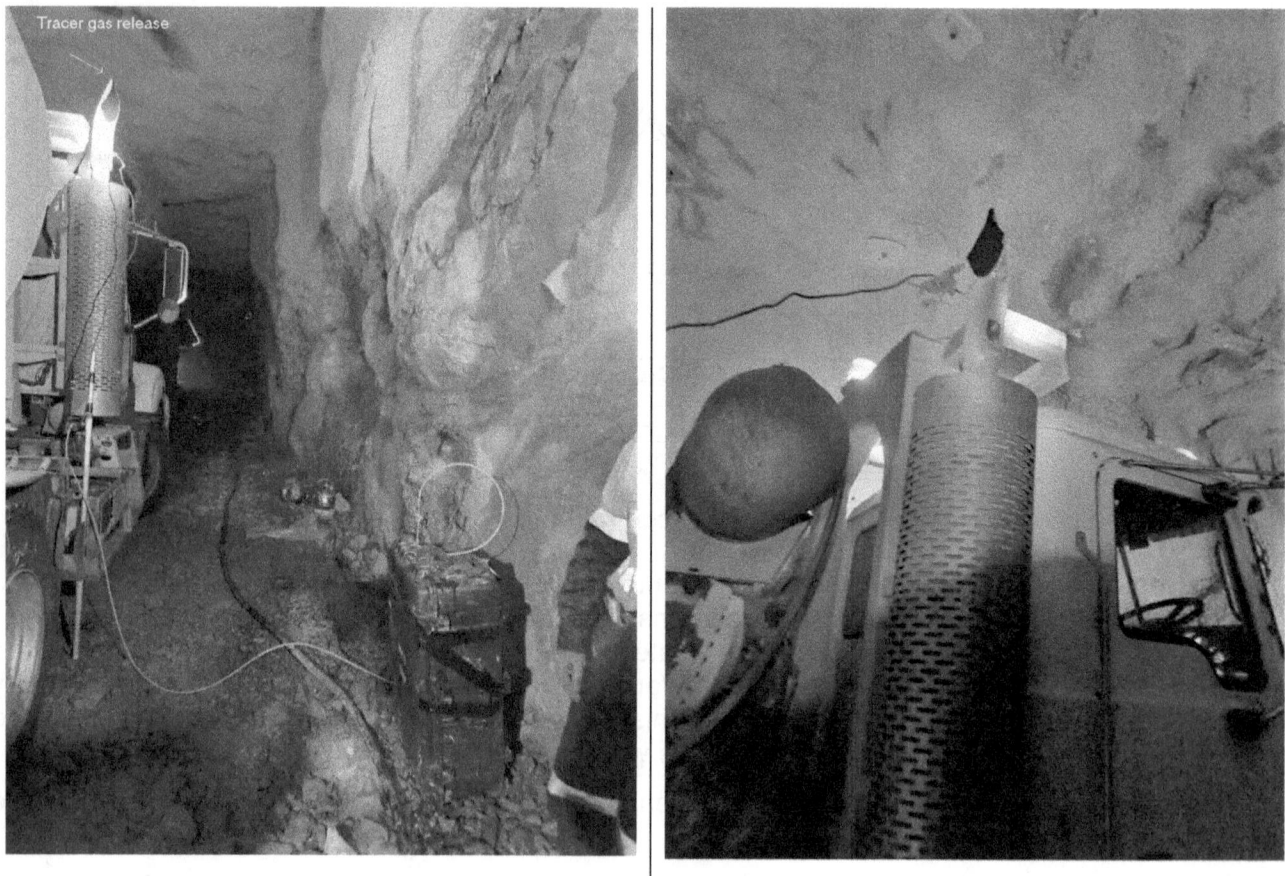

FIG 2 - Tracer gas release into the Agi truck exhaust.

RESULTS AND DISCUSSION

The tracer gas study of a number of underground mining activities, such as charging, bogging, hydro-scaling, shotcreting and truck driving, demonstrated that during those activities there were consistently relatively higher SF_6 concentrations measured during the hydro-scaling and shotcreting activities. Hence, the discussion in this paper will focus on the hydro-scaling and shotcreting activities.

Hydro-scaling and Shotcreting

During Hydro-scaling, the tracer gas was released into the exhaust of the Spraymec machine while it was scaling the loose rock with water and compressed air. The four Miran positions are shown in Figure 3 as M1, M2, M3 and M4.

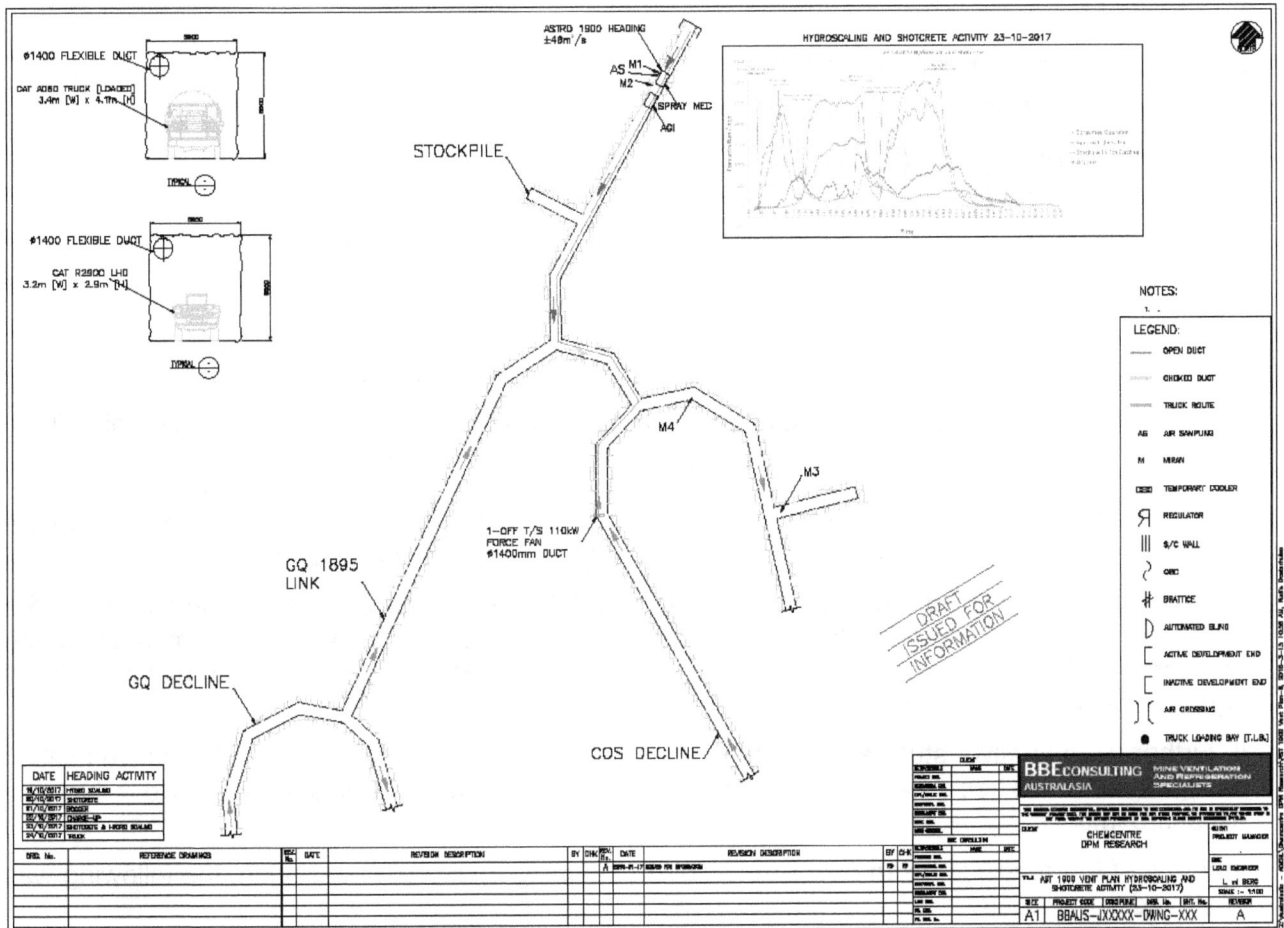

FIG 3 - Map of tracer gas study during the Hydro-scaling and Shotcreting activities.

During shotcreting, SF$_6$ was released at Agi truck first (Figure 3). Once the SF$_6$ concentration had reached a steady state air samples were taken near the Agi truck operator (M2 position). Spraymec operator stopped mid-activity. Then stopped SF$_6$ release and allowed to clear (~10 min). Resumed release of SF$_6$ at spraymec once shotcreting activity was resumed. Once the SF$_6$ concentration had reached a steady state, air samples were taken near the Spraymec operator (M1 position).

The tracer gas study observations (Figures 4) are specific to the ventilation configuration and the distance of the secondary ventilation duct to the face for the duration of these particular hydro-scaling and shotcreting activities. The results may differ for other ventilation configurations and cannot necessarily be considered universal. In this instance there was good ventilation in the face with the ventilation duct close to the face and good mixing (Table 1).

During hydro-scaling (Figure 5, extracted from Figure 4) the Spraymec and Agi operators received a similar SF$_6$ exposure from the Spraymec exhaust. Note: the Agi operator does not necessarily have to be present during this activity and this was merely a coincidence on the day.

During shotcreting the exhaust from the Agi truck mainly impacted the Agi operator with lower exposure to the Spraymec operator. In this instance the Spraymec operator exposure is only 60% of Agi operator exposure from the Agi truck. This suggests that having the ventilation duct close to the heading face ensures that the Spraymec operator benefits from the fresh air in the ventilation duct as it scavenges the Agi exhaust away from the heading face before it can impact the Spraymec operator.

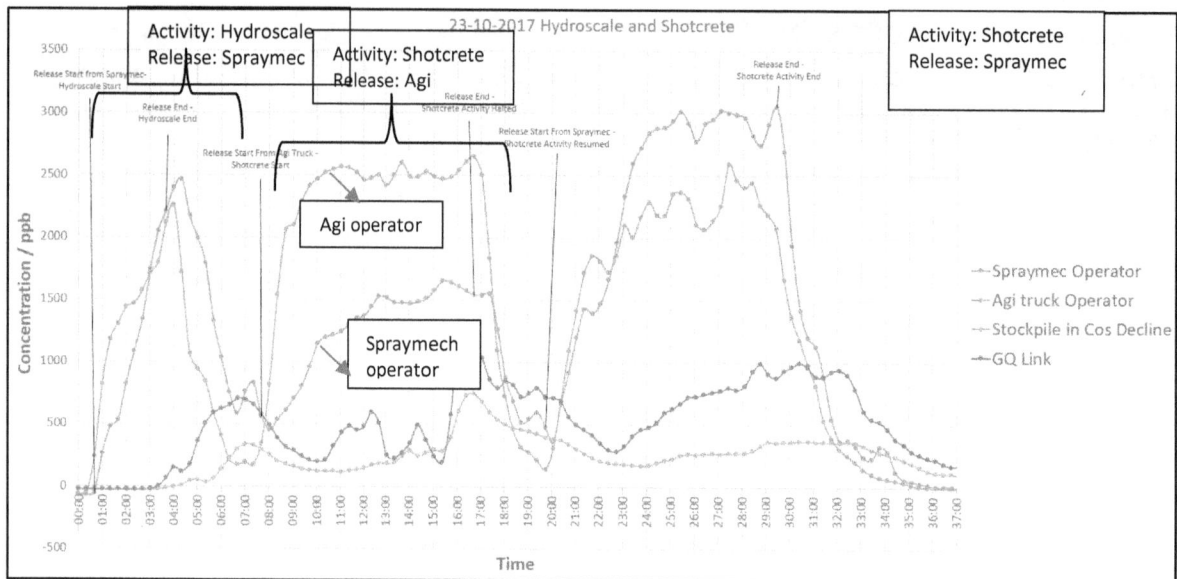

FIG 4 - Tracer gas study data during sequential Hydroscale and Shotcrete activity. M1 = blue line, M2 = orange line, M3 = grey line and M4 = green line.

However, during shotcreting the exposure from the Spraymec exhaust significantly impacted both the spraymec operator and Agi operator albeit slightly lower for the Agi operator.

The Agi operator received a similar exposure from both the Spraymec and Agi machines. Under the prevailing ventilation conditions, the Agi operator experienced approximately 10-15% more exposure compared to the Spraymec operator and would be at greater risk. It is worth noting that respiratory personal protective equipment is worn during shotcreting.

If further comparative studies can be undertaken with heading set-up in different ventilation configurations, this tracer gas approach can be used to optimise airflow in the heading to reduce the exposure impact.

In terms of the readings taken downstream of the heading in through ventilation, the following was observed:

- The stockpile is not significantly impacted by the activity. This is due to the small amount of fresh ventilation air entering the stockpile; and

- Any exhaust entering the stockpile does so by diffusion rather than active ventilation.

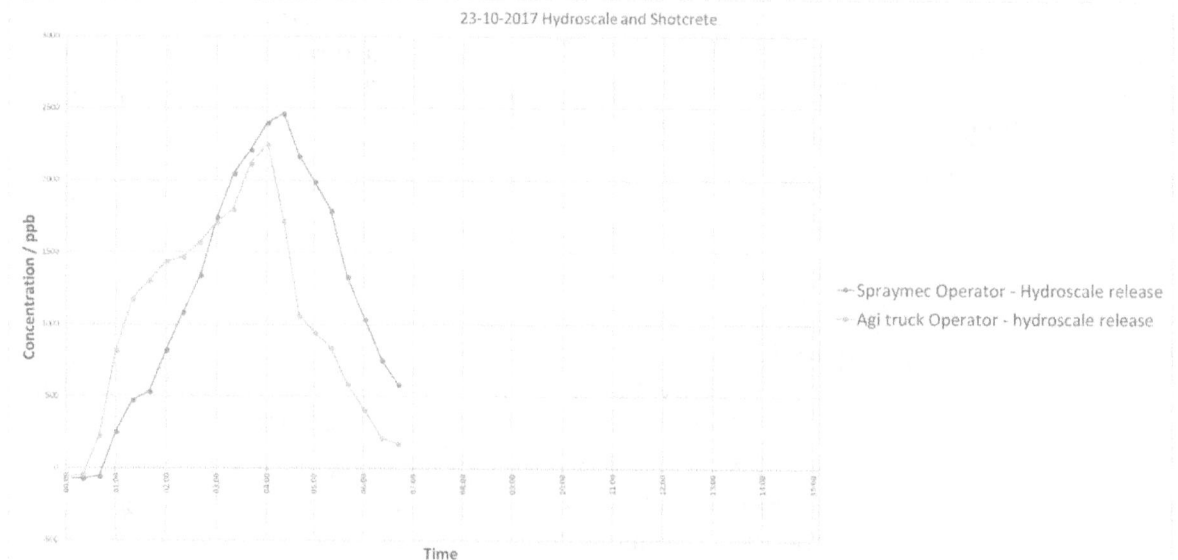

FIG 5 - Relative exposures (extracted from Figure 4) to the Spraymec and Agi truck operators during the Hydro-scaling activity.

Safe shelter area for observation personnel

It is worth noting that within the short timeframe of the activity the concentrations in the stockpile do not reach the peak values that are experienced by either operator and are also less than in the main return (GQ link). Because the stockpile was not activley ventilated at the time, the buildup of gas concenration relies on diffusion processes only and is therefore slow. This means that during a short duration activity an unventilated cuddy (as represented by the stockpile in this case) could be a natural 'place of safety' or shelter area for personnel that are in the general area but not involved with the actual activity. The data also shows that after the diesel sources have left the heading, concentratons are quickly diluted. In the case of the expriment (shown by Figure 4) this occurs within 2-3 minutes. After this period the concentrations are lower than in the cuddy, which then take some time to clear. This perhaps provides an opportunity to have administrative controls in place to exploit this phenomenon. For example, shelter in a cuddy (stockpile) during activity but leave stockpile soon after the diesel machines have vacated the heading.

The peaks observed at the heading (Figure 4) occur 6 minutes later at the through ventilation within QC link and the levels are approximately 20% of the levels at the heading.

Figures 5, 6 and 7 represent the exposure of the Spraymec and Agi truck operators during the hydro-scale and shotcrete activities from both the Agi truck and the spraymec machine with the trace lines superimposed on top of each other.

The tracer gas data (Figures 5, 6 and 7) shows that the Agi operator is exposed to similar amounts of SF_6 released from both pieces of equipment, which means he/she was at greater risk in this particular ventilation setup. The integrated total exposure to fumes is summarised in Table 2. This is the sum total from both exhausts over the exposure period. For this particular setup the Agi operator received approximately 12% more exposure to gas, but given the experimental errors from the Miran instruments ($\pm$ 10%) this may not be significant.

The spraymech operator receives almost 65% of his total exposure from his own machine and only 35% from the Agi truck. Therefore, any improvements to the Spraymec exhaust will be beneficial for the Spraymech operator. The Agi operator also gets as much as 50% of his exposure from the Spraymech. Thus, improvements to the Spraymech truck will have a benificial impact to both operators. Hence, a focus firstly on the Spraymech will have the greatest initial return on investment. With effective ventilation in place the concenration of both machines are rapidly diluted at similair rates.

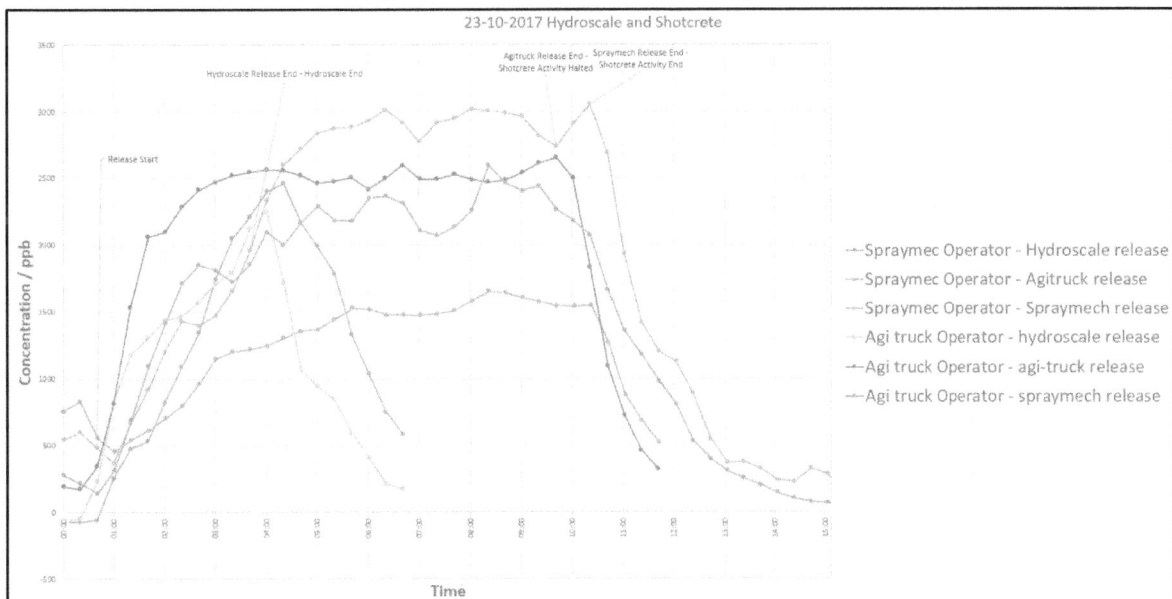

FIG 6 - Exposures (extracted from Figure 4) to the Spraymec and Agi truck operators during the Hydro-scaling and Shotcreting activities, normalised to the same time scale.

(a) Exposure of spraymec (grey) and Agi (orange) exhaust on the spraymec operator.

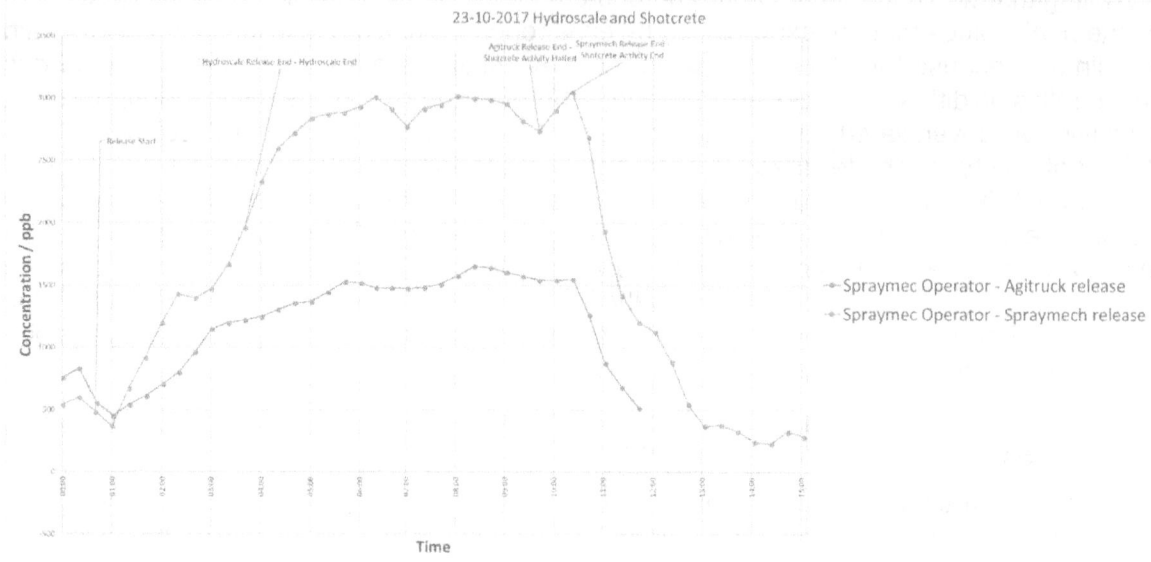

(b) Exposure of spraymec (green) and Agi (blue) exhaust on the Agi truck operator

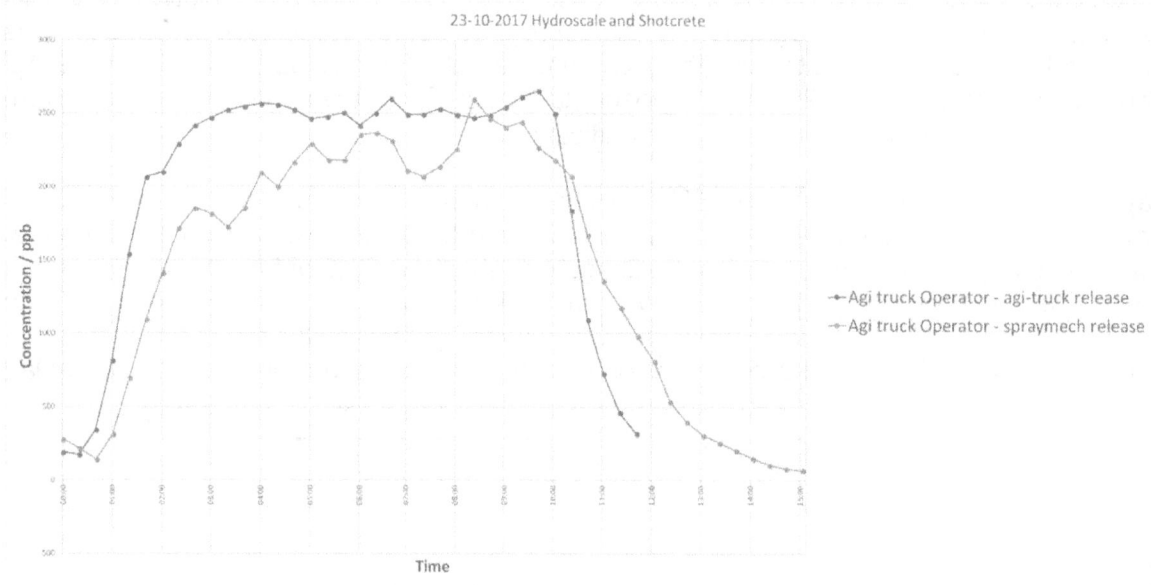

FIG 7 - Relative exposures (extracted from Figure 4) to the Spraymec and Agi truck operators during the Shotcreting activity.

In summary it can be concluded that the two operators experience the same nominal amount of exposure with both being impacted by each of the two machines.

Similar SF_6 flow profiles were observed for the hydro-scaling and shotcreting activities respectively performed on separate days in the same development heading under similar ventilation conditions (Table 1). The results obtained on separate days are within the Miran instruments' experimental error of 10% and hence the SF_6 profile data was reproducible.

TABLE 2 - Relative exposure calculated from integrated peak areas (as shown in Figure 4) experienced by the Spraymec and Agi truck operators during hydro-scaling and shotcreting activities.

Release Activity	Peak from Figure 4	Spraymec Operator, M1	Agi truck Operator, M2
Hydro-scale (Agi release)	1	27,603	22,054
Shotcrete (Agi release)	2	43,090	72,113
Shotcrete (Spraymec release)	3	79,734	65,548
Total			
Shotcrete	2 + 3	122,824	137,661
Hydroscale	1	27,603	22,054

Truck Exercise

During the truck activity, the tracer gas was set up to release into the tailpipe of the truck while the truck travelled from surface, down the decline past the fan that feeds the test level, got loaded and travelled back up past the test level again to surface (Figures 8 and 9).

FIG 8 - Tracer gas study during the Truck activity.

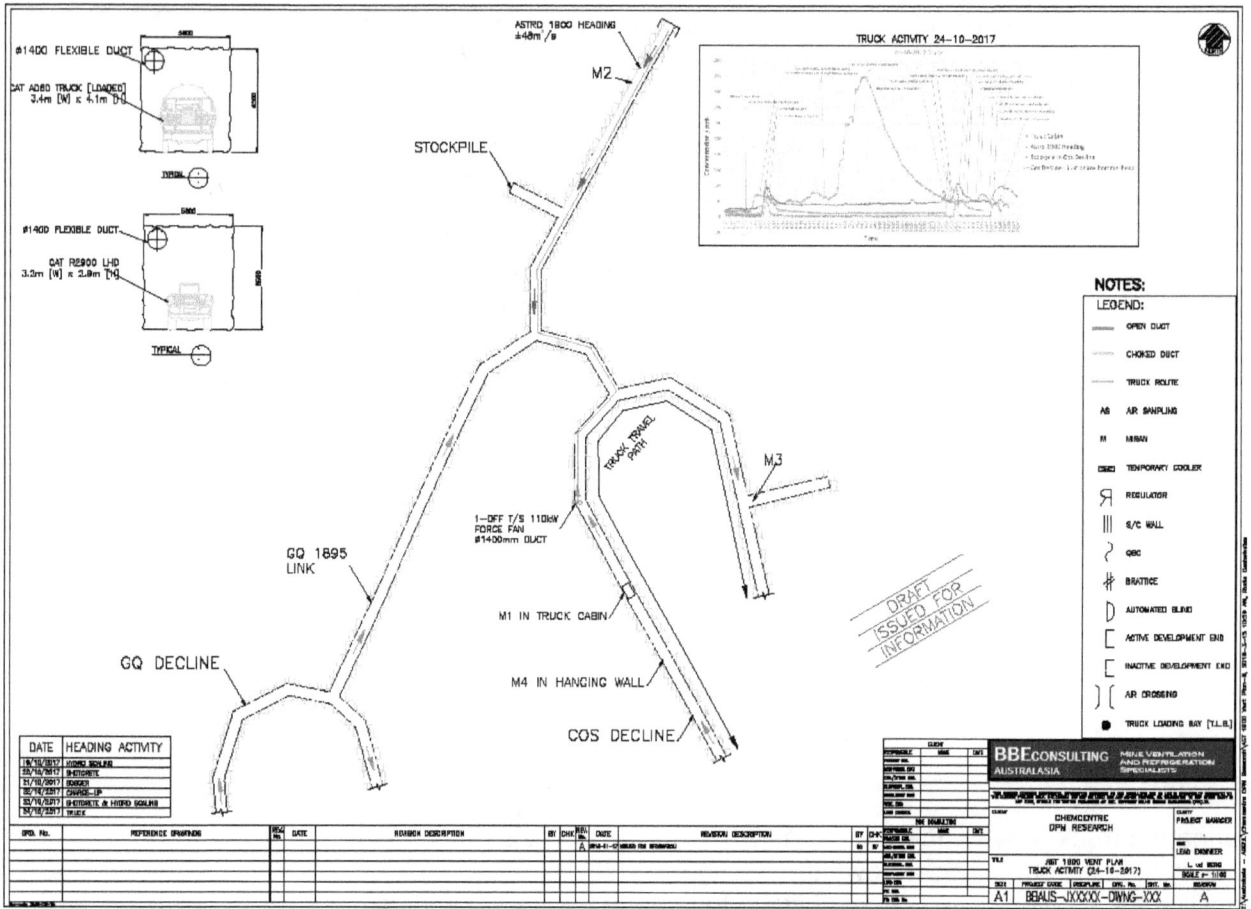

FIG 9 - Map of tracer gas study during the Truck activity.

The truck experiment (Figure 10) shows that SF_6 exposure levels to the truck driver are much lower when compared to operator exposure levels during shotcreting activities (Figure 4). However, care should be taken in comparing these directly since the SF_6 release rate would be constant whilst in reality exhaust rates between different pieces of equipment will be different. Subsequent studies should calibrate SF_6 release rates relative to exhaust rates of different diesel engine equipment to enable direct comparisons.

The levels of exhaust gas entering the truck cabin is only approximately 5% of that observed during the shotcreting activity. During the truck experiment, the truck driver opened the window (for a few minutes) while the track was stationary during loading and this resulted in a 9-fold increase in truck exhaust gas entering the cabin. It is worth noting that it took approximately 20 minutes after closing the window again before the levels in the cabin reached those prior to opening the window. This means that not only does the exposure levels increase, but period of exposure also increases when a window is opened. The benefits of ensuring the cabin remains isolated is clear.

A recommendation from this study is that if the truck driver needs to open the cabin window while stationary during loading, it is best that the window is left open while driving away for a certain amount of time to ensure faster clearance of exhaust from the cabin.

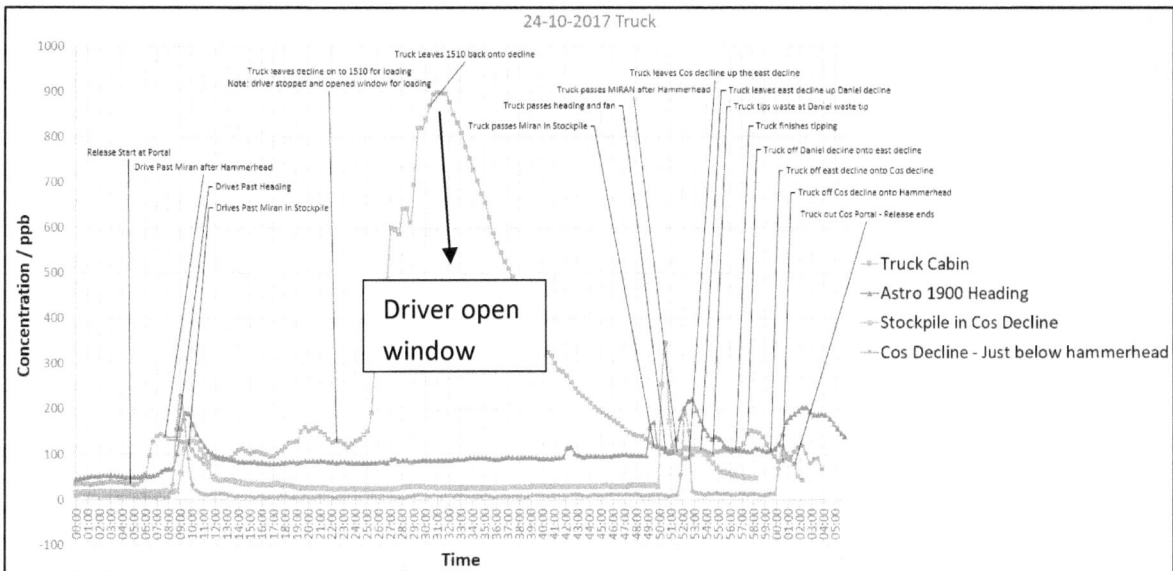

FIG 10 - Tracer gas study data during Truck activity.

Traverse Exercise

This study relates to the heading that was used for the tracer gas analysis of each activity described earlier. The objective of this particular experiment was to determine if the fresh air supplied by the duct is distributed uniformly (or not) along the heading cross section (left to right). For example, does the fresh air form a 'layer' along the sidewall where the duct is positioned or is it effectively distributed along the entire heading width. A second objective was to determine the penetration distance of fresh air into the blind heading from the duct discharge. For example, with the distance to the face does the fresh air reach the end of the heading or does the end experience a 'blind spot'.

During the traverse exercise, tracer gas was released into the secondary ventilation system near the outlet in the development heading in order to test the system effectiveness and side to side stratification. The 4 Miran detectors were placed at varying distances from the heading face with Miran 1 being closest to the face, 6.5 m from the face (Figure 11). The other Mirans were place at 5 m intervals from each other back from the face. Over a period of time the 4 Mirans were moved from one side of the heading to the other.

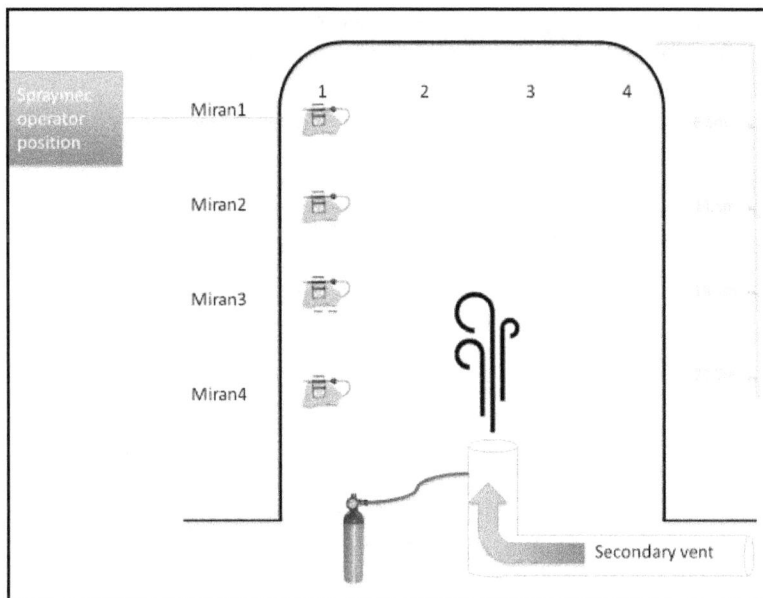

FIG 11 - Layout of Mirans and tracer gas release during the Traverse exercise. The Mirans are shown at position 1 ready to traverse across the heading from left to right.

As expected, Miran 1 (M1) which is closest to the face recorded the lowest concentration (Figure 12) and indicates that only about 83% of the fresh air from the vent bag reaches the face. The full amount of ventilation in the bag probably reaches a position of about 11.5m. Thus, there is a very rapid drop-off between 11.5m and 6.5m which also means that areas much closer to the face will probably have far less effective ventilation. Over time measurements were taken across the width of the heading. The similar concentrations suggest that there is little horizontal stratification across the heading despite the vent bag being near the right-hand side of the heading wall.

Future studies should investigate different ventilation configurations to determine if better scouring is possible.

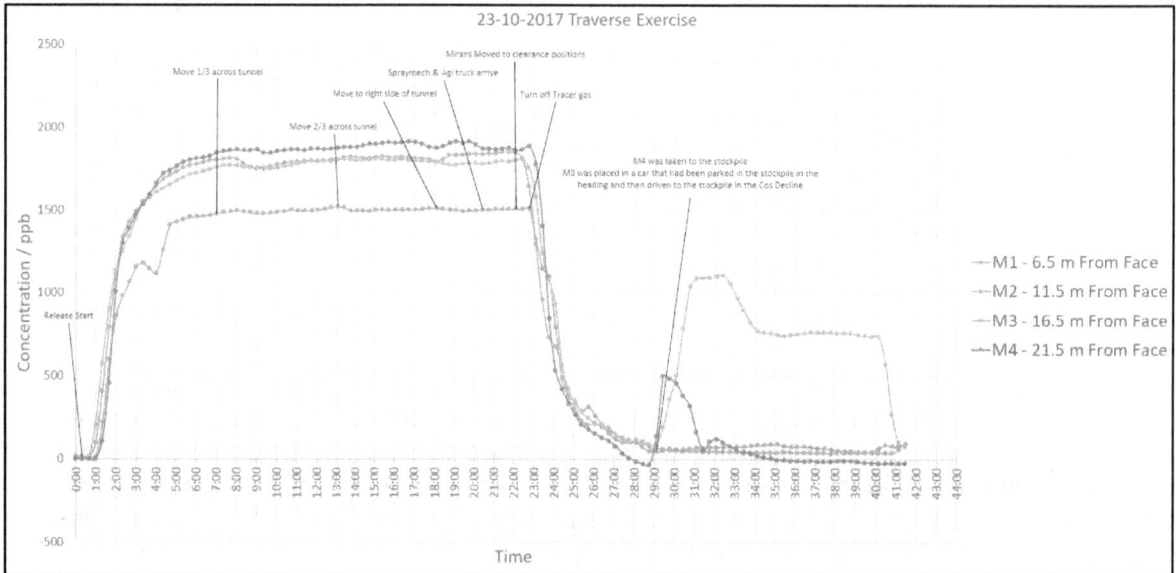

FIG 12 - Tracer gas study data during a Traverse exercise.

SUMMARY

The objective of this study was to undertake a pilot study to show how SF_6 tracer gas techniques can be applied to improve ventilation for better nDPM dilution over time. The findings are as followed:

- The tracer gas study of a number of underground mining activities, such as charging, bogging, hydro-scaling, shotcreting and truck driving, demonstrated that during those activities there were consistently higher SF_6 concentrations measured during the hydro-scaling and shotcreting activities.

- During shotcreting, the Agi truck operator experienced approximately the same exposure of SF_6 from the Agi truck and spraymec exhaust. In contrast, the spraymec operator received almost twice the exposure from the spraymec exhaust than from the Agi truck exhaust. The Agi operator in this instance was at greater risk.

  Hence, because the spraymec is the more significant contributor of exhaust to the operators it is recommended that a focus on improving systems around the spraymec will give the greatest initial return on investment.

- During a short duration activity in a development heading, an unventilated cuddy (as represented by the stockpile in this study) could be a natural 'place of safety' or shelter area for personnel that are in the general area but not involved with the actual activity at a development heading. This information can be utilised to better inform the planning of administrative controls to manage activities around other major diesel activities.

- The SF_6 results from the truck study suggest that the enclosed airconditioned cabin is very effective in managing exposure levels. However, the level of SF_6 exposure to the truck driver increases significantly when a window is opened (a 9-fold increase). Once the window is closed the clearance time is very slow. Thus, the opening of the window not only results in increased levels but also results in prolonged exposure to higher levels once the SF_6 is inside the cabin. The benefits of ensuring the cabin remains isolated is clear and some administrative controls need to be considered.

A recommendation from this study is that the truck driver should keep the window closed while stationary during loading. However, if the truck driver needs to open the window to communicate with the loader driver it is best that the window is left open while driving away for a certain amount of time to ensure faster clearance of exhaust from the cabin.

- A traverse exercise performed in a well ventilated development heading demonstrated that there was little horizontal stratification across the heading despite the vent bag being near the right-hand side of the heading wall. However, there was a very rapid drop-off in ventilation flow between 11.5m from the face and 6.5m from the face which means that areas much closer to the face will probably have far less effective ventilation.

Tracer gas effectively identified anomalies and concentration differences for different activities. Controlled experimental set-ups with different secondary ventilation configurations should be considered to allow comparative studies that will enable ventilation optimisation.

CONCLUSIONS

Tracer gas technology was applied successfully at Sunrise Dam underground gold mine to better understand and inform the following;

- SF_6 flow behaviour as a surrogate for diesel exhaust and relative source contribution to exposure of nearby equipment operators;

- The dispersal of gaseous and ultrafine particulate emissions from diesel exhaust, i.e. particularly nDPM, and the dilution efficiency of the mine ventilation with particular focus on the auxiliary ventilation at the face of a development heading;

- The impact of ventilation practises on the exposure levels; and

- The potential impact of nano-diesel particulate matter (nDPM) on air quality.

ACKNOWLEDGEMENTS

The research team and authors of this paper wish to acknowledge the funding support for this study from the Department of Mines, Industry Regulation and Safety (DMIRS) and from the Mineral Research Institute of Western Australia (MRIWA).

Also acknowledged is the in-kind support provided by AngloGold Ashanti, Barminco and the staff at the Sunrise Dam Gold Mine and from DMIRS, the Mining Industry Advisory Committee (MIAC) nDPM Work Group, and the Australian Institute of occupational Hygienists (AIOH) by providing scientific staff to participate in the scientific advisory panel for this project.

In-kind contribution to this project is also appreciated from ChemCentre and BBE Consulting Australasia.

REFERENCES

Heidari Nejad S, Takechi R, Mullins B J, Giles C, Larcombe A N, Bertolatti D, Rumchev K, Dhaliwal S, and Mamo J, 2014, The effect of diesel exhaust exposure on blood–brain barrier integrity and function in a murine model, J. Appl. Toxicol, 35: 41–47.

Nemmar, A, Al-Salam, S, Zia, S, Dhanasekaran, S, Shudadevi, M, Ali, B H, 2010, Time-course effects of systemically administered diesel exhaust particles in rats, Toxicology Letters, 194(3): 58-65.

Oberdörster, G, Sharp, Z, Atudorei, V, Elder, A, Gelein, R, Lunts, A, Kreyling, W, Cox, C, 2002, Extrapulmonary translocation of ultrafine carbon particles following whole-body inhalation exposure of rats, Journal of Toxicology and Environmental Health - Part A, 65(20): 1531-1543.

World Health Organization (WHO), Press release, 2012, http://press.iarc.fr/pr213_E.pdf.

Using ventilation to drive mine productivity

J Fernandez[1]

1.Managing Director, Zitrón Australia Pty Ltd, Perth WA 6166. Email: jfernandez@zitron.com.au

ABSTRACT

Ventilation is an essential component of any underground mine, playing a vital role in ensuring the health and safety of workers, and a mine's continuous operation.

At a time when mining companies are increasingly having to turn to deeper underground operations to access high-grade ore-bodies, the search is on for more efficient ventilation systems that will allow mining to be carried on at ever deeper levels, ensuring the protection of workers, the maximisation of productivity and the lowering of energy costs.

Mining companies are today looking to innovation to deliver new technologies and procedures that will enhance the profitability of their mines. With the energy costs associated with operating ventilation systems being a mine's second-highest cost of production after labour - between 25% and 50% (Cheryl and Robert 2009) of the total energy requirements of an underground mine - finding the right ventilation solution can be critical to a mine's performance.

The continued use of diesel-powered machinery in ever deeper underground operations is resulting in the need to deliver a greater volume of fresh air over longer distances, generating additional friction, which in turn generates heat and increases the operating pressures of the ventilation fans (resistance), resulting in more powerful ventilations systems being required.

The challenge is therefore to design and install more sophisticated and reliable ventilation systems that reduce air volumes and natural gas levels, and that significantly lower power consumption and thus deliver cost savings and corresponding reductions in greenhouse gas (GHG) levels.

Taking the time and making the investment to ensure proper planning and design of the ventilation system will ensure that it performs continuously at maximum efficiency, ensuring that the primary fans are able to operate within an optimum range that delivers the required airflow throughout the mine.

Continuous air quality and temperature monitoring are essential to provide miners with a safe and healthy workplace.

Effective Ventilation on Demand Systems (VoD) can save a significant amount of energy in a mine's operation, but come at a cost, both initial and ongoing. These systems will need to be fine-tuned and re-adjusted as the mine grows, and mine ventilation practitioners must be adequately trained to analyse the changing mine conditions and to successfully implement the required ventilation solutions.

In many instances, significant cost and energy savings can also be obtained by analysing and improving the current system on site.

With a growing number of mining companies opting to reduce the size and capabilities of their ventilation engineering departments and increasing their dependence on fan manufacturers, the experience of the fan manufacturer can deliver important benefits to projects.

To achieve optimal results, consideration should be given to the requirement for thorough testing at the factory. Fan suppliers should be able to test their fans at the factory at full power and simulating the mine's future requirements, prior to being despatched to site.

Key words: Energy savings, optimised ventilation systems, VoD, vertical fans, Factory Fan testing

INTRODUCTION

While the mining industry is facing internal and external pressures to reduce operating expenses (OPEX), the energy consumption of ventilation systems continues to increase, resulting in an associated increase in GHG emissions.

Despite improvements in battery technologies, most of the machinery used in underground mining operations continues to rely on diesel power. The move to electric propulsion is gaining pace, but it will be quite some time before it becomes the dominant power source.

Until that time comes, the continued use of diesel-powered machinery in ever deeper underground operations is resulting in the need to deliver a greater volume of fresh air over longer distances, generating additional friction, which in turn generates heat and increases the operating pressures of the ventilation fans (resistance), resulting in more powerful ventilations systems being required.

The challenge is therefore to design and install more sophisticated, efficient and reliable ventilation systems that reduce air volumes and natural gas levels, that significantly lower power consumption and thus deliver corresponding reductions in GHG levels.

With this objective in mind, I propose to analyse some important fan and ventilation system parameters that can contribute to achieving improved performance in each of these areas.

DEFINITION OF FAN PERFORMANCE REQUIREMENTS AND FAN TYPE

The starting point for any analysis is that a mine's ventilation system must be aligned to the mine's production requirements. Correct fan design duty specification and then fan selection is critical for the correct operation of any ventilation system.

An important consideration when providing a fan manufacturer with the design duty point for fan selection is that the size and structure of the mine varies over time which in turn changes the mine resistance, resulting in different fan duty point requirements (flow and pressure) over the Life of Mine (LOM) and, in some instances, for the mine's resistance levels in 25 years' time. Therefore, the duty point used for fan selection should be a succession of operating duty points rather than a single design duty point.

Maximum and minimum fan duties are generally provided to the fan manufacturers, establishing a fan duty envelope which is based on the mine modelling over its predicted LOM.

Effective guidelines for fan specification and tender adjudication are set out in papers published by Brake (2017) and Stachulak and Mackinnon (2002).

The fan supplier is responsible for selecting the most suitable fan type, with maximum efficiency and performance over the entire duty operating envelope, having regard to the mechanical and structural requirements of the fan installation and a mine's ambient operating conditions.

The potential erosion and corrosion of the main fan rotating parts must be studied, and then suitable blade materials selected.

There are several options available for the blade material, namely, stainless steel, special alloys, composite materials and carbon fibre. However, the most common materials used in blade fan applications are cast aluminium and cast iron. Aluminium material is usually preferred due to its lighter mass, however its resistance to wear is limited.

TABLE 1 – Aluminium Vs Steel blades

	Aluminium casting	Steel casting
Advantage	Light	High resistance to wear
Disadvantage	Low resistance to wear	Heavy

Zitrón has responded to this challenge by investing in a new 5-axis high-speed high-precision CNC machine centre, which is capable of machining complex aerofoil blades using superior quality materials such as forged aluminium rather than traditional casting blades.

Forged aluminium is ideal for applications where performance and safety are critical, but a lighter-weight metal is preferred. This option increases impeller blade resistance to erosion and corrosion, with a lower impeller weight (when compared traditional cast iron blades), thereby resulting in extended motor bearing life, among other benefits.

The use of CNC-machined blades provides other important advantages over cast blades, such as shorter lead times. The time to manufacture a blade is now defined in hours as compared to several weeks for traditional cast iron blades.

REDUCING ENERGY COSTS

The benefits of improving a ventilation system's energy efficiency are both financial and environmental. It is therefore vital that mine operators give due importance to ensuring that they put in place the optimum system for their mine.

Taking the time and making the investment to ensure proper planning and design of the ventilation system will ensure that it performs at maximum efficiency, ensuring that the primary fans are able to operate within an optimum range that delivers the required airflow.

There are two key factors to achieving an efficient, reliable and lower energy consuming ventilation system, namely, efficient fan design and an optimised ventilation system.

Testing the Fan efficiency

Achieving the necessary performance efficiencies requires a customised approach, where each fan that is installed in a mine is specifically designed to meet individual site-specific conditions, thereby ensuring that the ventilation system will be continuously running at its highest efficiency rate.

A typical design life for a primary fan installation is 25 years of continuous operation. This long-term durability requirement makes it vital that, prior to their installation, such fans be thoroughly tested in controlled environments while running at full power and at any possible ventilation circuit resistance.

Nowadays, using the available sophisticated simulation software, such as Vuma or Ventsim, the required system flows and pressures at various stages through the LOM can be easily predicted.

Maximum and minimum fan duties are normally provided to the fan manufacturers at the tender stage, therefore factory acceptance tests and site tests, to demonstrate the whole fan capacity are very important. The tests should also provide the mine operators with further useful information about the fan margins and its capability to adapt to un-planned changes in the mine conditions.

To permit such high capacity testing and facilitate the search for innovative engineering solutions Zitrón has invested in a cutting-edge test tunnel at its manufacturing operations in Gijon, Spain (Figure 1).

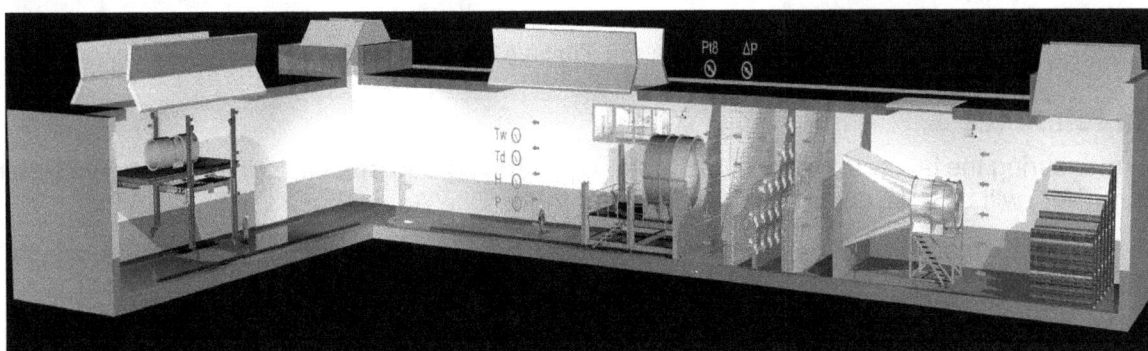

FIGURE 1 - Zitrón's test tunnel in Gijon

Measuring 100 metres in length and 52 square metres in cross-section, this is the largest certified test tunnel in the world, allowing axial fans up to 5.8 metres in diameter to be tested at full power, and up to 2,000 kW, whilst reproducing any real ventilation circuit resistance.

In situ testing has several limitations, with inevitably less accurate tolerance levels being commonly accepted. Also, in most instances it is not practical to alter the system resistance to measure the fan performance at several different duty points, or even close to the contracted duty. This may be one of the reasons why factory tests with stringent tolerance levels should be requested for mine primary fan projects.

The factory tests for a primary mine fan should not be limited to the typical mechanical-run tests that aim to demonstrate compliance with the specified vibration levels. Complete aerodynamic testing is very important when planning the purchase of a primary fan which should operate continuously during the whole LOM.

It is true that the fan factory aerodynamic test (Figure 2), as conducted in the testing laboratory, does not consider the possible 'system effect' once the fan is installed on site. However, the test allows for the complete fan performance testing, which in combination with the manufacturer's system calculations and fluid dynamic simulations will result in a very accurate prediction of any possible fan operating condition on site and allow for any necessary improvement of the system to be made on site as required.

FIGURE 2 – Axial fan at Zitrón's test tunnel in Gijon

Tests tolerance

When measuring the fan performance, the testing tolerance grade applied to the Factory Acceptance Testing (FAT) will be different to the grade applied to the Site Acceptance Testing (SAT).

There are three main standards generally followed by fan manufacturers and mine ventilation practitioners. These are:

- ISO 5801 (2007): Industrial fans – Performance testing using standardized airways. This standard is used for the Factory performance tests.

- ISO 5802 (2001): Industrial fans – Performance testing in situ. This standard is used for the Site performance tests.

- ISO 13348 (2008): Industrial Fans – Tolerances, Methods of Conversion and Technical Data Presentation. This standard sets out the applicable testing tolerance grades.

Please refer to table 2 below, showing the tolerance grades as defined in the ISO 13348.

TABLE 2 - (Extract from ISO 13348 Table 1: Guide for fan air and noise tolerance grades 1 to 4)

Tolerance grade (air and noise)	Typical application	Material of, and manufacturing processes used for, major aerodynamic components	Approx. min. power[a] kW
AN1	Mining (e.g main fan), process engineering, power stations (e.g. exhaust fan), wind tunnels, tunnels, etc.	Machined in some places, cast (high accuracy)	> 500
AN2	Mining, power stations, wind tunnels, tunnels, process engineering, air conditioning	Sheet or plastic material, partly machined, cast (medium accuracy)	> 50
AN3	Process engineering, air conditioning, industrial fans, tunnels, power station fans and industrial fans for harsh (abrasive or corrosive) conditions	Sheet material, cast (medium to low accuracy), special surface protection (e.g. hot-dip galvanizing, moulded plastics	> 10
AN4	Process engineering, ships fans, agriculture, small fans, power station fans and industrial fans for harsh (abrasive or corrosive) conditions	Sheet material, special surface protection (e.g. rubber coating), moulded or extruded plastics	-

[a] For each class, a recommendation has been given only for the lower power limit; an upper limit is not essential. For example, even if the power is greater than 500 kW, any one of the grades may be assigned.

While it is commonly accepted that tolerance levels (table 2 above) AN2 or AN3 will be applied for in-situ testing, more stringent levels can be assumed by the fan manufacturer when the fan is tested in a controlled environment.

Some of the mining fans tested by Zitrón in the test-tunnel can be measured to a tolerance grade of **AN1** (table 3 below).

Table 3 - Zitrón typical FAT tolerance grade

Parameter	Tolerance grade				Additional information
	AN1	AN2	AN3	AN4	
Volume flow rate	**±1%**	±2.5%	±5%	±10%	
Fan pressure	**±1%**	±2.5%	±5%	±10%	
Power	**+2%**	+3%	+8%	+16%	Negative deviations permissible
Efficiency	**-1%**	-2%	-5%	-12%	Positive deviations permissible

In some cases, Zitrón is also capable of offering the AN1 tolerance grade applied to the entire fan performance curve and not only to its near-peak efficiency portion (as it is allowed by the ISO 13348). This may be particularly important for mine fans required to operate over a wide range of their fan curve over their operating life, or for those cases when the mine duty has been incorrectly estimated.

The Zitrón FAT provides useful and accurate information about a fan's full capabilities as well as about its flexibility to adapt to unplanned changes in the operation conditions.

Optimised Ventilation System

Before opting to invest in the installation of more technological and expensive systems, mining companies should commit to analysing a mine's current ventilation design, looking for any optimisation that may be possible.

CFD Simulations

Computational Fluid Dynamics software (CFD) is used to identify improvements that can minimise system pressure loss (figure 3).

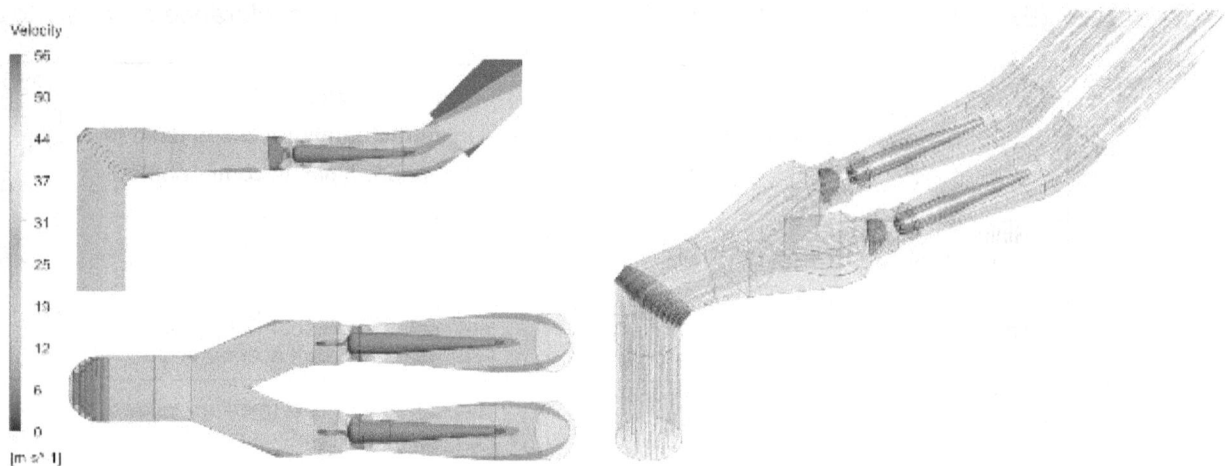

Figure 3 - System velocity path lines produced using Computational Fluid Dynamics software

It will allow the system designers to optimise the geometry of the key areas of the mine with the highest-pressure loss and thus energy consumption wastage.

Booster ventilation systems

In some cases, a mine's primary ventilation plan can be optimised by installing "booster" ventilation systems (figure 4), which involves several small standard auxiliary fans running in parallel, thereby reducing the total pressure to be overcome by the primary fans.

Figure 4 - Booster system based on small auxiliary fans

In the right environments, booster systems may allow for the required fresh air to be delivered over longer distances, thus avoiding the requirement to replace the existing main fans with more powerful units and producing a cost saving.

Mitigation of risks associated with starting up parallel fans

While the use of booster fans systems is widespread within the industry, there are some important considerations to take into account when planning a ventilation system based on fans running in parallel. Installing two or more booster fans in parallel may generate problems during the fan start up.

One of the traditional solutions would consist of the installation of a release valve or recirculation damper that allows for the fan flow recirculation, reducing the system resistance that the fans are working against. The valve can be remotely closed once the fans are running at full speed.

Please refer to figure 5 showing a damper door installed at the bulkhead in a ventilation system with two underground axial fans.

Figure 5 - Booster system with recirculation damper

Although the recirculation damper solution is well accepted within the industry, the author's experience has been that the use of Variable Speed Drives ('VSDs') is the best option in most cases.

There are several advantages of using VSD instead of Soft Starter or Direct On Line (DOL) systems (namely, reduced power line disturbances, lower power demand on start, controlled acceleration, adjustable torque limit, controlled stopping, amongst others) and avoiding potential problems during parallel fan operation is one of them.

When a fan enters the unstable part of its curve (dashed line) it will start to stall. This will occur when the resistance of the circuit is higher than the "stall limit" of the fan, see figure 6.

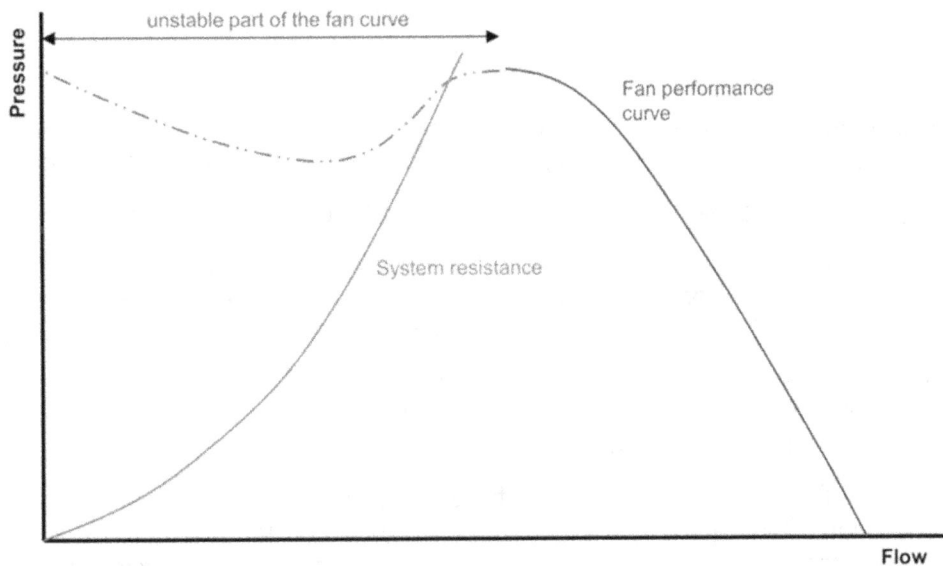

Figure 6 – Fan performance curve, Flow Vs Pressure

As can be seen in figure 7 below, when a single fan is running in the system, it will be delivering air flow Q1 at total pressure P1.

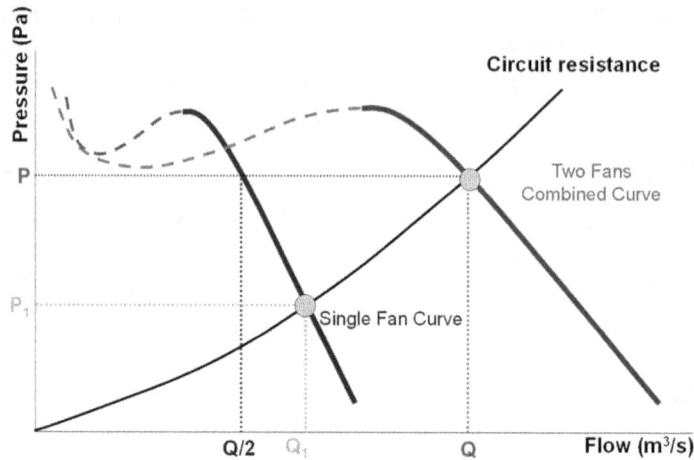

Figure 7 – Single fan and Two Fans combined curve

Running two or more fans in parallel will induce stalling earlier at a lower operating pressure. This is because successive fans must start up against the pressure already being developed by the running fan or fans, this pressure is called initial pressure, P0, see figure 8.

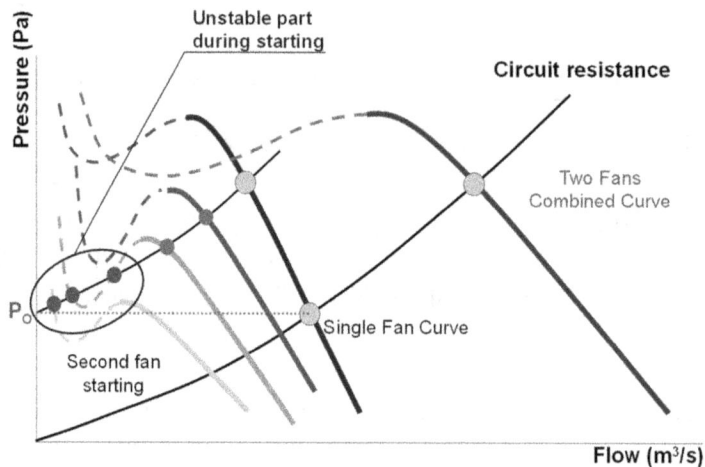

Figure 8 – Second fan starting

Running more fans in the system will develop a higher initial pressure in the system. This is the reason why running more fans in parallel will increase the risk of fan stall within the system.

The best way to mitigate or avoid this problem is to start all the fans simultaneously.

When fans are started across the line (Direct On Line), the starting current can be very high, up to 7 times the motor Full Load Current (FLC). If all the fans need to be started at the same time, the starting current will be very high and the power supply on site will in turn be highly oversized.

In order to avoid oversizing of the power supply, the use of VSDs is a very good option. VSD will limit the starting current to a fan by 1.1 > Motor FLC >1.2 on start up. In terms of electrical requirements on site, VSD will be the most cost-effective solution.

Moreover, the VSD will provide additional advantages, such as the ability to control the motor speed of any fan in operation. This will be very advantageous when one fan needs to be started while the remaining fans are already in operation. During the ventilation system's commissioning, a maximum "synchronizing" motor speed will be set up in the VSD. The synchronizing speed will be applied to all fans in the system, and it will be a value at which the initial system pressure is at an acceptable level to minimise risk of stall during the fan starting procedure.

Ventilation on Demand (VOD) systems

Once the existing ventilation system is optimised, the next option is to evaluate the implementation of VOD systems.

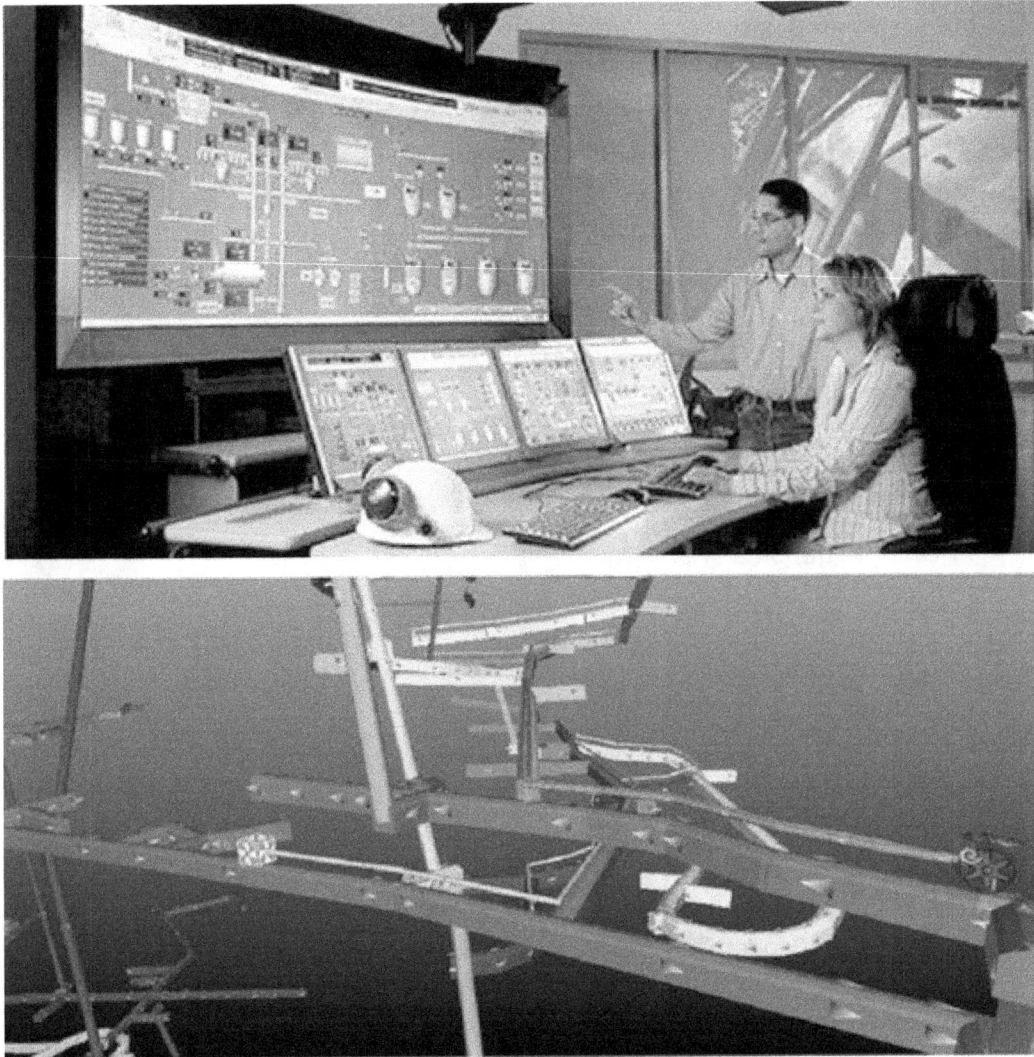

Figure 9 - Ventilation On Demand (VOD) Control Panel and Mine Layout

There is a widely-held belief in the mining community that more powerful fan systems will produce greater airflow. However, in many cases, this is either not true in any practical sense or is a very expensive way to achieve the required increase, as it involves both higher capital and operating costs.

A more sophisticated ventilation system based on real-time adjustment to the underground ventilation needs may help to reduce the air volumes required underground, which will in turn decrease power consumption, with a corresponding reduction in GHG.

VOD is an automated system that delivers fresh air to the underground areas only when and where it is needed, namely, when operating equipment and/or personnel are present.

By implementing a VOD system, the mine ventilation is provided with a real-time monitoring of the conditions at the working area, providing a safer environment for miners, generally through the availability of ample fresh air, and in some cases lower temperature and gas levels among others.

VOD systems will reduce the total air requirements and therefore the energy required to operate the fans will be lower. They have the potential to not only save power, but also equipment capital cost (smaller fans may be needed), as well as improve safety both in normal operations and also during emergency events such as a fire, when the ventilation system may be used for preventing the spread of smoke and enabling faster exhaust of hazardous fire gases.

The backbone of a typical VOD system may be as follows:

- **Main SCADA Control**: Allowing all inputs from the tracking system to be analysed and ventilation network equations solved in real-time.

- **Sensors and gas monitors**: Ensuring continuous monitoring of the conditions at any working area.

- **Automated flow regulators**: Enabling the dampers (blade doors) installed underground to control the resistance of the system and therefore the ventilation volumes.

- **Fan VSDs starters**: The Variable Speed drives (VSDs) will allow fan motor speed adjustment, therefore the control of volume flow delivered by the fans.

VOD Theory

The Ventilation On Demand basic principle to save energy is derived from the fan laws, derived using the Buckingham π theorem.

- Volume Flow, equation 1 $\quad q_{V2} = q_{V1} \times \left(\frac{n_2}{n_1}\right)^1 \times \left(\frac{d_2}{d_1}\right)^3$ (1)

- Pressure, equation 2 $\quad p_2 = p_1 \times \left(\frac{n_2}{n_1}\right)^2 \times \left(\frac{d_2}{d_1}\right)^2 \times \left(\frac{\rho_2}{\rho_1}\right)^1$ (2)

- Power, equation 3 $\quad P_{w2} = P_{w1} \times \left(\frac{n_2}{n_1}\right)^3 \times \left(\frac{d_2}{d_1}\right)^5 \times \left(\frac{\rho_2}{\rho_1}\right)^1$ (3)

Where:

- qv, is the volume of air
- n, is the rotational speed of fan
- d, is the diameter of fan impeller
- p, is the pressure generated by the fan
- Pw, is the power demanded by the fan impeller
- ρ, is the density of air

Since the flow is proportional to speed, pressure is proportional to the square of speed and power is proportional to the cube of speed, therefore as is shown in the table in Figure 10, a reduction of 20% in the flow will save close to 50% of the power required to operate the ventilation fans.

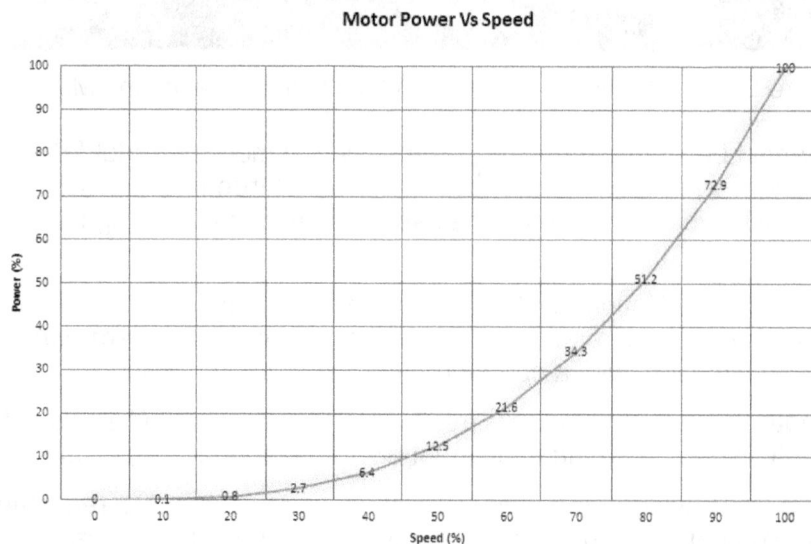

Motor Power Vs Speed

Figure 10 – Ventilation On Demand (VOD) Energy Saving Calculations

VOD systems offer the potential to achieve considerable energy savings. Correctly implemented, they can help to lower the total energy consumption demanded by the ventilation system by 25% to 50%.

However, a drawback of VOD is that its implementation is difficult and costly. Skilled personnel are required for both system integration in the mine and also during operation and maintenance of the equipment.

In addition, each mine is different and must be analysed individually and it is recommended to carry out a case study to decide which is the best option for each mine.

Alternative Solutions

The use of innovative but reliable systems may improve significantly the efficiency of the ventilation systems. Whilst horizontal twin fan installations (Figure 11) are the most commonly chosen option for underground environments, their use involves higher energy consumption, due to the need for 90 degree bends to connect the fans to the ventilation air rise.

The main reason to choose on a twin system is "reliability", as 60% of the system flow can be delivered by one fan only.

Figure 11 - Main Fans running parallel

Vertical Fans advantages

Some of the advantages of the vertical fan installations (Figure 12) can be found below:

1. Reduced footprint. The area required by a typical vertical fan installation is generally about one quarter of that with horizontal fan of the same capacity. That generally results in simplified installation and reduced constructions time and costs.

2. More Robust design. Vertical fans provide the ventilation system with an efficient aerodynamic, structural and civil works design.

3. Energy savings. The system shock losses are minimised as the fan is mounted directly on to the shaft collar and the air flow is straight through, eliminating the pressure loss generated in the ductwork connecting the shaft collar and the horizontal fans.

4. Reduced vibration levels. One of the most common reasons for fan down time is vibration. Vertical fans typically have strengthened designs where the load is distributed evenly over a strong circular section. The fan load transfer to the civil works (foundation) is more efficient in a vertical fan installation than in a horizontal one. The axial fan is symmetrical about its rotational axis, therefore the distortion resulting from load transfer in a radial direction is avoided.

5. Increased reliability. The potential issue due to fatigue is reduced, therefore the system reliability factor is significantly increased.

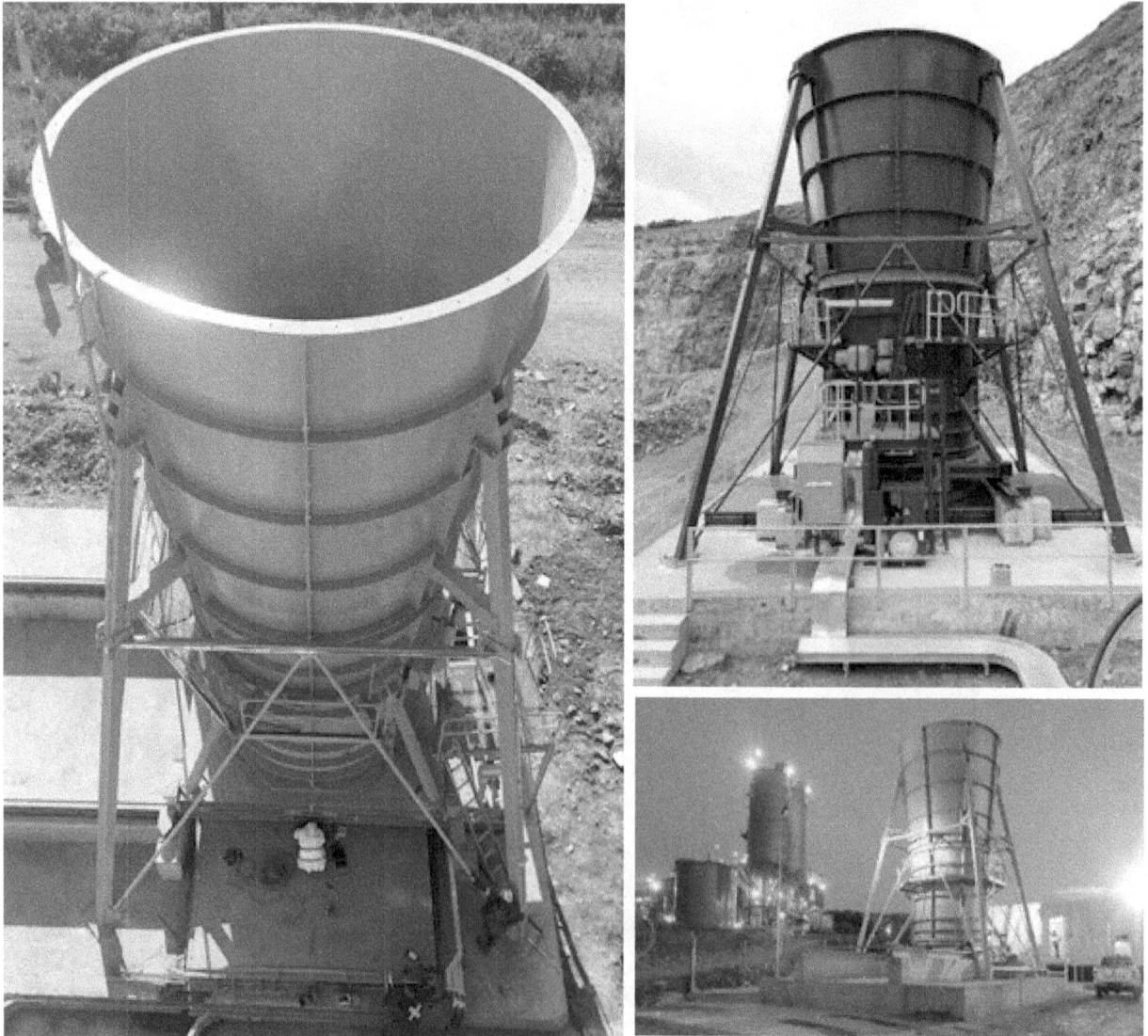

Figure 12 - Vertical mine main fans

Roll out system for Vertical Fans

Both mine operators and fan manufacturers need to understand the critical nature of a reliable ventilations system to the mining operation.

Although ventilation fans are not generally considered part of the mine production equipment, their reliability can seriously impact on mine production.

Zitrón offers a 'roll out system' (Figure 13) with rails that permits much easier and shorter interruptions to operations in the case of fan replacement and maintenance.

Figure 13 - roll out system

A whole primary fan replacement operation can be performed in 4 hours if a complete spare fan unit is available on site, as it is not necessary to remove the Fan evasé.

Another great benefit of the "roll out system" is that in the case of regular maintenance, the fan can be removed and access to the impeller is much easier. As a consequence, the cleaning of the impeller, the blade adjustment or any maintenance that needs access to the inner parts of the fan will be safer, easier and quicker.

Replacing the fan unit without using the roll-out system would involve the operation taking about 14 hours (assuming that a complete spare fan unit is available on site).

If the spares on site were one impeller and one motor (no fan casing), the time of disassembly and assembly of the impeller and motor should be also taken into account.

As is shown in the tables below, equipment replacement times are much quicker when using a roll-out system and the appropriate spare parts strategy is followed.

1. **Replacement time scenario A: Complete fan unit as spare (impeller, motor and casing).**

	No-Roll Out	Roll-Out
Fan replacement time	12 hours	4 hours

2. **Replacement time scenario B: Spare parts impeller and motor only.**

	No-Roll Out	Roll-Out
Impeller replacement time	22 hours	14 hours
Motor replacement time	24 hours	16 hours

Dugald River primary ventilation vertical fans - A Case study

Zitrón was recently challenged in Australia to deliver a ventilation system that would meet the highest technical requirements for air volumes, operational reliability and lowest running costs.

The project was for the provision of surface fans for the Dugald River underground zinc mine, operated by MMG Limited and located in Mt Isa, Queensland. It involved the installation of four (three in use and one spare) ZVNv 1-36-1400/8 (Figure 14) heavy duty Axial Vertical fans with impeller diameters of 3,600 mm and total installed power of 4,200 kW (1,400 kW each fan).

Figure 14 - Dugald River mine main fans

Zitrón's solution was based on using single vertical axial fans mounted directly on to the shaft collar, complemented by an innovative guide rail system that allows for a fast 'change-out' of the fan.

Extensive testing showed this to be the most efficient and effective solution for guaranteeing the required air volumes and operational reliability.

Zitrón worked with the client's ventilation engineers to analyse the changing mine conditions and develop an operating strategy that delivers increased system efficiency and savings of up to $1 million in annual operational expenditure, MMG Newsletter (September 2016).

CONCLUSIONS

Energy consumed by mine ventilation systems is higher than ever due to stringent specifications of the equipment used underground (primarily combustion engine emissions), to air quality standards and to mining at deeper levels which requires more powerful ventilation systems.

Continuous air quality and temperature monitoring are essential to provide miners with a safe and healthy workplace. Ventilation on Demand Systems may save a significant amount of energy in a mine's operation, but in most cases, significant energy savings can also be obtained by analysing and improving the current ventilation systems on site.

Fan factory testing, in combination with the manufacturer's system calculations and fluid dynamic simulations, will result in a very accurate prediction of any possible fan operating conditions on site and allow for any necessary improvement of the system to be made on site as required.

Consideration must be given to the requirement for full testing at the factory. Fan suppliers should be able to test their fans at the factory (prior to being despatched to site) at full power and simulating the mine future requirements. When testing the fans in a controlled environment, the most stringent testing tolerance grades can be accepted by the fan manufacturer and most importantly without the necessity to unnecessarily oversize the fan or increase its capital cost. This will ensure compliance with the project specifications and will help to mitigate any doubts or uncertainties about fan performance when testing on site.

REFERENCES

Brake,D J, 2013. Performance and acceptance testing of main mine ventilation fans, in Proceedings Australian Mine Ventilation Conference, pp 141–148 (The Australasian Institute of Mining and Metallurgy: Melbourne).

Stachulak, J S and Mackinnon, K A, 2001. Mine ventilation fan specification and evaluation, in Proceedings 7th Int Mine Vent Conf, pp 203–211

Cheryl L. Allen & Robert Da Prat, Variable Frequency Drives as Applied to Ventilation Systems (CIM Toronto May 2009)

MMG Limited, Australian operations newsletter, September 2016

The influence of atmospheric pressure change principles on gas sampling in mine closed fire zone

B W Lei[1], B Wu[2], H Y Wang[3] and H Q Zhu[4]

1. Dr. LEI, China University of Mining and Technology (Beijing), Ding No.11 Xueyuan Road, Haidian District, Beijing 100083 P. R.China. Email: leibws@163.com
2. Pro. WU, China University of Mining and Technology (Beijing), Ding No.11 Xueyuan Road, Haidian District, Beijing 100083 P. R.China. Email: wbelcy@vip.sina.com
3. Pro. WANG, China University of Mining and Technology (Beijing), Ding No.11 Xueyuan Road, Haidian District, Beijing 100083 P. R.China. Email: whyhyp@163.com
4. Pro. ZHU, China University of Mining and Technology (Beijing), Ding No.11 Xueyuan Road, Haidian District, Beijing 100083 P. R.China. Email: zhq@cumtb.edu.cn

ABSTRACT

Analysing the change of gas composition in mine closed fire area is the most important basis for evaluating the combustion state of fire area. However, due to the fluctuation of ground atmospheric pressure, the gas composition within a certain distance of the closed wall in mine closed fire area is affected by air leakage, which can not effectively reflect the true state of fire area. Grasping the law of ground atmospheric pressure change is the basis for drawing up the gas sampling scheme in mine fire area. Based on the annual atmospheric pressure hourly recorded data from six surface meteorological monitoring stations in Shanxi Province, China, this paper analysed the diurnal periodicity, annual periodicity, fluctuation amplitude and change rate of atmospheric pressure changes, and discussed and analysed the influence of surface atmospheric pressure variation law on gas sampling method and sampling timing in closed fire areas. The results show that: (1) the variation of atmospheric pressure does not have a diurnal periodicity, but the distribution of atmospheric pressure troughs and peaks is regular in every day; (2) The fluctuation of atmospheric pressure lead to the irregular expansion and contraction of the gas in the closed fire zone. Therefore, in order to obtain the real gas in the fire zone, it is necessary to monitor the pressure difference inside and outside the closed wall, and analyse the pressure difference inside and outside the closed wall synchronously when analysing the gas composition; (3) Continuous sampling and analysis equipment should be used to conduct gas analysis in closed fire area for correct analysis of combustion state in closed fire area.

INTRODUCTION

After the closure of the mine fire area, sampling and analysis of the gas in the fire area through the closed wall is an important method to judge the combustion state of the closed fire area, and the reliability of gas sampling is the premise to accurately analyse the combustion state of the fire area(Zhou, X., and Wu, B,1996) (Fauconnier,1992). The gas in mine closed fire area is not stationary. Because of the fluctuation of ground atmospheric pressure, the gas in closed fire area exchanges with the outside environment continuously through the gap of the closed wall, causing the gas in closed fire area to present the breathing state of 'expansion-contraction' (Gautier G,2004. When the gas in closed fire area expands, the gas in the fire area seeps out through the closed wall and closes the fire. When the gas in the zone shrinks, the outside environment leaks to the closed fire zone. Therefore, when the gas in the closed fire zone is expanding, gas sampling can more effectively reflect the real gas state in the closed fire zone(Cheng,2018).

The fluctuation law of surface atmospheric pressure is very complex. Currently, few researchers have studied its hourly variation law(Francart,1997). The understanding of periodic variation of atmospheric pressure is limited to diurnal and annual variation(Ren and Jiang,2005). When planning the time of gas sampling, the rescuers only consider that the change of atmospheric pressure is inversely proportional to the change of temperature. Therefore, the task of gas sampling is arranged in the period of the highest temperature of the day, and the time of gas sampling is fixed everyday for comparative analysis. However, besides temperature, atmospheric pressure variations are also affected by atmospheric circulation, atmospheric tides, earth movement and other factors. Abnormal weather such as cold wave outbreak and thunderstorm process can also destroy the periodic variation of atmospheric pressure. Therefore, the non-periodic variation of atmospheric pressure is much more complex than the periodic variation.

In order to effectively grasp the law of atmospheric pressure variation and provide references for the formulation of gas sampling measures in mine closed fire areas, based on the hourly atmospheric pressure data recorded in six areas of Shanxi Province, China, this paper made statistical analysis of the period of atmospheric pressure variation, fluctuation amplitude and rate of atmospheric pressure variation, and determined the principle of gas sampling in mine closed fire areas, so as to ensure the reliability of gas sampling results.

DATA SOURCE AND PREPROCESSING

Data source

Shanxi Province of China is the largest coal production area with a cumulative coal production of more than 16 billion tons, which accounts for more than a quarter of China's total annual coal production.It is a temperate continental monsoon climate with $34° \ 34' \ —40° \ 44'$ north latitude and $110° \ 14' \ —114° \ 33'$ east longitude. Coal resources in Shanxi are widely distributed. From north to south, the coalfields in Shanxi can be divided into Datong coalfield, Ningwu coalfield, Hedong coalfield, Xishan coalfield, Huoxi coalfield and Qinshui coalfield. Therefore, according to the classification of coal fields and considering the principle of average distribution in Shanxi Province, six sites (Datong, Ningwu, Lvliang, Taiyuan, Huozhou and Xiangyuan) were selected as typical representatives. The location information of each monitoring site is detailed in Fig. 1 and Table 11. The required atmospheric pressure hourly data were enquired from NMIC, short for National Meteorological Information Center (http://data.cma.cn/). The time and frequency of atmospheric pressure and temperature data acquisition were from January 1, 2017 to December 31, 2017.

1-Datong 2-Ningwu 3-Lishi 4-Taiyuan 5-Huozhou 6-Xiangyuan

FIG 1 – Geographical location map of Shanxi meteorological data monitoring station

Hourly data preprocessing of annual atmospheric pressure

Through the analysis of the meteorological data of Datong, Ningwu, Lvliang, Taiyuan, Huozhou and Yuanyuan, it could be found that due to unknown reasons of NMIC, the atmospheric pressure data are delayed 6 hours than the real time, so t is needed to adjust the atmospheric pressure data . There were 416 missing data in the six monitoring sites in total, the linear interpolation method was used to fill in the missing data since most of the missing data were single-point data. Based on the observation results, the annual atmospheric pressure hourly data trend charts in different regions are shown in Fig. 2.

TABLE 1 – Geographic location information of monitoring stations

Location	Station number	latitude (°)	longitude (°)	altitude of observation site （m）	altitude of barometric pressure sensor (m)
Datong	53487	113.25	40.05	1067.2	1068.4
Ningwu	53577	112.18	39	1437.4	1438.7
Lishi	53764	111.06	37.3	950.8	951.2
Taiyuan	53772	112.35	37.37	778.3	779.4
Huozhou	53869	111.42	36.35	550	551.2
xiangyuan	53884	113.02	36.31	877.9	879.1

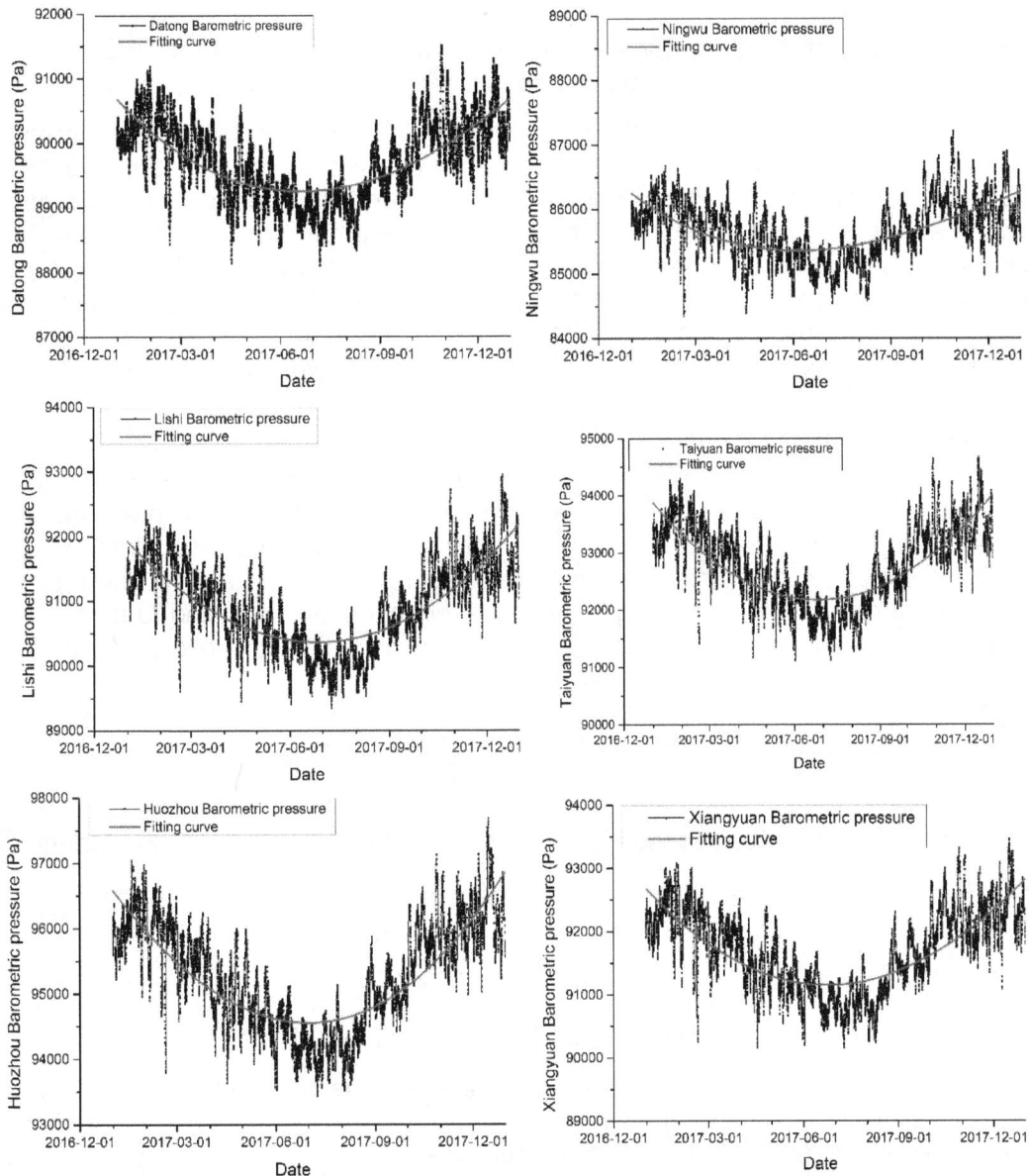

FIG 2 – Trends of atmospheric pressure hourly data in Shanxi Province

According to the analysis of Fig.2, the atmospheric pressures of the six locations of Datong, Ningwu, Lvliang, Taiyuan, Huozhou and Xiangyuan are relatively large, with fluctuations ranging from 2860Pa to 4250Pa. Based on the analysis of the annual trend of the above six regions, the atmospheric pressure change in Shanxi Province has obvious seasonal variation characteristics. The average atmospheric pressure is lower in summer, the average atmospheric pressure is higher in winter, and Huozhou has the largest annual amplitude, its value is 4250Pa.

The daily atmospheric pressure fluctuations of various locations in Shanxi Province are very severe, and extreme weather changes will cause rapid and dramatic changes in atmospheric pressure, such as the large-scale heavy snowfall in Shanxi Province on February 20, 2017 and the windy blue warning issued by the Shanxi Meteorological Bureau on April 17, 2017 ,the above-mentioned six regions suddenly experienced a rapid drop in atmospheric pressure. Among them, the atmospheric pressure in Huozhou was the most severe when heavy snowfall occurred, the atmospheric pressure in the region dropped continuously within 21 hours by 2220Pa.

RESULTS AND DISCUSSION

Time statistics of Atmospheric pressure peak

By extracting the peak data from the atmospheric pressure hourly data in Figure 2, the number of moments corresponding to the annual atmospheric pressure peaks in the six locations of Datong, Ningwu, Luliang, Taiyuan, Huozhou and Xiangyuan are analyzed, and a histogram is drawn, as shown in Fig. 3, where the horizontal axis represents the time of day, and the ordinate represents the quantity distribution of atmospheric wave peaks throughout the day.

It can be seen from Fig. 3 that during the day, the most frequent occurrence of the atmospheric pressure peak is from 6:00 to 8:00, and the peak occurrences in this period accounts for 23% to 30% of the total occurrences per day. The second most frequent time period which the atmospheric pressure peak waves occures is from 20:00 to 22:00, and the proportion is 22% to 26%. The results of this study are similar to those other researchers that the changes in atmospheric pressure are inversely proportional to the change in temperature. Therefore, when performing manual sampling analysis in closed fire areas, the sampling time is set from 11:00 to 17:00, so that the impact of air leakage from the closed wall on the fire zone gas can be avoided with a higher probability.

Atmospheric pressure rate analysis

Variation in atmospheric pressure fluctuations can cause changes in the pressure difference inside and outside the enclosed wall to cause air leakage. When the value of air pressure outside the enclosed wall is high, the external environment leaks air into the enclosed area, otherwise the enclosed area leaks air to the outside environment. Assuming that the airflow flowing through the enclosed wall is laminar flow, according to the gas equation of state and Darcy's law, the air exchange between the enclosed area and the external environment is as follows(Zhou, X., and Wu, B,1996):

$$Q = \frac{\Delta P_a}{\Delta t} \frac{V}{P_{s0}} \left[1 - \exp\left(-\frac{P_{s0}}{RV} \right) t \right] \qquad (1)$$

where Q (Pa) is the amount of air leakage from the enclosed wall caused by atmospheric pressure changes. ΔP_a is the amount of atmospheric pressure change in the external environment (Pa). When the outside air leaks into the enclosed area, it is proposed to be positive, and it is negative in the opposite case. P_{s0} is the initial pressure in the enclosed area (Pa). V is the volume of the enclosed area (m3). R is the total wind resistance of the closed wall ($N \cdot s/m^5$).

According to Eq. (1), the change in gas compression and expansion in the enclosed area is proportional to the rate of change of atmospheric pressure, independent of the absolute value of atmospheric pressure change. When the atmospheric pressure increases rapidly, the amount of air leakage from the outside to the enclosed area increases. When the atmospheric pressure drops rapidly, the gas in the enclosed area expands rapidly, and the amount of gas in the enclosed area increases to the outside. Gas sampling analysis at this time can better reflect the state of the gas in the enclosed area.

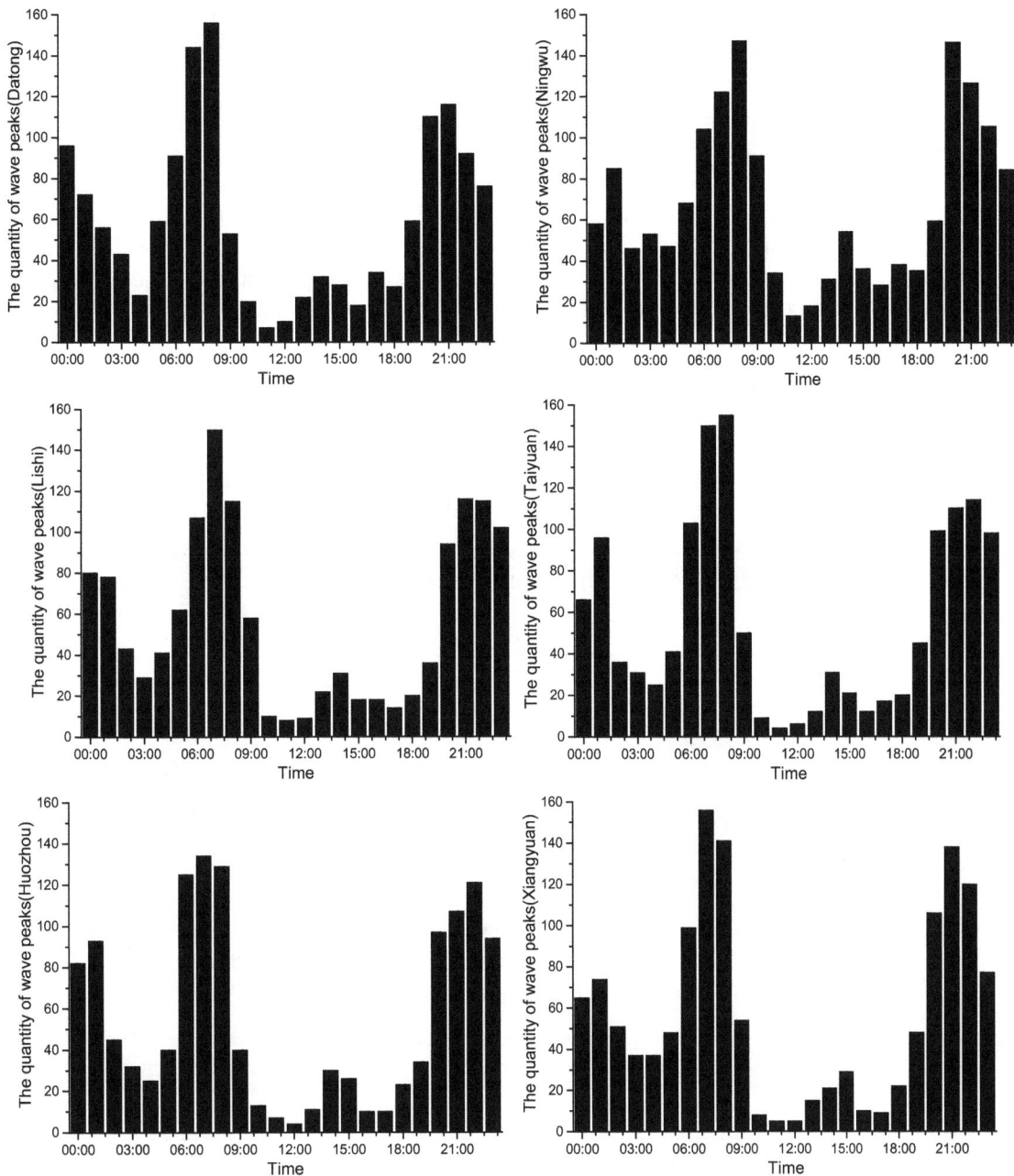

FIG 3 – Distribution of the number of atmospheric pressure peaks occurring in a day

By data-processing the atmospheric pressure hourly data in Fig. 2, the change characteristics of atmospheric pressure decay rate and increase rate are analyzed and the annual atmospheric pressure occurrence times at different decay speeds are counted, which are shown as Table 2. The number of occurrences of annual atmospheric pressure at different increasing rates are counted and are shown in Table 3.

By comparing and analyzing Tables 1 and 2, it could be drawn that the statistical law of the atmospheric pressure decay rate and increase rate is basically the same, that is, the atmospheric pressure change rate is mostly less than 50 Pa/h, followed by 100 Pa/h, and only few times that atmospheric pressure decay rate or increased rate are higher than 100Pa/h.

Periodic Analysis of Atmospheric Pressure

Spectral analysis of discrete data by Fourier transform is an effective method to analyze the periodic characteristics of signals. However, the spectrum analysis method is only suitable to the analysis of stationary and dynamic data. It is necessary to eliminate trends affection from non-stationary curves, and then perform Fourier transform analysis. After eliminating trends from the atmospheric pressure hourly data in the six sampling locations of Datong, Ningwu, Lvliang, Taiyuan, Huozhou and Handan, the time-varying data spectrum of the atmospheric pressure at each location are plotted in this paper, as shown in Fig.4.

It can be found that the main frequencies of atmospheric pressure data change are all near zero, that is, the periodic variation of atmospheric pressure is weak, and there is a small peak of 0.04162 in each region, which corresponds to the atmospheric pressure fluctuation period of 24 hours. The results show that the change of the ground pressure in Shanxi Province in China is dominated by non-periodicity, and the daily cycle changes are covered by non-periodicity. Therefore, in order to reduce the influence of non-periodic fluctuations of atmospheric pressure on gas sampling, it is preferable to use a gas continuous sampling device in the gas sampling analysis of mine fire area, such as a beam tube continuous analysis system to develop continuous sampling and analysis of the sampling site, and simultaneously monitor and record the change of pressure difference inside and outside the mine closed wall, so as to compare and analyze the pressure difference change during gas analysis to help determine the cause of gas change.

TABLE 2 – Statistics on the occurrence frequency of different attenuation rates at atmospheric pressure

	Atmospheric pressure decay rate（Pa/h）			
	$(-\infty,-200]$	$(-200,-100]$	$(-100,50]$	$(-50,0]$
Datong	2	14	209	855
Huozhou	0	18	19	933
Lishi	2	10	222	802
Ningwu	1	7	160	1059
Taiyuan	3	16	271	745
Xiangyuan	3	12	239	813

TABLE 3 – Statistics on the number of occurrences of different atmospheric pressure increase rates

	Atmospheric pressure increase rate（Pa/h）			
	$(0,50]$	$(50,100]$	$(100,200]$	$(200,+\infty)$
Datong	976	148	13	3
Huozhou	163	771	229	0
Lishi	921	160	11	2
Ningwu	1090	139	12	0
Taiyuan	877	195	19	3
Xiangyuan	872	215	22	1

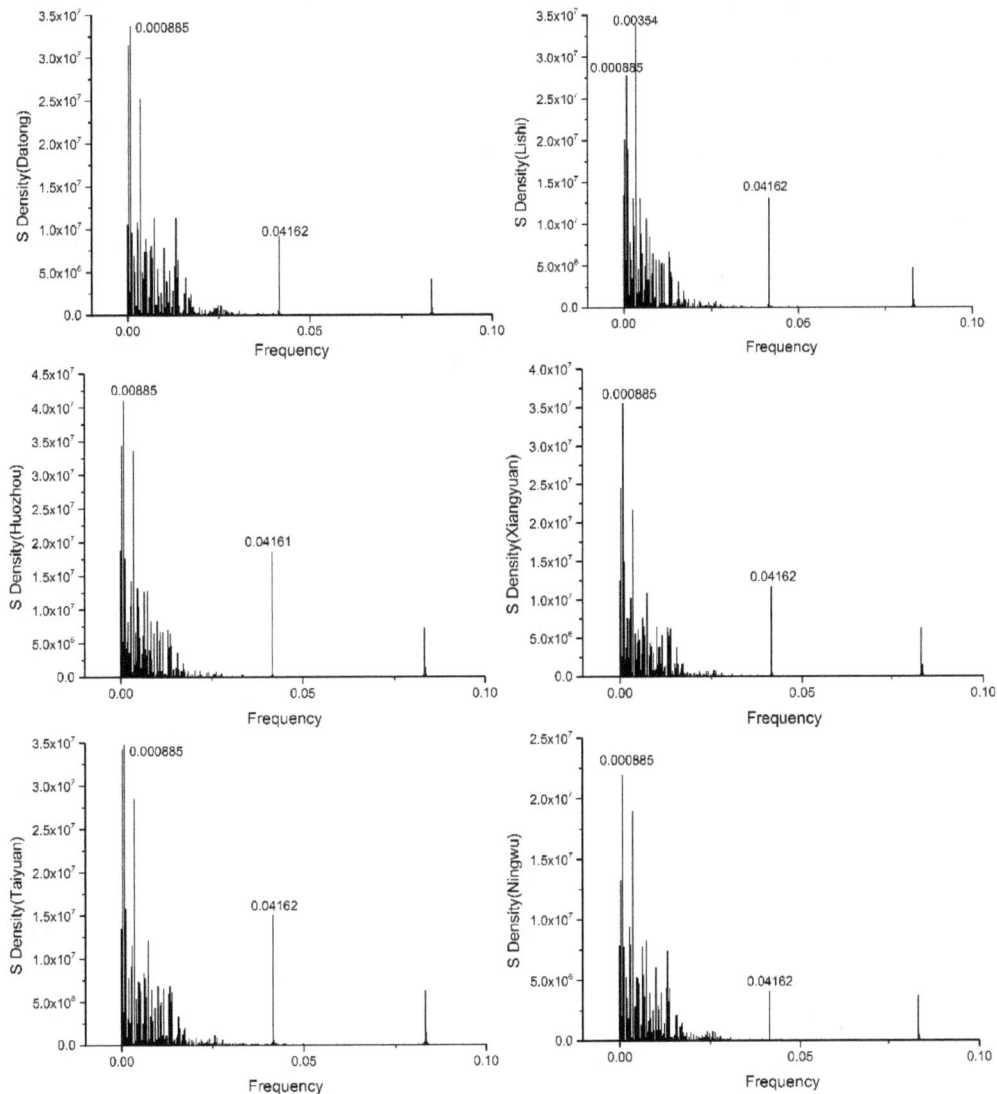

FIG 4 –   Hourly atmospheric pressure spectrum

CONCLUSION

(1) According to the characteristic analysis of atmospheric pressure hourly data, the annual cycle of atmospheric pressure changes regularly, and more than 50% of the atmospheric pressure- -wave peak are between 6:00~8:00 and 20:00~22:00 every day. Mine fire gas analysts are suggested to distribute the manual sampling time at 11:00~17:00 if they can't get continuous monitoring devices to conduct continuous sampling the mine closed fire area;

(2) The daily atmospheric pressure periodicity is weak, the irregular expansions and contractions of gas in the closed fire area are caused by the fluctuation of atmospheric pressure. Therefore, in order to obtain the real gas in the fire area, it is necessary to monitor the pressure difference and analyze the pressure difference inside and outside the closed wall synchronously when analyzing the gas composition; Because the amount of gas emitted from the enclosed area through the closed wall to the external environment is small, when the gas comes to the external environmen，it gets diluted greatly in a very short time, which is not conducive to the analysis of the gas composition in the fire area. Therefore, it is still necessary to collect the gas in the enclosed wall for component analysis.

(3) In order to eliminate the influence of atmospheric pressure fluctuation on gas migration, continuous sampling and analysis devices should be used to get samples from the closed fire zone gas when analyzing gas compositions in the closed area of mine fire zone.

REFERENCES

Gautier G , Bassanino M , Fernando T , et al. 2004 .Assessing the status of sealed fire in underground coal mines. Journal of Scientific & Industrial Research, 63(7):579-591.

Cheng, Jianwei. 2018. Explosions in Underground Coal Mines. (Springer International Publishing AG)

Fauconnier C J . 1992 .Fluctuations in barometric pressure as a contributory factor to gas explosions in South African mines. Journal of the South African Institute of Mining & Metallurgy, 92(5):131-147.

Francart, W., & Beiter, D. (1997). Barometric Pressure Influence in Mine Fire Sealing. In R. Ramani (Ed.), Proceedings of the 6th International Mine Ventilation Congress (pp. 341–342). New York, NY

Ren Jiang, Jiang Cheng- yu. 2005. The Relation of Gas to Gush in Excavated Roadway with Changes of Atmospheric Pressure and How to Handle It. Coal Technology, 24(5):1-3.

Zhou, X., and Wu, B. 1996. Theory of mine fire rescues and applications. (Coal Mining Industry Press: Beijing).

A study on the performance of novel blast-proof door under gas explosion

R K Pan [1*], T Qiu [2], J Wang [3], L G Zheng [4], X B Zhang [5] and M G Yu [6*]

1. Associate professor; Department of Safety & Science Engineering, Henan Polytechnic University, The Collaborative Innovation Center of coal safety production of Henan Province, Henan Polytechnic University, Henan key laboratory of prevention and cure of mine methane & fires; Henan, Jiaozuo 454000, China; prk2018@126.com
2. Department of Safety & Science Engineering, Henan Polytechnic University; Henan,Jiaozuo 454000, China; 1755958453@qq.com
3. Lecturer; Department of Safety & Science Engineering, Henan Polytechnic University; Henan, Jiaozuo 454000, China; wjhpu@hpu.edu.cn
4. Professor; Department of Safety & Science Engineering, Henan Polytechnic University; Henan, Jiaozuo 454000, China; zhengligang97@163.com
5. Department of Safety & Science Engineering, Henan Polytechnic University; Henan,Jiaozuo 454000, China; zhxbhpu@163.com
6. Professor; State Key Laboratory of Coal Mine Disaster Dynamics and Control, Chongqing University; Chongqing 400044, China; 13333910808@126.com

ABSTRACT

In the gas explosion mine disaster, the mine blast-proof door is often severely deformed and cannot be closed in time, so that the downhole airflow and the surface air are short-circuited by the wind, the mine can not carry out the wind, and the toxic gas can not be effectively discharged, causing a large number of casualties. To solve the defects of the original blast-proof door in applications, in this paper, a novel blast-proof door system is designed, which can be quickly reset after pressure relief of mine gas explosion. The scenario of a gas explosion in an underground coal mine was simulated. Through the experimental platform, the safety protection performance of the novel blast-proof door under different explosive equivalents was experimentally studied and analyzed. The performance parameters such as the pressure and stress acting on the novel blast-proof door were experimentally studied. The experimental results show that under the conditions of different explosive equivalents, the main part of the novel blast-proof door did not obviously deform, so the overall design of the novel blast-proof door is reasonable and has good safety protection performance. To further study the pressure distribution law and pressure-relief effect of the novel blast-proof door, the FLUENT software was used during the explosion. The research shows that the strength of the novel blast-proof door satisfies the design requirements, the main part has not been obviously deformed, and the entire novel blast-proof door system has excellent safety protection performance. The novel blast-proof door can solve the problem that the existing often be severe deformation in the gas explosion and not timely closed. So the investigation provides a novel designing ideas, research methods and theoretical support for enriching the technical means to prevent gas explosion disasters. It has important practical significance for improving the mine disaster prevention technology level and minimizing the loss of gas explosion disaster.

Keywords: novel blast-proof door; safety protection performance; experiment; numerical simulation

INTRODUCTION

Coal is the most important energy source for the Chinese economy. Other rapidly growing economies in Asia and Africa also increasingly rely on coal to satisfy their growing appetite for energy (Edenhofer, 2015). According to the statistics, coal is the main basic energy and chemical raw material in China and accounted for 62 percent of the energy consumption in China in 2016(Chen et al., 2019); in recent years, with the significant increase in coal production, the number of major accidents and deaths in coal mines has been declining (Chen and Hao 2012). However, as the number one killer of coal mine safety, gas explosion, has caused many casualties and much property damage to the country and society. Among the potential accidents, that can lead to fatal injuries in underground coal mines, gas explosion accidents have been recognized as the leading threat to miners' safety in China (Meng et al., 2019).Therefore, to protect the personal safety of closed buildings and mines, various blast-proof doors have been designed (Chun et al., 2018; Zhao and Qian, 2012; Meng, et al., 2016; Tavakoli and Kiakojouri, 2014; Chen and Hao, 2013). A mine blast-proof door is a type of safety equipment that can prevent the destruction of the main ventilator when gas and coal dust explode; the door opens when the main ventilator is out of operation and prevents gas accumulation in the underground chamber and main return air duct.

At present, some achievements have been made in the research on the safety protection performance and safety of blast-proof doors; for example, Hsieh et al.(Hsieh, Hung and Chen, 2008) conducted experimental and numerical simulations on the explosion-proof performance of the stiffened door structure. The research shows that the structure has good impact resistance. Harold Brode H L, Louca L A, Remennikov A M, Yuen S C K, et al.(Brode, 1955; Louca, Pan and Harding, 1998; Remennikov, 2003; Yuen and Nurick, 2005) studied different failure modes of the blasting plate on the stiffened plates and the relationship between deformation and tear, which provided data support for future research. Choi Y et al.(Choi et al., 2016) performed experimental research and an analysis on the detonation characteristics of the blast-proof door. The experimental results show that the blast-proof door is bent under the action of the explosion shock wave, and an elliptical opening is formed at the blast-proof door. Li Xiudi et al.(Li et al., 2016) used the ANSYS/LS-DYNA software to establish a numerical calculation model. The results show that the response of the protective door is related to the high impulse of the thermobaric bomb explosion shock wave; Veeredhi L S B et al. (Veeredhi and Ramana, 2015) used computational methods to study the degree of damage that could be caused by blast-proof doors subjected to explosive loads under blast loading. The results show that the blast-proof door structure with cap reinforcement is better in practical applications. Lee J et al. (Lee and Choi, 2018) studied the propagation of pressure waves in the tunnel caused by an explosion near the blast-proof door, and they explained the propagation of pressure waves through experiments and simulations. Chen W et al. (Chen and Hao, 2012) used the finite element program Ls-Dyna to analyze the explosion-proof load capacity and energy absorption capacity of the blast-proof door. The results show that the panel with this new structure can withstand higher explosive loads. Chen L et al. (Chen, Fang and Zhang, 2008) analyzed the dynamic response of a reinforced-concrete dome blast-proof door under blast loading. The results show that the strain rate strongly affects the response of the reinforced concrete blast-proof doors under blast loading. Koh C G et al.(Koh, Ang, and Chan, 2003) conducted a dynamic analysis of the application of the shell structure in the blast-proof door and provide numerical examples of benchmark problems and their application on the blast-proof door.

In recent years, there are many studies on the occurrence law and mechanism of the gas explosion, but there are few studies on the interaction between a shock wave of gas explosion and blast-proof door, which is not conducive to the design of blast-proof door. Therefore, aiming at the structure and main problems of original blast-proof doors, a novel blast-proof doors which can be reset in time in case of catastrophe has been developed on the basis of previous studies. Through experiments and numerical simulation, the dynamic response characteristics of the novel blast-proof door under the action of gas explosion shock wave, the pressure distribution law of gas explosion shock wave under pressure-relief condition and the pressure-relief effect are studied. It is of practical significance to improve the technological level of mine disaster prevention and to minimize the loss of gas explosion disasters.

DESIGN OF THE NOVEL BLAST-PROOF DOOR STRUCTURE

Figure. 1 shows the system diagram of a novel blast-proof door. A self-reset ventilating shaft novel blast-proof door is fitted to the side of the return air wellhead. During normal ventilation, the novel blast-proof door is fixed on the side of the ventilating shaft. When an explosion occurs in the ventilating shaft, the original blast-proof door is destroyed by the shock wave or deformed and difficult to reset and seal. Under these circumstances, the novel blast-proof door is automatically reset through the lead rail to cover, seal and fix the original ventilating shaft cover, to avoid the short-circuit of air flow in the ventilation system and ensure normal ventilation underground. In the case of anti-wind rescue, the novel blast-proof door is reset and locked, and after the anti-wind end is normal, the anti-wind lock device is opened, and the novel blast-proof door is returned to the side of the ventilating shaft.

On the basis of fully considering the factors such as how to make use of the original explosion-proof door structure, reduce cost and improve flexibility, this study designs the whole structure of explosion-proof door, the quick opening subsystem, the quick reset subsystem, the electronic control subsystem and the sealing subsystem of explosion-proof door. The new explosion-proof door not only has the basic functions of the original explosion-proof door, but also meets the special requirements of rapid reset. At the same time, the design of the explosion-proof door not only considers the cover, sealing and fixing of the original air shaft cover, but also considers the movement, docking and locking. After the original explosion-proof door was damaged by the impact of gas explosion, the new explosion-proof door can move the reserve explosion-proof door in place quickly and fix and lock reliably when the mine is backwind

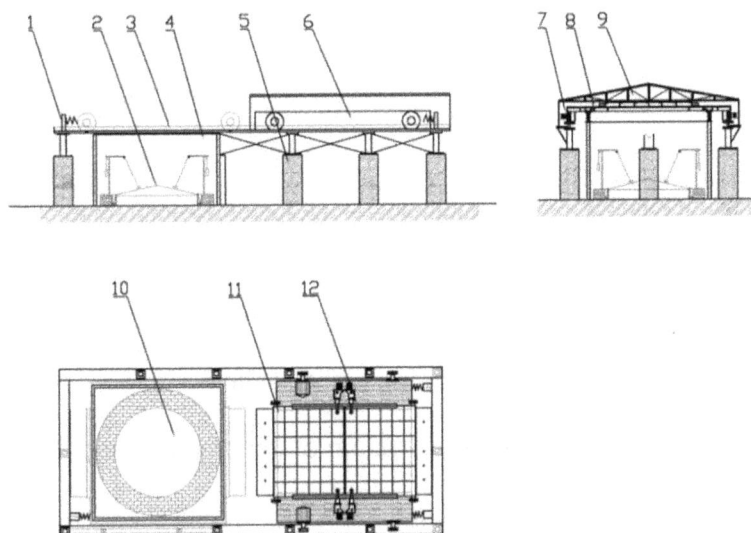

1-Electrical control device; 2-Ventilating shaft blast proof door;3-Novel blast proof door lead rail; 4-Concrete square box;5-Novel blast proof door bracket; 6- Novel blast proof door; 7- Scooter beam walking device; 8- Scooter sealing device;9- Canopy; 10-Ventilating shaft wellhead; 11-Ventilating shaft cover and bearing device; 12-Connecting rod locking device.

Figure. 1 System diagram of a novel blast-proof door

EXPERIMENTAL SECTION

Experimental conditions and processes

The pressure-relief test system of the novel blast-proof door that we built was used in the experiment, as shown in Figure. 2 (a), the location of the measuring point is shown in Figure 2 (b). The ratio of the system to the field size is 1:4, in which the diameter of the wellbore is 1.2 m, the size of the square sealing pool is 1.875mx1.875mx1.025m, and the size of the single novel blast-proof door is 0.684mx1.3m. The explosive gas in the airbag is a mixture of acetylene and pure oxygen, and the gun is inserted into the explosive gas balloon, sealed and placed at the bottom of the wellbore. After preparation, the ignition control switch is activated, and detonation is begun to verify the overall safety protection performance and pressure-relief effect of the newly designed novel blast-proof door. The experimental explosion equivalents are gradually performed in increasing order. The experimental conditions are shown in Table 1.

(a) Experimental reality

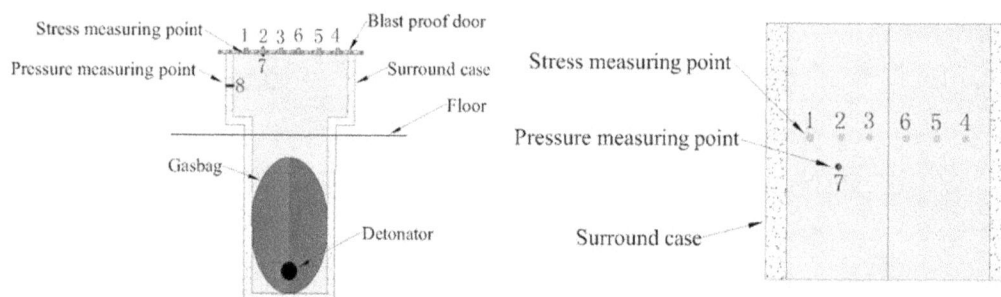

(b) Measuring point position

Figure. 2 pressure-relief test system of the novel blast-proof door

Table 1 Experimental conditions

Number	1	2	3	4	5	6	7	8
Mixed gas volume (l)	5	10	20	20	20	25	25	45
Explosive equivalent (g)	2.13	4.27	8.53	8.53	8.53	10.66	10.66	19.17

Experimental results and discussion

Figure. 3 shows the trend of the stress and pressure change of the novel blast-proof door in the 8th experiment. Figures. 3 (a) - (c) show that the stress values of the corresponding points (1-4, 2-5, and 3-6) of the left and right novel blast-proof doors are similar in size and development trend, which indicates that the stress of the left and right novel blast-proof doors is symmetrical. Compared with other locations, the stresses at points 1 and 4 near the rotating axis are larger. The main reason is that after the gas explosion, the explosion shock wave rapidly impacts both sides. Figure. 3 (d) shows that the explosion shock wave can be sensed in advance because the 8th measuring point is below the 7th measuring point, but because the 7th measuring point is located on the blast-proof door and directly faces the explosion shock wave, the pressure detected at the 7th measuring point is slightly higher than that at the 8th measuring point. The maximum pressure of this experimental test is 0.89 MPa, and the maximum stress is 168 MPa, which is lower than the allowable stress of 245 MPa.

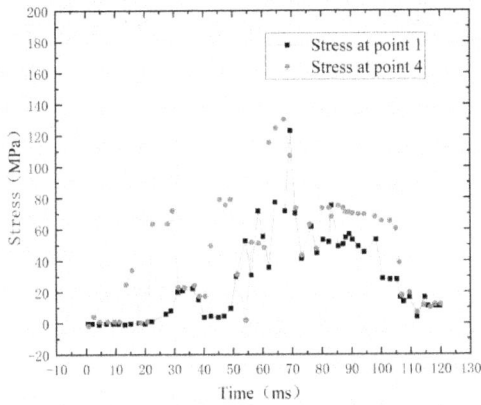

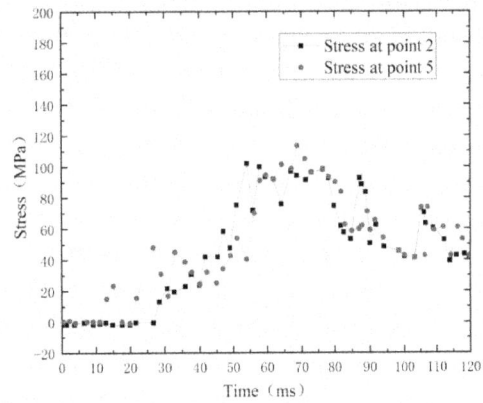

(a) Stress comparison at points 1 and 4 (b) Stress comparison at points 2 and 5

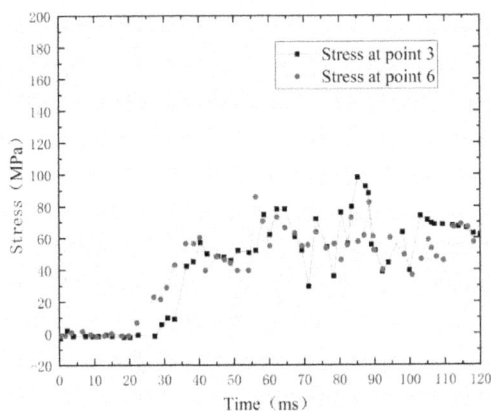

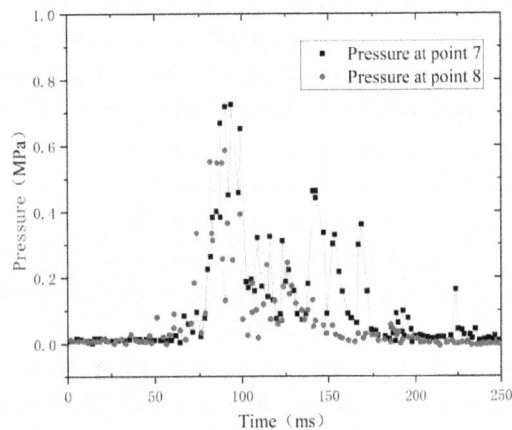

(c) Stress comparison at points 3 and 6 (d) Pressure comparison at points 7 and 8

Figure. 3 Stress and pressure variation trend of novel blast proof door

Figure. 4 shows the stress comparison between the left and right novel blast-proof doors in the 8th experiment. The 1st, 2nd, and 3rd measuring points and the 4th, 5th, and 6th measuring points reach the maximum stress at successively earlier time. Due to the gas explosion, the gas near the explosion source quickly hits outward, which results in the fastest time for the 1st and 4th measuring points to reach the maximum stress at the left and right ends. Since the novel blast-proof door is a split-type structure, when the explosion shock wave reaches the novel blast-proof door, first, the pressure-relief is opened from the middle as shown in Figures. 7-10, the 3rd and 6th measuring points first begin to relieve pressure; then, the 2nd and 5th measuring points begin to relieve pressure; finally, the 1st and 4th points begin to release pressure. Therefore, the maximum stress values reached by the 1st, 2nd, and 3rd measuring points and the 4th, 5th, and 6th measuring points successively decrease.

Table 2 shows a summary of the experimental results of the novel blast-proof door. In the 5th-8th experimental data tests, the explosion equivalent increases from 8.53g to 19.17g, the maximum pressure increases from 0.34 MPa to 0.89 MPa, and the maximum stress increases from 89 MPa to 168 MPa. Thus, the trend is that the maximum pressure and stress increase with the increase in explosive equivalent. The explosion equivalent is only related to the total mass of the fuel, and a greater total mass corresponds to a larger explosion equivalent. Since the first five explosion equivalents are small, the novel blast-proof door components have no obvious deformation. When the explosion equivalent is increased to 10.66g, the weight of the weight plate and the bearing weld are obviously deformed in the sixth experiment. In the seventh experiment, the welding point of the counterweight plate is broken, and the safety rope limit is broken. In the eighth experiment, the explosion equivalent reached 19.17g. The welding seam of the novel blast-proof door was deformed, and the connecting spring of the limit chain was pulled off. In general, in the 5th-8th experimental data test, the stress level is lower than the allowable stress of 245 MPa, which indicates that the strength of the novel blast-proof door component conforms to the design requirements, and the main part does not have obvious deformation.

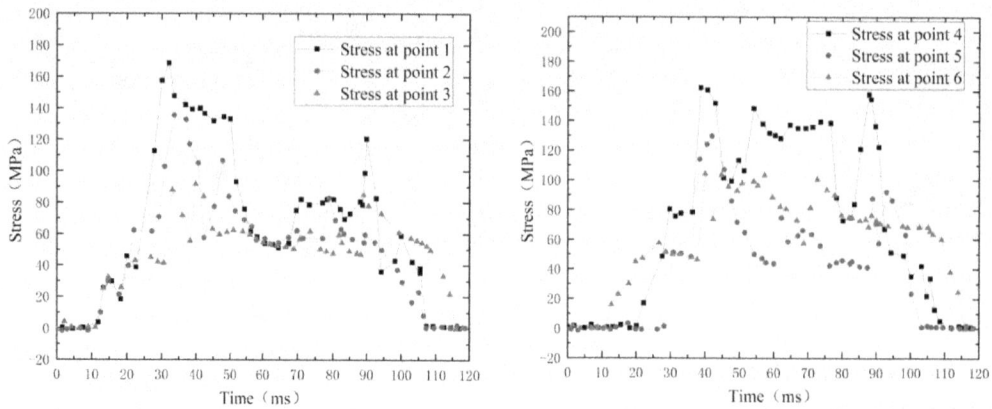

(a)Comparison of stress at points 1, 2, and 3 (b) Comparison of stress at points 4, 5, and 6

Figure. 4 Stress comparison between the left and right novel blast-proof doors

Table 2 Summary of the experimental results of the novel blast-proof door

Number of experiments	Mixture gas volume (l)	Explosion equivalent (g)	Maximum pressure (MPa)	Maximum stress (MPa)	Blast proof door opening and resetting	Component deformation
1	5	2.13	Not detected	Not detected	Reset after opening 1/3	No obvious deformation
2	10	4.27	Not detected	Not detected	Reset after turning on 90°	No obvious deformation
3	20	8.53	Not detected	Not detected	Reset after turning on 90°	No obvious deformation
4	20	8.53	Not detected	Not detected	Reset after turning on 90°	No obvious deformation
5	20	8.53	0.34	89	Reset after turning on 90°	No obvious deformation
6	25	10.66	0.53	113	Reset after turning on 90°	Partially deformed
7	25	10.66	0.72	128	Reset after turning on 90°	Partially deformed
8	40	19.17	0.89	168	Reset after turning on 180°	Partially deformed

SIMULATION SECTION

Simulation conditions

Owing to the limitation of experimental conditions, the test range of pressure and stress is limited. To further study the pressure distribution law and pressure-relief effect of gas explosion shock wave under pressure-relief condition of the whole novel blast-proof door. In this paper, the FLUENT software is used to simulate the gas explosion pressure field in the wellbore. The pressure distribution and pressure-relief effect on the novel blast-proof door are analyzed.

Figure. 5 shows the physical model of a gas explosion in the ventilating shaft novel blast-proof door. Figure. 5(a) shows the novel blast-proof door, where the diameter of the wellbore is 4.8 m, the size of the square sealed pool is 7.5 m × 7.5 m × 4.1 m, and the size of a single novel blast-proof door is 2.735 mx5.2 m, the thickness is 3 mm, the door frame is 50 angle steel, the quality of a single novel blast-proof door is 480 Kg, and the weight is 775 Kg, to simplify the model, the total depth of wellbore and sealing pool is 10m. (For the sake of simplicity of wellbore sketch and simulation results, the counterweight is not drawn in Figure. 5, but it is taken into consideration in the calculation of the movement of novel blast-proof door); Figure. 5(b) shows the 1:1 calculation model established according to (a), and the calculation network is a divided grid. To improve the accuracy of the pressure calculation, the computational grid near the wall is locally encrypted. In the Figureure, ac and bc indicate the double opening of the doors.

The wellbore and square sealed tank are filled with a mixture of CH4 and air. The volume concentration of CH4 is 9.5% of the stoichiometric ratio, and the mass concentration of O2 is 21%. At the initial moment, the local high-temperature method is used to ignite and detonate at the bottom of the wellbore. The explosion shock wave flow after detonation is an unsteady turbulent flow, the turbulence model uses the k-ε turbulence equation, this model is defined as:

$$\rho \frac{D_k}{D_t} = \frac{\partial}{\partial x_i}\left[\left(u + \frac{u_t}{\sigma_k}\right)\frac{\alpha \partial_k}{\partial x_i}\right] + G_k + G_b - \rho\varepsilon - Y_M$$

$$\rho \frac{D_\varepsilon}{D_t} = \frac{\partial}{\partial x_i}\left[\left(u + \frac{u_t}{\sigma_k}\right)\frac{\alpha \partial_k}{\partial x_i}\right] + C_{1\varepsilon}\frac{\varepsilon}{k}(G_k + C_{3\varepsilon}G_b) - C_{2\varepsilon}\rho\frac{\varepsilon^2}{k}$$

In the above equation, G_k denotes the turbulent kinetic energy caused by mean velocity gradient, G_b denotes the turbulent kinetic energy caused by buoyancy effect, and Y_M denotes the effect of turbulent expansion on the total dissipation rate. $u_1 = \rho C_u \frac{\varepsilon^2}{k}$ denotes the turbulent viscous coefficient. In FLUENT, as the default constant, $C_{1\varepsilon}$=1.44, $C_{2\varepsilon}$=1.92, C_u=0.09. The Plante numbers corresponding to turbulent kinetic energy k and dissipation rate are σ_k=1.0, σ_ε=1.3, respectively.

The wall surface uses the standard wall function, The standard wall function is treated by Launder and Spalding's near-wall method:

$$U^* = \frac{1}{k}\ln(Ey^*)$$

Where $U^* \equiv \frac{U_P C_u^{1/4} k_P^{1/2}}{\tau_w/\rho}$, $Y^* \equiv \frac{\rho C_u^{1/4} k_P^{1/2} y_P}{u}$, k is Von Karman, the value is 0.42; E is the experimental constant, the value is 9.81; U_P is the average velocity of the fluid at point P; k_P is the turbulent energy at point P; y_P is the distance from point P to the wall; u is the dynamic viscosity coefficient.

The methane combustion chemical reaction uses the EBU (vortex diffusion) model, which is suitable for turbulent combustion, and the average fuel consumption rate can be expressed as follows:

$$R_F = -C_R\rho\left(\frac{\varepsilon}{k}\right)\left(Y_F^{'2}\right)^{1/2}$$

Where R_F is the fuel burning rate per unit space; C_R is the model coefficient, $C_R \approx 6$; k is the turbulent flow energy; ε is the turbulent flow energy dissipation rate; k is the turbulent flow energy; $Y_F^{'}$ is the pulsation value of the fuel mass fraction, $Y_F^{'} = Y_F - \overline{Y_F}$; $\overline{Y_F}$ is the Reynolds average of the fuel mass fraction.

Finally, iteratively solves the PISO algorithm with better convergence, and the iteration time step is 1s-5s.

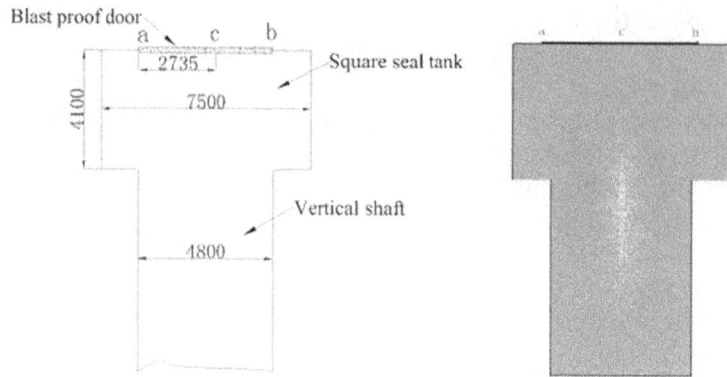

(a) Schematic diagram (b) Calculation model

Figure. 5 Physical model of a gas explosion

Simulation results and discussion

Figure. 6 shows the pressure results of a single novel blast-proof door obtained with the FLUENT software, which shows a similar pressure change trend as the experimental results. The explosion shock wave reaches the novel blast-proof door at 24 ms. The novel blast-proof door was not opened at this moment, and the maximum pressure was only 26KPa. Since the initial pressure of the pressure wave on the novel blast-proof door is small, the angular acceleration of the novel blast-proof door is small when it is just opened, and the opening speed is slow. With the advance of time, the pressure gathers around the novel blast-proof door, and the pressure on the novel blast-proof door gradually increases. After that, the angular acceleration increases as the pressure increases, and the opening speed increases faster. From 24 ms to 38 ms, the maximum pressure on the novel blast-proof door increases from 26 kPa to 782 kPa, the average pressure on the novel blast-proof door increases from 23 kPa to 467 kPa, and the novel blast-proof door opening angle is also increased from 0° to 12°. At 38 ms, the pressure on the novel blast-proof door reaches the maximum. After 38ms, since the novel blast-proof door opening exceeds 12°, the accumulated pressure is released enough and the pressure begins to gradually decrease. Finally, the novel blast-proof door is opened at 54 ms, and the pressure-relief process is completed, and the pressure-relief effect is best.

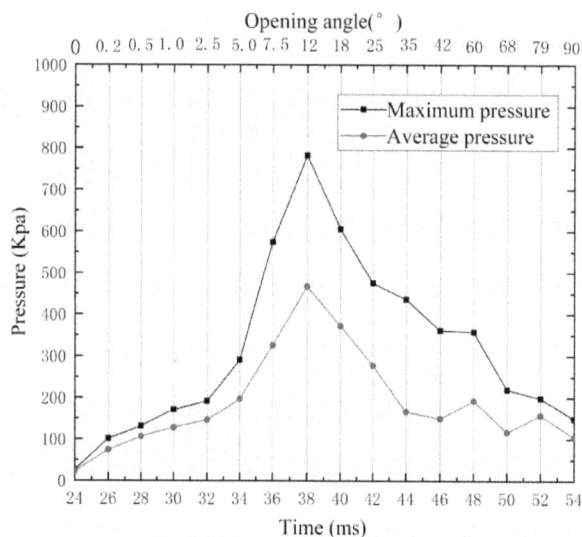

Figure. 6 Pressure results of a single novel blast-proof door

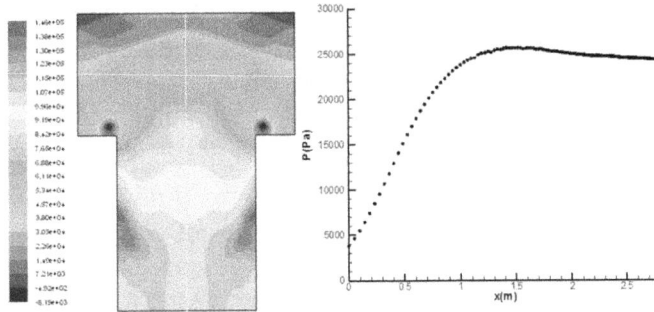

(a)Pressure field (b) Pressure distribution

Figure.7 Pressure state at 24 ms with 0°opening angle of the venting doors

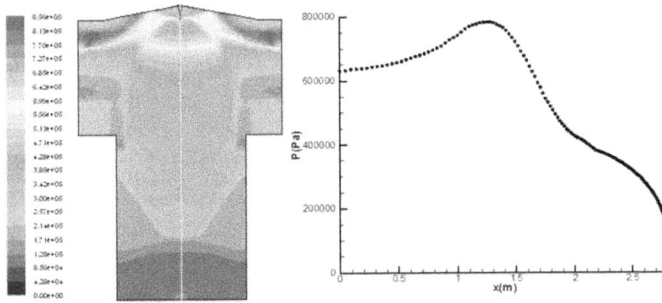

(a)Pressure field (b) Pressure distribution

Figure.8 Pressure state at 38 ms with 12°opening angle of the venting doors

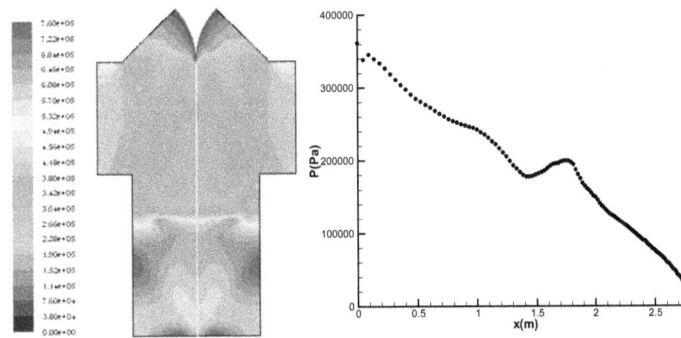

(a)Pressure field (b) Pressure distribution

Figure.9 Pressure state at 46 ms with 42°opening angle of the venting doors

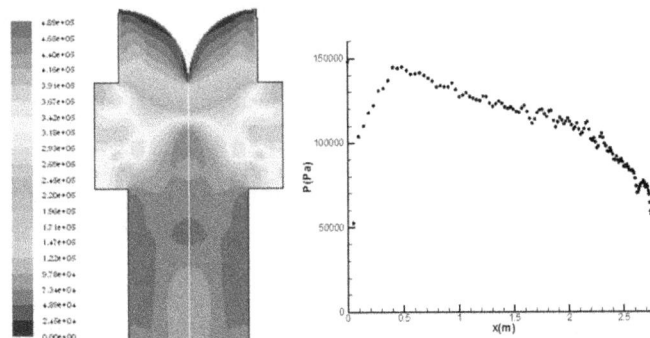

(a)Pressure field (b) Pressure distribution

Figure.10 Pressure state at 54 ms with 90°opening angle of the venting doors

Figures. 7-10 show the results of the flow field analysis at different time. The pressure distribution in the wellbore during the entire explosion process is shown in detail: (a) shows the pressure field of the gas explosion shock wave, and (b) shows the pressure distribution of the single novel blast-proof door. This process can be approximately divided into three stages:

Prestart stage: The prestart stage is from the moment of ignition to 24 ms. Figure. 7 shows the pressure distribution on the single novel blast door at 24 ms. After the ignition, the initial pressure of the explosion shock wave that reaches the novel blast-proof door is small, the pressure is mainly concentrated near the explosion source. In the middle area near the two novel blast-proof doors, the pressure is higher and the pressure on both sides is lower, at the moment, the explosion-proof doors have not been opened.

Start-up stage: The duration from 24 ms to 54 ms is the start-up stage. Figures. 8-9 show the pressure distribution on the single novel blast door at 38 ms and 46 ms. At the beginning of the novel blast-proof door, the shock wave is impacted upward from the explosion source. Due to the small opening angle of the novel blast-proof door, the airflow accumulates, that the pressure on the novel blast-proof door increases, and the pressure near the explosion source decreases. After 38 ms, the pressure on the novel blast-proof door begins to gradually decrease because the novel blast-proof door opens more than 12°, and the accumulated pressure was sufficiently released. When the shock wave passes through the obstacle, the shock wave intensity near the novel blast-proof door is strengthened due to the excitation effect (Xu and Yang, 2004). At this time, the novel blast-proof door opening angle continues to increase, the explosion shock wave near the novel blast-proof door is given priority release, the pressure decreases, and the unexploded shock wave remains near the explosion source, which results in greater pressure near the explosion source. After the novel blast-proof door is opened, the pressure in the middle of the two novel blast-proof doors decreases rapidly and is lower than the pressure on both sides due to the pressure-relief between the two novel blast-proof doors.

Decompression stage: after 54 ms is the decompression stage. Figure. 10 shows the pressure distribution at 54 ms and 90 degrees of the opening angle of the novel blast-proof door. At this time, the novel blast-proof door is in the late stage of opening because the pressure increases, the angular acceleration increases, and the opening speed rapidly increases. The shock wave near the novel blast-proof door is quickly released, the shock wave near the explosion source continuously rushes upward, and the pressure-relief effect is best.

CONCLUSION

(1) This study designed, manufactured and tested a novel blast-proof door that can be quickly reset under catastrophic conditions. The research results show that the novel blast-proof door is feasible and has good safety protection performance. It enriches the theory of gas explosion prevention and control, and innovates in research content.

(2) From the experimental research and analysis of the pressure-relief characteristics of the novel blast-proof door, the experimental results show that the stress level of the main part of the novel blast-proof door is lower than the allowable stress of the material, and there is no plastic deformation. At the same time, some problems were exposed. In the seventh experiment, the limit safety rope was broken, and in the eighth experiment, the connecting spring of the limit chain was broken, which showed that the limit of explosion-proof door by the safety rope or chain could not meet the strength requirements

(3) To further study the pressure-relief process and deformation of the novel blast-proof door, the numerical simulation method is adopted. The simulation results show that the simulation results have similar pressure changes with the experimental results, and the design of the novel explosion-proof door is reasonable and the overall reliability meets the requirements. This study increases the novel blast-proof door system, which can better guarantee the safety of the mine after the explosion than the traditional blast-proof door. The explosion-proof performance of this novel blast-proof door system must be further studied.

ACKNOWLEDGMENTS

This work was carried out with funding from the National Key R&D Program of China (Grant No. 2018YFC0808103), National Natural Science Foundation of China (Grant No. 51304070, 51674103) and supported by the Science Research Funds for the Universities of Henan Province. The authors wish to thank these organizations for their support. They also wish to thank the readers and editors for their constructive comments and suggestions to improve the manuscript.

REFERENCES

Brode, H L, 1955. Numerical Solutions of Spherical Blast Waves, Journal of Applied Physics, 26(6): 766-775.

Chen, H, Qi, H and Long R, et al., 2012. Research on 10-year tendency of China coal mine accidents and the characteristics of human factors. Safety Science, 50(4): 745-750.

Chen, L, Fang, Q and Zhang Y, et al., 2008. Rate-sensitive numerical analysis of dynamic responses of arched blast doors subjected to blast loading, Transactions of Tianjin University, 14(5): 348-352.

Chen, W and Hao, H, 2012. Numerical study of a new multi-arch double-layered blast-resistance door panel, International Journal of Impact Engineering, 43(none): 16-28.

Chen, W S and Hao, H, 2013. Numerical Studies on the Blast-Resistant Capacity of Stiffened Multiple-Arch Panel, Key Engineering Materials, 535-536: 514-517.

Chen, X J, Li, L Y and Wang, L, et al., 2019. The current situation and prevention and control countermeasures for typical dynamic disasters in kilometer-deep mines in China, Safety Science, 115: 229-236.

Choi Y, Lee, J and Yoo, Y H, et al., 2016. A Study on the behavior of blast proof door under blast load, International Journal of Precision Engineering and Manufacturing, 17(1): 119-124.

Chun, Y W, Song, C Z and Wan, Z Z, et al., 2018. Multi-objective explosion-proof performance optimization of a novel vehicle door with negative Poisson's ratio structure, Structural and Multidisciplinary Optimization, 58(4): 1805–1822.

Edenhofer, O, 2015. King Coal and the queen of subsidies, Science, 349(6254): 1286-1287.

Hsieh, M W, Hung, J P and Chen, D J, 2008. Investigation on the blast resistance of a stiffened door structure, Journal of Marine Science and Technology, 16(2): 149–57.

Koh, C G, Ang, K K and Chan, P F, 2003.Dynamic Analysis of Shell Structures with Application to Blast Resistant Doors, Shock and Vibration, 10(4): 269-279.

Lee, J and Choi, Y, 2018. Effects of a Near-Field Explosion in a Tunnel Behind a Blast Proof Door, International Journal of Precision Engineering and Manufacturing, 19(4): 625-630.

Louca, L A, Pan, Y G and Harding, J E, 1998. Response of stiffened and unstiffened plates subjected to blast loading, Engineering Structures, 20(12): 1079-1086.

Li, X D, Geng, Z G and Miao, C Y, et al., 2016. Failure mode of a blast door subjected to the explosion wave of a thermobaric bomb, Journal of Vibration and Shock, 35 (16): 199-203.

Meng, F, Zhang B and Zhao, Z, et al., 2016. A novel all-composite blast-resistant door structure with hierarchical stiffeners, Composite Structures, 148: 113-126.

Meng, X F, Liu, Q L and Luo, X X, et al., 2019. Risk assessment of the unsafe behaviours of humans in fatal gas explosion accidents in China's underground coal mines, Safety Science, 210: 970-976.

Remennikov, A M, 2003. A review of methods for predicting bomb blast effects on buildings, Journal of Battlefield Technology, 6(3): 155–161

Tavakoli, HR and Kiakojouri, F, 2014. Numerical dynamic analysis of stiffened plates under blast loading, Latin American Journal of Solids and Structures,11(2): 185–99.

Veeredhi, L S B and Ramana R N V, 2015. Studies on the Impact of Explosion on Blast Resistant Stiffened Door Structures. Journal of The Institution of Engineers (India): Series A, 96(1): 11-20.

Xu, J D and Yang G Y, 2004. Numerical simulation of the barricade encouraging effect in the process of gas explosion propagation. Journal of China Coal Society, 29(1): 53-56.

Yuen, S C K and Nurick, G N, 2005. Experimental and numerical studies on the response of quadrangular stiffened plates. Part I: Subjected to uniform blast load, International Journal of Impact Engineering, 31(1): 55-83.

Zhao, H and Qian, X, 2012. Simulation Analysis on Structure Safety of Two Typical Refuge Chamber Shell Forms Under Explosion Load, Procedia Engineering, 45(2): 910-915.

Impact of where a measurement is taken on the value of a heat stress index

M A Tuck[1]

1. Associate Professor, Federation University Australia, Ballarat VIC 3353.
   m.tuck@federation.edu.au

ABSTRACT

Heat stress is an issue that affects many mines worldwide and a range of indices are used to determine the thermal comfort or safety of these environments. Heat stress indices generate numeric values that are used to evaluate the thermal safety of hot and humid environments in mines. Heat stress indices are generated from actual measured values of properties such as air dry and wet bulb temperature, air velocity, radiant temperature and others. It is also recognised that across an airway in a mine, the temperature, humidity and air velocity are not constant but vary in their values. In determining a heat stress index for a location, how much error is introduced by arbitrarily selecting a location to take measurements? Should all measurements be taken at the same point for a given location? This paper seeks to answer these questions.

INTRODUCTION

Heat affects many mines worldwide. Heat affects the productivity, health and safety of a mine's workforce. Over the years a number of indices have been developed to determine the thermal safety and thermal comfort of the working environment, some of which are used in mining operations. To generate numerical values for these indices actual measured values of properties such as air dry and wet bulb temperature, air velocity, radiant temperature and others are used. In underground mine airways these properties are however not constant but are variable in terms of both time and location within the particular cross-section. This means that the measured value for the heat stress index will depend on where and when the measurement is taken for the particular location under investigation.

This paper seeks to answer the following;

1. Does the location of a measurement in an airway impact on the calculated value of a heat stress index?

2. How much error is introduced depending on the location comparted with using an average air velocity value?

3. Does the choice of heat stress index have an impact on the level of error?

This paper is going to concentrate on changes in air velocity across an airway cross section and assumes that the temperatures and humidity remain constant to provide a simplified analysis to answer the above questions.

BACKGROUND INFORMATION

The velocity data for this study was obtained from Hardcastle, Grenier and Butler (1991) who describe work undertaken using electronic vane anemometers. As a part of this study, measurements were undertaken underground in a uranium mine in three sections of straight airway, free of obstructions and changes in cross sections. The three cross sections are illustrated in figure 1, 2 and 3. Standard traverses were made at all three stations to determine the average air velocity. In addition point measurements were made to develop the velocity contours shown in figures 1 to 3, the area for each station was also measured. The average air velocity measurements were as follows using the Davis #1 instrument, station 1: 0.47 m/s, station 2: 2.98 m/s and station 3: 0.93 m/s. Using these average velocities and selecting values of other velocities from the station velocity contours as reported in Tables 1 to 8 the Basic Effective Temperature (BET) and simple Air Cooling Power were calculated for the average and spot velocities. The % difference between the Effective

Temperature and Air Cooling Power values were also calculated and reported in Tables 1 to 8 for a number of air dry bulb and wet bulb combinations at an atmospheric pressure of 100 kPa.

The Effective temperature and Air Cooling Power were calculated using the equations outline by Pickering and Tuck (1997). The Basic effective temperature can be calculated using the following equation for air velocities in the range 0.5 - 3.5 m/s:

Basic Effective temperature = BET °C

Dry bulb = td °C

Wet bulb = tw °C

V = air velocity m/s

Works for range of velocities from 0.5 to 3.5 m/s, if V > 3.5 m/s assume V = 3.5 m/s, if V < 0.5 m/s assume V = 0.5 m/s.

$$BET = \frac{4 \times (4.12 - x1) + x2}{1.65176} + 20$$

Where

$$x1 = \frac{8.33 \times (17 \times x3 - (x3 - 1.35) \times (tw - 20))}{(x3 - 1.35) \times (td - tw) + 141.6}$$

$$x2 = \frac{17 \times ((td - tw) \times x3 + 8.33 \times (tw - 20)}{(x3 - 1.35) \times (td - tw) + 141.6}$$

$$x3 = 5.27 + 1.3 \times V - 1.15 \exp(-2V)$$

This study has focused on a simple Air Cooling Power index described by Wyndham (1974). It is a rational index and so is based on the metabolic heat balance given by:

$$M = Br + Rad + Con + Evap + Cond + Ac$$

Where

M = metabolic heat generation

Br = respiratory heat exchange

Rad = radiative heat transfer

Con= convective heat transfer

Evap = evaporative heat transfer

Cond = conductive heat transfer

Ac = heat storage/accumulation in the body

Under hot conditions found in mines, the conductive component is very small as is the respiratory heat exchange, as such these are assumed to be zero. To ensure health and safety, the core temperature of the body should not be allowed to rise, therefore Ac in the equation is also zero. This means that the heat balance equation becomes:

$$M = Rad + Con + Evap$$

Equations, as given below, can be written for each of the radiative, convective and evaporative components to determine the right hand side of the equation which represents heat transfer to the environment or ventilation airflow, Wyndham (1974), Gibson (1976).

$$Rad = 17 \times 10^{-8} \times \left(\left(\frac{t_r}{2} \right) + 290.7 \right)^3 \times (t_{sk} - t_r)$$

$$Con = 8.32 \times \left(\frac{P}{101.3} \right)^{0.6} \times v^{0.6} \times (t_{sk} - t_r)$$

$$Evap = 15140 \times \left(\frac{P}{101.3} \right)^{0.6} \times v^{0.6} \times \left(\frac{e_s}{P} \right)$$

$$e_s = 0.6105 \times exp \left(\frac{17.27 \times t_{sk}}{237.3 + t_{sk}} \right) - 0.6105 \times exp \left(\frac{17.27 \times t_w}{237.3 + t_w} \right)$$

$$+ 6.6 \times 10^{-4} \times P \times (t_d - t_w) \times \left(1 + 1.15 \times 10^{-3} \times (t_d - t_w) \right)$$

Where

P = atmospheric pressure kPa, t_{sk} = skin temperature °C, t_r = radiative temperature °C, t_d = dry bulb temperature °C, t_w = wet bulb temperature °C, v = air velocity m/s.

According to Wyndham [8] a reasonable assumption is that in hot environments the skin temperature can be assumed to be 35°C and the radiative temperature is equal to the dry bulb temperature.

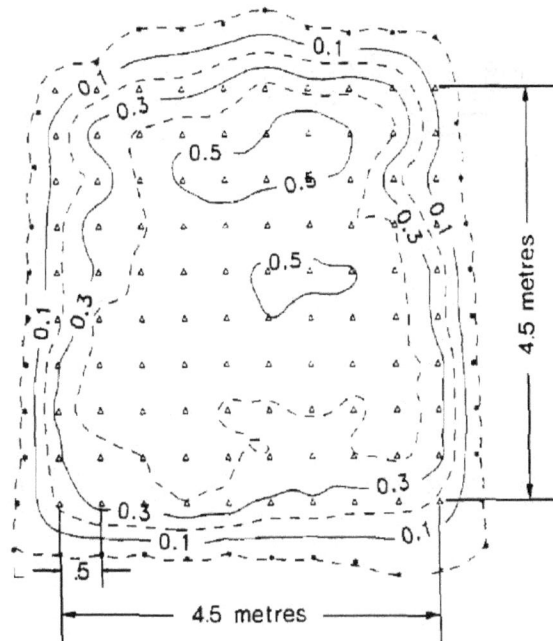

FIG 1 – Velocity contours at location 1 after Hardcastle, Grenier and Butler (1991)

RESULTS

The results are shown in tables 1 to 8 for the following dry/wet bulb temperature combinations, 25/20°C, 25/25°C, 30/20°C, 30/25°C, 30/30°C, 35/20°C, 35/25°C, 35/30°C. In addition, it is assumed that the atmospheric pressure is 100 kPa, skin temperature is 35°C and the radiant temperature equals the dry bulb temperature.

DISCUSSION

Inspection of tables 1 to 8 reveals a number of interesting points.

1. Firstly the values of the heat stress indices are position i.e. velocity dependent

2. The error or percentage difference from the value for the average airflow depends on the heat stress index used. The tables show that the error is larger when the ACP is used compared to using the BET. Another point to note from the tables is that for location 1, the error with the BET is zero. This is because each of the BET values is calculated at an air velocity of 0.5 m/s as the formula for BET works only in the range 0.5 to 3.5 m/s. All velocity

values for location 1 are below 0.5 m/s so a value of 0.5 m/s is used in the calculation. The same is true at the opposite end of the velocity spectrum for location 2 where the value of BET for the velocity of 4 m/s is calculated using 3.5 m/s.

3. Similar trends are shown for both BET and ACP in that the error is highest at low velocities and reduces with an increase in velocity. However this only applies because the average air velocity in all cases excluding location 2 is close to the maximum velocity.

4. As the wet bulb depression reduces, i.e. the relative humidity increases, the error in both BET and ACP reduces. However this effect is small.

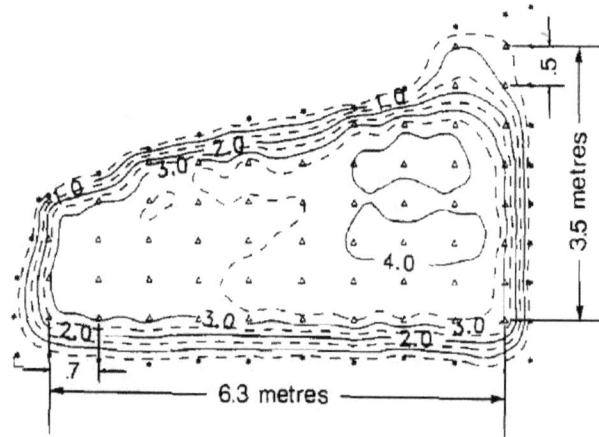

FIG 2 – Velocity contours at location 2 after Hardcastle, Grenier and Butler (1991)

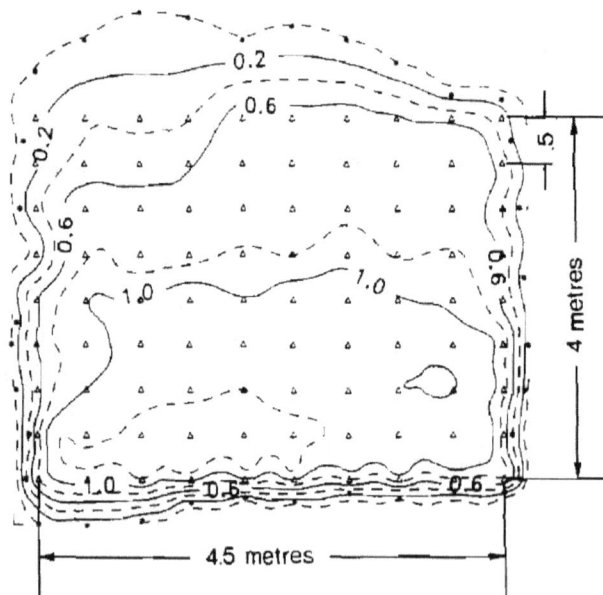

FIG 3 – Velocity contours at location 3 after Hardcastle, Grenier and Butler (1991)

The results show that a considerable error can be introduced depending on the position. As such it is important to measure the air flow properties at the location where workers are likely to be working in the airway cross section to accurately determine the risk of the heat exposure. This however does pose another question. Looking at the velocity contours in figures 1 to 3 it can be seen that there will be a wide variation on the velocity flowing across a human body located at any location in the cross sections, which velocity should be chosen?

Table 1: Effective Temperature, Air Cooling Power and difference between calculated values and the value calculated based on the average velocity for locations 1, 2 and 3. Air dry bulb temperature 25°C, Air wet bulb 20°C.

Location	Average velocity m/s	Velocity m/s	BET °C	% difference from average velocity BET °C	ACP W/m²	% difference from Average velocity ACP W/m²
1	0.47		20.11		445.1	
		0.1	20.11	0.00	204.5	54.06
		0.2	20.11	0.00	285.6	35.83
		0.3	20.11	0.00	351.2	21.10
		0.4	20.11	0.00	408.4	8.25
		0.5	20.11	0.00	460.1	3.37
2	2.98		15.22		1251.9	
		1.0	18.78	23.39	673.0	46.24
		2.0	16.84	10.64	995.6	20.47
		3.0	15.19	0.20	1256.8	0.39
		4.0	14.42	5.26	1484.6	18.59
3	0.93		18.34		646.3	
		0.2	20.11	9.65	285.6	55.81
		0.6	19.8	7.96	507.8	21.43
		1.0	18.78	2.40	673.0	4.13

Table 2: Effective Temperature, Air Cooling Power and difference between calculated values and the value calculated based on the average velocity for locations 1, 2 and 3. Air dry bulb temperature 25°C, Air wet bulb 25°C.

Location	Average velocity m/s	Velocity m/s	BET °C	% difference from average velocity BET °C	ACP W/m²	% difference from Average velocity ACP W/m²
1	0.47		22.65		334.2	
		0.1	22.65	0.00	160.7	51.92
		0.2	22.65	0.00	219.2	34.41
		0.3	22.65	0.00	266.5	20.26
		0.4	22.65	0.00	307.8	7.90
		0.5	22.65	0.00	345.7	3.44
2	2.98		16.42		916.1	
		1.0	21.08	28.38	498.6	45.57
		2.0	18.62	13.40	731.2	20.18
		3.0	16.37	0.30	919.6	0.38
		4.0	15.26	7.06	1083.9	18.32
3	0.93		21.27		479.3	
		0.2	22.65	6.49	219.2	54.27
		0.6	22.29	4.80	379.5	20.82
		1.0	21.08	0.89	498.6	4.03

Table 3: Effective Temperature, Air Cooling Power and difference between calculated values and the value calculated based on the average velocity for locations 1, 2 and 3. Air dry bulb temperature 30°C, Air wet bulb 20°C.

Location	Average velocity m/s	Velocity m/s	BET °C	% difference from average velocity BET °C	ACP W/m²	% difference from Average velocity ACP W/m²
1	0.47		22.77		427.8	
		0.1	22.77	0.00	183.7	57.06
		0.2	22.77	0.00	266.0	37.82
		0.3	22.77	0.00	332.5	22.28
		0.4	22.77	0.00	390.6	8.70
		0.5	22.77	0.00	443.1	3.58
2	2.98		19.98		1246.5	
		1.0	21.97	9.96	659.0	47.13
		2.0	20.86	4.40	986.4	20.87
		3.0	19.97	0.05	1251.4	0.39
		4.0	19.56	2.10	1482.6	18.94
3	0.93		22.06		632.0	
		0.2	22.77	3.22	266.0	57.91
		0.6	22.58	2.36	491.5	22.23
		1.0	21.97	0.41	659.0	4.27

Table 4: Effective Temperature, Air Cooling Power and difference between calculated values and the value calculated based on the average velocity for locations 1, 2 and 3. Air dry bulb temperature 30°C, Air wet bulb 25°C.

Location	Average velocity m/s	Velocity m/s	BET °C	% difference from average velocity BET °C	ACP W/m²	% difference from Average velocity ACP W/m²
1	0.47		25.32		316.6	
		0.1	25.32	0.00	139.8	55.84
		0.2	25.32	0.00	199.3	37.05
		0.3	25.32	0.00	247.5	21.83
		0.4	25.32	0.00	289.6	8.53
		0.5	25.32	0.00	327.6	3.47
2	2.98		21.95		909.5	
		1.0	24.4	11.16	484.0	46.78
		2.0	23.06	5.06	721.2	20.70
		3.0	21.93	0.09	913.1	0.40
		4.0	21.39	2.55	1080.5	18.80
3	0.93		24.52		464.5	
		0.2	25.32	3.26	199.3	57.09
		0.6	25.11	2.41	362.7	21.92
		1.0	24.4	0.49	484.0	4.20

Table 5: Effective Temperature, Air Cooling Power and difference between calculated values and the value calculated based on the average velocity for locations 1, 2 and 3. Air dry bulb temperature 30°C, Air wet bulb 30°C.

Location	Average velocity m/s	Velocity m/s	BET °C	% difference from average velocity BET °C	ACP W/m²	% difference from Average velocity ACP W/m²
1	0.47		28.63		182.2	
		0.1	28.63	0.00	86.7	52.41
		0.2	28.63	0.00	118.9	34.74
		0.3	28.63	0.00	144.9	20.47
		0.4	28.63	0.00	167.7	7.96
		0.5	28.63	0.00	188.2	3.29
2	2.98		24.99		502.6	
		1.0	27.71	10.88	272.7	45.74
		2.0	26.28	5.16	400.8	20.25
		3.0	24.97	9.89	504.6	0.40
		4.0	24.32	2.68	595.0	18.38
3	0.93		27.83		262.1	
		0.2	28.63	2.87	118.9	54.64
		0.6	28.42	2.12	207.1	20.98
		1.0	26.28	5.57	272.7	4.04

Table 6: Effective Temperature, Air Cooling Power and difference between calculated values and the value calculated based on the average velocity for locations 1, 2 and 3. Air dry bulb temperature 35°C, Air wet bulb 20°C.

Location	Average velocity m/s	Velocity m/s	BET °C	% difference from average velocity BET °C	ACP W/m²	% difference from Average velocity ACP W/m²
1	0.47		24.89		409.7	
		0.1	24.89	0.00	161.9	60.48
		0.2	24.89	0.00	245.4	40.10
		0.3	24.89	0.00	313.0	23.60
		0.4	24.89	0.00	371.9	9.23
		0.5	24.89	0.00	425.2	3.78
2	2.98		23.31		1240.9	
		1.0	24.42	4.76	644.5	48.06
		2.0	23.79	2.06	976.8	21.28
		3.0	23.3	4.59	1245.9	0.40
		4.0	23.09	0.94	1480.6	19.32
3	0.93		24.47		617.0	
		0.2	24.89	1.72	245.4	60.23
		0.6	24.78	1.27	474.3	23.13
		1.0	24.42	0.20	644.5	4.46

Table 7: Effective Temperature, Air Cooling Power and difference between calculated values and the value calculated based on the average velocity for locations 1, 2 and 3. Air dry bulb temperature325°C, Air wet bulb 25°C.

Location	Average velocity m/s	Velocity m/s	BET °C	% difference from average velocity BET °C	ACP W/m²	% difference from Average velocity ACP W/m²
1	0.47		27.4		298.1	
		0.1	27.4	0.00	117.8	60.48
		0.2	27.4	0.00	178.5	40.12
		0.3	27.4	0.00	227.7	23.62
		0.4	27.4	0.00	270.6	9.23
		0.5	27.4	0.00	309.4	3.79
2	2.98		25.51		902.9	
		1.0	26.86	5.29	468.9	48.07
		2.0	26.11	2.35	710.7	21.29
		3.0	25.5	5.06	906.5	0.40
		4.0	25.23	1.10	1077.3	19.32
3	0.93		26.92		448.9	
		0.2	27.4	1.78	178.5	60.24
		0.6	27.27	1.30	345.1	23.12
		1.0	26.86	0.22	468.9	4.46

Table 8: Effective Temperature, Air Cooling Power and difference between calculated values and the value calculated based on the average velocity for locations 1, 2 and 3. Air dry bulb temperature 35°C, Air wet bulb 30°C.

Location	Average velocity m/s	Velocity m/s	BET °C	% difference from average velocity BET °C	ACP W/m²	% difference from Average velocity ACP W/m²
1	0.47		30.54		163.4	
		0.1	30.54	0.00	64.6	60.47
		0.2	30.54	0.00	97.9	40.09
		0.3	30.54	0.00	124.8	23.62
		0.4	30.54	0.00	148.3	9.24
		0.5	30.54	0.00	169.6	3.79
2	2.98		28.67		494.9	
		1.0	30.03	4.74	257.0	48.07
		2.0	29.29	2.16	389.6	21.28
		3.0	28.66	4.56	496.9	0.40
		4.0	28.37	1.05	590.5	19.32
3	0.93		30.1		246.1	
		0.2	30.54	1.46	97.9	60.22
		0.6	30.42	1.06	189.2	23.12
		1.0	30.03	0.23	257.0	4.43

CONCLUSIONS AND FURTHER WORK

The results show that a considerable error can be introduced depending on the position and hence the velocity. The impact of this does depend on the heat stress index applied and in particular on limitations to the calculated range for indices such as BET. The differences are greatest at lower air velocities and reduce as air velocity increases. The effect of relative humidity is similar to the effect of velocity albeit that it is not as pronounced.

This shows that evaluating a heat stress index will be position dependant in an airway. Also given the size of a human body this poses another issue. Where in the airway should velocity be measured to give a good indication of the thermal strain that the body is being subjected to?

The analysis presented is developed from a limited data set. Further testing using a wider dataset needs to be undertaken. This can be using published data such as that presented by Kohler and English (1983) or by establishing a dataset from underground measurements. The later would be preferable as it would enable the combined effects of variation in air velocity and dry/wet bulb temperatures to be investigated.

ACKNOWLEDGEMENTS

The author would like to acknowledge the support of Federation University Australia in researching and writing this paper.

REFERENCES

Gibson K. L. (1976) The computer simulation of climatic conditions in underground mines. PhD Thesis, University of Nottingham.

Hardcastle S. G., Grenier M.G. and Butler K. C. (1991) Electronic vane anemometry finding a suitable replacement of mechanical analog devices for mine airflow assessment. Proc 5th US Mine Vent Symp. Wang Y. J. (ed) Morgantown W.V. June 3-5. pp 482- 493. SME Littleton CO.

Kohler J. L. and English L. M. (1983) Determination of velocity measurement correction factors and guidelines. USBM Research Report J0308027

Pickering A. J. and Tuck M. A. (1997) Heat: Sources, Evaluation, Determination of heat stress and heat stress treatment. Mining Technology, June 1997, 79, No 910, p 147-156. Doncaster, UK

Wyndham C. H. (1974) The physiological and psychological effects of heat. Chapter 7 The ventilation of South African Gold Mines, Burrows J (ed), pp 93 – 137. The Mine Ventilation Society of South Africa, Johannesburg

A novel approach for predicting the incubation period of spontaneous combustion of coal based on thermogravimetric analysis (TGA)

X X Zhong[1] and F Hou[2]

1. Professor, China University of Mining and Technology, Xuzhou 221116,
   Email: zhxxcumt@cumt.edu.cn
2. Student, China University of Mining and Technology, Xuzhou 221116,
   Email: 1297929816@qq.com

ABSTRACT

Accurately predicting the incubation period of spontaneous combustion (IPSC) is of great significance to the prevention of coal self-ignition. In order to overcome the disadvantage of long test period of the adiabatic oxidation methods, in this paper, the composite reaction behavior of coal with oxygen is carried out in a thermogravimetric analyzer under different constant temperature conditions and multiple heating rates (1, 2,4 and 8 K·min⁻¹) with temperature rising from ambient temperature to 1073K. A novel approach for predicting the IPSC of coal is then proposed. Firstly, based on Starink iso-conversion method and its transformation formulas, the relationships are determined between apparent activation energy, pre-exponential factor and the extent of conversion, as well as the reaction model under different extent of conversion of coal in the water evaporation stage and the oxygen absorption stage. Then, the relationships between temperature and the extent of conversion in the water evaporation stage and the oxygen absorption stage are obtained respectively based on the constant temperature experiments. Finally, the results are substituted into the prediction model which is transformed from Arrhenius equation to calculate the reaction time of coal in the water evaporation stage and the oxygen absorption stage respectively, and the sum of these two parts of time is the IPSC of coal. This approach has the advantages of short test period and good repeatability.

INTRODUCTION

Coal spontaneous combustion happens during coal mining, storage, and transportation, (Pandey J et al., 2015; Deng, J et al., 2015; Lu Y et al., 2015) which leads to huge economic losses and resource waste as well as seriously threatens the safety of workers because of harmful substances and the gas or coal dust explosion evoke in the process (Song, Z Y et al., 2014). The incubation period of spontaneous combustion (IPSC) is one of the major parameters to characterize the coal spontaneous combustion. It refers to the time from coal is exposed to the air to spontaneous combustion, which is an important theoretical basic for guiding the fire prevention works in mines (Wang D M et al., 2011).

Statistics, analogy, experimental simulation and model solving methods are the most commonly used for predicting the IPSC of coal presently. The statistical and analogous approaches are applied on the mining site, the results of which usually take 'month' as units and have a poor precision (Zhang G S et al., 1990). They are generally as a reference to evaluate the accuracy of predicted results in laboratory (Li X C et al., 2014; Dan Z et al., 2017). The experimental simulation methods simulate the whole process of coal spontaneous combustion under laboratory adiabatic conditions, so the test period is usually long (tens of days or months). They can also be divided into two categories, large-scale and small-scale simulation experiments, in accordance with the coal consumption.Large-scale simulation experiments can truly reflect the process of coal spontaneous combustion (Stott J B, 1987; Smith A.C. et al., 1991; Cliff D, 1998; Wen H, 2004; Xu J C et al., 2001; Deng J et al., 1999). However, a large number of coal consumption (90kg~3000t) as well as the disadvantages of long test periods, much workload, high cost and poor repeatability which limits the widely application of this method. Small-scale coal spontaneous combustion experiments are favored by some scholars (Lu W et al., 2006; He Q L et al., 2004; Dai G L et al., 2005) which has the advantage of high speed, low cost and easy control. But it is necessary to dry the coal sample at 105~110 °C before the experiments which ignores the influence of water on the IPSC of coal. Thus the predicted results are quite different from the actual values.

The model solving method is a kind of method that combines theory and experiment and the IPSC of coal can be tested quickly by it. The models established by the Soviet scholar И.B Kalengin (1984) and Chinese scholars Xu J C et al. (2000), Deng J et al. (2001) are the most influential, and many scholars (Li L et al., 2010; Liang Y T et al., 2014; Yu M G et al., 2001a; Yu M G et al., 2001a; Yang Y L et al., 2014; Liu J et al., 2006; Chen W S et al., 2005; Qu L N et al., 2018) have proposed some new models based on them. The reaction rate and exothermic intensity in these models are usually calculated by the Arrhenius equation while the methods for determining the kinetic parameters are different. For example, Li L et al. (2010) obtain the activation energy of three typical coal samples based on the thermal equilibrium equation of coal oxidation and the adiabatic oxidation experiments; Yu M G et al. (2001a; 2001b) obtain the heat of coal oxidation in air by DSC method; Liu J et al. (2006) drawn the conclusion that the oxidizing decomposition of coal can be described as a first-order chemistry reaction, and obtained the activation energy and IPSC according to the Costs-Redfen kinetic analysis method; Chen W S et al. (2005), Qu L N (2018) based on the results of Liu J et al. (2006) obtained the activation energy of different temperature stages. The methods for determining the kinetic parameters above are all the based on the 'single heating rate' experiments. However, as the kinetic analysis methods are continuously improved and optimized (Zhang Y, 2016; Rotaru A, 2012; Dupuy J, 2015, Qi X Y et al., 2016, Ma L Y et al., 2017), the single heating rate methods is no longer applicable now, and replaced by the iso-conversional methods of multiple heating rates.

In summary, the approach of large-scale coal spontaneous combustion simulation is the most accurate in the existing methods for predicting the IPSC of coal while the disadvantage of long testing period restricts the wide application of this method. The model solving method can achieve the aim of testing the IPSC of coal quickly, but the present methods of determining the kinetic parameters need further improvement and optimization. In order to solve the problems, this paper have proposed a novel approach for predicting the IPSC of coal based on thermogravimetric analysis (TGA).

PREDICTION PRINCIPLE

Coal oxidative decomposition is a typical gas-solid reaction. Kinetics deals with measurement and parameterization of the process rates. Ignoring the influence of pressure, the rate can be parameterized in terms of two major variables: temperature T and the extent of conversion α. The temperature dependence of the process rate is typically represented through the Arrhenius equation and the dependence on the extent of conversion by the reaction model, f(α) (Vyazovkin, S et al., 2011), yielding the following equation:

$$\frac{d\alpha}{dt} = Aexp\left(-\frac{E}{RT}\right)f(\alpha) \quad (1)$$

Where, α is the extent of conversion; t is the time, s; dα/dt is the process rate; A is the pre-exponential factor, s^{-1}; E is the activation energy, $kJ \cdot mol^{-1}$; R is the universal gas constant, R=8.314 $J \cdot mol^{-1} \cdot K^{-1}$; T is the system temperature during the coal reaction process, K; f(α) is reaction model. It is worth emphasizing that the experimentally determined kinetic parameters are called 'apparent' in the following text. Rearranging the Eq. (1) and integrating between 0~t and 0~α on both sides of the equal sign, obtaining the formula:

$$t = \int_0^t dt = \int_0^\alpha \frac{exp\left(\frac{E}{RT}\right)}{Af(\alpha)} d\alpha \quad (2)$$

The Eq. (2) is used for predicting the IPSC and it also applies to the combustible substances except coal. The relationship of E, A, f(α), T with respect to α are unknown. And the aim of kinetics analysis in this study is mainly to identify these unknowns.

For the experiments that the temperature is controlled by thermogravimetric analyzer in accord with a program set up by an operator, the temperature changes linearly with time so that β=dT/dt=constant, where β is the heating rate. Rearranging Eq. (1) and the integral with respect to time is usually replaced with the integral with respect to temperature leads to:

$$\int_0^\alpha \frac{d\alpha}{f(\alpha)} = g(\alpha) = \frac{A}{\beta} \int_{T_1}^{T_2} exp\left(-\frac{E}{RT}\right) dT \quad (3)$$

Where T_1 and T_2 are the initial temperature and final temperature of the reaction stage respectively; $g(\alpha)$ is the integral of $f(\alpha)$. For different kinetic reaction mechanisms the reaction model and its integration are different. The integral term $\int_{T_1}^{T_2} exp\left(-\frac{E}{RT}\right) dT$ in Eq. (3) does not have an analytical solution. And a number of approximate solutions are offered. Among them, the Straink method is more accurate than others (Vyazovkin, S et al., 2011). Taking the Straink approximation into Eq. (3) and obtaining another expression of the kinetic equation:

$$\ln\left(\frac{\beta}{T^{1.92}}\right) = -1.0008\left(\frac{E}{RT}\right) + ln\left(\frac{AR^{0.92}}{g(\alpha)E^{0.92}}\right) - 0.312 \quad (4)$$

It can be seen from the Eq. (4) that when α=constant during a plurality of program-controlled linear heating experiments (ie. multiple β), a line can be obtained by fitting the scatter plots of $\ln(\beta/T^{1.92})$ with respect to 1/T. The apparent activation energy and pre-exponential factor at different extent of conversion can be calculated by the equation:

$$E_\alpha = -\frac{KR}{1.0008} \quad (5)$$

$$A_\alpha = \exp(C + 0.312)\frac{g(\alpha)E^{0.92}}{R^{0.92}} \quad (6)$$

Where K and C are the slope and intercept of the line respectively. The apparent activation energy and pre-exponential factor at other extent of conversions can be obtained by the same method. When solving the pre-exponential factor by Eq. (6), the reaction model must be determined firstly. The commonly used reaction models have been given in the literature (Hu R Z et al., 2008), and 20 kinds of gas-solid reaction models (Appendix 1) are selected as the scope of determining the coal-oxygen reaction model. The Eq. (4) can be rearranged mathematically to obtain the equation:

$$\ln\left(\frac{g(\alpha)}{T^{1.92}}\right) = -1.0008\left(\frac{E}{RT}\right) + ln\left(\frac{AR^{0.92}}{\beta E^{0.92}}\right) - 0.312 \quad (7)$$

In Eq. (7), when α=constant, for any of the reaction model in Appendix 1, another line can be obtained by fitting the scatter plots of $\ln(g(\alpha)/T^{1.92})$ with respect to 1/T. And the apparent activation energy (expressed by E_x and the value of x ranges from 1 to 20) which is corresponding to the selected reaction model can be obtained by Eq. (5). It is assumed that the selected reaction model is exactly the actual reaction model of coal-oxygen reaction, then the value of E_x should be equal or similar to E_α that obtained by Eq. (4). Thus the reaction model can be determined by comparing the value of E_x and E_α.

The kinetic parameters (E_α, A_α, $f(\alpha)$) under different extent of conversion in coal spontaneous combustion process can be obtained by the kinetic analysis above, and the function of their change with respect to the extent of conversion can be obtained by curve fitting. But the temperature diversification during the process is nonlinear and irregular, and cannot be directly obtained by theoretical analysis. Studies have shown that coal has the characteristics of reacting in stages with the temperature increasing (Yu M G et al., 2009). Therefore, in this paper constant temperature experimental methods have been proposed to determine the relationship between temperature and the extent of conversion. First of all, let the coal samples fully react constant temperature to obtain the equilibrium thermogravimetry (TG). Then, converting the equilibrium TG into the corresponding extent of conversion. Finally, obtaining the function of temperature versus the extent of conversion by fitting.

EXPERIMENTAL

Coal sample

The experimental coal sample in this study was taken from the 23107 working face of No. 13 coal seam of Xiegou (XG) Colliery in Shanxi Province which belongs to bituminous coal. On-site statistics showed that the incubation period of coal spontaneous combustion is about one month. Fresh raw coal was sealed at the production site and sent to the laboratory for storage under an inert atmosphere. Before the experiments, the collected raw coal is processed to remove the surface layers and then crushed in an oxygen-free glove box. Screened particles ranging between 0.075 mm to 0.15 mm in size which are commonly used in the literature (Ma L Y et al., 2017) are used in this study. The elemental and industrial analysis of the testing coal sample are shown in Table 1.

TABLE 1 –The elemental and industrial analysis of the coal sample

The elemental analysis results （%）		The industrial analysis results （%）	
C	79.27	M_{ad}	1.15
H	4.68	A_d	15.34
N	1.30	V_{daf}	44.39
S	0.20	F_{Cd}	47.08
O	14.51	-	-

Testing procedure

The experiment in this paper consists of two parts, program-heating experiment and constant temperature experiment. The main instrument is SDT-Q600 synchronous thermal analyzer produced by American TA Company which is used to measure and record the continuous thermogravimetric during sample reaction. The auxiliary instrument is MF-4 gas mixture, pure nitrogen bottle and pure oxygen bottle for formulating a gas stream with certain oxygen concentration.

The gas flow rate of the temperature programmed experiment are set to 100ml·min^{-1}. The initial temperature is ambient temperature, and rose to 1073K at the heating rate of 1, 2, 4, 8K·min^{-1} respectively. Each experiment is weighed 10mg coal sample and repeated three times. The constant temperature experiment also set the gas flow rate to 100ml·min-1 and the initial temperature to ambient temperature. When selecting the constant temperatures, the results of the temperature programming experiment needs to be referred to. In this paper, 303 K, 313 K, 323 K, 333 K, 343 K, 353 K, 393 K, 413 K, 433 K, 453 K, and 473 K are selected as constant temperatures. Each experiment is weighed 10 mg of sample. Three replicate experiments are performed for each set of experiments.

THE IPSC PREDICTION OF XG SAMPLE

In view of the influence of oxygen concentration on the IPSC of coal, this study only uses the gas flow with 21% oxygen concentration (equivalent to air but here obtained by 79:21 volume ratio of nitrogen and oxygen) as an example to illustrate the novel approach. The influence of different oxygen concentration on the IPSC of coal will be explored in subsequent studies.

Programmed-heating experimental results and analysis

Fig.1 illustrates the TG profiles characterizing the combustion of XG raw coal at different heating rates (1, 2, 4 and 8 K·min^{-1}). The results show that the TG changing trends with the rise of temperature are similar. Based on the previous literatures and a series of TG profiles obtained in this study, the coal reaction process is divided into three stages: water evaporation stage, oxygen absorption stage, and combustion stage. There are several characteristic points existing on the TG profiles, as shown in Fig.2, M_0, M_1, M_2, and M_3 are the initial or final TG of the respective stages, and T_0, T_1, T_2, and T_3 are the corresponding temperatures. The characteristic temperature T_2 is also referred to the initial combustion temperature. And the time of the temperature rises from T_0 to T_2 is just the IPSC of coal which includes the water evaporation stage and the oxygen absorption stage. Comparing the profiles in Fig.1, it is found that when the initial TG (M_0) under all heating rates are the same, M_1 and M_2 are almost equal too, in spite of the TG changing process are different. Therefore, the conclusion that the total amount of TG changed during the water evaporation stage (M_0-M_1) and the oxygen absorption stage (M_2-M_1) are substantially equal under different heating rates can be drawn. Based on the above analysis, a novel research idea that predicting the reaction time of the water evaporation stage and the oxygen absorption stage respectively and then the sum is the IPSC of coal has been proposed.

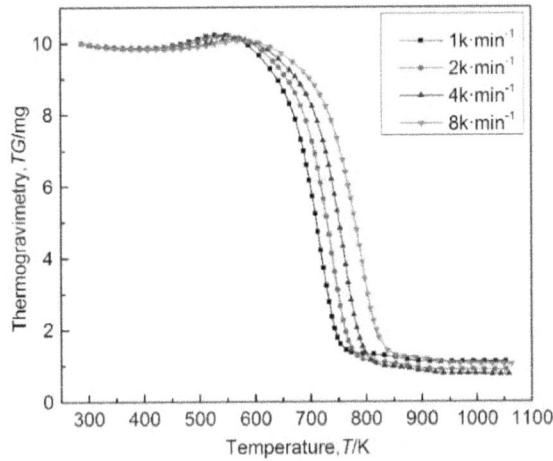

FIG 1 –The curves of TG vs. T at multiple temperature-increasing rates

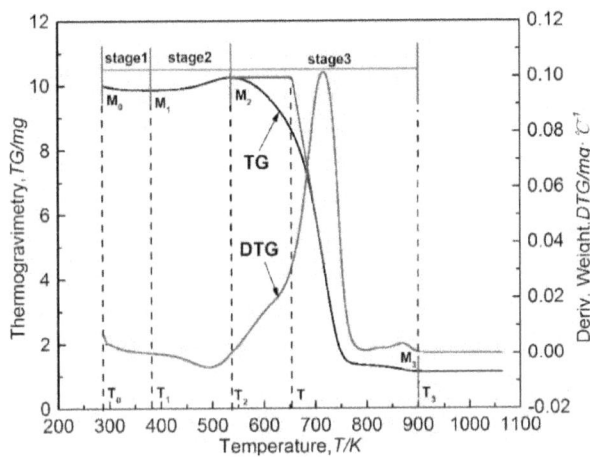

FIG 2 –Diagram of stage division and characteristic points

Based the results of the program-heating experiments, the relationship of kinetic parameters change with respect to the extent of conversion in the water evaporation stage and the oxygen absorption stage can be obtained respectively using iso-conversional methods. First of all, converting the TG to the corresponding extent of conversion by the equation:

$$\alpha = \left| \frac{M-m}{\Delta M} \right| \qquad (8)$$

Where M is the initial TG, m is the changing TG, ΔM is the total amount of TG changed of the two stages. After the conversion, the extent of conversion in each stage varies from 0 to 1. Next, fitting the scatter plots of $\ln(\beta/T^{1.92})$ with respect to 1000/T (in order to make the calculated apparent activation energy's unit is kJ·mol⁻¹, the slope value obtained here is reduced by 1000 times) to obtain the lines corresponding the extent of conversion in a range of 0.05~0.95 with a step of 0.05, as shown in Fig.3.

Then, the apparent activation energy (E_α) at each extent of conversion is calculated from the slope of the line by Eq. (5), and the results are shown in Fig. 4. Finally, fitting the scatter plots of the apparent activation energy versus extent of conversion and obtaining the functional expressions..

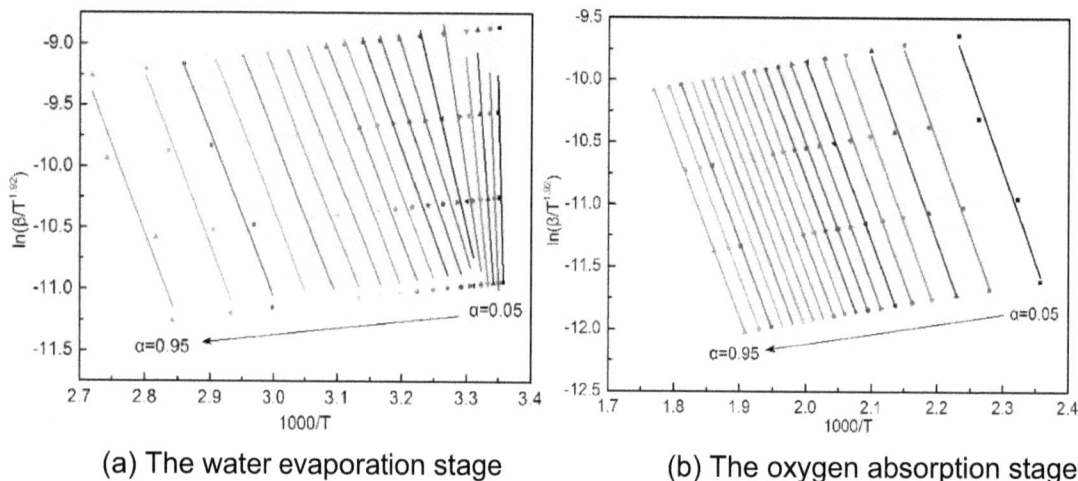

(a) The water evaporation stage   (b) The oxygen absorption stage

FIG 3 –The results of fitting $\ln(\beta/T^{1.92})$ vs. 1000/T

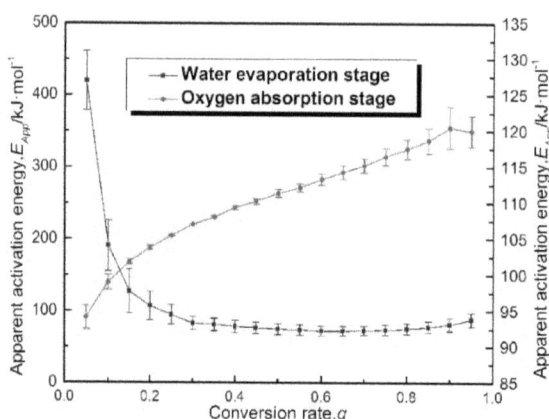

The results are shown in Table 2

FIG 4 –The profiles of E_α vs. α

TABLE 2 –The expressions of E_α vs. α

Corresponding stage	E_α/ kJ·mol-1	Domain
Water evaporation stage	$E_\alpha=1082.95\times\exp(-\alpha/0.041)+95.53$	0,0.3
	$E_\alpha=-37.94\times\alpha+94.02$	0.3,0.55
	$E_\alpha=9.3\times10\text{-}4\times\exp(-\alpha/0.097+71.91)$	0.55,1
Oxygen absorption stage	$E_\alpha=-33.51\times\exp(-\alpha/0.64)+127.25$	0,1

For any of the extent of conversion above, fitting the scatter plots of $\ln(g(\alpha)/T^{1.92})$ with respect to1000/T and obtaining the apparent activation energy (E_x). The reaction models have been determined by comparing the value of E_x and E_α, and the results are shown in Fig. 5.

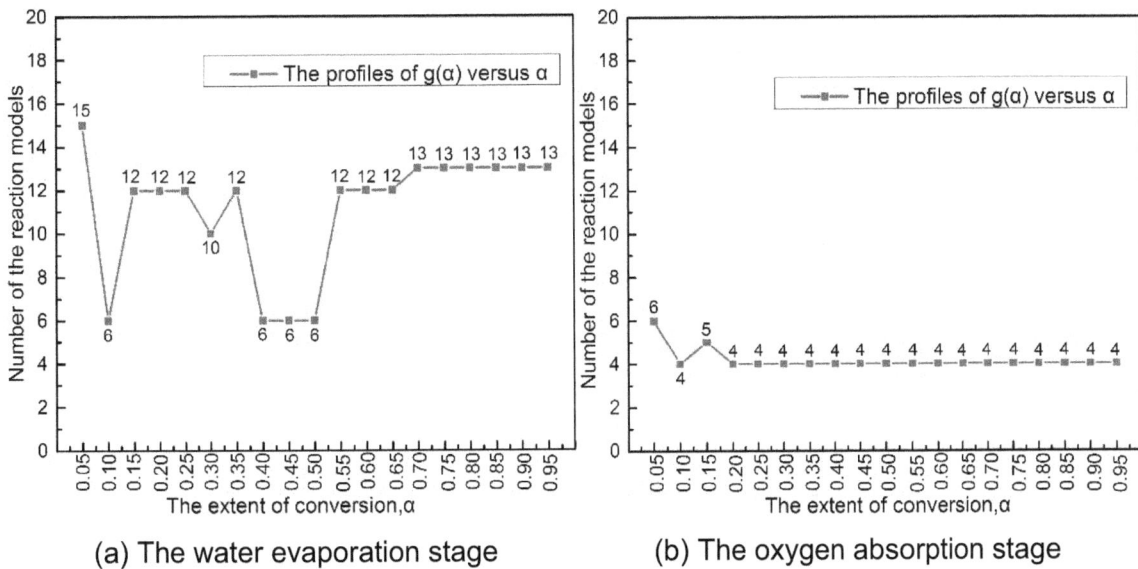

(a) The water evaporation stage   (b) The oxygen absorption stage

FIG 5 –The profiles of g(α) vs. α

According to the apparent activation energy and reaction models obtained, the pre-exponential factors at each extent of conversion have been calculated by Eq. (6), and the logarithm of the results are shown in Fig. 6. In view of the exponential change of the pre-exponential factor with the extent of conversion, the accuracy of the fitting is poor. Studies have proved the kinetic compensation effect of carbon oxidation reaction throughout conversion for all samples (Yip K, 2011). Thus, the kinetic compensation effect, that is, the linear relationship between lnA and E_α, is used to obtain the expressions of pre-exponential with the extent of conversion. Its mathematical expression is:

$$\ln A = aE_\alpha + b \quad (9)$$

Where a and b are the compensation coefficients, and the unit of a is $mol·KJ^{-1}$. Fitting the scatter plots of lnA with respect to E_α to obtain the value of a and b. Then the expressions of pre-exponential with the extent of conversion are obtained by the expressions of E_α, and the results are shown in Table 3.

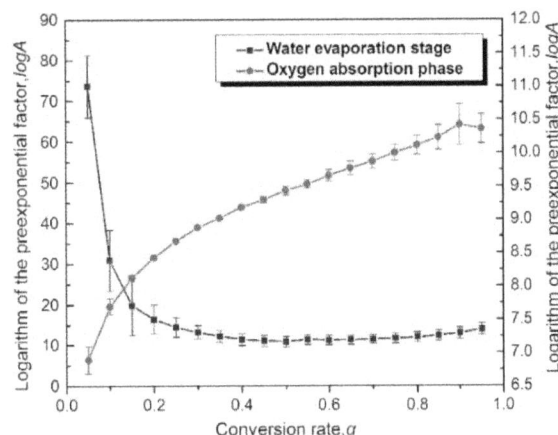

FIG 6 –The profiles of logA vs. α

Constant temperature experimental results and analysis

Fig.10 illustrates the profiles of TG and temperature versus time of XG raw coal in constant temperature experiments. From the reaction time, the results show that the TG reaches equilibrium in a short time (about 30 minutes) in the water evaporation stage but a long time in the oxygen absorption stage. From the perspective of TG change, the results show that the value of equilibrium TG decreases with the increasing of temperature in the water evaporation stage but increases in the oxygen absorption stage. The reasons for this result are as follows. In the process of the water evaporation, the coal sample can reach the set temperature in a very fast time due to the small amount. And the higher the temperature, the more thorough the water evaporation. However, in the oxygen absorption stage, because of the complicated microscopic pore of the coal sample and the porosity increases after the water evaporated which leads it is difficult for the oxygen around to diffuse into the pores and react with the active groups on the surface. Macroscopically, the reaction time of the oxygen absorption stage is long. Converting the equilibrium TG to the corresponding extent of conversion by Eq. (8) and obtaining the relationship between temperature and the extent of conversion by fitting, the results are shown in Table 4.

TABLE 3 – The expressions of A vs. α

Corresponding stage	A/S⁻¹	Domain
Water evaporation stage	$A=\exp[34.58+446.39\times\exp(-\alpha/0.041)]$	0,0.3
	$A=\exp(33.96-15.64\times\alpha)$	0.3,0.55
	$A=\exp[24.84+3.84\times10{-4}\times\exp(-\alpha/0.097)]$	0.55,1
Oxygen absorption stage	$A=\exp[26.31-10.02\times\exp(-\alpha/0.64)]$	0,1

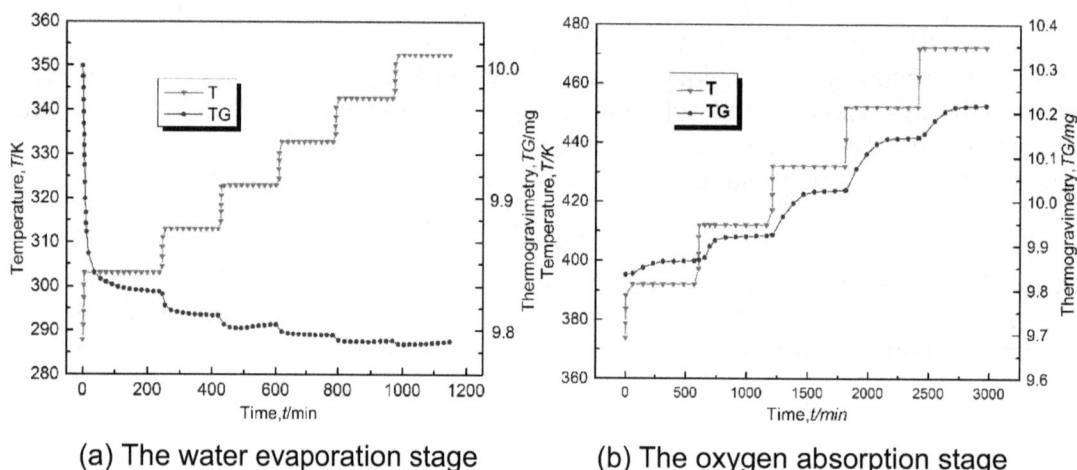

(a) The water evaporation stage   (b) The oxygen absorption stage

FIG 10 – The profiles of T vs. t and TG vs. t in constant temperature experiments

TABLE 4 – The expressions of T vs. α

Corresponding stage	T/K	R^2
Water evaporation stage	$T=1.4\times10{-7}\times\exp(\alpha/0.05)+298.51$	0.99222
Oxygen absorption stage	$T=-166.966\times\exp(-\alpha/1.128)+525.72$	0.99297

Results of the IPSC

According to the results above, the reaction time of the water evaporation stage and the oxygen absorption stage can be calculated respectively by Eq. (2). The results show that the reaction time of the water evaporation stage is 14.63d and the reaction time of the oxygen absorption stage is 13.58. Thus the IPSC of the tested coal is 28.21d, which is close to the on-site statistical result which demonstrates the accuracy of this novel method.

CONCLUSION

In view of the limit of the present methods, a novel approach for predicting the IPSC of coal have been proposed based on the iso-conversional kinetic analysis. The program-heating experiments at multiple heating rates and constant temperature experiments are carried out. The experimental result showed that the total amount of TG changed during the water evaporation stage and the oxygen absorption stage are substantially equal under different heating rates, thus an approach is proposed to predict the reaction time of the two stages respectively and eventually the IPSC of coalare. The result of the constant temperature experiments shown that the reaction time of the water evaporation stage is longer than that of the oxygen absorption stage. The IPSC of the tested coal is 28.21d, which is close to on-site statistics, which proves the accuracy of this novel method.

REFERENCES

Pandey J, Mohalik N K, Mishra R K, 2015. Investigation of the Role of Fire Retardants in Preventing Spontaneous Heating of Coal and Controlling Coal Mine Fires, Fire Technology, 51(2):227-245.

Deng, J, Xiao, Y, Li, Q W, Lu, J H, Wen, H, 2015. Experimental studies of spontaneous combustion and anaerobic cooling of coal, Fuel, 157:261-269.

Lu Y, Qin B, 2015. Identification and control of spontaneous combustion of coal pillars: a case study in the Qianyingzi Mine, China, Natural Hazards, 75(3):2683-2697.

Song, Z Y, Kuenzer, C, 2014. Coal fires in China over the last decade: A comprehensive review, International Journal of Coal Geology, 133:72-99.

Wang D M, 2011. Mine Fires, 103 p (China University of Mining and Technology Press: Xu Zhou).

Zhang G S, 1990. Estimation Method of Natural Fire Period of Coal Seam and Its Extension Way, Mine Safety, 06:61-63+18.

Li X C, 2014. Pore-scale numerical study on the self-heating process of packed coal stockpile, PhD thesis (unpublished), Dalian University of Technology.

Dan Z, 2017. Analysis of the Spontaneous Combustion Period of Lignite in Pingzhuang Hongmiao Coal Mine, Modern Mining.

Stott J B, 1987. A 'full-scale' laboratory test for the spontaneous heating of coal, Fuel, 66(7):1012-1013.

Smith A.C., Miron Y, Lazzara CP, 1991. Large-scale studies on spontaneous combustion of coal, US Bureau of Mines, Report of Investigation, 346.

Cliff D, 1998. Large scale laboratory testing of the spontaneous combustibility of Australia coals, in Queensland Mining Industry Health & Safety conference, Bresbance, Australia: Queensland miningcounci1, 175-179.

Wen H, 2004. Experiment simulation of whole process on coal self-ignition and study of dynamical change rule in high-temperature zone, JOURNAL OF CHINA COAL SOCIETY, 06:689-693.

Xu J C, Xue H L, Wen H, Li L, 2001. Analysis of Factors Affecting the Composite Thermal Effect of Coal Oxygen, China Safety Science Journal, 11(2):31.

Deng J, Xu J C, Zhang Y D, Li L, 1999. Experiment and numerical analysis of the shortest spontaneous ignition period of coal, JOURNAL OFCHINA COAL SOCIETY, 03:52-56.

Lu W, Wang D M, Zhou F B, Dai G L, Li Z H, 2006. Tendency of Spontaneous Combustion of Coal Based on Activation Energy, Journal of China University of Mining &Technology, 02:201-205.

He Q L, 2004. Study on Experiment and Computer Simulation of Coal Oxidation of Low Temperature and Whole Spontaneous Process, PhD thesis (unpublished), China University of Mining &Technology.

Dai G L, Wang D M, Lu W, Zhou F B, 2005. Adiabatic study on low-temperature self-heating oxidation of coal, Journal of Liaoning Technical University, 04:485-488.

И.B Kalengin, 1984. Coal spontaneous combustion and its prediction, pp 201-205, (Coal Industry Press: Beijing).

Xu J C, Xu M G, Deng J, Ge L M, 2000. Research on Prediction Technology of Natural Fire Period Based on Analysis of Coal-Oxygen Compound Process [J]. FIRE SAFETY SCIENCE, 03:21-27.

Deng J, 2001. Coal spontaneous combustion prediction theory and technology, (Shanxi Science and Technology Press: Xi An).

Li L, Jiang D Y, B B Beamish, 2010. Calculation of ignition times under adiabatic conditions by activation energy, JOURNAL OFCHINA COAL SOCIETY, 35(5):802-805.

Liang Y T, Song S L, Lou H Z, Lin Q, 2014. An analytic solution of coal spontaneous combustion period calculation model, Journal of China Coal Society, 40 (9):2110-2116.

Yu M G, Wang Q A, Fan W C, Liao G X, 2001a. Study on Prediction of Natural Fire Period of Coal Seam, Journal of China University of Mining &Technology, 30(4):384-387.

Yu M G, Huang Z C, Yue C P, 2001b. Mathematical model for solving the shortest spontaneous ignition period of coal, Journal of China Coal Society, 26(5):516-519.

Yang Y L, Li Z H, Hou S S, Gu F J,2014. The shortest period of coal spontaneous combustion on the basis of oxidative heat release intensity, International Journal of Mining Science and Technology, 24(01):99-103.

Liu J, Chen W S, Qi Q J, 2006. Study on spontaneous combustion period of coal based on activation energy index, Journal of Liaoning Technical University, 25(2):161-163.

Chen W S, Wu Q, 2005. Discussion on the relationship between spontaneous combustion tendency of coal and spontaneous ignition period and activation energy, Mine Safety, 36(12):53-55.

Qu L N, 2018. A study on the prediction method of coal spontaneous combustion development period based on critical temperature, Environmental Science and Pollution Research, 1-13.

Zhang Y, 2016. Evaluation of the susceptibility of coal to spontaneous combustion by a TG profile subtraction method, Korean Journal of Chemical Engineering, 33(3):1-11.

Rotaru A, 2012. Thermal analysis and kinetic study of Petroşani bituminous coal from Romania in comparison with a sample of Ural bituminous coal, Journal of Thermal Analysis & Calorimetry, 110(3):1283-1291.

Dupuy J, Leroy E, Maazouz A, 2015. Determination of activation energy and preexponential factor of thermoset reaction kinetics using differential scanning calorimetry in scanning mode: Influence of baseline shape on different calculation methods, Journal of Applied Polymer Science, 78(13):2262-2271.

Qi X Y, Li Q Z, Zhang H J, Xin H H, 2016. Thermodynamic characteristics of coal reaction under low oxygen concentration conditions, Journal of the Energy Institute, 90(4).

Ma L Y, Wang D M, Wang Y, Dou G L, Xin H H, 2017. Synchronous thermal analyses and kinetic studies on a caged-wrapping and sustained-release type of composite inhibitor retarding the spontaneous combustion of low-rank coal, Fuel Processing Technology, 157:65-75.

Vyazovkin, S., Burnham, A. K., Criado, J. M., Perez-Maqueda, L. A., Popescu, C., Sbirrazzuoli, N., 2011. ICTAC Kinetics Committee recommendations for performing kinetic computations on thermal analysis data, Thermochimica Acta, 520(1-2), 1-19.

Hu R Z, 2008. Thermal Analysis Kinetics (2nd Edition), Science Press.

Yu M G, Lin J J, Lu C, Chen L, 2009. Experimental study on characteristic of coal's constant temperature oxidation at different temperature, JURNAI OF HENAN POLYTECHNIC UNIVERSIIY (NATURAI SCIENCE), 28(3):261-265.

Yip K, Esther N, Chun-Zhu L, 2011. A mechanistic study on kinetic compensation effect during low-temperature oxidation of coal chars, Proceedings of the Combustion Institute, 33(2):1755-1762.

APPENDIX 1

APPENDIX 1 –The 20 gas-solid reaction models

	Reaction model	Code	$f(\alpha)$	$g(\alpha)$
1	One-dimensional diffusion	D1	α^2	$\frac{1}{2}\alpha^{-1}$
2	Two-dimensional diffusion	D2	$\alpha + (1-\alpha)\ln(1-\alpha)$	$[-\ln(1-\alpha)]^{-1}$
3	Two-dimensional diffusion	n=2	$\left[1-(1-\alpha)^{\frac{1}{2}}\right]^2$	$(1-\alpha)^{\frac{1}{2}}\left[1-(1-\alpha)^{\frac{1}{2}}\right]^{-1}$
4	Three-dimensional diffusion	D3,n=2	$\left[1-(1-\alpha)^{\frac{1}{3}}\right]^2$	$\frac{3}{2}(1-\alpha)^{\frac{2}{3}}\left[1-(1-\alpha)^{\frac{1}{3}}\right]^{-1}$
5	Three-dimensional diffusion	D4	$1-\frac{2}{3}\alpha-(1-\alpha)^{\frac{2}{3}}$	$\frac{3}{2}\left[(1-\alpha)^{-\frac{1}{3}}-1\right]^{-1}$
6	Three-dimensional diffusion	D3	$\left[(1-\alpha)^{-\frac{1}{3}}-1\right]^2$	$\frac{3}{2}(1-\alpha)^{\frac{4}{3}}\left[(1-\alpha)^{-\frac{1}{3}}-1\right]^{-1}$
7	Avrami–Erofeev	A3,n=1/3	$[-\ln(1-\alpha)]^{\frac{1}{3}}$	$3(1-\alpha)[-\ln(1-\alpha)]^{\frac{2}{3}}$
8	Avrami–Erofeev	A2,n=1/3	$[-\ln(1-\alpha)]^{\frac{1}{2}}$	$2(1-\alpha)[-\ln(1-\alpha)]^{\frac{1}{2}}$
9	Avrami–Erofeev	A1	$-\ln(1-\alpha)$	$1-\alpha$

10	Avrami–Erofeev	n=3/2	$[-\ln(1-\alpha)]^{\frac{3}{2}}$	$\frac{2}{3}(1-\alpha)[-\ln(1-\alpha)]^{-\frac{1}{2}}$
11	Avrami–Erofeev	n=2	$[-\ln(1-\alpha)]^2$	$\frac{1}{2}(1-\alpha)[-\ln(1-\alpha)]^{-1}$
12	Avrami–Erofeev	n=3	$[-\ln(1-\alpha)]^3$	$\frac{1}{3}(1-\alpha)[-\ln(1-\alpha)]^{-2}$
13	Avrami–Erofeev	n=4	$[-\ln(1-\alpha)]^4$	$\frac{1}{4}(1-\alpha)[-\ln(1-\alpha)]^{-3}$
14	Mampel Power	R1,n=1	α	1
15	Mampel Power	n=3/2	$\alpha^{\frac{3}{2}}$	$2/3\,\alpha^{-1/2}$
16	Mampel Power	n=2	α^2	$\frac{1}{2}\alpha^{-1}$
17	Contracting sphere	R3,n=1/3	$1-(1-\alpha)^{\frac{1}{3}}$	$3(1-\alpha)^{\frac{2}{3}}$
18	Contracting sphere	n=3	$3(1-(1-\alpha)^{\frac{1}{3}})$	$(1-\alpha)^{\frac{2}{3}}$
19	Contracting cylinder	R2,n=1/2	$1-(1-\alpha)^{\frac{1}{2}}$	$2(1-\alpha)^{\frac{1}{2}}$
20	Power law	F2	$(1-\alpha)^{-1}-1$	$(1-\alpha)^2$

Health Hazard Management – Respirable Dust/Diesel Emissions

Robust control of diesel engine particulate emissions in underground coal mining

N Coplin[1]

1. Engineering Manager, Orbital Australia, Balcatta WA 6021. Email: ncoplin@orbitalcorp.com.au

ABSTRACT

The need to protect workers from diesel particulate matter (DPM) had led the underground coal mining industry to install disposable filter systems on their vehicles. While the disposable filters are efficient at removing significant DPM, the following major issues have arisen:

- High cost of operations. Disposable filters cost $250-300/each and need to be changed at least once per shift resulting in an estimated cost of up to $164M/year to the NSW underground coal mining industry in replacement filters alone.

- Improper installation, damaged seals and failure to install a new filter when the old filter is removed means that workers are still being exposed to excessive amounts of DPM.

ACARP, the Australian Coal Association Research Program, funded project C25073 to evaluate the suitability of conventional on-road heavy duty diesel wall-flow filter systems to be adapted for use in underground coal operations at a proof-of-concept (PoC) level.

One of the key benefits with the use of a wall-flow DPF system is its tamper-proof design mitigating the risk of operating unfiltered diesel plant in poorly ventilated areas. Elimination of the need for continual replacement of disposable filters also provides the underground coal sector with significant operational savings estimated to be up to 80% of the incumbent technology.

The follow-on ACARP project C26070 has progressed the industrialisation of the wall-flow DPF system with additional site work, assessment of system robustness and the embedding of real-time, and near-real-time, monitoring technology.

INTRODUCTION

The project aimed to develop a wall-flow DPF type system to replace existing wet-element disposable filter systems used in typical Load Haul Dump (LHD) vehicles in underground coal operations.

The overall objectives were to:

- Prove a wall-flow DPF system could reduce Diesel Particulate Matter (DPM) emissions to a lower level than the current disposable filter systems;

- Prove that the wall-flow DPF system could be engineered to meet coal sector regulations such as explosion proofing, surface temperature and exhaust gas temperature limits;

- Prove that sufficient regeneration of the wall-flow DPF system would occur during typical operations so as to extend the interval between filter cleaning, thereby reducing maintenance costs. This was first verified on the engine dynamometer using the transient LHD test cycle developed by the project, and then later confirmed in the field at site tests.

A summary of the results that show that the first two objectives were achieved was outlined in an earlier AusIMM paper at the 2017 mine ventilation conference (Coplin, 2017) and detailed in the publish ACARP C25073 report (Coplin, Chicka, 2018). The detail in this paper will focus on demonstrating how the third objective is achieved and monitored.

BACKGROUND

The LHD platform selected for demonstration was the Coaltram model, with an electronic Caterpillar C7 engine and coal sector approved Diesel Exhaust System (DES) satisfying industry mandated requirements for explosion protection.

The simplified system schematics in FIG 1 compare the layout of a typical conventional disposable wet-element exhaust filtration system to that of the wall-flow DPF system. With the conventional system, exhaust from the engine is cooled through a water-filled exhaust scrubber before being filtered through the disposable filter element. The disposable filter is typically a paper, or coated paper element (similar to an engine air intake filter), which traps the particulates in the exhaust stream. An oxidation-only catalyst is sometimes also included upstream of the scrubber to reduce gaseous pollutants such as carbon monoxide (CO) and unburnt hydrocarbons (HC).

Because of the disposable filter's construction, the exhaust gas has to be first significantly cooled; however the cooling process does carryover substantial amounts of water both as liquid and in the form of saturated exhaust. This carryover water (and not the diesel particulates) can be a significant cause of premature filter blockage. Regular and early replacement of disposable filters is a substantial operating cost burden for the industry, further compounded by the costs of disposal and associated environmental handling. Whilst the current technology works to reduce particulates, the operational challenges can in instances lead to incorrectly fitted or operation without filters, resulting in a direct impact on ambient mine air quality during operations.

With the wall-flow DPF system, the wall-flow filter is installed before the wet exhaust scrubber since it requires heat for its effective operation. To satisfy tailpipe gas temperature requirements, the water scrubber is still required but is staged following the filter.

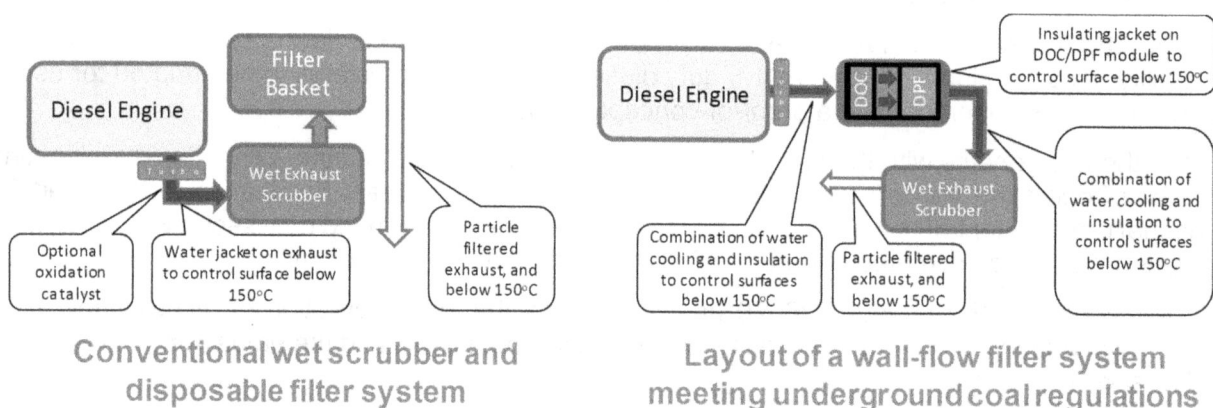

Conventional wet scrubber and disposable filter system

Layout of a wall-flow filter system meeting underground coal regulations

FIG 1 – Comparison of conventional wet scrubber, disposable filter system and wall-flow DPF systems

The wall-flow filter system technology discussed in this paper is detailed in FIG 2. The filter component is the second stage of the assembly and is a porous ceramic element where particulates are trapped as the exhaust passes through the structure. The combustion exhaust gas is forced to pass through the wall structure (hence the name wall-flow) because the cell structure of the substrate alternately blocks entry and exit flow channels. This wall-flow filtration method is understood to offer significant reduction in ultrafine particles compared to flow-through-filter (FTF) systems, such as metal monolith constructed substrates.

The underground coal sector is highly regulated compared to other mining environments, both in Australia and internationally. The primary reasons for this superior regulatory oversight relates to the explosion risk posed by an ambient operating environment that may contain elevated methane levels and deposition of coal dust on equipment. A methane rich environment may be explosive, but can also affect the operation of diesel equipment which "breathes" this enriched air. However, it is the deposited coal dust that poses an immediate explosion risk at an ignition point well below that of ambient methane. Consequently, regulations (AS3584.2:2008, MDG43) are in place to limit the temperature of surfaces or other ignition sources to below 150 °C, with additional constraints placed on other ignition sources such as electrical equipment (for example Standards Australia, AS4871.xx and AS60079.xx).

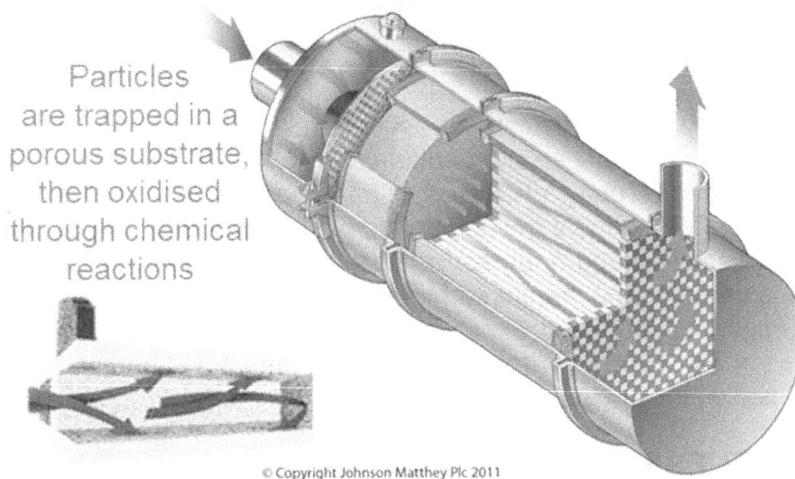

FIG 2 – Outline of how a wall-flow diesel particulate filter works

The consequence of these regulations is that the coal mine itself can not be used as a development environment for new equipment, as may be the case in other mining operations where these operational risks are not present.

The approach to development, including the generation of representative underground coal LHD engine test cycles was described in the previous AusIMM paper published in 2017 (Coplin, 2017). The methodology used provided robust correlation between in-service performance and work undertaken on an engine dynameter.

The earlier ACARP project evaluated several diesel particulate system configurations, with data from three of these to be reported in the earlier publication. Both the 10.5" and 9.5" systems used the same cell density, chemical formulation and precious metal washcoat loading levels; the only difference being diameter of the substrate.

Even though the 10.5" substrate was a little more challenging to package in the design constraints of the Coaltram LHD, this was selected as the basis of the design to proceed with for industrialisation as its larger capacity provided improved particulate storage capacity – important during extended idle periods when the system would not be able to regenerate. FIG 3 shows the system as installed on a Coaltram LHD for site trials.

FIG 3 – Installation of the wall-flow DPF on a Coaltram LHD for surface trials at site

The particulate and gaseous emission performance of the 10.5" DPF was measured to be compliant with MDG43 year 2020 limits for carbon monoxide (CO), oxides of nitrogen (NO_x), hydrocarbon (HC) and $PM_{2.5}$.

MDG43 standards apply to total particulate measurements, $PM_{2.5}$. This includes all the compounds deposited on the sampling filter paper and is different to NIOSH5040 methods which focus on the elemental carbon component. Total particulates, rather than speciation for elemental carbon, is used by engine and exhaust aftertreatment developers during lab based testing for a number of reasons, including single step sampling and analysis, rapid turn around of results in-house, and improved repeatability of measurement. The measure of total particulates is valid in that all the sample is from the exhaust of the engine under test, unlike in a working mine environment where some particulates may be attributable to other sources unrelated to diesel exhaust and hence the use of EC allows for traceability to what is diesel sourced.

SYSTEM ROBUSTNESS

The removal of stored particulates, predominantly carbon based soot, would be a substantial challenge if temperature alone was used to facilitate oxidation into CO_2. Many modern day wall-flow DPFs include a catalytic washcoat to assist in reducing the oxidation temperature of the trapped soot. This substantially lowers the exhaust feedgas temperature requirement from over 550 °C to typically 350-400 °C. Whilst better, this is still a substantial temperature for diesel engines to reach – compounded by the typically low duty cycle operation of an LHD in coal mining operations, with many LHDs spending more than 50% of their daily duty at idle.

To further lower the temperature required for soot oxidation, it was proposed to use technology which not only included a catalytic DPF, but also a pre-catalyst for preparation of the feed gas chemistry delivered to the DPF. The particular wall-flow technology used in this research work is proprietary to Johnson Matthey Catalysts Inc. and is effective in enabling oxidation of stored soot at temperatures as low as 250 °C (ref. Johnson Matthey website). The specific wall-flow technology tested was a part of the CCRT (catalytic continually regenerating trap) product family.

Whilst retention of exhaust heat for the purpose of improving wall-flow DPF performance is desirable, both the management of thermal loadings and exhaust gas composition chemistry were issues that required engineering solutions to be developed and are discussed in this paper.

Exhaust Back Pressure Stability

The robustness of a wall-flow DPF system can be characterised by a number of parameters, but the easiest tell-tale is whether the system reaches a stabilised pressure condition measured as the differential pressure across the system over a consistent test cycle regime. What this signifies is that the DPF system is storing and regenerating (oxidising) soot produced by the engine in a stabilised manner.

The basis of the testing undertaken to qualify stability was the custom developed transient cycle. The custom transient cycle exposes the DPF to significant periods of idle (non-regenerating periods to load the DPF with soot), typical periods where operational tasks may be undertaken (where the temperature and feedgas composition will provide increased soot rates, but may also provide the conditions suitable for oxidation of stored soot), and periods of idle between tasks, where the aftertreatment system is allowed to cool down and become inefficient (so as to be conservatively biased). The test cycle loops every 30 minutes, for a total duration of 100 hours. The use of a transient capable dynamometer facility not only allows the testing to be completed accurately, it also allows for each cycle repeat to be identical. The process is substantially automated including the acquisition of data – eliminating variations that can occur infield.

FIG 4 compares the performance of an insulated 10.5" system over cycles starting at 1 hour, 50 hours and 100 hours into the test, and at the end after the DPF is fully regenerated by a period of high load operation to burn-off any stored soot.

The data show that DPF back pressure has peaked and stabilised by the 50 hour point and maintains those conditions through to the 100 hour point. The burn-off is seen to be effective in restoring DPF performance to original pre-sooting conditions.

Similar back pressure stability is seen with the engine operated on Eromanga UMF (Underground Mining Fuel), despite this fuel having higher levels of sulfur (100ppmS) compared to automotive specification fuel meeting the National Fuels Standard 2000, which has only 10ppmS specified for diesel. The issue identified with UMF was that the high sulfur content though not affecting backpressure, did result in contamination of the catalytic substrates reducing oxidation effectiveness and thus emission compliance of the system with respect to both gaseous and particulate emissions. The effects of the high sulfur contamination where found to be reversible with high exhaust temperatures when the engine was operated at peak torque for a sustained period. More detail about the performance of the wall-flow DPF system will be published in the final report for the ACARP C26070 project.

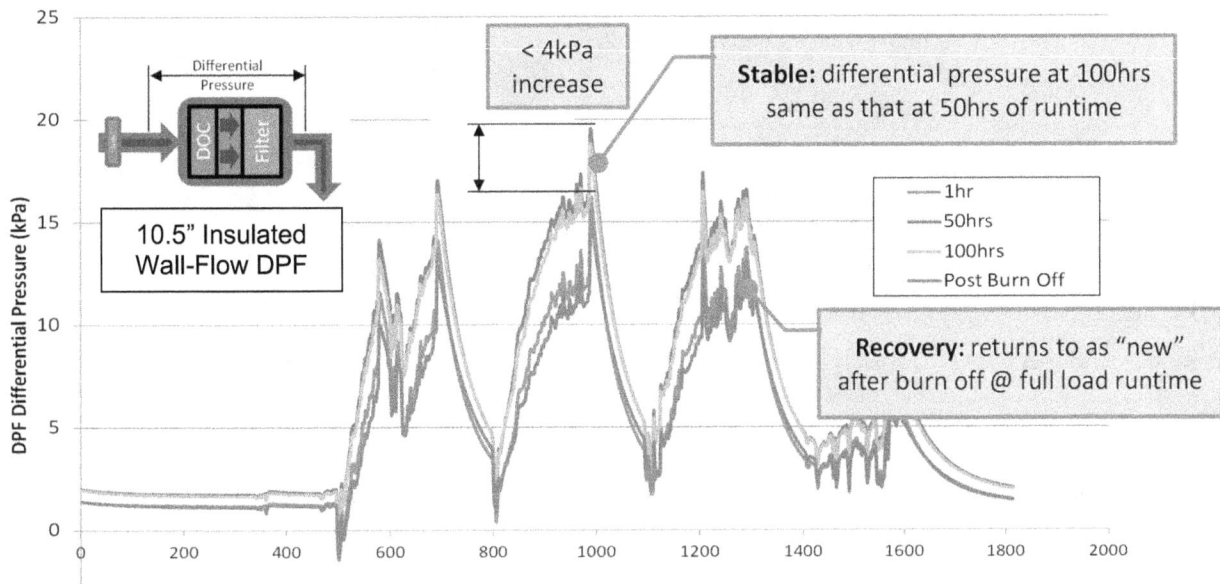

FIG 4 – Comparison of DPF differential pressure and stabilisation
over industry developed transient test

Demonstration of Thermal Control

Conventional practice in the underground coal sector is to use a water jacket around exhaust components as the means of reducing surface temperatures. The use of a water jacket poses compromises with regard to the implementation of a wall-flow DPF system. Firstly, the water jacket not only cools the surface of the pipe, it also cools the temperature of the exhaust feedgas. If the gas temperature upon exit from the engine's turbo is already on the low side of optimal, it will only get lower with water-cooled pipework. Secondly, exhaust heat lost to the water system needs to be dissipated in the LHD's radiator necessitating this to be larger than needed to cool just the engine.

Using a non-conducting insulation barrier around the exhaust pipework and DPF unit has benefits with regard to heat retention in the exhaust gas and minimisation of heat to the coolant circuit. Hotter exhausts may however increase the evaporation rate, and hence water consumption rate, of the wet exhaust scrubber.

The proof-of-concept design sought to evaluate and optimise the properties of the insulation barrier, with the ultimate goal of demonstrating that surface temperatures could be controlled with insulation as the primary measure.

The industrialised version of the wall-flow DPF system uses a mix of water-cooled and insulated features. The insulated features need to be protected from coal dust contamination and this is done by housing the insulated DPF core in a solid housing.

Through extensive testing it was verified that DPF core could be thermally controlled through insulation alone, just as it had been demonstrated during the proof-of-concept work. Since the insulation prevented exhaust and catalytically generated heat from dissipating radially from the whole body of the DPF it was found to concentrate at the connection points into and out of the DPF core. At these points, bolted flange joints are used to interconnect exhaust pipework. These pipes could be completely water-cooled, but for reasons of optimising heat loss from the exhaust, water-cooling is focused mostly to specific joints. FIG 5 shows which features of the Coaltram specific DPF system were thermally managed by water-cooling and which by insulation alone.

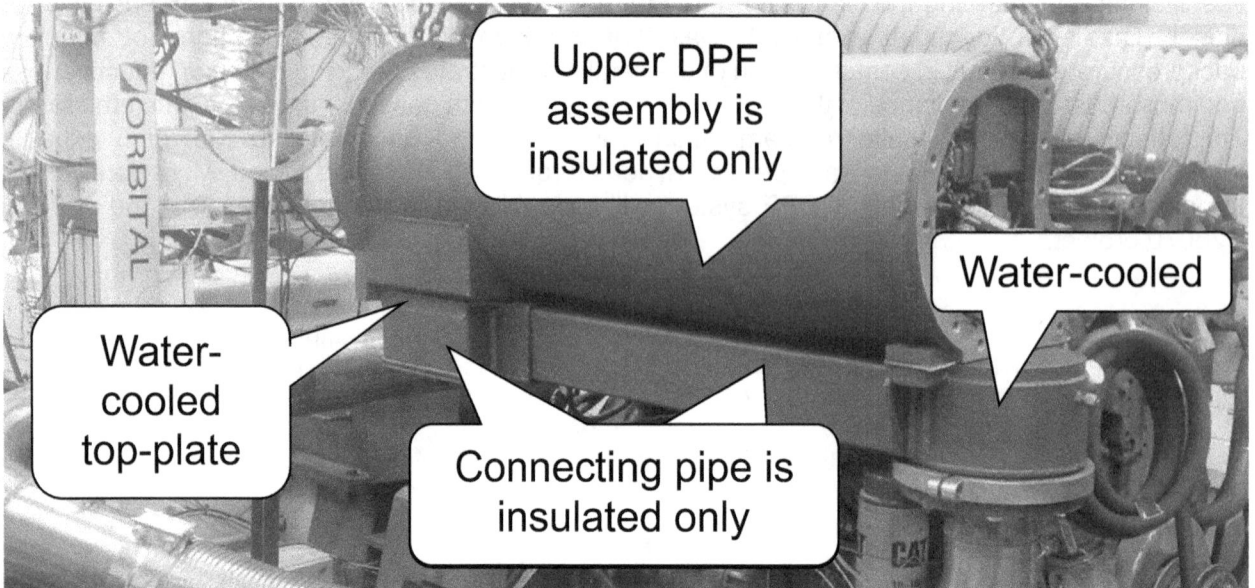

FIG 5 – Outline showing which DPF housing features were water-cooled and which were insulated

The flange joints themselves are optimised to direct heat into the water-cooling. This has necessitated design detail on the non-cooled side of the joint to minimise radial flow of heat from inner flowing exhaust gases and selection of exhaust gasket materials to optimise transfer.

Regulations prescribe that surface temperatures are not only controlled at nominal operating conditions, but also under adverse conditions of elevated ambient methane levels. Elevated ambient methane is combusted by the engine to produce more power and consequently higher exhaust temperatures. The equipment is generally fitted with methanometers which shut down the engine at ambient concentrations 1.25%; hence this is the capped test limit. However, 1.25% ambient methane in the intake air represents a significant excess fuel loading for the engine.

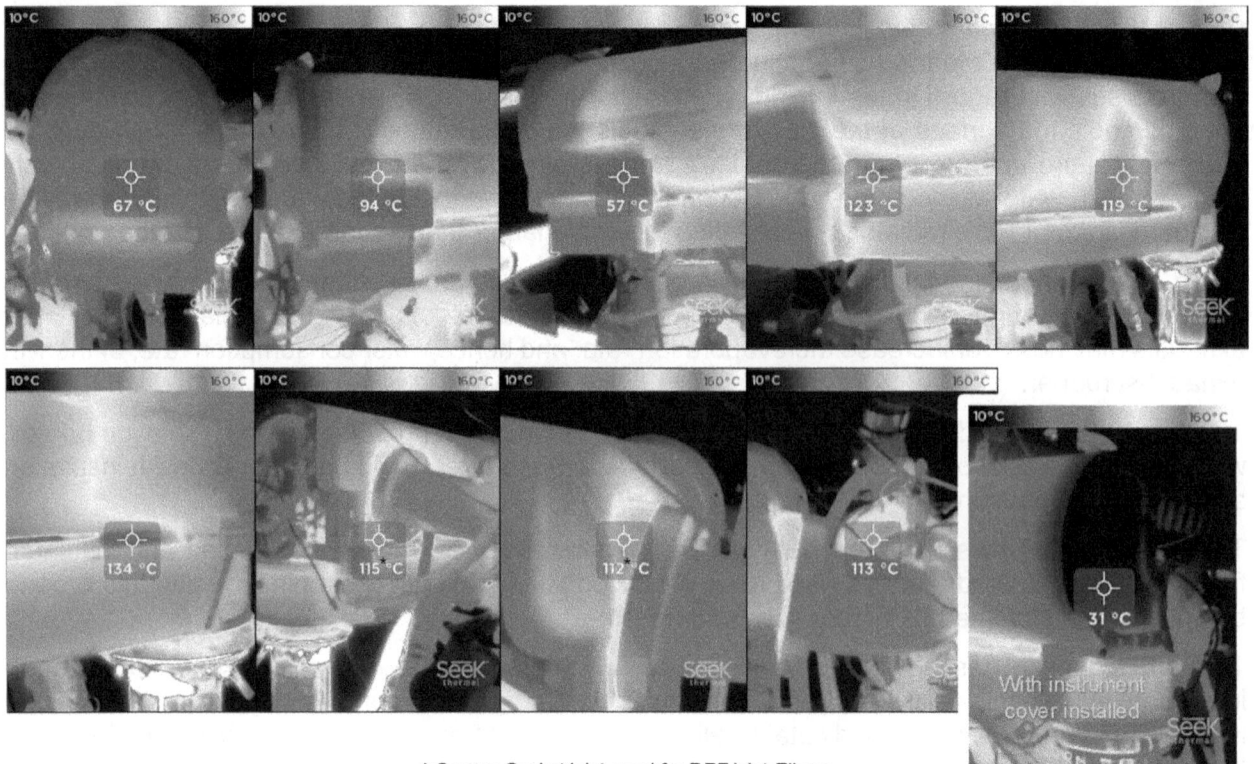

* Copper Gasket joint used for DPF inlet Elbow

FIG 6 – SEEK thermal images for 10.5" Industrialised DPF at 1500rpm Full-Load

FIG 6 shows thermal images captured by a SEEK thermal camera for the industrialised DPF assembly. Whilst arguably not as accurate as thermocouple results (refer the ACARP project report; Coplin, Chicka, 2017), they provide a more complete impression of the overall level of thermal management achieved. The full-scale readings on the thermal images taken whilst the engine was operating at 1500rpm full-load (with no methane injection) indicate peak temperatures of between 98 and 134 °C, well below the mandated 150 °C limit. These results are with a copper gasket at the DPF inlet flange which offers improved thermal transfer compared to an alternate mica-based gasket (see FIG 7). Note: no water-filled exhaust scrubber is fitted to the system during dynamometer testing.

It should be noted that the electronic DPF monitoring system is housed in an Ex 'd' enclosure integrated into the end of the DPF module. The reason for this is to integrate the sensing of pre-DPF and post-DPF temperatures and pressures within the one assembly. From within this Ex 'd' enclosure, direct access to the exhaust stream before and after the DPF is possible. However, the ramifications of this close integration are that the heat transfer to the electronics within the enclosure is a real consideration. Also seen in FIG 7 is the effectiveness of the insulation and isolation of the Ex 'd' instrumentation enclosure (blue region at end of the DPF module). In addition to thermal insulation from the DPF core, the water-cooled exhaust outlet feature stabilises heat transfer from the outlet sensors to the enclosure. Inside the enclosure, peak temperatures no higher than 76°C have been recorded during testing, including hot soak after engine shutdown; well below the 125°C rating of the electronic systems used.

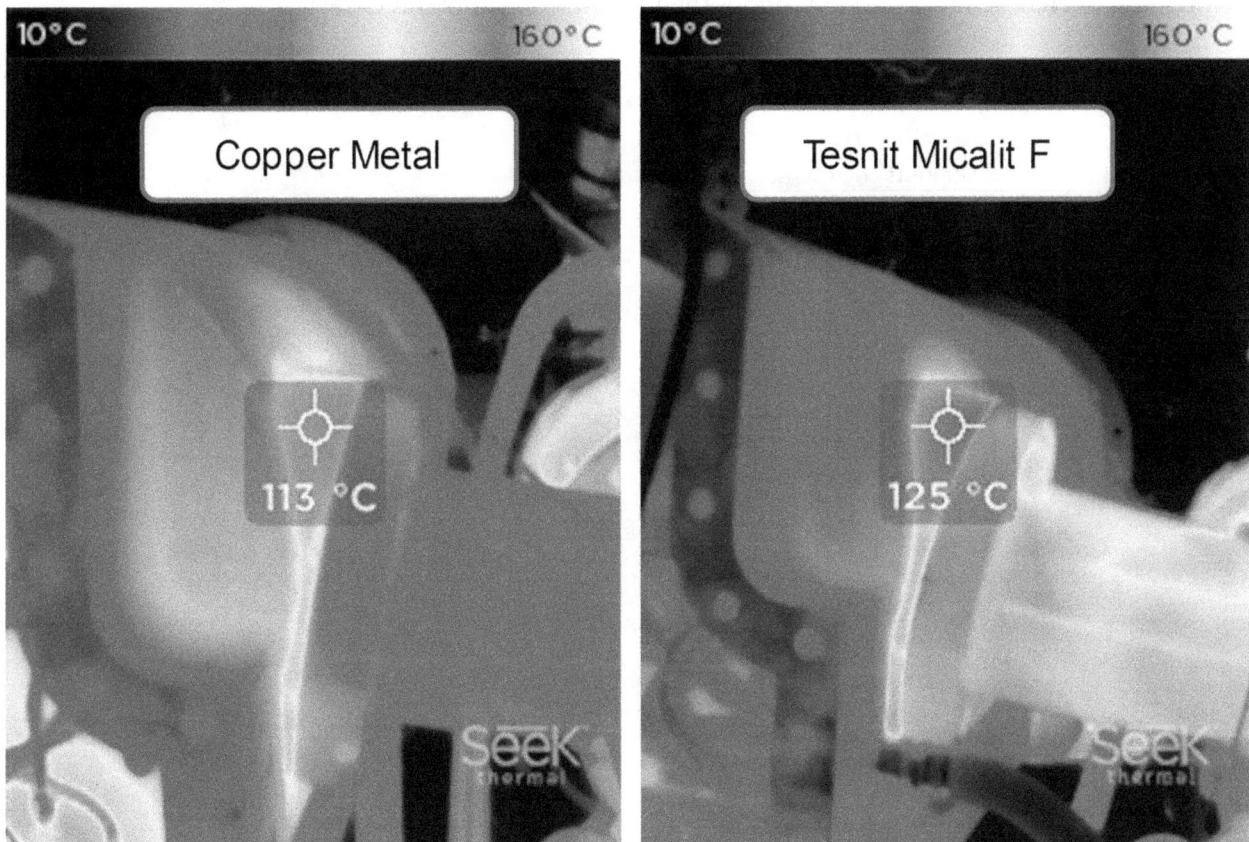

FIG 7 – SEEK thermal images for the DPF inlet flange comparing thermal performance of different gaskets

Monitoring the Performance and Robustness of the System

Whilst testing results have shown that a DPF system can be designed to be inherently robust within prescribed operating boundaries, the use of embedded electronic sensors allows for the vigilant monitoring of both the operational input factors and performance output results.

Conventional sensors deployed to monitor DPF performance include temperature and pressure sensing. Pressure sensing is best done as a differential measurement across the DPF substrate rather than as an absolute reference. Temperature sensing includes both monitoring of gas conditions as well as external surfaces which must comply with stringent limits because of the ignitability of coal dust and ambient methane. The introduction of electronic sensing direct to high temperature exhaust is a new challenge area for DPF system integrators; one which has not been required previously as in the past it has been sufficient to sample at exhaust conditions post wet-scrubber where they are cooler.

The next level of system monitoring comes with the use of sensor technology which can measure exhaust emissions real-time, or near real-time. Technology to do this has reached maturity and has been progressively introduced into the automotive and heavy on-highway sectors over the last few years. For diesel engines, NO_x sensors are used to assess the variation in emissions both before and after the DPF aftertreatment system. These NO_x sensors utilise a development of the lambda sensor (relative air-fuel ratio sensor) technology which has been in service since the 1970's. This sensing technology has converged to virtually all manufacturers using a similar principle. NO_x sampling can be undertaken effectively in real-time, although specific algorithms and numerical processing is required to filter for stabilised readings and tuned to each application. FIG 8 shows representative data from the NOx sensor when all readings are recorded. With onboard processing of these readings, filtering allows only the stabilised results to be considered as part of any monitoring assessments. FIG 9 shows the FIG 8 data after filtering and also compares the results to those obtained at selected steady state operating conditions when the DPF system was tested fitted to an engine dynamometer facility. Good correlation is seen, providing confidence in the ability to realtime assess NOx compliance, but also predict DPF regeneration system performance as the NO2 component of NOx is integral to the regeneration performance of the CCRT.

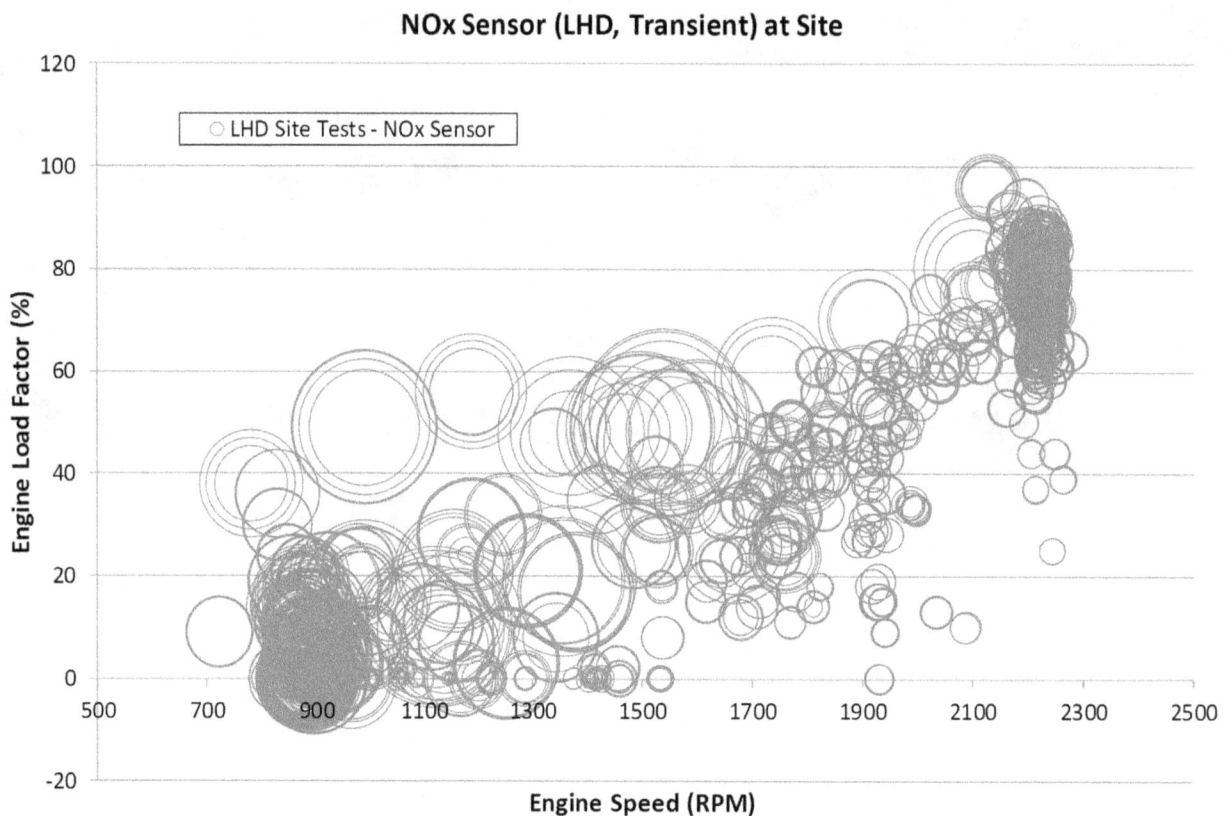

FIG 8 – All NOx sensor data reported by during operational duty of the LHD

PM sensing is comparatively new and a particular technology approach has yet to dominate. None of the currently available PM sensing technology is truly real time, with some requiring sampling periods of stabilised engine operation to work effectively. For the trials with this wall-flow DPF system, a resistive PM sensor technology was used adapted from a system implemented for on-road heavy duty engine compliance on-board diagnostics. Resistive PM sensing makes use of the conductivity of carbon as a means of quantifying PM concentration during deposition. This sensing technology can be described as only being near-realtime in that a discrete period of sampling is required to determine PM concentration and a separate discrete period is required to regenerate (or reset) the sensor for its next sample. This time base is of the order of fractions of a minute to a couple minutes, making it unsuitable for instantaneous readings. However, the use of the sensor is beneficial in determining the integrity of the DPF. If the DPF is undamaged, it will collect PM. If the DPF substrate has failed, either through cracking or thermal instability, PM will increase significantly. The sensor will detect this dramatic change in PM as a result of a significant change in deposition rate, or sensor cycle time.

Whether real-time, or near-real-time, the use of on-board embedded exhaust emissions sampling is set to provide a more rapid notification of changes to baseline performance than conventional periodic sampling and compliance checking which can be spaced days, weeks or months apart. The biggest challenge in implementing this real-time technology comes not from the sensing science; but the regulations around the use of non-conventional electronics in the highly regulated underground coal mining sector.

CONCLUSIONS

The emission performance results previously demonstrated by the proof-of-concept wall-flow DPF system developed under ACARP Project C25073 were repeated with the industrialised DPF system. These emission levels included DPM sufficiently low to pass the MDG43 year 2020 (0.025g/kWh) standards, including allowance for deterioration factors (DF); and Gaseous emission levels sufficiently low to pass the MDG43 year 2020 standards for Carbon Monoxide (CO), Oxide of Nitrogen (NOx), Nitrogen Oxide (NO) and Hydrocarbons (HC), including allowance for deterioration factors.

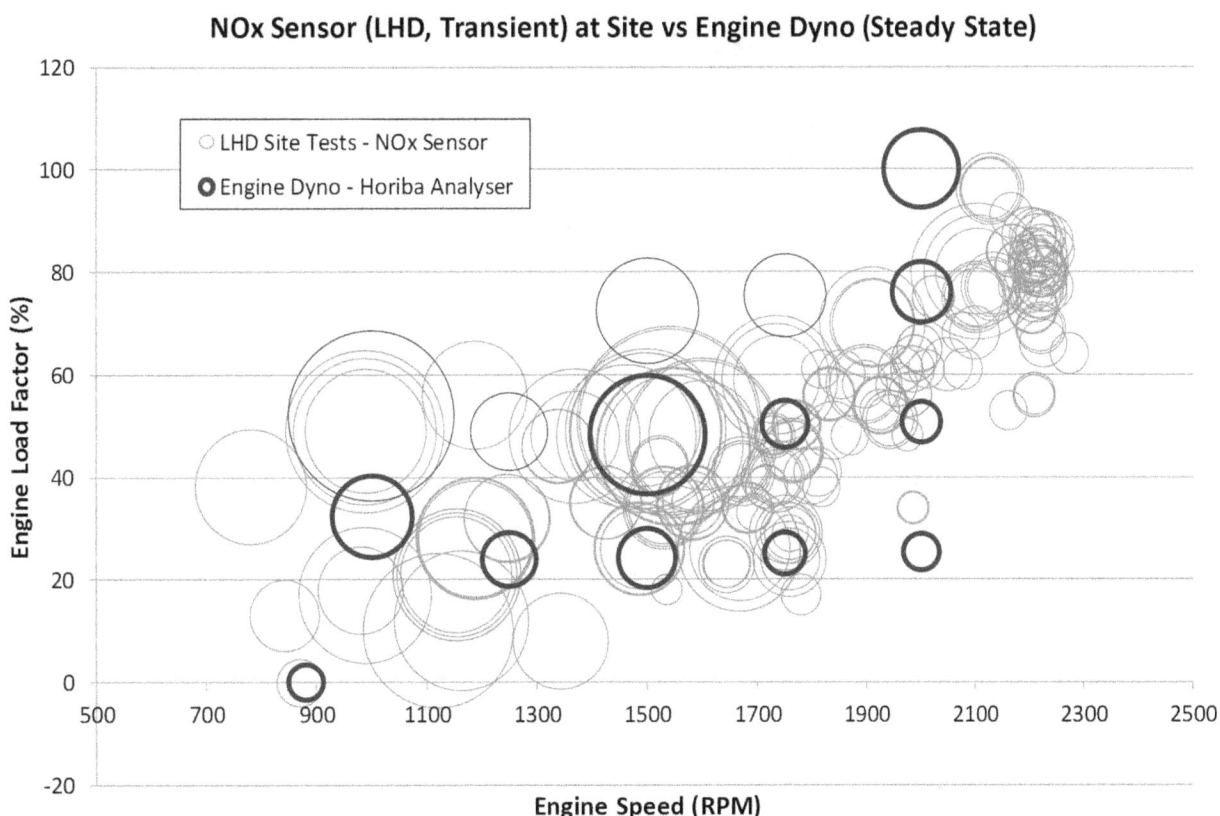

FIG 9 – Comparison of NOx to engine dyno data after filtering strategy
applied by the on-board monitor

The industrialisation of the system was focused primarily on improving and validating robustness of the selected wall-flow DPF technology:

- The larger of DPF substrate options from the proof-of-concept work was used for the industrialised design as it provided greater capacity for soot storage under adverse engine operating conditions, namely extended idle where exhaust gas temperatures were not sufficient to ensure continual catalytic regeneration was attainable;

- Robustness in the control of DPM with the continual catalytic regeneration as seen through sooting tests based on a developed light duty cycle which showed a stabilised exhaust back pressure for the system only a few kPa higher than the starting condition of an unloaded (soot free) DPF. Whilst stabilised back pressure results could be achieved with the use of Eromanga UMF fuel, this high sulfur fuel was seen to adversely affect emission results and would not be recommended for use with the selected CCRT technology;

- Thermal control of the system was re-validated using the industrialised design. The industrialised design packaged the DPF core inside a pressure rated metal assembly, with insulation methods proposed from the earlier proof-of-concept work. This not only controlled surface temperatures but allowed exhaust feed-gas temperatures to be held higher than they would be with a water-cooled approach to surface temperatures. Water-cooled flanges were still required in discrete locations to direct heat losses and control surface temperature;

- Electronic monitoring technology was developed and packaged into an Ex 'd' rated enclosure mounted integrally with the DPF enclosure. This close-coupling of the DPF with the monitoring system allowed for measurement of exhaust gas temperatures, pressures and selected regulated emission components (NOx and PM) to be accommodated without the additional risks associated with the remote mounting this sensing equipment. The use of electronic realtime, or near-realtime, monitoring of DPF and tailpipe emissions has the potential to substantially improve in-service compliance with occupational air quality standards applicable to enclosed environments found in underground mining operations.

ACKNOWLEDGEMENTS

Orbital would like to acknowledge ACARP (formerly the Australian Coal Association Research Program) for the funding provided to undertake this work program. ACARP is a unique and highly successful mining research program that has been running in Australia since it was established in 1992. It is 100% owned and funded by all Australian black coal producers through a levy of five cents per tonne paid on saleable coal. ACARP's research covers a wide range of important areas including all aspects of the production and utilisation of black coal including health, safety and the environment.

ACARP Contact:	Patrick Tyrrell
Industry Monitor/s:	Shayne Gillette (Centennial Coal)
	Greg Briggs (Centennial Coal)
	Andy Withers (Peabody)
	Steve Coffee (South32)
	Dr Bharath Belle (Anglo-American)

REFERENCES

Coplin, 2017, Australian Mine Ventilation Conference 2017, AusIMM, Paper 84

Coplin, Chicka, ACARP C25073 Final Report 2017

Johnson Matthey Catalysts Inc. 2017, http://www.jmdpf.com/diesel-particulate-filter-DPF-passive-systems-johnson-matthey (viewed 12 May 2017)

NSW Department of Resources and Energy, 2015, Mechanical Design Guideline, MDG43 – Technical standard for the design of diesel engine systems for use in underground coal mines

Standards Australia, 2008, AS/NZS 3584.2:2008 – Diesel engine systems for underground coal mines - Explosion protection

Standards Australia, 2013, AS/NZS 4871.6:2013 – Electrical equipment for mines and quarries – Diesel powered machinery and ancillary equipment

Psychrometric evaluation of workplace impacts upon change to battery mobile equipment from diesel

K Macdonald[1], C McGuire[2], W Harris[3] and D Witow[4]

1. Mine Ventilation Engineering Graduate, Hatch Ltd, Mississauga Ontario Canada.
   Email: katrina.macdonald@hatch.com
2. Mine Ventilation Engineer, Hatch Ltd, Mississauga Ontario Canada.
   Email: chris.mcguire@hatch.com
3. Gas Handling & Ventilation Specialist (MAusIMM), Hatch Ltd, Brisbane Queensland 4000.
   Email: wendy.harris@hatch.com
4. Mine Ventilation Systems Lead, Hatch Ltd, Sudbury Ontario Canada.
   Email: darryl.witow@hatch.com

ABSTRACT

Mine access construction cost is often the largest component or even majority of a mining project's total capital, and ventilation flow capacity typically defines the quantity and size of these developments. With a strong shift towards new technology in mobile equipment electrification, Owners are seeking savings in mine access brought on by reductions in total mine ventilation requirements. Control of heat underground is a key and often defining consideration for underground airflow demand. With new battery electric equipment, engineers are using energy balance and efficiency calculations to approximate heat generated in the workplace. The moisture produced by diesel engines contains latent heat, versus the solely sensible heat produced by battery machines; these impact physiological heat for workers differently. Psychometric analysis is used to improve the comparison of diesel to battery mobile equipment to capture the net effect on workers underground. The analysis and discussion presented in this paper aims to improve confidence for projects evaluating or adopting this new technology.

INTRODUCTION

In 2019, a joint effort by Caterpillar and Hatch published results from a test that compared heat from equivalent diesel and battery powered underground loaders during a simulated production cycle (Macdonald *et al*, 2019). Measurements in the local ventilation airstream were taken and calculations performed to quantify the heat and water vapour released from the machine in each of the battery-powered and diesel-powered scenarios. Given that wet bulb and dry bulb measurements were taken, the latent heat of vaporisation in the diesel exhaust was quantified in addition to the sensible heat.

From this test data, it was estimated that the diesel vehicle produced seven to nine times more heat than the equivalent battery vehicle (Macdonald *et al*, 2019). This ratio oversimplifies the impact felt by workers, since it does not distinguish the portion of heat released as latent heat by the diesel vehicles. The evaporated moisture results in increased humidity of the workplace, which directly impacts the body's ability to regulate internal temperature. Therefore, it is interesting to consider not just the total heat output, but the impact on both the temperature and humidity of the workplace.

This paper will refer repeatedly to the previous work undertaken by the authors in 2018 for publish in 2019. The analysis of the test data at the time was focused on the total heat output from the vehicles, not the impact on workplace conditions, to which the authors shift their focus in this current work. Note that the previous paper placed significant emphasis on the interaction between the ventilation air stream and the rock strata, while this interaction will be negated in the sensitivity analyses presented in this paper for simplicity.

This paper uses two metrics to provide an indication of workplace heat; the wet bulb temperature (WB) and the wet bulb globe temperature (WBGT). The WBGT, commonly used in North America for workplace temperature monitoring, combines the effects of the dry bulb temperature (DB) and humidity of the airstream on the worker's risk of heat stress.

Notably, for the same heat and moisture input to a workplace, the impact on the WB and WBGT metrics is not constant for different inlet air conditions. The nonlinearity of psychrometric functions results in a variable impact to workers depending on the conditions of the mine air entering the workplace. It is interesting to consider whether the benefits of a switch to battery-electric equipment are eroded in particular situations (e.g. mines with particularly dry or humid intake air). These effects are explored in a sensitivity analysis.

COMPARISON OF DIESEL AND BATTERY-ELECTRIC VEHICLES

Electric motors are significantly more efficient than an equivalent diesel engine. While diesel engine efficiency is typically around 33%, electric motors are typically around 95% efficient. As such, for the same task an electric motor will require less input energy and have lower energy losses to inefficiencies. This waste energy reports to the surroundings as heat, increasing the temperature of the surrounding rock, the mobile equipment itself, and the ventilation air stream. A theoretical comparison of diesel engine and electrical motor efficiencies predicts a three-fold higher heat generation by a diesel unit compared to a battery-electric unit. Thus, for the same activities a battery-electric fleet will produce significantly less heat than a diesel fleet.

Another important distinction between diesel engines and electric motors is the production of water vapour. The combustion process in diesel engines produces water. Due to the high gas temperature the water in the engine exhaust stream is in vapour form and thus contains significant energy in the form of latent heat of vaporisation. As such, the heat from a diesel engine reports as a combination of sensible and latent heat. In comparison, electric motors produce no water vapour and therefore report all heat as sensible heat.

Diesel combustion produces water at a rate of 1.1 to 1.5 litres of water per litre of diesel combusted on a liquid-to-liquid basis (Kibble, 1978; McPherson, 1993). However, in-field studies have observed higher influx of water vapour in exhaust air downstream of diesel vehicles. This is believed to be the result of water released by engine cooling systems and exhaust scrubbers along with enhanced evaporation in local hotspots created on the rock and vehicles surfaces. Ratios as high as 10 litres of water per litre of diesel consumed have been reported (McPherson, 1986; Mousset-Jones, 1987). A ratio of 5 litres of water vapour emitted to the exhaust stream per litre of diesel fuel combusted is commonly used in mining heat models.

MEASURED HEAT AND MOISTURE OUTPUT

Previous work by Macdonald *et al* (2019) conducted in-field testing of a diesel underground loader and an equivalent battery-electric underground loader. Both units were subjected to identical simulated mucking cycles. Vehicle parameters and entering/exiting airstream parameters were monitored over the course of the test. The results of this study indicated the heat released from the diesel machine was on the order of seven to nine times that of the battery powered unit with heat emitted by the diesel unit estimated as 320 to 350 kW and heat emitted by the battery estimated as 36 to 50 kW. This is significantly higher than the three-fold increase predicted by theoretical comparison of engine/motor efficiencies. The additional reduction in heat generated by battery-electric units may be the results of differences in the units' torque converters/power control, ancillary equipment coupling/control, and regenerative braking capabilities. Details of the test work and analysis are published in the North American Mine Ventilation Symposium proceedings (Macdonald, McGuire, Armburger, and Baumann, 2019).

Comparison of inlet and outlet air stream psychrometrics also enabled the estimation of water vapour produced over each test. During the diesel test, water vapour production was observed at approximately 25 mL/s, or a ratio of about 4.5 litres of water per litre of diesel consumed. This is comparable with typical measured values as discussed above. Water vapour emissions into the exhaust air over the battery-electric test was observed to be less than 0.6 mL/s, which can likely be attributed to regular evaporation from the rock surfaces. As such, the diesel unit heat load was emitted as approximately 15-20% latent heat while the battery-electric unit heat load was emitted essentially as pure sensible heat.

WORKPLACE CONDITIONS

A worker's ability to regulate their body temperature is dependent on the surrounding air DB temperature, humidity, air velocity and other factors. Thermoregulation of the body relies on the release of body heat to the surrounding environment via radiation and evaporation of perspiration from the skin. In humid conditions evaporation is suppressed and control of body temperature is more difficult. Similarly, increased airflow improves evaporation and improves a worker's ability to regulate their body temperature. Thus, several indices have been developed to indicate the perceived temperature of the workplace including WB temperature and WBGT.

The WB temperature is that which is measured by a thermometer covered in a wet cloth. The WB temperature describes the temperature the air would reach if cooled to a relative humidity of 100% by evaporation. So, at high relative humidities the WB approaches the DB temperature and at low relative humidities the WB is below the DB temperature. The WBGT temperature indicates the combined impact of the DB temperature, WB temperature, and infrared radiation measured by the globe thermometer temperature. Indoors or underground where radiation can usually be regarded as negligible, the WBGT can be approximated as the sum of 70% of the WB and 30% of the DB (i.e., the WBGT will lie between the WB and DB temperatures, closer to the WB). Both WB and WBGT will give an indication of the felt impact on workers.

To understand the felt impact of diesel and battery-electric units, their impact on DB, WB and WBGT is considered. Energy released as sensible heat will impact the air DB temperature, while the total enthalpy increase (as both sensible and latent heat) will impact the air WB temperature. For two equal heat sources, one entirely sensible heat and one a mixture of sensible and latent heat, the purely sensible heat source will result in a greater increase in DB temperature while both will produce roughly the same change in WB temperature. Therefore, although a diesel engine will produce more total heat, the release of some of that heat as latent heat will help depress the DB temperature relative to what would be seen if the heat were released purely as sensible heat. Furthermore, the change in WB or WBGT will depend on the inlet airstream conditions. The enthalpy of the inlet air stream will impact the temperature rise associated with a certain input of sensible/latent heat.

During the diesel/battery LHD study, psychrometric conditions were monitored at the inlet and outlet air streams. These temperatures have been corrected to reflect the temperature local to the vehicle (i.e., prior to interaction with the rock strata). The outlet temperature was observed to increase by 8.1 °C WBGT over the diesel test and 1.2 °C WBGT over the battery test, or 5.4 °C WB and 0.8 °C WB, respectively. Thus, the relative impact felt in the workplace was six to seven times larger for the diesel unit compared to the battery unit. This can be compared to the seven to nine-fold difference observed in heat load of the diesel and battery units. Note that these temperature values are specific to the intake air conditions during the field test and cannot be said to reflect a "general" or "typical" impact for the range of inlet air temperature and humidity conditions experienced in various underground mines.

SENSITIVITY ANALYSIS

In order to provide more general discussion from the previous test work, the measured heat (enthalpy) and water emissions were applied to a variety of inlet airstream conditions. This effectively simulates the operation of these two diesel and battery-electric loaders in mines with differing climates than that of the test location. Figure 1 shows the impact of varying the relative humidity of the inlet air stream while maintaining a constant inlet air DB temperature and pressure. Similar results can be seen by varying any of the psychrometric properties (any parameter that affects the enthalpy of the inlet air). For Figure 1, the temperature and barometric pressure are held at a constant 23.6 °C DB and 116.0 kPa respectively, to match the conditions measured during the actual test period.

Diesel equipment was predicted to result in a 5 to 8 °C WB or 7 to 10 °C WBGT temperature rise, while battery equipment was predicted to only result in a 0.5 to 1.5 °C WB or 1 to 2 °C WBGT temperature rise. A larger temperature rise was observed for dry, low-enthalpy inlet air conditions as compared to wet, high-enthalpy inlet air conditions for both equipment types and both temperature metrics. This reflects the fact that the constant heat released by the equipment is a smaller proportion of the enthalpy of the entering air, and therefore has a smaller impact on the resultant exhaust temperature. This exemplifies the impact of mine-specific conditions showing that hot/humid mines may see less temperature rise associated with underground equipment than cool/dry mines.

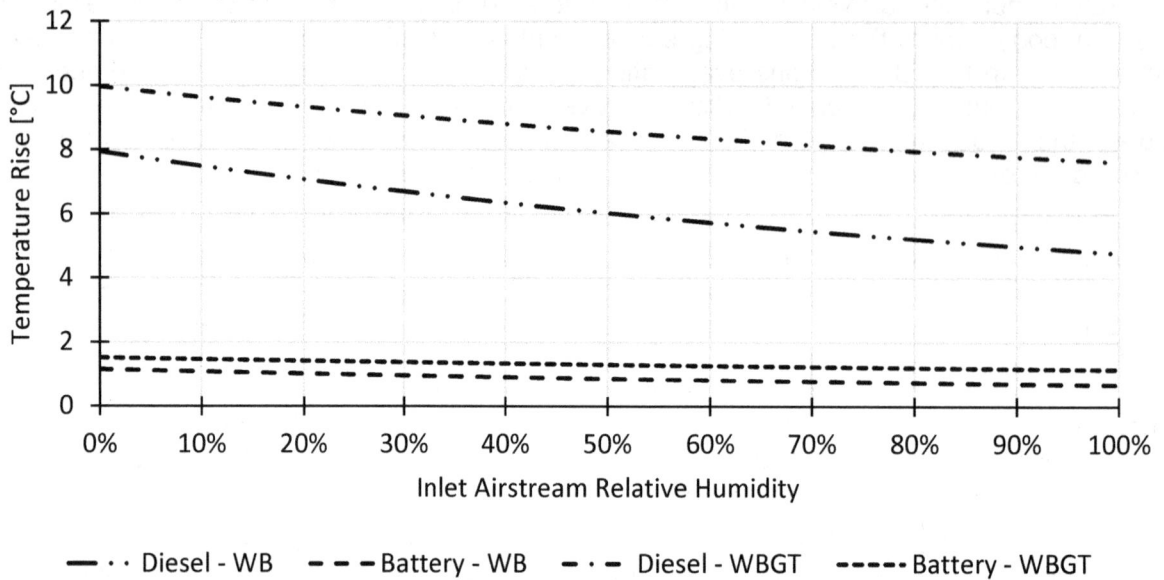

Figure 1: Exhaust Temperature Rise for Varied Inlet Air Stream Relative Humidity

To explore how these inlet conditions impact the relative performance of diesel and battery equipment, the ratio of the diesel temperature rise divided by the battery temperature rise was calculated for all of the sensitivity cases. These calculated ratios are compared to the equipment heat output ratio in Figure 2.

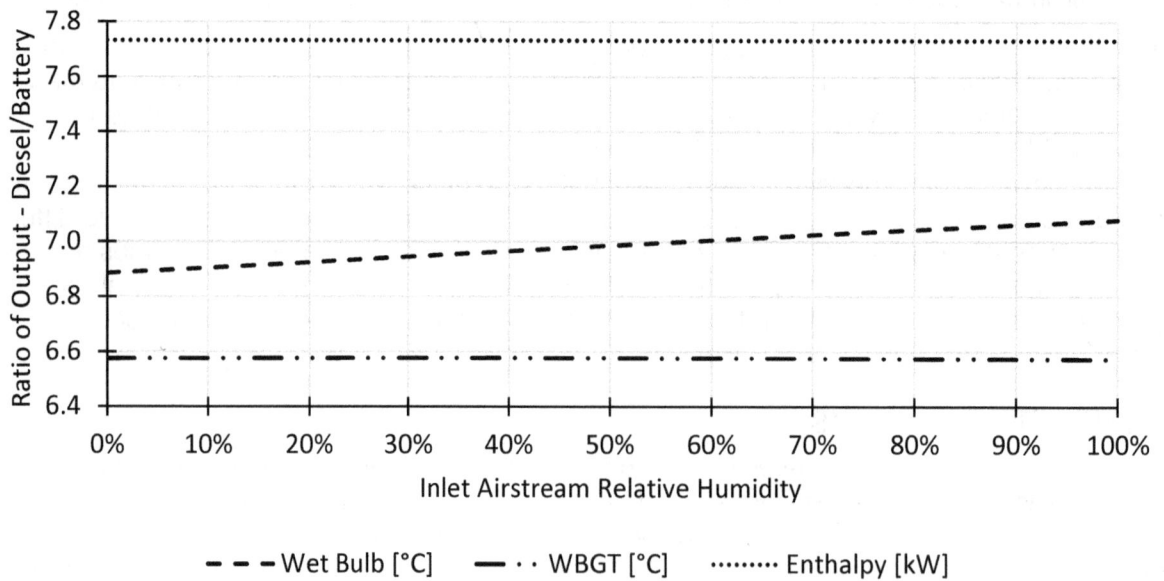

Figure 2: Ratio of Diesel and Battery Equipment Temperature Rise in the Workplace

For all inlet conditions it can be seen that the relative impact of diesel vs. battery equipment felt by workers, as indicated by the relative temperature increase, is much less than the relative heat loads of the two units. Comparison of the equipment heat loads gives a ratio of approximately 7.7 times higher heat for diesel equipment as compared to battery equipment; yet, comparison of the workplace temperature impact gives a ratio of 6.5 to 7.1 times higher diesel workplace impact as compared to that of battery equipment. This means that the true benefit of conversion to battery-electric equipment could be 8-11% lower on a WB basis and about 15% lower on a WBGT basis than the benefit predicted by gross heat loads alone. As such, the ventilation savings of a battery-electric fleet compared to a diesel fleet may be over-stated if relative fleet heat loads are considered alone. The ventilation required to maintain a safe workplace underground should consider the impact felt by workers of equipment type.

This shows that it is crucial to have a thorough understanding of the ventilation requirements during early mine design phase, instead of applying blanket conversion factors on some assumptions relative to a diesel fleet. Take for example the scenario above, where a mine is developed and later finds that the benefit for a battery conversion was overestimated by 15%. For this very crude example, suppose that the outcome was that the ventilation system needed to be upgraded to provide 15% more air flow to counter the increased heat load and maintain the mine throughput with safe workplace temperatures. From the fan affinity laws, we understand that a 15% increase in airflow requires a 52% increase in ventilation system power, which could become cost prohibitive. Similarly, dropping cooler outlet temperatures may also be difficult due to limitations with coolant fluid or air condensate freezing. Early understanding of the expected heat loads may have led to different decisions in the mine design which improve ventilation efficiency, optimisation of cooling system design within the ventilation network, etc.

Another point of interest exemplified by Figure 1 and Figure 2 is the variation of relative WB and WBGT temperature rise with inlet conditions. Figure 1 shows that both WB and WBGT metrics are impacted by inlet conditions for diesel and battery equipment; however, it can be observed that the WB temperature rise in the diesel case falls more rapidly with increased inlet relative humidity compared to that of the battery unit. This shown more clearly in Figure 2 by the variation of the diesel/battery ratio on a WB basis with inlet conditions. The WBGT temperature rise was found to be proportionally consistent between diesel and battery equipment, resulting a roughly constant ratio as shown in Figure 2. This highlights the importance of considering the workplace temperature metric in use when considering the ventilation savings offered by battery-electric equipment. Specifically, a hot/humid mine using a WB basis will observe a relatively larger benefit if switching from diesel to battery equipment.

CONCLUSIONS

When considering a conversion to a battery-electric equipment fleet, it is important to consider the impact felt by workers instead of just the raw heat output from the equipment. The felt impact (on a WB basis) in the workplace for either battery or diesel is shown to be variable by up to 30% depending on the mine intake air conditions. Additionally, it is shown that the benefit to the workplace from switching to a battery fleet is as much as 15% less in terms of impact felt by workers than by comparison of the equipment heat generation alone. This can translate into errors in the mine ventilation and cooling selection if a simple heat ratio is applied to mine design and workplace conditions are not properly considered. A thorough understanding of the ambient conditions and underground heat loads is required to quantify this benefit.

This work emphasises the fact that there is no suitable "one size fits all" approach to defining ventilation requirements in battery-electric mines. There isn't a single "diesel to battery conversion factor" that can be generally applied, and even the selection of the defining thermal metric for the workplace will impact the potential savings. When regulatory air volume requirements associated with diesel equipment fleets are removed, the ventilation engineer's task gets significantly more challenging. Each mine must be considered on a case by case basis, with unique constraints and challenges defining key decisions on ventilation and cooling system design.

REFERENCES

Kibble, J D, 1978. Some notes on mining diesels, *Mining Technology*, October:393-400.

Macdonald, K, McGuire, C, Armburger, J and Baumann, J, 2019. Direct comparison of heat produced by battery vs. diesel LHD, in *Proceedings of the 17th North American Mining Ventilation Symposium* (ed: A Madiseh, A Sasmito, F Hassani and J Stachulak), pp 404-413 (Canadian Institute of Mining, Metallurgy and Petroleum).

McPherson, M J, 1986. The analysis and simulation of heat flow into underground airways, *International Journal of Mining and Geological Engineering*, 4:165-196.

McPherson, M J, 1993. *Subsurface Ventilation and Environmental Engineering* (Chapman & Hall).

Mousset-Jones, P et al, 1987. Heat transfer in mine airways with natural roughness, in *Proceedings of 3rd US Mine Ventilation Symposium*, pp 42-52 (Penn State).

Working toward a better understanding of diesel engine exhaust emissions (DEEE) for underground mines in Western Australia

J Oding[1]

1. District Inspector of Mines, Department of Mines, Industry Regulation and Safety (the Department), East Perth 6004. Email:junior.oding@dmirs.wa.gov.au

ABSTRACT

Western Australian underground mines are dependent on diesel equipment for drilling, loading and hauling ore and waste rock and transporting the workforce and supplies. Underground mine atmospheres are controlled by a combination of mechanical ventilation, dust suppression and vehicle exhaust controls to maintain contaminant levels below health-based exposure standards and as low as reasonably practicable (ALARP). Ongoing reliance on diesel powered equipment in Australia presents the potential for increased concentrations of contaminants in the working atmosphere resulting from vehicle emissions.

This paper describes the changing technologies used to control underground atmospheres, evolving exposure standards and sampling method developments with particular emphasis on those relating to diesel exhaust emissions and the impact on workforce exposures.

Historical and current workforce exposures to diesel engine exhaust emissions (DEEE) focus on personal exposure to respirable, ultrafine diesel particulate matter (DPM). As the majority of these particles have an aerodynamic diameter less than 100 nanometres (nm) they are referred to as nano-diesel particulate matter (nDPM).

The paper highlights a number of important issues that need to be considered and understood when assessing worker exposure to DEEE and how to ensure that existing controls are still effective. Briefly, the issues include:

1.   the components of the mixture of DEEE

2.   guidelines for atmospheric contaminants to protect the general public compared with occupational exposure standards to protect workers

3.   existing emission standards in Australia compared with international standards

4.   use of real-time monitoring equipment to assist with determining control effectiveness

5.   use of tailpipe measurements to monitor reduced function of in situ controls

6.   comparison of background levels of DEEE in the central business district (CBD) with exposure trends for mine workers since 2002.

INTRODUCTION

In 2012, the International Agency for Research on Cancer reclassified diesel engine exhaust as a category 1 carcinogen to humans based on evidence that exposure is associated with an increased risk for lung cancer (IARC, 2012). According to Cancer Council Australia (2019), DEEE is the second most common cancer causing agent to which workers are exposed in Australia.

Ideally, the DEEE from all diesel engines should only consist of water, carbon dioxide and nitrogen. However, many hazardous and carcinogenic components are present in DEEE. It is a mixture of particulate matter, aerosols and gaseous components. Table 1 shows the typical diesel exhaust gas composition produced by diesel engines.

TABLE 1 – Typical diesel exhaust gas composition

Components		Typical component concentration range in diesel exhaust gas	Component concentration in natural dry ambient air
Nitrogen	N_2	75 - 77 %-vol	78.08 %-vol
Oxygen	O_2	11.5 - 15.5 %-vol	20.95 %-vol
Carbon dioxide	CO_2	4 - 6.5 %-vol	0.038 %-vol
Water	H_2O	4 - 6 %-vol	-
Argon	Ar	0.8 %-vol	0.934 %-vol
	Sub-total	> 99.7 %-vol	
Nitrogen oxides	NOx	1000 - 1500 ppm-vol	0 %
Sulphur oxides*	SOx	30 - 900 ppm-vol	0 %
Carbon monoxide	CO	20 - 150 ppm-vol	0 %
Total hydrocarbons	THC (as CH_4)	20 - 100 ppm-vol	0 %
Volatile organic compounds	VOC (e.g. aldehydes)	20 - 100 ppm-vol	0 %
Particulates*	DPM	20 - 100 mg/m³, dry, 15% O_2	
%-vol: concentration, percentage, volume basis ppm-vol: concentration, parts per million, volume basis * varies depending on composition of fuel and age and condition of engine			

(Source: CIMAC, 2008)

No exposure standard has been set for the DEEE mixture as there are many components that vary considerably from engine to engine, and between physical environments. The adopted time-weighted average (TWA) guideline for mine workers in Western Australia (WA), and many other jurisdictions around the world, is 0.1 mg/m³ of DPM, analysed as elemental carbon (EC). This is measured by the National Institute for Occupational Safety and Health (NIOSH) *5040 Analytical Method* (Bugarski, 2017).

A number of DEEE control systems are employed by the WA mining industry, such as electronic engine management systems, exhaust gas recirculation systems (EGR) and fuel additives. This is in addition to installation of exhaust catalysts and/or filters. In some circumstances, controls are combined to minimise exhaust emissions or to optimise the fuel and air ratio mixture for power efficiency (Bugarski, 2011). Several published studies have shown additives may resolve one problem, but can create other problems. For example, emission of heavy metals or other contaminants have been reported from use of some fuel additives (Zeller, 1992; Miller 2007; Bünger, 2012; and MECA, 2007). However, optimally the focus is for engine manufacturers to implement new technology diesel engines (NTDE) with reduced exhaust emissions (IRSG, 2012).

The Department requires all underground mine operators to have a site diesel emission management plan (DPM Plan) that works in conjunction with the Ventilation management and Health and hygiene management plans. The challenge for all mine operators is to determine how well the controls described in the DPM Plan are working, and whether they are effective in controlling DEEE to maintain the exposure risk to ALARP.

BACKGROUND

In April 2013, the Mining Industry Advisory Committee (MIAC) endorsed the Department guideline, *Management of diesel emissions in Western Australian mining operations*, for anyone planning or conducting underground mining where diesel engines are being used, or are likely to be used. Following this, the Department implemented a DPM monitoring program at a number of mine sites. Measurements of DPM, using mass-based, real-time monitors that were compliant with the NIOSH 5040 analytical method, were taken by inspectors during regulatory site inspections of underground mines. Concurrently, many underground mining operations proactively commenced developing their site-based DPM Plan and the Department reported significant reductions in average DPM measurements.

Exposure standards and targets

General population target

The National Environment Protection Council (NEPC) was established under the *National Environment Protection Council Act 1994* and has two primary functions:

1. to make National Environment Protection Measures (NEPMs)
2. to assess and report on the implementation and effectiveness of NEPMs in participating jurisdictions. This includes Western Australia.

The NEPMs are designed to assist in protecting or managing particular aspects of the environment. This includes the development of standards on ambient air toxins and air quality goals (under review in 2019). The NEPC (2016) goals are shown in Tables 2 and 3.

TABLE 2 – National Environmental Protection (Ambient Air) Measure

Pollutant	Averaging period	Maximum concentration standard	Maximum allowable exceedances
Carbon monoxide	8 hours	9.0 ppm	1 day a year
Nitrogen dioxide	1 hour	0.12 ppm	1 day a year
	1 year	0.03 ppm	None
Photochemical oxidants (as ozone)	1 hour	0.10 ppm	1 day a year
	4 hours	0.08 ppm	1 day a year
Sulphur dioxide	1 hour	0.20 ppm	1 day a year
	1 day	0.08 ppm	1 day a year
	1 year	0.02 ppm	None
Lead	1 year	0.50 $\mu g/m^3$	None
Particles as PM_{10}	1 day	50 $\mu g/m^3$	None
	1 year	25 $\mu g/m^3$	None
Particles as $PM_{2.5}$	1 day	25 $\mu g/m^3$	None
	1 year	8 $\mu g/m^3$	None

(Source: NEPC, 2016)

TABLE 3 – NEPM (Ambient Air Quality)

Pollutant Goal	Averaging period	Maximum concentration
Particles as $PM_{2.5}$	1 day	20 $\mu g/m^3$ by 2025
	1 year	7 $\mu g/m^3$ by 2025

(Source: NEPC, 2016)

Occupational exposure standards

Occupational exposure standards for airborne contaminants are specified by the Mines Safety and Inspection Regulations 1995 (MSI Regulations), which directly reference the *Workplace exposure standards for airborne contaminants* (Safe Work Australia, 2013) for most identified hazardous substances. This includes guidance on interpretation and application of the standards. Occupational exposure standards are distinctly regulated in the MSI Regulations for respirable and inhalable particulates that are not otherwise classified by Safe Work Australia.

The exposure standards are based on an eight hour day and five day working week (Safe Work Australia, 2018), which is not representative of extended mining operation rosters. The Quebec Model is one of several methods for adjusting exposure standards for extended shifts, and reduces the exposure standard based on a reduced recovery time and extended exposure for agents that produced chronic or latent health effects (Drolet, 2015).

Typical gaseous airborne contaminants from DEEE are shown in Table 4, along with the current occupational exposure standards. These standards are enforced by the Department under the *Mines Safety and Inspection Act 1994*.

TABLE 4 – Occupational exposure standards for gaseous components of DEEE

Pollutant	Time weighted average (8hrs)	Short term exposure limit (STEL)
Carbon monoxide	30 ppm	NA
Carbon dioxide	5000 ppm	30,000 ppm
Nitrogen dioxide	3 ppm	5 ppm
Nitric oxide	25 ppm	NA
Sulphur dioxide	2 ppm	5 ppm
Ammonia	25 ppm	35 ppm

(Source: Hazardous Chemical Information System, Safe Work Australia, 2018)

The Australian Institute of Occupational Hygienists (AIOH) DPM Position Paper (AIOH, 2007; 2013) proposed the use of a Time Weighted Average (8 hours) exposure standard of 0.1 mg/m^3 measured as submicron elemental carbon (EC) for diesel particulate (DP). The Department adopted this as an interim standard and Safe Work Australia references the interim DPM exposure values in their 2018 standard.

Engine emission standards

Currently in Australia, emission standards are only applicable to new road vehicles. These standards have been in place since 1970. However, the United States of America and European Union have established emission standards to reduce emission rates of contaminants and improve engine efficiencies over time by setting and iteratively reducing these emission standards. The sequence of reduction of emission standards has been demonstrated previously by Dallman (2016).

Predicted concentrations using calculations based on the *Donaldson engine horsepower and exhaust flow guide* (Donaldson, 2016) are presented in Table 5. Using manufacturer specifications and assuming the selected mining diesel engines conform to the Euro 5 emission standards, the predicted contaminant concentrations should be maintained at levels compliant with occupational exposure standards. A further assumption in these calculations is that ventilation dilution rates of 0.05 m^3/s/kW are maintained.

TABLE 5 – Predicted contaminant concentrations (based on Donaldson, 2016)

Exhaust Temp	C'	100	(In the work place)		Tailpipe @ US/EU Standards			Diluted with Vent (In the work place)		
		To meet std DP of 12hrs – 0.07mg/m³	Diluted Cooled exhaust gasses		Exhaust Gas concentration (g/kWh)			MSIR [r.10.52] 0.05 m³/s/kW		
		Tailpipe DP Conc	MSIR [r.10.52]	MSIR [r.10.52]	3.5	0.4	0.02			
			CO (1500)	NOx(1000)	CO	NO	PM	CO	NO	PM
Type	kW	mg/m³	ppm	ppm	ppm	ppm	mg/m³	ppm	ppm	mg/m³
Truck	565	2.08	49.99	33.32	514.84	53.78	3.33	13.94	1.46	0.11
LHD	333	1.37	75.86	50.57	339.26	35.44	2.20	13.94	1.46	0.11
Truck	515	1.89	54.84	36.56	469.28	49.02	3.04	13.94	1.46	0.11
Truck	450	1.75	59.46	39.64	432.83	45.21	2.80	13.94	1.46	0.11
LHD	352	1.48	70.39	46.93	365.59	38.19	2.37	13.94	1.46	0.11
LHD	310	1.30	79.93	53.29	321.97	33.63	2.09	13.94	1.46	0.11
LHD	256	1.07	96.79	64.53	265.88	27.77	1.72	13.94	1.46	0.11
LHD	235	1.16	89.44	59.62	287.75	30.06	1.86	13.94	1.46	0.11
Truck	600	1.25	83.09	55.40	309.71	32.35	2.01	13.94	1.46	0.11
Truck	439	1.36	76.13	50.76	338.03	35.31	2.19	13.94	1.46	0.11
LHD	305	1.25	82.82	55.21	310.73	32.46	2.01	13.94	1.46	0.11
LHD	263	1.48	70.14	46.76	366.91	38.33	2.38	13.94	1.46	0.11
LHD	257	1.10	94.30	62.87	272.90	28.51	1.77	13.94	1.46	0.11
LHD	123	0.60	173.39	115.60	148.42	15.50	0.96	13.94	1.46	0.11

IMPORTANCE OF MONITORING DIESEL EXHAUST

There are a number of methods of monitoring diesel exhaust. Traditionally mass-based DPM measurements are used for both tailpipe and occupational exposures.

Few mining operations undertake tailpipe DPM monitoring. Table 6 shows a typical mass-based tailpipe DPM monitoring result. The results demonstrate variability even between identical models of machinery.

TABLE 6 – Tailpipe DPM monitoring – mass concentration based measurements

Underground vehicles	Engine hours	Minimum (mg/m³)	Maximum (mg/m³)	Average (mg/m³)
A1	23,403	0.4	17.7	4.6
A2	8,157	-	0.0	0.0
B1	9,914	3.4	18.9	9.4
B2	22,401	10.4	34.5	18.5
B3	1,810	0.2	3.4	1.2
B4	1,893	13.5	54.6	28.2
C1	3,873	1.5	34.6	10.7
C2	416	1.6	50.0	9.5
C3	3,760	3.9	33.1	9.4
D1	312	2.0	14.5	5.6
D2	64	2.4	18.0	5.7
D3	2,820	0.8	47.9	13.4
D4	40	0.6	51.3	12.9

Note: Where letters are the same, this identifies the same asset make and model of vehicle.

New technology diesel engines have reduced the size and mass of DPM, but an increase in particle numbers due to optimised combustion conditions in low emission engines, has been reported (Kittelson, 1997; Ghose, 2015; Stafford, 2008). It is suggested that monitoring based on number concentration may allow for better understanding of the health impacts of diesel particulates.

Department inspectors found no correlation could be established between mass concentration measurements using the monitoring equipment and NPET number concentrations when measurements are made at the engine exhaust tailpipes.

Department inspectors observed that workplace environmental conditions such as temperature and humidity affect the number and mass concentration of airborne particulates. This should be factored when converting mass concentration to an equivalent number concentration when measuring workplace DPM concentrations.

Portable and in-situ real-time monitoring instruments to measure mass concentration and number concentration (different instruments) are available. Operators are advised to adopt a particular type of instrument and continue to monitor using the same instrument to establish a baseline emission profile for that particular diesel equipment.

Real-time monitors identify where high levels of contaminants occur and assist in identifying defective equipment. This can facilitate earlier identification for rectification or replacement of defective controls. Efficiency of the DPF can also be compared with baseline values. Monitoring has been undertaken at some mining operations showing that the filtration efficiencies reduce to below 50%, with regular maintenance. Inspectors have observed that efficiencies approach 0% in cases where the DPF has not been maintained or monitored.

DATA COLLECTED

Department inspectors took a series of spot measurements in several underground mines, at various locations and activity levels using real-time monitors that report DPM particle counts. Background levels in Perth during off-peak traffic periods are compared with levels recorded in underground mines.

Portable real-time mine site DPM measurements

Many underground operations have implemented measures to limit exposure to DPM through the development of a DPM Plan. The Department provides guidance for mining operations during their inspections on measures to manage the hazard by encouraging the monitoring of the performance of the DPF. This includes using real-time monitoring devices that measure in both mass and number concentrations, at the exhaust tailpipe and the workplace.

From 2017, inspectors conducted site monitoring using equipment based on DP number concentrations. The main findings were:

- average number concentrations were observed with a range of 200% to 1200% of the time-weighted average (TWA) of 0.1 mg/m³ (EC), even when diesel particulate filters (DPF) have been installed in fleet equipment at inspected minesite workplaces.

- on levels with no equipment working, elevated particulate number concentrations were observed due to accumulation of particulates in areas with series ventilation circuits.

The results of the monitoring at various mining operations, locations, ventilation rates, activities and types of diesel equipment is set out in Table 7.

TABLE 7 – Monitoring results using a real-time particle counter

Activity	Monitor location	Air flow (m³/s)	Engine kW	Average number /cubic cm	Average diameter (nm)	Average mass (µg/m³)
Load-haul-dump truck (LHD)	In drive	14	242kW	109,000	74	342
LHD	In cabin	14	242kW	67,000	71	63
LHD	In drive	20.5	220kW	885,000	59	589
Shotcrete machine	In drive	28	74.9kW	149,000	60	119
LHD	In cabin	13	345kW	50,000	50	30
LHD	In drive	13	345kW	170,000	61	172
LHD	In cabin	36	333kW	79,000	73	152
Shotcrete machine	In drive	13	Combined 412kW	1,191,000	66	1,271
LHD	In drive	13	256kW	134,000	56	96
LHD	In drive	19	263kW	158,000	53	89
LHD	In drive	20	333kW	1,133,000	30	69
LHD	In drive	18	263kW	110,000	74	198
LHD	In drive	38	263kW	477,000	67	538
Charging machine	In drive	16.2	74.9kW	90,000	74	144
LHD	In drive	16.8	123kW	401,000	67	455
Charging machine	In drive	13.8	74.9kW	103,000	61	97
Trucks	Decline	87	- 567kW	248,000	39	55
Trucks	Decline	87	567kW	274,000	39	56
LHD/truck	Access	47	275kW, 567kW	70,000	42	23

Activity	Monitor location	Air flow (m³/s)	Engine kW	Average number /cubic cm	Average diameter (nm)	Average mass (μg/m³)
LHD/truck	Access	47	275kW, 567kW	71,000	43	24
LHD	Decline	26	290kW	517,000	45	95
LHD	Decline	26	290kW	613,000	44	94
LHD	In Drive	15	(310kW)	90,000	51	39
General run	All areas inspected	NA	NA	85,000	55	73
General run	All areas inspected	NA	NA	270,000	84	725
General run	Level s	NA	NA	152,000	59	192
General run	All areas inspected	NA	NA	90,000	51	59
General run	All areas inspected	NA	NA	143,000	51	60

New technology diesel engines produce up to 10^9 particles/cubic centimetres concentration. Of the DPF tested and submitted to the Department, results have shown that they are capable of reducing the concentration down to 10^4 particle/cubic centimetres concentration at the exhaust tailpipe (DieselNet 2002; Arndt, 2013; Mammoth, 2013). Arndt (2013) found a correlation between particle mass and particle number, based on engine power and kilometres travelled when monitoring exhaust from tailpipes.

Background DPM measurements in Perth CBD (off-peak)

To compare measurements in an urban environment with those observed underground, background DPM particle number readings were taken in Perth during off peak traffic conditions. A portable real-time particle monitor was carried along a random route in the Perth CBD. Measurements ranged between 1,000 and 10,000 with multiple short-term peaks (Figure 1) corresponding to vehicles passing near the equipment. The highest peaks occurred at a bus stop at point G when a diesel powered bus stopped and idled while passengers entered and exited the bus.

These peak values at point G exceed 220,000 particles per cubic centimetre and are comparable to measurements reported in the underground mine in the vicinity of active load haul dump trucks. It appears that prevailing winds effectively reduce the background air concentrations to around 1,000 particles per cubic centimetre.

Further work is required to evaluate any health consequences from these short-term peaks.

HISTORICAL DPM EXPOSURES (PERSONAL)

The Department has been recording occupational exposure levels in the workplace for all mines in WA since 1977. The CONTAM database system was established so sites would conduct personal exposure monitoring for dust and respirable silica and then focus on measures to reduce unacceptable exposures. Over the years, more contaminants were added to the system, and the requirement to identify and investigate exceedances, has been further emphasised. More recently, the methodology evolved towards a risk based approach and migration of the CONTAM database into the Safety Regulation System (SRS) Health and Hygiene database.

The occupational exposure monitoring of DPM for underground mining operations in WA submitted to the Department is shown in Figure 2. This shows the number of samples taken per year against the number of exceedances per year. It also indicates that with the considerable increase in sampling frequency and focus on improving control measures, the rate of exceedances of the adopted exposure standard has significantly decreased from around 65% of all samples in 2003 to only 4.5% in 2018.

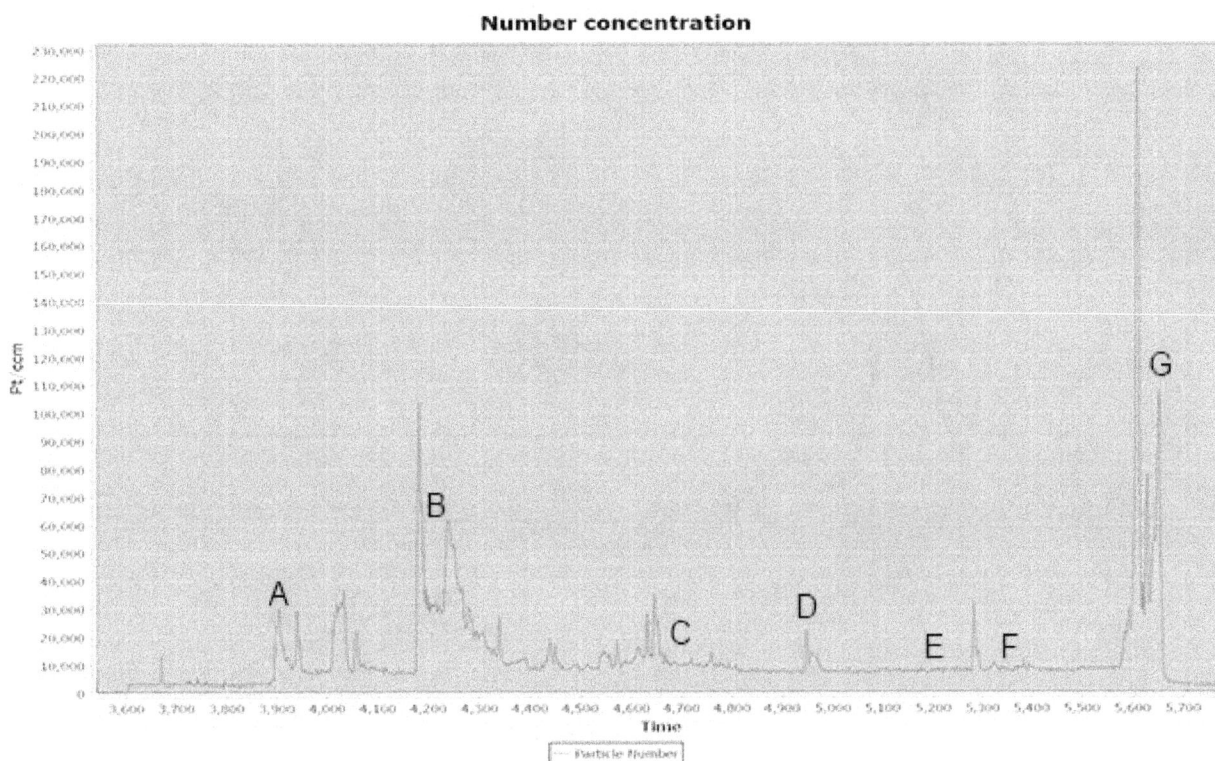

FIGURE 1 – Real-time monitoring data in Perth streets

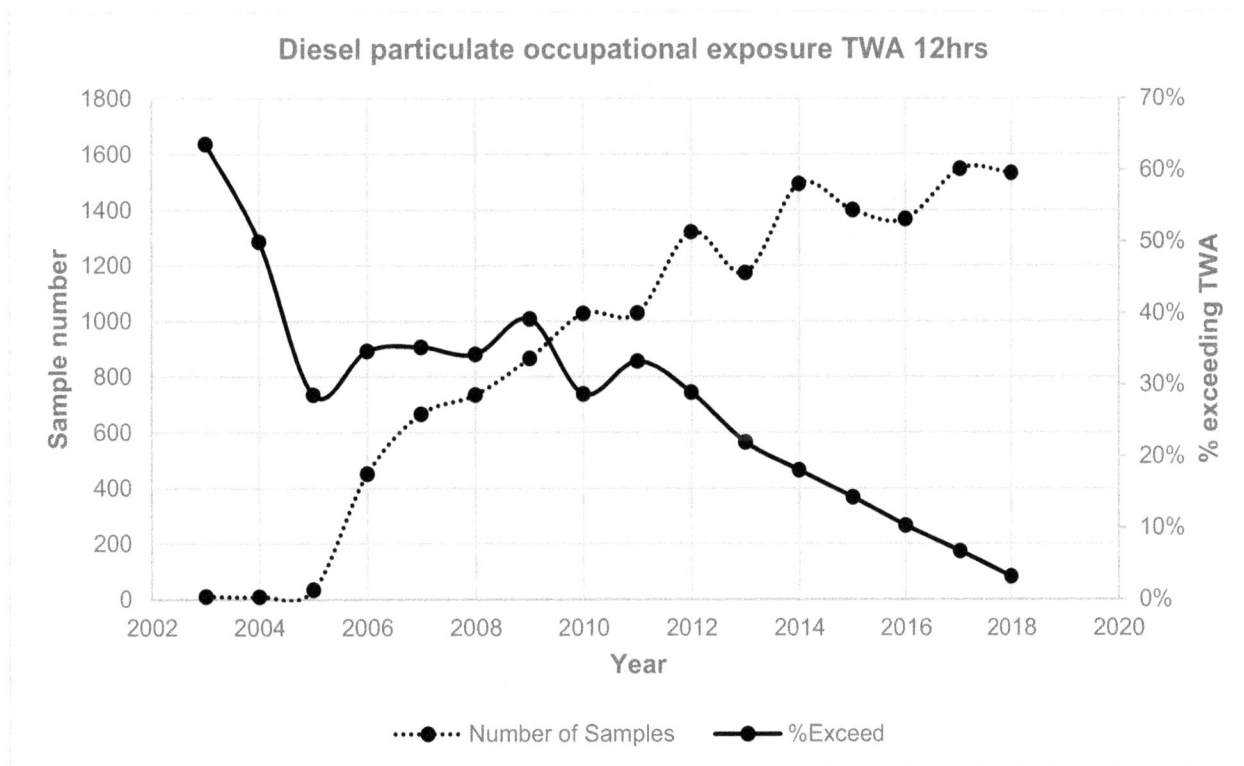

FIGURE 2 – Personal samples and number of exceedances
(Source: Department CONTAM and SRS)

Figure 3 shows annual average personal exposure levels. The annual average DPM (EC) concentration shows a decline from almost 120 µg/m³ in 2003, with downward trending towards

20µg/m³ for 2018. The guideline is 0.1mg/m³ (100 µg/m³) measured as EC using NIOSH 5040 analysis method.

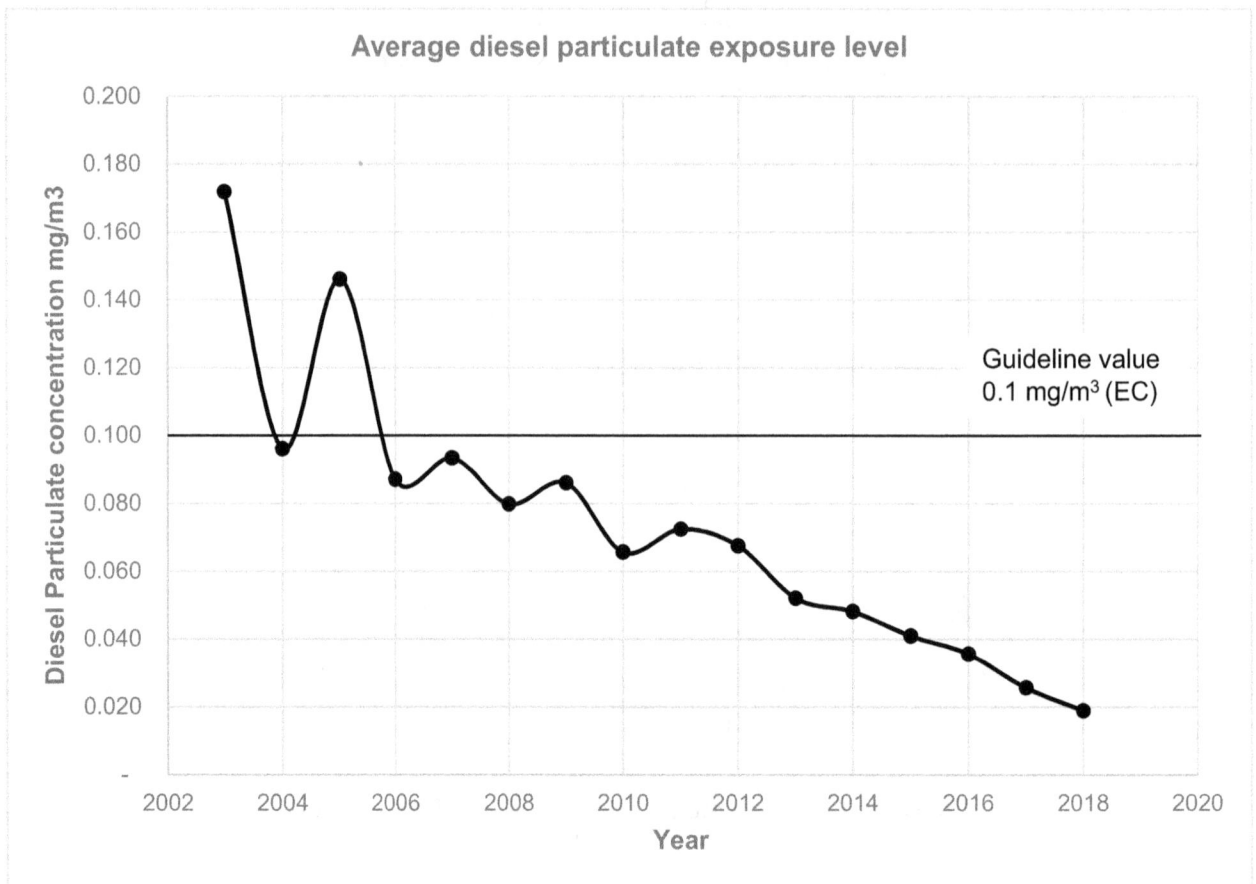

FIGURE 3 – Average personal exposure levels to DPM (EC)

CONTRIBUTORY FACTORS TO HIGH DP EXPOSURES

Exposure of workers to DEEE may be increased when:

- Series ventilation is used which adds contaminants cumulatively to the main intake air. Inspectors have measured the ambient background DPM concentrations at levels up to three times above the TWA guideline at a lower working level before any mining activity has commenced in that work area. This often corresponded to series primary ventilation systems and was not observed for parallel ventilation systems. In parallel systems contaminated air from the level is exhausted and not allowed to contaminate the intake which supplies the air to the deeper workplaces.

- There is inadequate or no performance monitoring of the DPF and other DEEE control systems.
  If the controls are not monitored to determine its effectiveness, durability and robustness then the outcome of the management plan will fail or be ineffective. This is reflected by intermittent or regular exceedances in the personal monitoring results submitted to SRS.

- A failure of DEEE controls will allow contaminants to penetrate into the work environment and expose workers.

- Regular real-time monitoring equipment is not used in the workplace to identify when controls are failing and initiate early remedial work.

- DPM management plans do not include light vehicles and ancillary diesel equipment. Measurements made by inspectors indicate very high DEEE from some light vehicle diesel engines.

CONCLUSIONS

The key message in this paper is that it is necessary to better understand DEEE in each workplace and adopt the Plan-Do-Study-Act process (Deming, 1950).

In planning, mine operators need to understand

- the DEEE components, not limited to engines, controls, emission of gasses and particulates and
- the implications of ventilation dilution rates at each stage of the mine, with respect to DEEE.

In doing, mine operators need to employ robust and effective controls and to implement action plans to identify when these controls begin to fail.

In studying, to ensure the plan is meeting the standards for workplace exposures.

In acting, defects from the standard set by the plan should trigger a response to rectify any defects.

REFERENCES

Arndt, M, 2013. PN to PM correlation [online], presentation for AVL TechDay. Available from: https://www.avl.com/documents/10138/0/Arndt+-+PM+vs+PN.pdf/e2fa6032-7784-496e-b47d-0842fa574e6d (Joint Research Centre, Ispra, Italy).

Australian Institute of Occupational Hygienists (AIOH), 2013. Position paper: Diesel particulate matter and occupational health issues [online]. Update of Position paper 2007. Available from: https://gastech.com/files/dpm/Diesel-Particulate-Matter-and-Occupational-Health-Issues.pdf.

Bugarski, A, Janisco, S, Cauda, E, Noll, J, Mischler, E, 2011. Diesel aerosols and gases in underground mines: guide to exposure assessment and control [online]. Available from: https://www.cdc.gov/niosh/mining/UserFiles/works/pdfs/2012-101.pdf (Centers for Disease Control and Prevention, NIOSH).

Bugarski, A, Cauda, E, Barone, T, Vanderslice, S, 2017. Diesel particulate matter exposure and concentration monitoring in underground mines: practices and challenges [online]. Conference presentations, 23rd annual Mining Diesel Emissions Council (MDEC) Conference. Available from: http://www.mdec.ca/2017/S7P2_Bugarski.pdf (Ontario, Canada).

Bünger, J, Krahl, J, Schröder, O, Schmidt, L, Westphal, G, 2012. Potential hazards associated with combustion of bio-derived versus petroleum-derived diesel fuel [online], Critical Reviews in Toxicology, 42(9):732-50. Available from: https://www.ncbi.nlm.nih.gov/pubmed/22871157.

Cancer Council Australia, 2019. Diesel [online]. Available from https://www.cancer.org.au/preventing-cancer/workplace-cancer/diesel.html.

International Council on Combustion Engines / Conseil International des Machines à ski Combustion (CIMAC), 2008. Guide to diesel exhaust emissions control of NOx, SOx, particulates, smoke and CO2 [online]. Available from: https://www.cimac.com/cms/upload/Publication_Press/Recommendations/Recommendation_28.pdf.

Dallman, T, Menon A, 2016. Technology pathways for diesel engines used in non-road vehicles and equipment [online]. Available from https://www.theicct.org/sites/default/files/publications/Non-Road-Tech-Pathways_white-paper%20paper_vF_ICCT_20160915.pdf (International Council on Clean Transportation, Washington DC, USA).

Deming, W, 1950. PDSA Cycle (Plan-Do-Study-Act) [online], The Deming Institute. Available at: https://deming.org/explore/p-d-s-a

Department of Mines, Industry Regulation and Safety (DMIRS), Government of Western Australia, 2013. Management of diesel emissions in Western Australian mining operations – Guideline [online]. Available from: http://www.dmp.wa.gov.au/Documents/Safety/MSH_G_DieselEmissions.pdf

DieselNet, 2002. Diesel exhaust particle size [online], DieselNet Technology Guide. Available from: http://courses.washington.edu/cive494/DieselParticleSize.pdf.

Donaldson, 2016, Engine horsepower and exhaust flow guide [online]. Available at: http://anyflip.com/etjx/xvpn/basic

Drolet, D, 2015. Guide for the adjustment of permissible exposure values (PEVs) for unusual work schedules – Technical guide (4th edition) [online]. Available from: http://www.irsst.qc.ca/media/documents/pubirsst/t-22.pdf (Institut de recherche Robert-Sauvé en santé et en sécurité du travail (IRSST), Quebec, Canada).

Ghose, C, 2015. Small in size, big problem [online], Down to Earth. Available from: https://www.downtoearth.org.in/coverage/small-in-size-big-problem-13399

International Agency for Research on Cancer (IARC), 2012. IARC monographs on the evaluation of carcinogenic risks to humans, Volume 105: Diesel and gasoline engine exhausts and some nitroarenes [online]. Available from: https://www.iarc.fr/media-centre-iarc-news-58/

International Agency for Research on Cancer (IARC) Review Stakeholder Group (IRSG), 2012. A global and historical perspective on the exposure characteristics of traditional and new technology diesel exhaust: Impact of technology on the physical and chemical characteristics of diesel exhaust emissions – Report [online]. Available from: https://www.concawe.eu/wp-content/uploads/2017/01/irsg_diesel_report_may_7_2012_final-2012-02105-01-e.pdf

Kittelson, D, 1998. Engines and nanoparticles: a review [online], Journal of Aerosol Science, 29(5/6):57–588. Available from: http://dns2.asia.edu.tw/~ysho/YSHO-English/1000%20CE/PDF/J%20Aer%20Sci29,%20575.pdf

Mammoth Equipment and Exhausts, 2013. Diesel particulate filter presentation [online]. Available from http://www.mammothequip.com.au/images/DPF%20Presentation%20-%20Email%20Version%202013.pdf

Manufacturers of Emission Controls Association (MECA), 2007. Emission control technologies for diesel-powered vehicles – white paper [online]. Available from: http://www.meca.org/galleries/files/MECA_Diesel_White_Paper_12-07-07_final.pdf

Miller A, Ahlstrand, G, Kittelson, D, Zachariah, M, 2007. The fate of metal (Fe) during diesel combustion: morphology, chemistry, and formation pathways of nanoparticles [online]. Available from: https://www.cdc.gov/niosh/mining/UserFiles/works/pdfs/tfomf.pdf (Centers for Disease Control and Prevention, NIOSH).

NEPC, 2016. National Environment Protection Council (Western Australia) Act 1996 (WA) - National Environment Protection (Ambient Air Quality) Measure (NEPM), https://www.legislation.gov.au/Details/F2016C00215

Safe Work Australia, 2013. Guidance on the interpretation of workplace exposure standards for airborne contaminants [online]. Available from https://www.safeworkaustralia.gov.au/system/files/documents/1705/guidance-interpretation-workplace-exposure-standards-airborne-contaminants-v2.pdf.

Safe Work Australia, 2018. Hazardous chemical information system (HCIS) [online]. Available from: http://hcis.safeworkaustralia.gov.au/

Safe Work Australia, 2018. Workplace exposure standards for airborne contaminants [online]. Available from: https://www.safeworkaustralia.gov.au/system/files/documents/1804/workplace-exposure-standards-airborne-contaminants-2018_0.pdf.

Stafford, N. Newer diesel engines emit more harmful nanoparticles (2008) [online], Chemistry World. Available from:

https://www.chemistryworld.com/news/newer-diesel-engines-emit-more-harmful-nanoparticles/3002891.article

Zeller, W, Westphal, T, 1992. Effectiveness of iron-based fuel additives for diesel soot control [online]. Available from: https://www.cdc.gov/niosh/mining/works/coversheet500.html (Centers for Disease Control and Prevention, NIOSH).

Study on coordinated control and intelligent regulation for the disaster smoke flow in the coal mines

K Wang[1], H Hao[1], S Jiang[1], Z Wu[1] and H Shao[1]

1. Key Laboratory of Coal Methane and Fire Control, Ministry of Education.
2. School of Information and Electrical Engineering.
3. School of Safety Engineering.
4. State Key Laboratory of Coal Resources and Safe Mining; China University of Mining & Technology, Xuzhou, China.

ABSTRACT

After a disaster occurred in a coal mine, the serious deaths were caused by the smoke flow. In order to control the smoke flow during a disaster in coal mines, the distributed facilities were pre-installed in the key branches in the ventilation network, the project for coordinated control during the emergency rescue was put forward. The smoke flow control effects of distributed modular facilities in different scenarios were studied, and a collaborative optimal method to install the ventilation facilities was proposed. A new type of ventilation facility with fire prevention, adjustable for the roadway deformation and continuous regulation for the section of the roadway was developed. The key components of spare high pressure gas cylinder and battery were adopted after power outages, and the double insurance was realized by the automatic switching technology. The coordinated control and intelligent regulation system was controlled by remote instruction through the ground under normal condition. Multiple factors of cross-perception technology was developed, such as the communication state, temperature, smoke, the concentration of the CO, O2, CO2 and other gases, and the situation of a disaster was judged under the condition of abnormal communication interrupt. By decision tree learning method, a tree model of multiple factors was established, and through the data mining and multiple factors of cross-analysis. Based on the judges, the coordinated control and intelligent regulation system makes independent decisions.

Key words: disaster smoke flow; coordinated control; intelligent regulation; ventilation network; independent decision

INTRODUCTION

At the present time, approximately 90% of coal is produced in under-ground mines in China (Lin, 2014). Exogenous fire has the characteristics of sudden occurrence and serious damage in the underground coal mine, especially once the fire breaks out in the main intake airflow roadway, improper handling will cause a large number of casualties and economic losses (He, 2008). More importantly, the spread of some poisonous and harmful substances like CO, HCl and CO2 endanger miners' health and safety. In addition, the disordered mine local ventilation system resulted from the belt fire can trigger secondary accidents like gas or dust explosion, severely threatening the operation safety of a coal mine (Li et al, 2013, 2012). According to Japan, USA and UK fire statistics, very few people among the number of deaths died directly from the fire, up to 78.9% of the people indirectly died from the smoke plume (Rickard, 2009; 2013; Charles, 2012; Rowland, 2011). During emergency fire rescue, many mine exogenous fires cause increased accidents and serious losses because the conductors make wrong decisions by not estimating enough of the toxic airflow and harmful gas spreading state (Vauquelin, 2009; Tilley, 2013; Wang, 2012; 2013). Exogenous fire accidents have happened in coal mine airflow roadways over the last five years in China, as shown in Table 1.

After disaster occurs in a coal mine, efficient evacuation and escape guidance are important to ensure the safety of miners (Wang, et al, 2015). The speed of personnel evacuation, the environment of the escape roadway, the distance to the safety exit, laws of smoke transport, and other factors were comprehensively analyzed (Dong-Ho 2006). The selection of the optimal path in the complex network of the roadways can improve the escape efficiency.

Table 1. Exogenous fire accidents occurred in coal mine airflow roadways
over the last four years in China

Accident date	Mining company	Specific location	Direct cause	Deaths
2012-01-27	Tang-nei coal mine in Guang-xi	Belt roadway	Belt fire	2
2012-05-11	Jia-zhuang coal mine in He-bei	Electrical cavern	Air compressor fire	5
2012-08-29	Qing-nian coal mine in Liao-ning	Intake roadway	Cable fire	5
2012-09-22	Long-shan coal mine in Hei-longjiang	Intake roadway	Cable fire	12
2013-02-28	Ai-jiagou coal mine in He-bei	+ 750 m intake roadway	Air compressor fire	13
2014-12-12	Xin-he coal mine in Shan-xi	No. 501 belt roadway	Belt fire	4
2015-11-20	Xing-hua coal mine in Hei-longjiang	No. 30 coal belt roadway	Belt fire	22
2016-7-4	Tian-yi coal mine in Liao-ning	400 m intake roadway	Air compressor fire	11
2016-11-29	Jing-you coal mine in Hei-longjiang	Intake roadway	Belt fire	22
2017-3-9	Dong-rong coal mine in Hei-longjiang	Vice shaft	Belt fire	17

McGrattan (2010) introduced the coal mine fire detection sensor and the air flow remote control system based on fire simulation data. Perera (2011) developed a remote monitoring and control system, which can simulate underground ventilation in real time. Charles (2012) introduced a ventilation network space navigation control system, which is used to simulate the evacuation route during a cataclysm in an underground coal mine. There are many scholarly studies regarding the automation of the entire ventilation system but a lack of special research regarding the roadway of a fire control system in a coal mine.

An emergency rescue system in coal mines includes: disaster forecasting and early warning, emergency rescue plan preparation, implementation and other emergency relief operations. Through identification, analysis and evaluation of the hazard sources in a site, the scope of the disaster spread can be forecast and an emergency rescue plan can be formulated (Edwards, 2009; Gurina, 2010). Used the volume rendering technique to make a three dimensional simulation model of smoke flow spread, visualized by using a time sequence, and then revealed the plume migration law. There is little research on mine personnel escape during fire emergency rescue but plenty of research on the model of personnel evacuation. Machacek (2013) proposed the Magnetic Field Model, which takes the relevant person as an object in a magnetic field. The power by which the person moves forward is the magnetic field force and the related traveling control equation. The model is based on the coulomb theorem. Cecelia (2005) studied the discrete time dynamic network model during the evacuation process. Analyzing the characteristics of crowd movement and the phenomenon of blocking in the process, they described the personnel evacuation model as a network model. The Exodus Building, a model of fire scene path selection using real fire data from the USA and UK, can analyze the probability of the evacuation direction of people in the fire evacuation process (Yuan, 2015).

Extinguishing to make people withdraw is the key to emergency rescue after the occurrence of a mine fire, and the safety of personnel during evacuation is particularly important. Thus, we need to study multiple information integration of the fire smoke flow parameters and personnel distribution in a complex ventilation network. Roadway fire simulation can analyze the personnel evacuation in advance and make a safer evacuation route and rescue plan. With the spread of fire smoke, toxic and harmful gas will pollute more of the roadway. We must establish an effective airflow control scheme and provide a good condition for people to escape. When fire occurs in a complicated ventilation network in a coal mine, after the remote emergency rescue system changes the ventilation network structure, the fire smoke flow and distribution of harmful gases should be simulated by FDS. The numerical simulation results can guide the configuration of a remote emergency rescue system for mine fire automation and be applied to personnel escape route design.

ANALYSIS OF BELT FIRE RESCUE MODEL

Mine Fire Rescue Expert Decision Support System

Decision Support Systems (DSS) is the fastest growing branch of information systems. It is based on management science, operational research, cybernetics and behavior science. Using computer and information technology, policymakers orient semi-structured or unstructured systems to make decisions. Wang D. (Wang, et al, 1996) designed the Mine Fire Rescue Decision Support System (MFRDSS) based on DSS technique. The main functions are to simulate a mine fire by computer, choose the best route to escape, control the airflow to avoid disasters, and calculate the branch air volume of the ventilation network. A new type of MFRDSS was developed at Liao-ning Engineering Technology University, and applied in the No.2 coal mine of Jin-chuan Company (Jia, 2006). The software can simulate the airflow state in the ventilation network, and guide the escape routes during a coal mine fire. Expert Systems (ES) emphasize widely available specialized knowledge in the specific domain, ES attempts to simulate or replace experts, automatically solving problems for high performance using a formally established knowledge database.

To apply rescue decisions in practice, automatic control technology and computer technology are combined. Automatic mine fire remote control rescue air-doors are developed, using optical fiber communication technology to realize the remote monitoring. Fusing the ventilation environment parameters monitoring and ES-DSS technology, the MFR-ES-DSS independent rescue system was set up. New ideas for the problem solving of DSS have guided the development of MFR-ES-DSS. Through man-machine interaction, the motility and science of mode thinking could be enhanced for policy-makers, and the decision quality and speed could be improved.

Belt fire rescue model

According to the existing theoretical research on coal mine fires, ventilation facilities would be pre-installed in key roadways by the structural characteristics of the ventilation network, and the facility status can be changed to control the smoke plume by using a remote monitoring system. Some of the air doors are in a normally open state while others are in a normally closed state. When disaster occurs, the smoke plume control system is activated, the air-doors in the normally open state should be immediately shut down to stop the intake flow from the belt roadway to the mining area. So that the miners in the mining area can avoid the disaster of poisonous gas. While the air-doors in the normally closed state are immediately opened to directly import the smoke plume into the air-return roadway. All of the air-doors in the underground coal mine component of the rescue system can be controlled from the ground monitoring center. The remote control system provides a safe environment for firefighting and personnel evacuation. According to the belt fire models statistics in ventilation systems, the fire smoke plume control methods can be divided into two models, as shown in Fig. 1.

The fire smoke plume control system can a control smoke plume into a small range and import it into the return air roadway. However, when the rescue system starts, with the influence of the gas emission and airflow pressure, the ventilation network structure and the fan operating conditions change. The airflow with smoke being directed into the air-return roadway, causes the air volume to fall sharply in the mining area, which may also cause secondary disasters. Fig. 1a shows wind speed sensors of FS1 and FS2 that are set to monitor the airflow volume of the key branches during the fire rescue. The smoke sensors YW1, YW2, YW3 and YW4 monitor the fire. After the belt fire begins, air-doors FM1, FM2 and FM3 are closed, and air-doors FM4 or FM5 are open. When the air volume requirement of the branch changes, we can adjust air-doors FM4 or FM5 to different opening degrees, and the system can control the air-doors to open and close step by step. We must adjust the air volume to guarantee that the smoke plume dose not rolling back and import the smoke into the air-return roadway, and the air volume also provides the conditions for firefighting. Ensure the fresh air volume in the non-smoke area for personnel to escape from the auxiliary transport roadway. To ensure that miners can quickly escape into the auxiliary transport roadway, a small escape door is inset on the air-doors to help the miners escape from the smoke area to the non-smoke area after the air-doors closed. Fig. 1b shows that when a fire breaks out, air-door FM1 is closed, preventing the smoke plume from entering into the mining area. FM2 is opened to import the smoke into the air-return roadway, and when the air volume requirement of the branch changes, we can adjust the air-doors (FM2) to different opening degrees, and the system can control the air-doors to open and close step by step. However, as there is no fresh airflow into mining area during the fire emergency rescue, the miners must escape into the refuge chamber and leave the coal mine after the fire is put out. The refuge chamber can be disassembled according to the mine barrier structure, and divided into various different types that can hold 30-100 people. The permanent refuge station can achieve the goals of providing continuous oxygen and communication to the inside of the station by drilling throughout the rock to the ground. However, the refuge chamber or rescue capsule is a closed space for the underground workers who cannot evacuate to provide a avoid disaster in time and win rescue time during a coal mine accident.

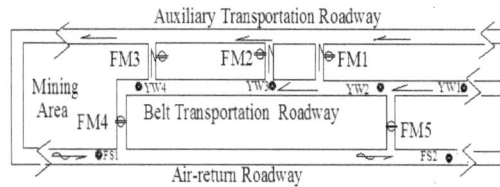

（a） Belt and auxiliary transportation in different roadway

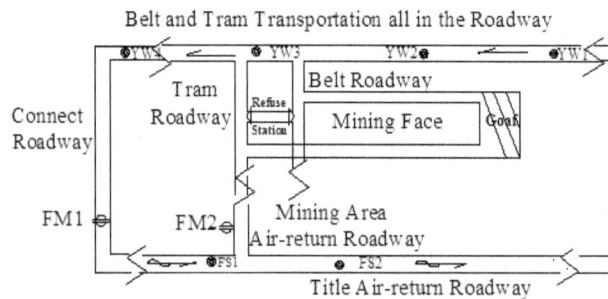

（b） Belt and tram transportation all in one roadway

Fig. 1 Belt fire rescue models in different ventilation system

Results of numerical simulation

Fire dynamics simulation software used areas from high buildings, factories and others to promote to the tunnel, underground railway and so on. Many successful fire process simulations made the abstract fire become a concrete image inquiry. Combining the ventilation system and the simplified physical model of the Da Liu-ta coal mine, after fire occurred in the transport belt of the main roadway in the No. 1 plate area, the fire smoke spread in the roadway very quickly, leading to reduced visibility. If the miners fail to escape before the smoke spread to the coal face, then the escape rate will fall, even until it is impossible to escape due to the high temperature toxic and harmful gases conditions. To analyze the control method of fire smoke flow, numerical simulation is mainly used to analyze the distribution law of smoke flow, temperature and visibility.

1) Air flow control facilities in the model

To grasp the method that is effective in controlling the smoke flow, we add switched air-doors to control the route of smoke flow. Comparing the numerical simulation results, we set the air-doors FM1, FM2, FM3, FM4 as open in the contact lane between the coal belt and the auxiliary transport roadway and set the air-doors FM5, FM6 as closed in the contact lane between the coal belt and the return air roadway. We close the air-doors FM1, FM2, FM3, FM4 when a fire occurs; if the fire location is near the front of the belt roadway, we open the closed door FM6; if the fire location is near the end of the belt roadway, we open the closed door FM5. With the control method, fire smoke control and personnel escape safety can be realized. In the simulation process, through setting the time to control the opening and closing of the air-doors, we observe the smoke flow migration before and after the emergency rescue system changes the ventilation. The model of the ventilation roadway fire smoke control system simulation is shown in Fig. 2.

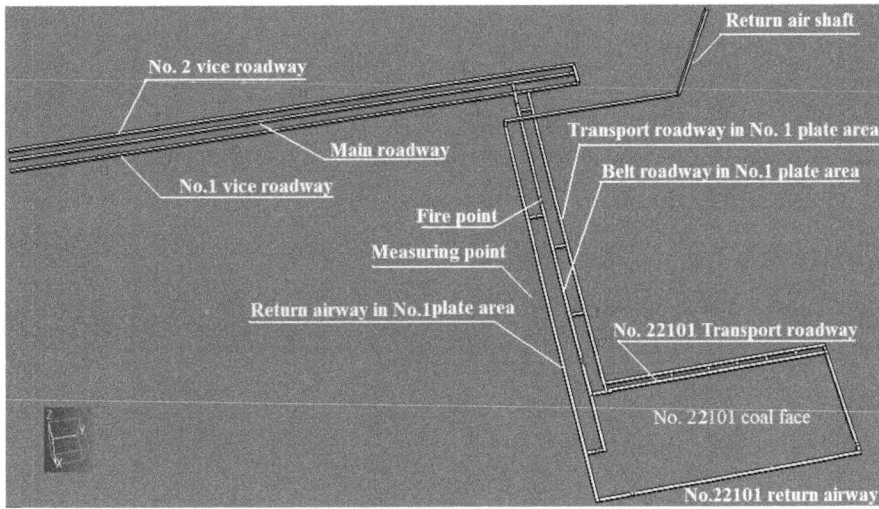

Fig. 2 The model of the roadway fire smoke control system

2) Velocity analysis of the smoke spread

The high concentration of smoke produced by the fire not only affects the breathing of miners in distress but also reduces the visibility of the escape path, seriously hindering the escape efficiency. Therefore, by analyzing the spread speed and path of high concentration smoke in the ventilation, we can find the methods of smoke flow path control and escape route selection for the miners, the results were shown in fig. 3.

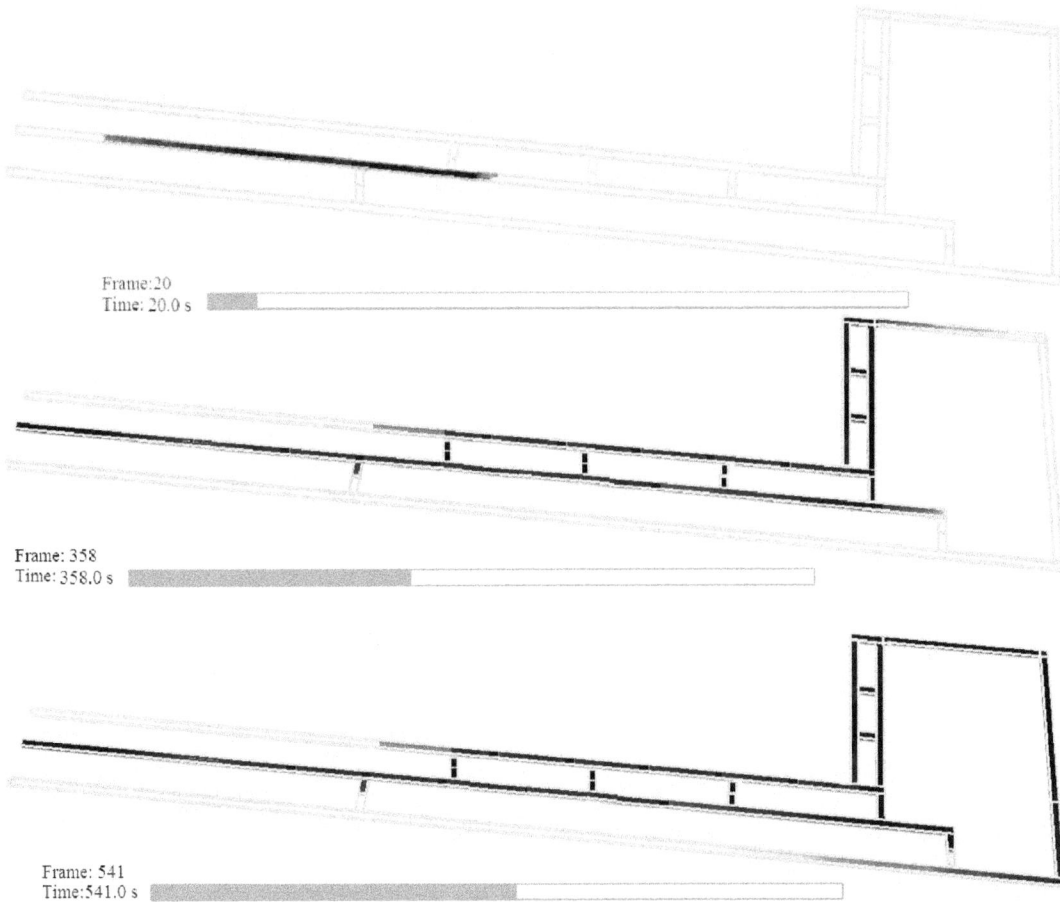

Frame:700
Time: 700.0 s

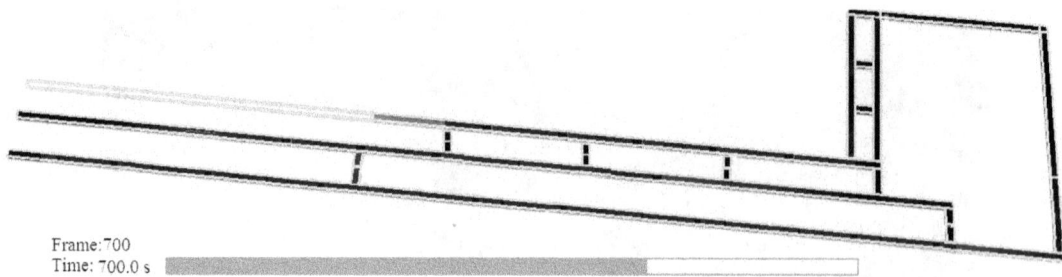

(a) Spreading law of smoke for the front belt roadway fire without a smoke control system

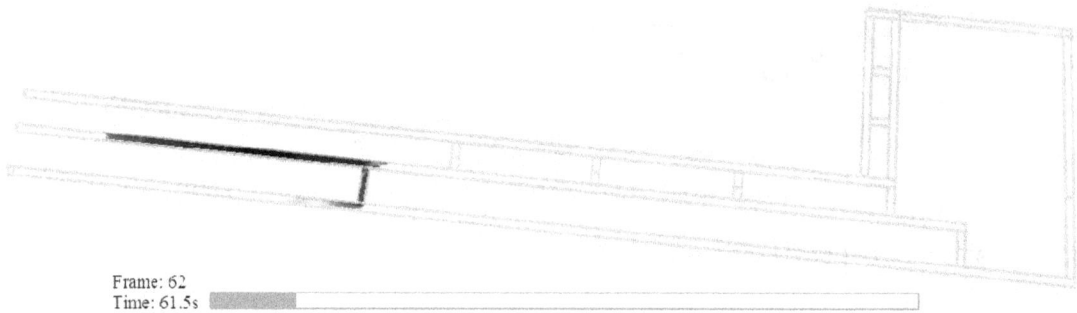

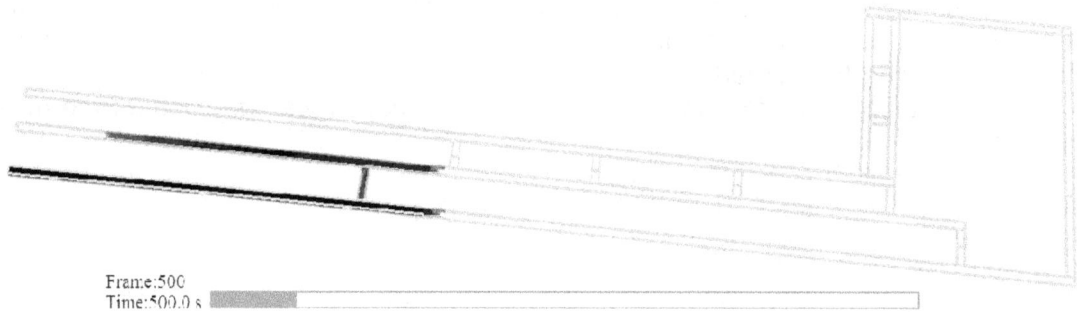

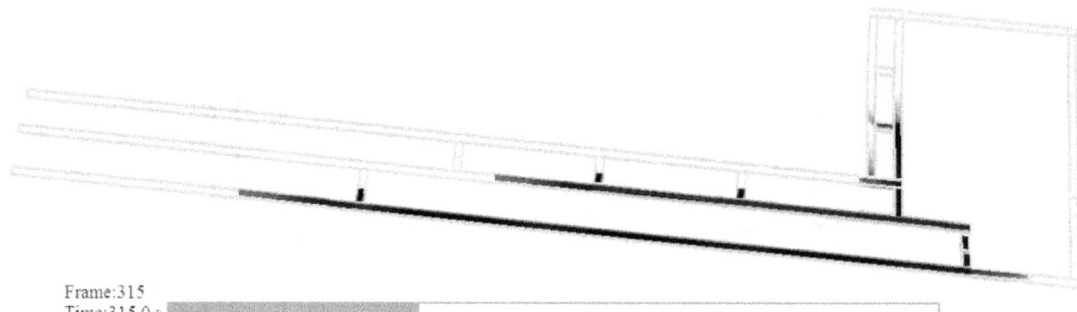

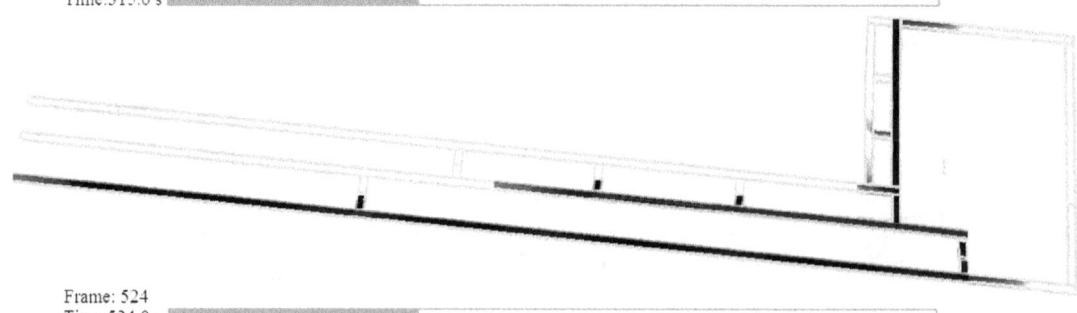

Frame: 62
Time: 61.5s

Frame:500
Time:500.0 s

(b) Spreading law of smoke for the front belt roadway fire with a smoke control system

Frame:315
Time:315.0 s

Frame: 524
Time:524.0 s

(c) Spreading law of smoke in the tail of the belt roadway fire with a smoke control system

Fig. 3. Spreading law of fire smoke at different times in the ventilation

Fig. 3 shows the process of smoke spreading after a roadway fire. In the early stage, the smoke quickly gathers at the top of the fire source, then rapidly spreads, following the air flow in the roadway, with a small amount of smoke rolling back in the process. Fig. 3a shows that without a smoke control system, after fire occurs for 20 s, the smoke roll back phenomenon appears in the main transport roadway in the No. 1 plate area. Once it happens, the fire burns more and more intensely, the smoke spreads and the air flow is polluted. After the smoke enters the auxiliary transport roadway in the No. 1 plate area via cross-heading, it will continually spread to the coal face, seriously threaten the safety of miners. At 358 s, thick smoke spreads into the coal face; thus, miners need to escape from the coal face before 350 s. Personnel escape is almost impossible after 541 s because all the coal face and the air intake roadway are covered by thick smoke. At approximately 700 s, the whole No. 1 plate area is polluted by smoke, which makes firefighting and personnel evacuation in the area difficult.

To create conditions for firefighting and personnel escape, we avoid the smoke flow spreading to the miners in a crowded area and transport it to the return air roadway via control measures. We consider the buffer time of the fire monitoring alarm; after 60 s of fire occurrence, air-doors FM1, 2, 3, 4 are closed by the simulation system. If the fire occurs in the front belt roadway, air-door FM6 is opened; if it occurs in the tail of the belt roadway, air-door FM5 is opened to guide the smoke flow into the return air roadway. Fig. 3b shows a fire that occurs in the front belt roadway; the smoke flow control system automatically opened FM6 after 60 s, and the smoke was transported into the return air roadway at 61.5 s. Due to the large amount of smoke flow into the return air roadway, a little smoke flowed into the belt roadway downstream. The fire smoke was transported into the return air roadway stability at 500 s and unable to enter other areas; thus, the miners could escape through the auxiliary transport roadway.

To check the response time of the smoke control system, Fig. 3c shows the fire in the tail of the belt roadway, with the smoke flow control system automatically opening FM5 after 200 s. Much smoke transported into the return air roadway at 315 s. Little smoke flowed into the belt and auxiliary transport roadway in the coal face due to the delays of smoke control measures. Under short circuit ventilation conditions, smoke spread at a low velocity in the coal face and did not affect personnel escape. The fire smoke transported into the return air roadway stability at 524 s; thus, the miners could escape through the auxiliary transport roadway. Short circuit ventilation can lead to an increase in air quantity in a fire roadway, with the roll back of smoke becoming smaller; thus, the condition is good for firefighting.

APPLICATION OF EMERGENCY RESCUE SYSTEM

Based on the numerical simulation and theoretical analysis, the belt fire emergency rescue system of the No. 1 plate in the Da Liu-ta coal mine is established. The system-related control equipment and monitoring software successfully fuse the fire plume control and personnel escape information into one platform. The emergency rescue system can ensure the realization of a remote start under disaster conditions, integrating the smoke flow path and the personnel position data into the emergency rescue software in the ground monitoring center. During an emergency rescue, it can monitor and control the wind quantity of each branch in the underground ventilation and we can see the miners' escape trajectory in real time. It effectively improves the efficiency of firefighting and the success rate of personnel escape.

Smoke Flow Control System

Electricity black-out must be carried out during the mine fire because the fire would damage the communication cable, reducing the reliability of the monitoring system. We must guide the smoke flow control and the personnel escape route with the key monitoring data during an emergency rescue(Zheng, 2011). The smoke flow remote control system includes: a local workable air-door, communication control station, sensors and solenoid valves, ground monitoring center, computer foreground form and background processing software, etc. In a fire emergency rescue, the underground equipment must operate steadily and remote control in the ground monitoring center must be realized. Combined with the advantages and disadvantages of using the air-doors, a portal structure that can overcome roadway deformation was designed, being effectively used in an underground mine. To realize the remote control of the air quantity of ventilation branches, the air-door structure was designed with an adjustable opening (adjustable open and closed area). 5 sensors were installed on the door structure to monitor the opening position, using a three-position five-way solenoid valve to control the door. The mine-used intrinsically safe controller and related equipment were designed based on the PLC (Programmable Logic Controller). An optical communication module, which uses optical fiber to realize remote communication, was built inside. UPS (Uninterruptible Power System) power, which provides a continuous power supply for the emergency rescue system for more than 5 hours, was also built inside. A high pressure cylinder was used as a standby gas source to provide power for adjusting the air-doors' opening area. According to the characteristics of the smoke flow remote control system, the ground center station was designed and developed. The functions are gathering information from each substation, processing the data, exchanging data with the computer, sending control commands to each station, etc. The control panel, rescue start and recovery switch buttons, buttons for adjusting and controlling the air-doors' opening degree, the switch of the doors state and the communication state light display, etc. were designed according to the method of controlling the air quantity of ventilation branches. The air quantity control button of the ground center station can send a command to the controller of the FM5 or FM6 air-door; after receiving a command, the station can control solenoid valves, causing them to open and close the air-door step by step; the position sensors then feedback information to the ground station. The hardware of the remote control system of the fire smoke flow in the roadway is shown in Fig. 4.

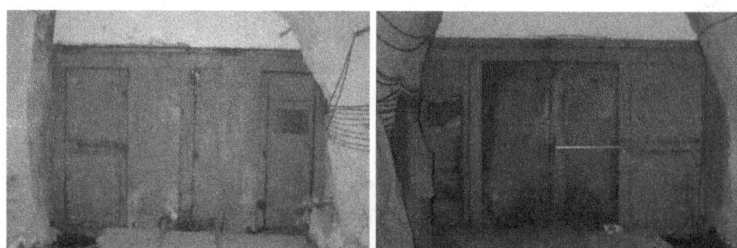

(a) Physical maps of air-doors

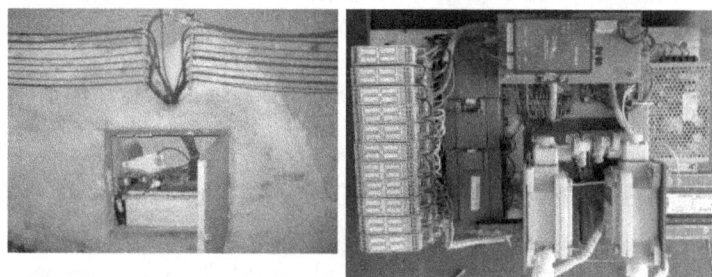

(b) Physical maps of controller

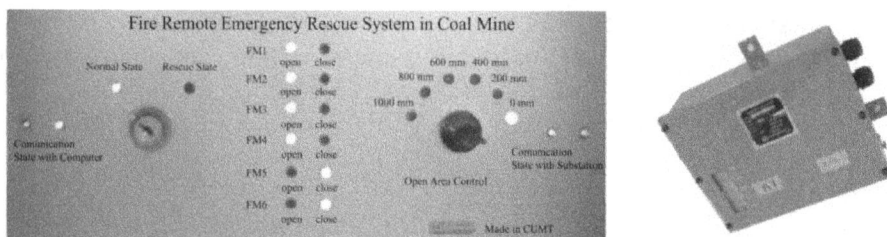

(c) Control panel of central station and personnel location base station

Fig. 4 Physical maps of the fire remote emergency rescue system in coal mines

The Da Liu-ta coal mine built an underground disaster wireless emergency communication system based on a mesh network last year. By reforming the existing personnel location system, the personnel location base station was connect to an emergency communication system to realize real-time tracking of the personnel escape trajectory under fire conditions. Finally, mixing the personnel position system and the fire smoke flow control system together, a unified regulation platform, which improved the efficiency of a fire emergency rescue and the success rate of personnel escape, was set up.

The mine fire smoke control system can analyze all the monitoring parameters through a background software, according to the personnel escape trajectory; the control method of the smoke flow path and air quantity must ensure the safety of personnel. Changes of the ventilation network structure, branch resistance, characteristics of ventilator operating conditions and ventilation network would be embedded into the background database. After a fire occurs, the real-time air quantity of a branch is monitored by the air speed sensors FS1 and FS2, which provide the data support for the smoke flow control. Using the key branches' air quantity, the real-time quantity of each branch can be obtained by performing a ventilation network calculation, which can guide the fire emergency rescue. Through the numerical simulation and the field test of the fire smoke remote control system, the ideal air quantity of the key branches during a fire emergency rescue can be calculated. According to the ideal air quantity, we set up the threshold of the key branches; when the error between actual monitoring and the ideal air quantity is over 10%, the system can automatically adjust the air quantity of the key branch. The visual form of the remote fire smoke control and personnel escape information monitoring system is shown in Fig. 5.

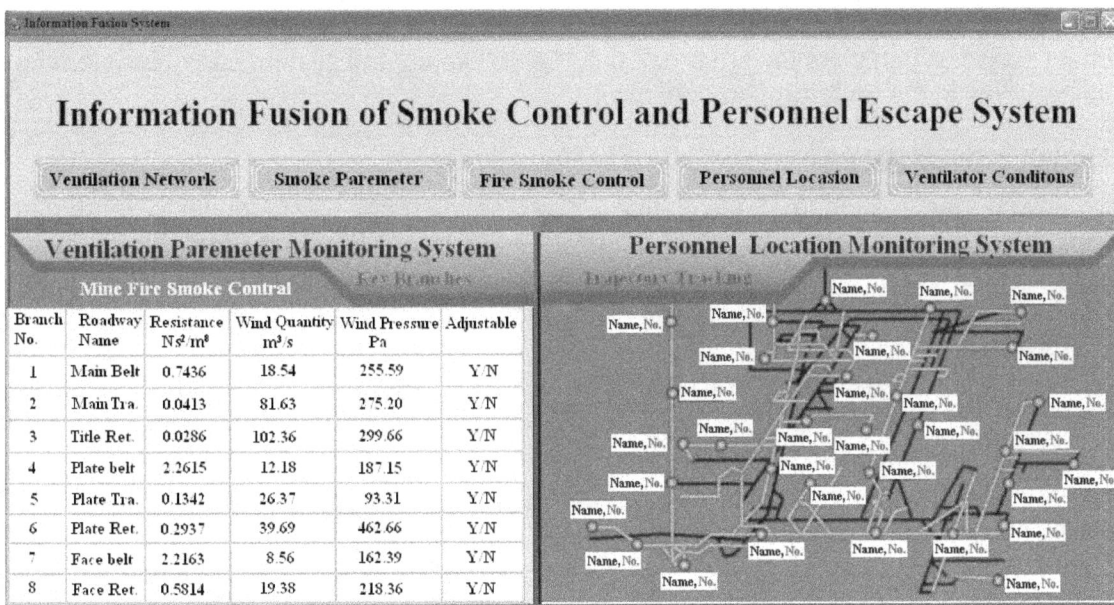

Fig. 5 Visual form of remote fire smoke control and personnel
escape information monitoring system

Analysis for application effect

Since the application of the remote fire smoke control and personnel escape monitoring system in the No. 1 plate area in the Da Liu-ta coal mine, simulated fire drills have been conducted three times. The results of field measurement, simulation results, and the results displayed in visual form all verify a good coupling, which means that the system can be applied well in a fire emergency rescue. During the exercise, air-doors could be opened and closed according to a predetermined function, and the smoke flow could also move according to the path designed. Because of the difficulty of simulating fire and the possibility of causing a disaster, in the exercise, the ventilation network structure changed without the influence of fire heating air pressure and the air quantity of each branch distribution was studied. In each exercise, after starting the fire smoke control system and achieving the intended function, the air quantity of the ventilation branches was determined and compared with the air quantity during a normal ventilation period. The system was used to monitor the air quantity of the key branches, and the solution to the air quantity of each branch was determined via a ventilation network calculation. The linear coefficient of each sensor was calibrated, ensuring the best state for fire emergency rescue.

CONCLUSION

1) Through analyzing the hazard of belt fire to ventilation system in No.1 plate area in Da Liu-ta coal mine. According to the characteristics of fire smoke control in the ventilation network, a method was proposed that fuse the fire plume control and personnel escape information into one platform.

Improved coupling the smoke flow control system to ventilation system during fire emergency rescue, which provides a theoretical basis and technical assurance for firefighting and personnel escape.

2) According to the fire situation results, Personnel escape is almost impossible after 541 s because all the coal face and the air intake roadway are covered by thick smoke. At approximately 700 s, the whole No. 1 plate area is polluted by smoke, which makes firefighting and personnel evacuation in the area difficult. The coal face temperature became stable at approximately 60 ℃, but the escape route would be blocked by high temperature, with the toxic smoke being the key harmful factor. The miners must withdraw from the transport roadway in the coal face within 280 s and from the auxiliary transport roadway in the No. 1 plate within 700 s to escape successfully. However, the length of the coal face was more than 500 meters; thus, miners cannot escape from the working surface within 280 s.

3) Combined numerical simulation results and the fire smoke control ideas, software and hardware of the fire smoke control system are developed. The air quantity of the belt and auxiliary transport roadways changed to within reasonable limits, providing good conditions for firefighting and personnel escape. When the mine fire break out, the smoke control system can guarantee the safety of the miner' lives in the underground coal mine.

ACKNOWLEDGMENTS

This work was partly supported by National Key Research and Development Program of China (2018YFC0808100), supported by the Fundamental Research Funds for the Central Universities (Grant No. 2018QNA03).

REFERENCES

Lin, W.J., 2014. Innovation and development of underground coal mining technology. Coal Eng. 46, 4-8 (in Chinese).

He X.J., Jiang S.G., Wu Z.Y., 2008. Mine fire disaster pneumatic remote control emergent rescue system in west wing transportation roadway of Long Dong coal mine. Coal Science and Technology. 36, 53-54 (in Chinese).

Li C.P., Li Z.X., Cao Z.G., 2013. 3D simulation modeling techniques of fire fume spread process for underground mines. Journal of China Coal Society. 38, 257-263 (in Chinese).

Li C.P., Li Z.X., Cao Z.G., 2012. Holistic modeling and three-dimensional visualization of complex fields for underground mines. Journal of University of Science and Technology Beijing. 34, 744-749 (in Chinese).

H. Rickard, I. Haukur, 2013. Heat release rate measurements of burning mining vehicles in an underground mine. Fire Safety Journal. 61, pp. 12-25.

Charles D.L., Inoka E.P., 2012. Evaluation of criteria for the detection of fires in underground conveyor belt haulage ways. Fire Safety Journal. 51, pp. 110-119.

Rowland J.H., Verakis H., Hockenberry M.A., Smith A.C., 2011. Effect of air velocity on conveyor belt fire suppression systems. Trans Soc Min Metall Expl. 328, pp. 493-501.

O. Vauquelin, G. Michaux, C. Lucchesi, 2009. Scaling laws for a buoyant release used to simulate fire-induced smoke in laboratory experiments, Fire Safety Journal.44, pp. 665-667.

N. Tilley, B. Merci, 2013. Numerical study of smoke extraction for adhered spill plume sin-atria: impact of extraction rate and geometrical parameters, Fire Safety Journal. 55, pp. 106-115.

Wang K, Jiang S.G., Zhang W.Q., 2012. Remote Control Technology and Numerical Analysis of Airflow in Coal Mine Fire Disaster Relief Process. Journal of China Coal Society. 37, 1171-1176 (in Chinese).

Wang K, Jiang S.G., Zhang W.Q., 2013. A study on numerical of smoke flow regulation during remote emergency rescue of roadway fire disaster. Disaster Advances. vol.6, pp. 331-340.

Dong-Ho Rie, 2006. A study of optimal vent mode for the smoke control of subway station fire. Tunnelling and Underground Space Technology, 21, pp. 3-14.

K.Mc.Grattan, R.Mc.Dermott, S. Hostik, J. Floyd, 2010. Fire dynamics simulator (Version5) user's guide, National Institute of Standards and Technology, Gaithers burg, pp. 230.

Perera, I. E., Litton, C. D., 2012. Impact of air velocity on the detection of fires in conveyor belt haulage-ways. Fire Technology. 48, pp. 405-418.

Charles, D.L., Inoka, E.P., 2012. Evaluation of criteria for the detection of fires in underground conveyor belt haulage ways. Fire Safety Journal. 51, pp. 110-119.

Edwards J. C., Wang C. C., 1999. CFD Analysis of mine fire smoke spread and reverse flow conditions. Tien J C. Proceedings of the 8th US Mine Ventilation Symposium. Rola: University of Missouri-Rolla Press, pp. 417.

Gurina, E. I., 2010. Modeling of the operation of a mine counter rotation fan by means of the fluent suite. Journal of Engineering Physics and Thermophysics. 83, pp. 985-990.

Z. Machacek, R. Hajovsky, 2013. Modeling of temperature field distribution in mine dumps with spread prediction. EKTRONIKA IR ELEKTROTECHNIKA Vol.19, pp. 53-56.

Cecelia W. B., 2005.A Case Study Of An Anti-Tbrrorism Evacuation Model Of A Magnetic Levitation Station In An Urban Environment [D]. Dissertation, Morgan state University, December, 2005.

Yuan L.M., Alex C.S., 2015. Numerical modeling of water spray suppression of conveyor belt fires in a large- scale tunnel. Process Safety and Environmental Protection,Vol. 95, pp. 93-101.

Wang DM, Wang SS, Guo JY, 1996. A mine fire rescue decision making support system. Journal of china coal society. 21, 624-629.

Jia JZ, Liu J, Zhao QL, 2006. Study of mine fire rescue decision support system for No.2 mine area of Jinchuan company. China safety science journal. 16, 131-135.

Zheng X.P., Cheng Y., 2011. Modeling cooperative and competitive behaviors in emergency evacuation: A game-theoretical approach. Computers &Mathematics with Applications. 62, 4627-4634.

Intelligent mining at Huangling Mine, with special mention of ventilation safeguard system

Z Yan[1], J C Tien[2] Y Wang[3] and X Chang[4]

1, 3, 4 Xi'an University of Science and Technology, Shaanxi, China
2 Monash University, Melbourne, Australia (Retired)

ABSTRACT

The development of computing, monitoring and control technologies in the past several decades has resulted in an increasing interest in the potential of mining automation in China's massive mining industry. Eliminating human from mining process and enhancing operational efficiency are the major goals. Over 90% of China's coal is mined underground where ventilation is essential to maintain productive and sustainable condition, it is expected that these new technologies will further improve coal mining safety and productivity. Over 70 companies have started using this so-called "intelligent" or "unmanned mining", among them, Huangling Mining Co. which has been experimenting this system at two underground coal operations in Shaanxi Province since 2015. Results have shown great potential for this technology to improve mine safety and increase efficiency. This paper introduces this intelligent system and addresses lessons learned at Huangling.

INTRODUCTION

China is the largest mining country in the world. Driven by its unprecedented industrial growth the past several decades, its mining industry has been challenged to provide much needed minerals and fuels to feed its massive power, manufacturing and construction industries. Among the minerals produced, coal continues to dominate to provide much needed energy to fuel its economical engine.

Coal consumption in China rose every year the last two years (2017 and 2018), constituting around 58% its total energy mix in 2018. Although its share of total energy decreases as cleaner energy sources gained ground (Anon., 2019b), it is still the primary source of energy in China (Figure 1).

As of 2018, there are over 5,800 coal operations, of which 1,200 produces more than 1.2 Mt annually, or 80% of China's total coal production (Personal Communication, 2019). Despite increased and shifting focus on the use of "green energy" in recent years, with nearly 1,000 GW in operation, China accounts for about half the world's coal-fired power, with the United States (259 GW) and India (221 GW) a distant second and third, according to Global Coal Plant Tracker (Hood, 2019). This energy mix is not to expected to change significantly in the near future.

More than 90% of China's coal operations is mined underground, a well-planned and well-maintained ventilation system is essential. There has been much improvement in China's coal mining safety overall over the last decade, fatality rate per million tones (Mt) was 0.093 in 2018. Many factors contributed to this significant change, among them, increased safety public awareness, improved mechanization, use of detection and monitoring instrumentations, and enhanced enforcement. To further improve ventilation safety, China has been pushing toward more mechanization and automation underground, this slowly evolved into unmanned mining, or intelligent coal mining.

DEVELOPMENT OF INTELLIGENT UNMANNED MINING IN CHINA

The increasing use of sensors and monitoring system in Chinese mines led to the next natural path: an intelligent or unmanned mining working environment. This development started around 2010, first with a systematic use of remote monitoring and control at the mining face (1.0), followed by automatic face-straightening (2.0) the following three-five years; and an "integrated mining face" (3.0) at the present time, hoping will finally lead to a totally unmanned working face (4.0). Mining automation consists of three major components: equipment, software and communication systems, there have been developments in all three areas in China's mining industry. The following discussion will focus only on underground coal operations.

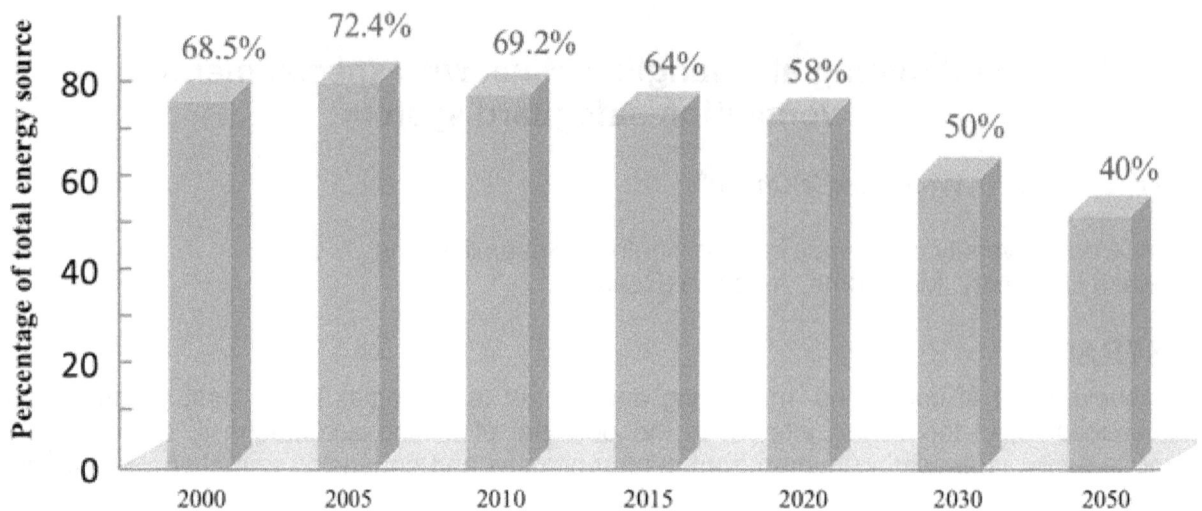

FIG 1 – Coal contributes roughly 58% of China's primary energy in 2018, although this share is expected to decrease in the coming years.

An integrated mining face would start with a geological model based on exploration data and modified interactively only with on-site data collected through geological radar, magnetic CT scan and airway laser scanning, they are further refined according to local rock characterization, a final working face model is constructed using real-time proximity recognition and surrounding environmental attributes. Challenges at this stage is to accurately develop shearer cutting and tramming control as well as advance roof control technologies based on real-time high-precision 3D geological model and equipment positioning data. All extraction operations, shield movement, AFC and stage-loader operations, air velocity, dust and methane emission data can be monitored and controlled remotely from surface (Figure 2).

FIG 2 – Over 90% of China's coal is mined underground. Photo shows a shearer cuts coal in a thick coal seam underground.

Based on most recent information, up to 70 longwall faces in major coal mining companies such as Shenhua, Shaanxi Coal and Chemical Industry Group Co., Ltd (Shaanxi-Coal), Yan-kuang (Yanzhou Mining), etc. have been working to develop intelligent longwall mining faces, with various stages of implementation at varying degrees of success. Among the companies, Huangling Mines #1 and #2 at Shaanxi-Coal have been among the most successful ones due to increased investment and concerned effort in the senior management. Huangling system is staffed with one miner in the production area and two monitors at the surface control room monitor and control underground mining operations. This paper will introduce the system, experience the past several years and lessons learned at Huangling Mine #1.

HUANGLING MINE #1

Huangling Mining Group Co., Ltd, is a subsidiary of the state-owned Shaanxi Coal and Chemical Industry Group Co., Ltd. The Company, in mid Shaanxi approximately 200 km north of Xi'an (Figure 3), has two underground coal operations with a combined annual production of 14 million tones (Mt) in 2017 (6 Mt for Mine #1 and 8 Mt for Mine #2).

FIG 3 – Huangling Mining Complex is approximately 200 km northwest of Xi'an

Coal seam at Mine #1 is bituminous coal with a heating value of 5,600 Kcal (or 22218 btu). At an average mining depth of 323 m, its average minable seam is 2.02 m in thickness. Mine is accessed through three horizontal adits, which are also served as intake airways (there are also #2 and #3 Slope Intakes), with exhausting air returned through #2 and #3 Return Slopes where main fans are located (Figure 4).

Huangling Mine #1 is ventilated using a multi-fan exhausting system. A total of 13,550 m^3/min fresh air is being brought in through three horizontal adits at the east side of the property (7,550m^3/min), #2 Intake Airway (5,000 m^3/min) and #4 Intake Airway (1,000 m^3/min), and exhausted through two exhausting fans (as of Jan 31, 2019):

> #2 Slope Return Fan: BD-II-8, No 27 or 2.7-m in diameter;
>
> > 10,107 m^3/min at – 2,550 Pa. with 355 KW power
>
> #3 Slope Return Fan: FBCDZ, No 27, or 2.7 m in diameter;
>
> > 11,009 m^3/min at – 2,300 Pa. with 355 KW power.

The coal seam has a relative methane emission rate of 7.7 m^3/t and absolute methane emission at 89.17 m^3/min. The seam also emits 1.89 m^3/t of CO_2, and absolute emission rate of 21.92 m^3/min. Huangling is classified as a gassy mine[1].

Although Chinese ventilation regulations require that all coal mines install safety monitoring systems to collect real-time data in underground airways and working faces, accuracy of measurements, data transmission and data-sharing capabilities, anti-jamming, vibration, methane and dust

---

[1] Mines with a methane emission exceeding 40 m^3/min are classified as gassy.

management are still not be able to meet system requirement, which present challenges to the development of an intelligent mining system. More work is still needed to improve data accuracy, transmission speed, and system integration, optimization and stability to ensure consistent system performance for potential advance detection and production planning.

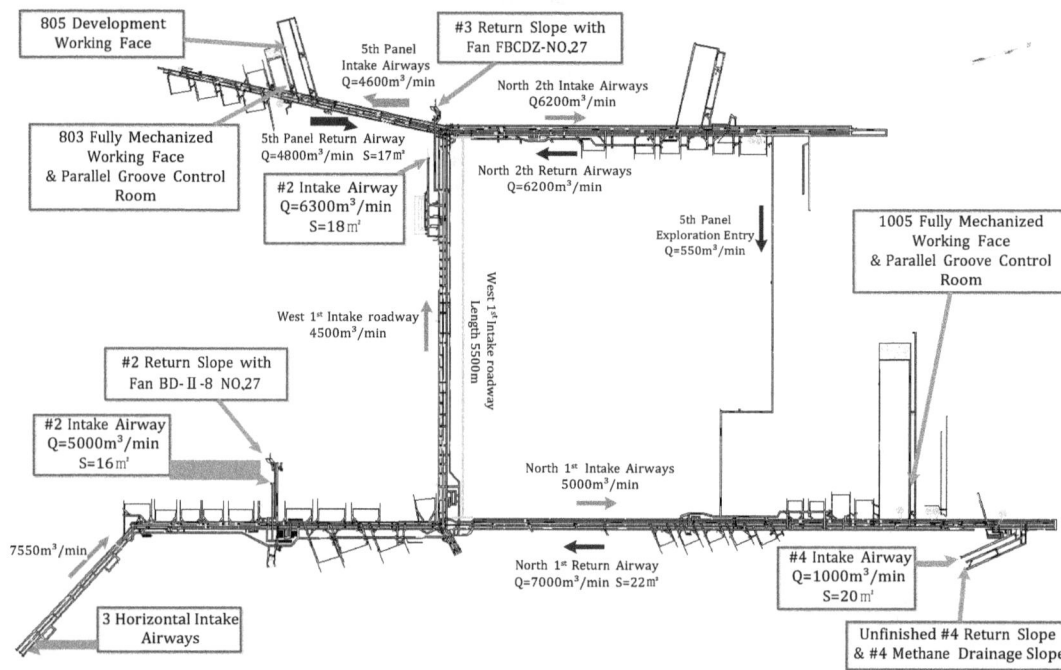

FIG 4 – Drawing showing Huangling #1 Mining Layout and Ventilation System

INTELLIGENT UNMANNED MINING SYSTEM AT HUANGLING

The Intelligent Unmanned system at Huangling is an integrated system in an advanced network environment, key functions include automatic AV monitoring and control, communication and supervision hoping to achieve fully mechanized remote control, and in most cases, a semi-automatic mining at this point. The System coordinates and controls all mining activities – coal extraction, shield support movement and AFC haulage using an SAS memorized cutting control subsystem, a hydraulic support control subsystem, a SAV video supervision subsystem, SAP integrated hydraulic oil feeding subsystem, transporting subsystem and several other specialized subsystems to control the entire mining, transporting and supporting activities (Figure 5)

During coal extraction, the shearer, AFC, shield supports, and stage loader be effectively coordinated and interlocked to allow predetermined and memorized cutting and tramming sequence be implemented with remote intercept/control as an occasional assist when necessary.

It is essential that a set of standard data exchange and technical protocol be strictly adhered to ensure accurate and efficient data collection, integration, storage, retrieval and exchange.

Monitoring System at Gateroads

A critical component of an Intelligent System is the successful implementing its monitoring and control center at gateroads. Depending on specific requirement, the Center supports and controls both a fully automated control mode and individual control mode, the former can fully control mining extraction remotely at the Control Console (Figures 6-8), while the latter allows each function (e.g., shearer, shield, AFC, stage loader, face pumping stations, etc.) be individually controlled.

The Control Console is designed and can monitor and control all ventilation and production activities to ensure all mining equipment/facilities operate safely and efficiently, automatic.

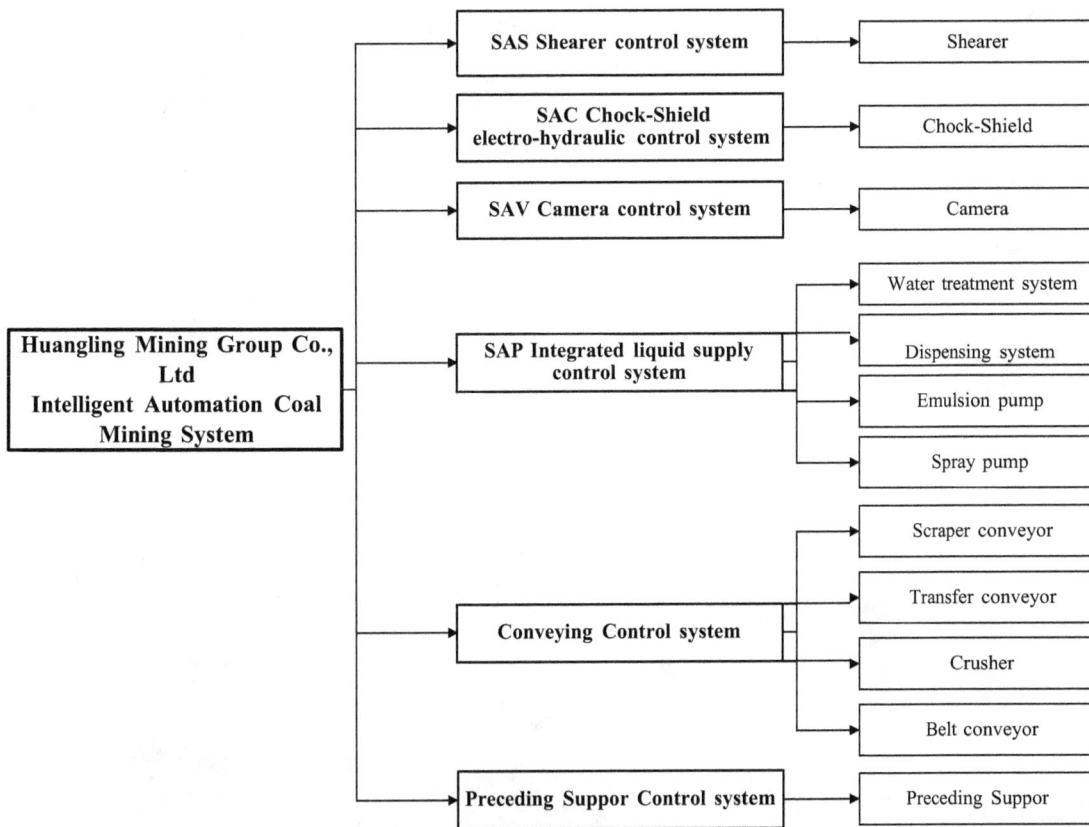

FIG 5a – Block diagram showing the intelligent control subsystems.

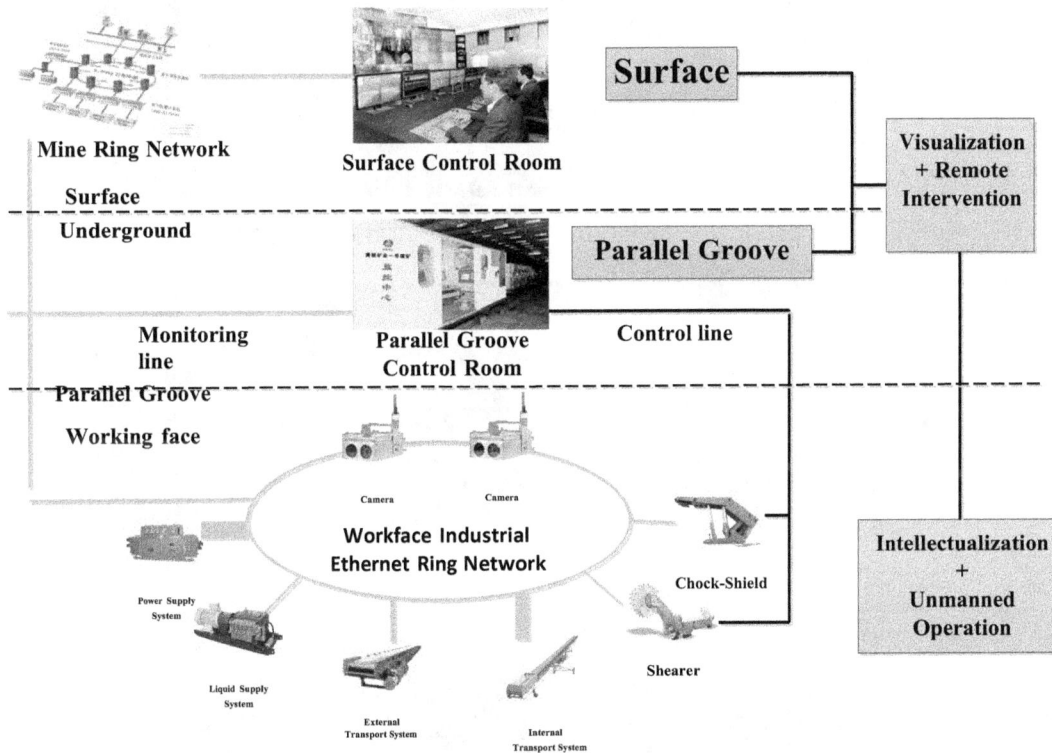

FIG 5b – Schematic diagram showing the intelligent control subsystems for both surface and underground subsystems.

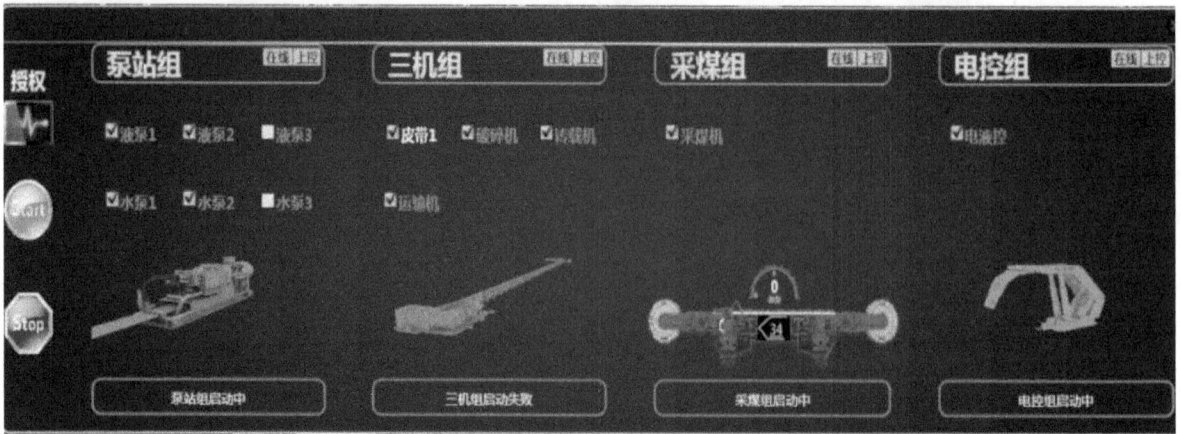

FIG 6 – Photo Showing a "One-button Control" Console

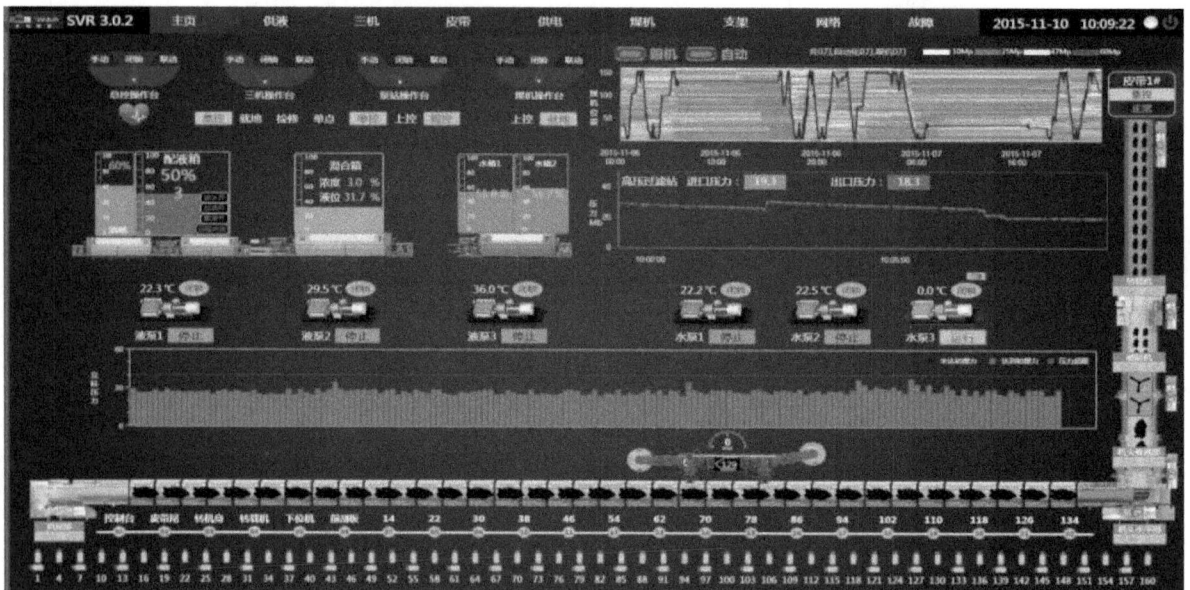

FIG 7 – Displaying face activities at the Console.

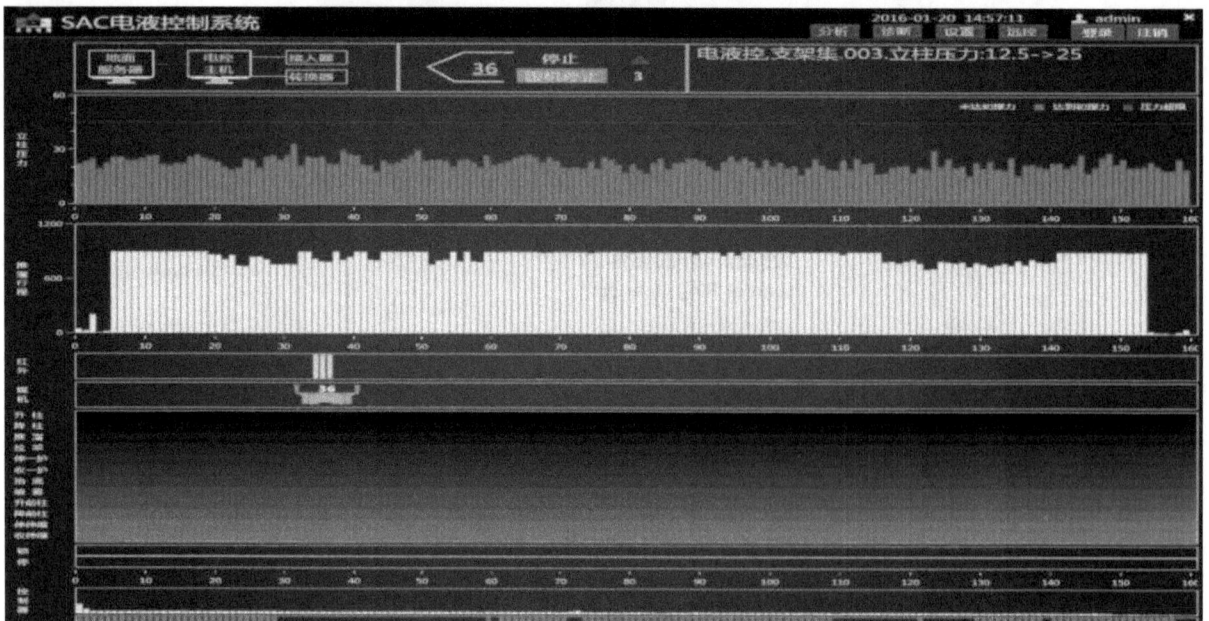

FIG 8 – Shear's cutting activities are displayed at the Console.

Surface Control Center and 3D Real-time Image Display

All activities are controlled by the operators on the Surface Command Center (Figure 9), all operational commands are sent to the shearer, electrohydraulic system, hydraulic oil pumping stations through a 10-GB Ethernet system, all operational data and commands are performed and displayed in real-time. Streaming media is used to reduce the load to the looped network bandwidth at the mine while accessing cameras at the face, making it possible to transmit AV data between underground and surface Command Center for decision-making, the system can also be accessed using a mobile phone. Three-dimensional virtual reality imagines of respective equipment (shearer, AFC and hydraulic supports) are displayed in real-time. All production data, including cutting, transporting and support movement can all be displayed clearly (Figure 10).

Ventilation Modeling and Analysis

To provide adequate ventilation in workings all airways is key to all underground operations, this is further enhanced through the use of Intelligent ventilation and safety management subsystem, which can either perform basic ventilation data analysis or targeted general calculation, or both (Figure 11).

An essential feature of the ventilation monitoring and control system is its ease of use with all important information displayed in real-time and in color. First, a network calculation is performed using real-time data and interactively updated based on a predetermined ventilation schematics map, calculated quantity for different airways are displayed in both tabulated form and block diagrams, or a "Ventilation Energy Consumption Diagram", commonly used in China. Each "block" (or rectangle) is the graphical depiction of a particular airway, with its width, height and block area representing air quantity, pressure drop and total energy consumption; this is designed for easy visual inspection and analysis. Corrective actions or changes based on changing airway resistance values, excessive pressure variations in specific areas, and individual airway energy consumption, can be made remotely on surface when necessary.

FIG 9 – Surface Control Room where operators can monitor extraction activities 24/7, can take over control of operations remotely, if necessary.

FIG 10 – A 3D real-time image showing shear cutting is displayed.

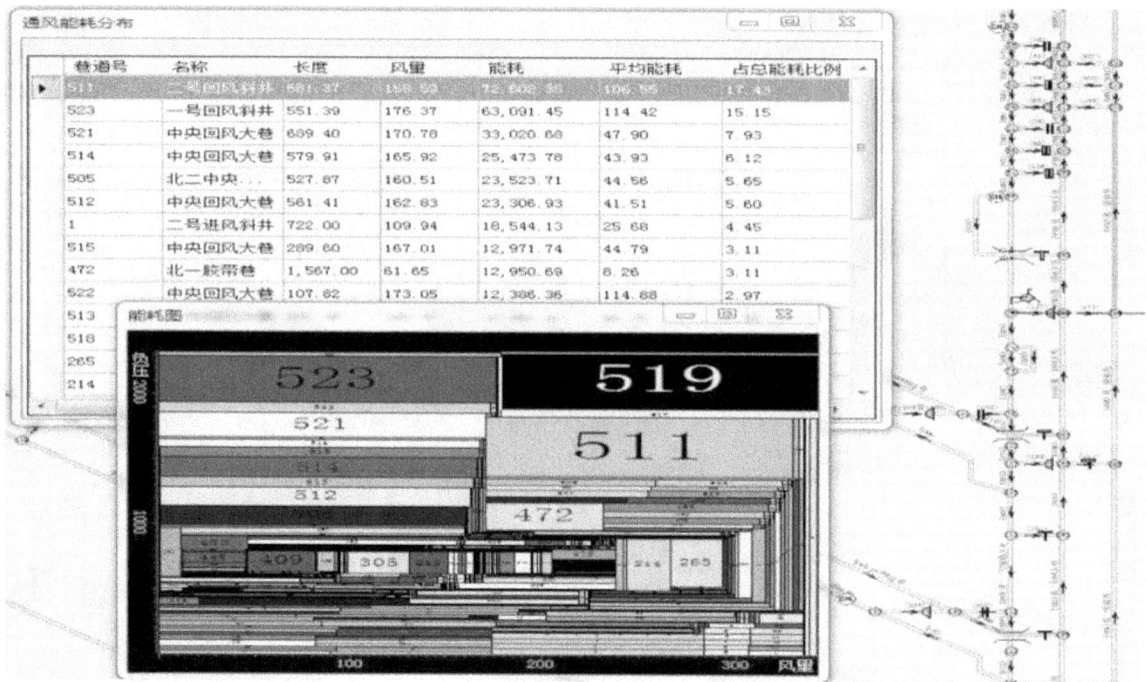

FIG 11 – Screen showing ventilation network schematics, real-time ventilation data in tabulated forms and block diagrams, or a "Ventilation Energy Consumption Diagram". Each "block" (or rectangle) is the graphical depiction of an individual airway, with its width, height and block area representing air quantity, pressure drop and total energy consumption. Blocks with different colors with color for easy identification.

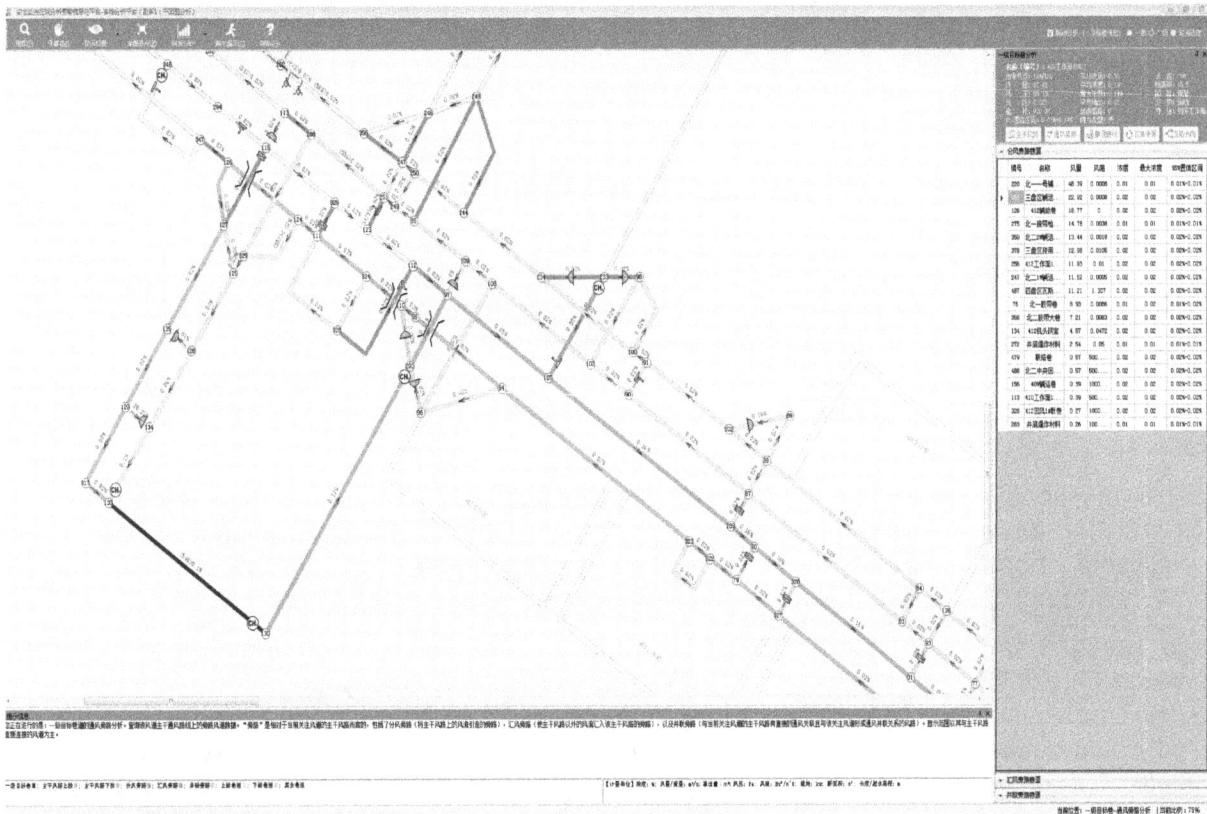

FIG 12 – Screen showing ventilation network schematics, real-time ventilation data in tabulated forms

Gas Monitoring Subsystem

Huangling Mine #1 is classified as a gassy coal operation, methane monitoring and control operation is a critical part of the ventilation management subsystem. The subsystem is capable of monitoring and discriminate "normal" and "abnormal" methane emission through real-time measurement. An early AV warning system with predetermined concentration range and color code based on real-time measured and calculated methane data has significantly facilitated monitoring and forecasting capabilities (Figure 13).

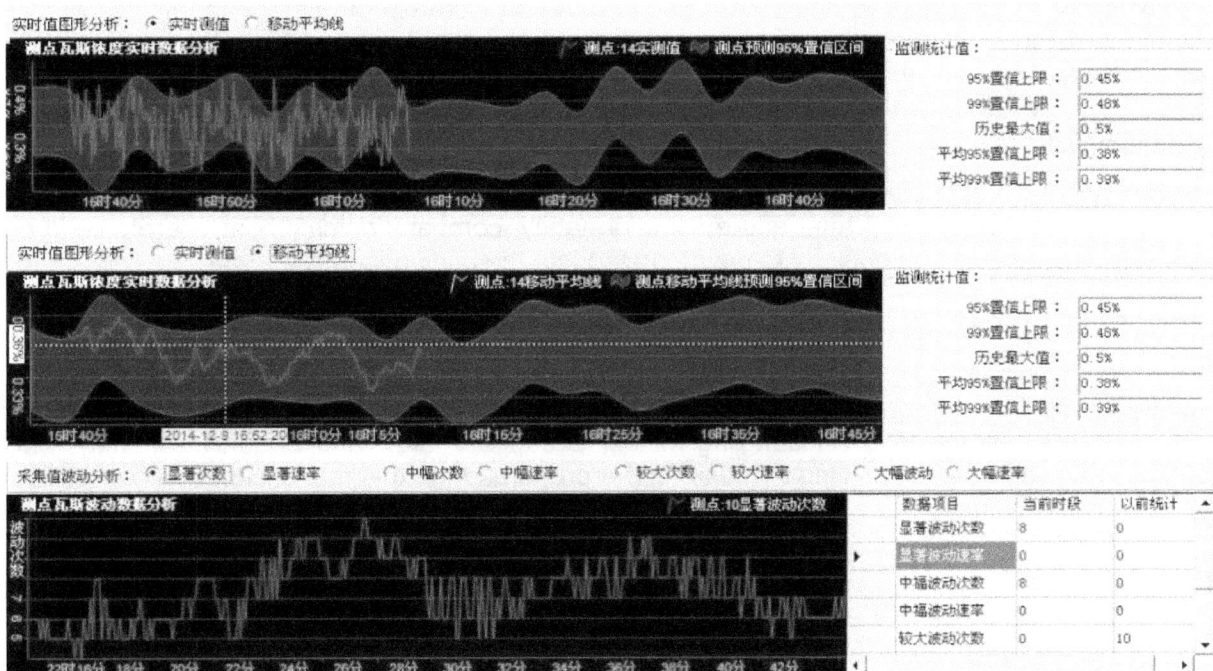

FIG 13 – Screen showing time-dependent methane concentration series.

The key to a successful program is to be able to provide sound technical support, both software and hardware, around the clock. The support team includes well-trained IT personnel as well as mining personnel with practical experience to provide seamless interactions to allow real-time information transmission, and to provide timely support in case of emergencies.

OPERATING COST COMPARISON: INTELLIGENT FACE VS. TRADITIONAL LONGWALL FACE

Despite the significant interest and application of autonomous technology, autonomous or unmanned mining is still relatively new in China. Much progress has been made but more needs to be done. Case study at Huangling suggests that, with proper/favorable geological conditions, mega mining companies with strong financial backing have been producing welcome results. As this type of operation is still new rapidly evolving in China, accurate production and financial data are lacking.

Results at Huangling show that the an unmanned mining face has been averaging 8.5 passes per 8-hr shift, producing 170,000 mt of raw coal per month. This doubles the amount of coal mining in a traditional longwall face with similar conditions, resulting in an additional coal production of 1 Mt annually or roughly an additional revenue of 280 M yuan (US$41.18 M2) per year.

Face workers have also been significantly reduced, from an original 9-man crew (3 shearer operators, 5 shield support operators, 1 AFC operator) to only one worker at the face, resulting in an 133 tonne/man. Due to automation, the number of shield operators at both the headgate and tailgate was halved, from original 8 on site to 4 operating off the face (2 each at the headgate and tailgate, 1 operator and 1 assistant operator.), resulting in an annual labor saving of 8 M yuan (US$1.18 M).

SUMMARY

Following the huge interest and application of mechanization and automation technology over the years in China, its mining industry has shown increasing interest to move toward intelligent, autonomous, or unmanned mining. Today, over 70 such longwall faces have been tried by several major mining companies in China, such as Shenhua, Yang-Kuang, and Shaanxi-Kuang, with varying degrees of success, and Shaanxi-Kuang is perhaps the most successful in this endeavor. Experience at Huangling suggests that, its effort as of this point has been concentrating mostly on how to make this system work, and as of today, the system has met its goal of developing a functional intelligent unmanned mining system.

To fully justify an intelligent system, there needs to be an increase of operational efficiency, or the ratio of the output gained relative to the input needed to run the operation by considering total cost, which includes capital, operational expenses and effort, while output consists revenue, product quality, quantity and growth. Huangling Mining has been the pioneer of this new technology and has obtained.

Preliminary results at Huangling suggest that given proper geological conditions, a strong technical and management support system, and most importantly, company commitment, there should be a promising future for this technology to be integrated into the mining industry.

ACKNOWLEDGEMENT

The authors would like to express their gratitude to the management of Shaanxi Coal and Chemical Industry Group Co., Ltd for their support of this project. Special thanks are due to Mr. Jindao Fan, former Chairman of the Board, Huangling Mining Group Col, Ltd. and Chuan Li, Manager, Mine Ventilation Department, Huangling #1 Mine, in the preparation of this manuscript and permission to publish project findings.

REFERENCES

Anon. (2017) "Status and Prospect of China's Coal Industry," Shaanxi Institute of Dev' and Reform, Xi'an, Shaanxi Prov., https://www.sohu.com/a/215735638_99917590

Anno. (2019a) "Coal Industry Annual Report, 2018," China Coal Mining Association.

---

2 Assuming an exchange rate of US$1 = 6.8 yuan)

Anno. (2019b) "China's 2018 Coal Usage Rises 1%, but Share of Energy Mix Falls," Reuters, Feb. 28, http://www.mining.com/web/chinas-2018-coal-usage-rises-1-share-energy-mix-falls/

Anon. (2019c) "World's Most Automated Underground Mine," Website: http://www.north-

parkes.com/improvement/worlds-most-automated-underground-mine

Anon. (201x) "Longwall Mining – Automation," University of Wollongong Underground Coal Website; http://www.undergroundcoal.com.au/longwall/automation.aspx

Anon. (201X) "Mining Safety and Automation," CSIRO Website: https://www.csiro.au/ en/Research/EF/Areas/Coal-mining/Mining-safety-and-automation

Anon. (2019) "Benefits of Increased Productivity Fuels Mining Automation Market," Transparency Market Research, SB Wire, April 2, http://www.sbwire.com/press-releases/mining-automation-market/release-1186415.htm

Hood, Mariowe (2019) "Push for More Coal Power in China Imperils Climate," Phys.org, March 28; https://phys.org/news/2019-03-coal-power-china-imperils-climate.html.

McAree, Ross (2018) "Mining Automation: Enhancing Precision," AusIMM Bulletin, December, https://www.ausimmbulletin.com/feature/mining-automation-enhancing-precision/

Personal Communication (2019) with Xintan Chang, Xi'an University of Science and Technology, April 15.

Heat and Refrigeration

Predicting annual underground thermal flywheel effects

M D Griffith[1] and C Stewart[2]

1. Senior Software Applications Engineer, Howden Australia, Brisbane QLD 4101. Email: martin.griffith@howden.com
2. Operations Director, Howden Australia, Brisbane QLD 4101. Email: craig.stewart@howden.com

ABSTRACT

The intake air for an underground mine varies daily with the rise and fall of the sun, and annually with the changing of the seasons. This study analyses the effect on the mine working environment of the annual variation of air intake temperature and humidity. A method is proposed to model the seasonal temperature variation in all parts of a cold climate mine and the results compared to a comprehensive data set of observed temperature. A Thermal Flywheel effect is observed, whereby temperature variations from the surface are damped and delayed deeper in the mine, due to the thermal capacitance of the rock strata.

INTRODUCTION

A key consideration in developing a mining ventilation system is managing heat. Heat is emitted into a mine from a range of sources, including rock strata heat and operating machinery. Air temperature also increases with depth, due to auto compression. Modelling and designing environmental conditions in a mine is important to ensure temperatures remain at safe and productive levels for all mine workers and machinery.

However in addition to identifying the heat sources, properly modelling their variation and effect can be difficult. For example, a truck operating underground will produce a variable amount of heat and moisture, depending on whether it is idling or in motion or whether driving up or down an incline. Another variable is the input temperature of the air from the atmosphere. The temperature and humidity of the air entering the mine will vary on daily and annual cycles. For hot climate mines, the mine is often designed on the basis of atmospheric temperatures equating to the 95% percentile or more of annual temperature daily maximums. For a cold climate mine, the mine modeller may need to consider the application of heaters at the air intakes during the winter.

Adding to the difficulty in considering the range of input temperatures to a mine is what is known as the thermal flywheel effect (Danko, Bahrami, Asante, Rostami, & Grymko, 2012). In an underground mine, the temperature of the rock strata responds to the temperature of the air. Therefore, when hot air enters a mine, the rock may absorb the heat; then when cooler air enters the mine, the hot rock may release the energy stored during the hotter part of the cycle. In this way the rock acts as a heat capacitor, affecting the temperature in an airway in two ways. Firstly, the temperature variations in the mine are reduced as compared to the atmospheric variation; secondly, a phase lag is present between the airway temperature and the atmospheric temperature, meaning the peak and trough of the airway temperature occur sometime after the peak and trough of the atmospheric temperature variation.

The variation of atmospheric temperatures and the thermal flywheel effect pose a challenge to the Gibson function method (Gibson, 1976) (McPherson, 1993) commonly used for rock-air heat transfer calculations for underground mines. An assumption of the Gibson function is that air at a constant temperature has been passing through the airway for the entirety of the time since the airway was mined. In the presence of annual and daily variations this assumption is not accurate. While it is possible to assume a mean value of air temperature, it is not necessarily a reasonable simplification of the dynamics of the system. Danko (Danko, Bahrami, Asante, Rostami, & Grymko, 2012) presented simulation results demonstrating the annual thermal flywheel effect on a single 6km long airway, comparing simulation results with and without accounting for the temperature variation history. The study found significant differences when comparing their results taking account of temperature history, with the Gibson function-based heat simulations of the Ventsim, VUMA, and CLIMSIM mining ventilation softwares.

This study goes beyond simulating a single airway and attempts to model the variation of heat throughout a complex mine model, accounting for seasonal variations of atmospheric temperatures. The model uses a modified Gibson function to calculate the rock temperatures as they respond to changing air intake temperatures and then stores and uses the results for subsequent periods of time. The results of the flywheel model are compared with a standard Gibson function-based heat simulation and validated against observed temperatures recorded at the mine upon which the model is based.

METHOD

The numerical mine model tested in this study was built and simulated using VentSim™ Design (Version 5.2). To simulate the air flows and pressures in the mine network, the software uses a network airflow solver based on the gradient method of Todini & Pilati, 1987. Thermodynamic simulations are performed using a discrete cell calculation method, where airways are divided into smaller segments, allowing heat effects to be calculated using the Gibson algorithm for that portion and transported to the next segment (Stewart, 2014). The heat simulation incorporates heat exchange between the air and the rock strata, heat from air auto-compression and heat from pre-configured sources, such as diesel and electric engines and refrigeration units.

The heat simulation uses rock strata heat transfer based on a calculation of the rock surface temperature. The rock surface temperature is calculated from a dimensionless temperature gradient, G, which can be obtained from Gibson's algorithm (McPherson, 1993). In summary, the rock surface temperature is a function of the virgin rock temperature, the temperature and flow quantity of the air in the airway, the age and size of the airway and the rock material properties, comprising the density, specific heat, heat conductivity and thermal diffusivity. Underlying the algorithm is the problematic assumption of constant airway temperature over the airway's history.

In order to model the heat taking into account the variation over an annual cycle, there are a number of potential approaches to resolving the problem.

- Method 1: A traditional approach using a single heat simulation could be run, with mean, maximum or 95th percentile atmospheric temperature inputs to produce an average result that may satisfy conditions most of the time. Or

- Method 2: a series of single heat simulations could be run for each month of the year, adjusting the atmospheric temperatures accordingly. This will create a model of the heat variation, however, each simulation will calculate rock surface temperatures based on an assumption of constant atmospheric air conditions.

- Method 3: A third flywheel option is presented in this paper, where the history of the temperature variation is utilised in each subsequent simulation by means of a modified Gibson's function. In this method, a heat simulation is run, where the age of the rock is made equal to one month, and the virgin rock temperatures set to the current rock skin temperatures determined from an initial heat simulation run with the mean values of the intake temperatures. For each subsequent simulation, the age of the rock is kept the same, the virgin rock temperatures are updated according to the new rock skin temperatures and the intake air temperature is changed according to the seasonal variation. This means each step of the flywheel cycle is affected by both the changing intake temperatures and the rock temperature changes of previous steps.

The proposed method 3 may not be valid for a full range of conditions as the Gibson algorithm can return a substantial difference between virgin rock temperature and rock surface temperature after one month. Where virgin rock temperature is very high compared to air temperature for example, the result should be treated with caution. A further problem is the dependence on the period interval chosen; choosing a period interval of less than one month will produce a smaller difference between virgin and surface rock temperatures, which will in turn affect the results. The current method of selecting one month intervals was determined on an adhoc basis using comparison of the method with published results and with experimental observation, as outlined later in this paper. In addition, models that are subject to strong natural ventilation pressures may exhibit large variations in heat and airflows from one heat simulation to the next, even with constant air intake temperatures. A model should be tested for such conditions before considering results of the flywheel simulation.

The methods were tested against real mine data (mine name supressed for confidentiality). Figure 1 shows an image of the mine's ventilation network. The mine in question is a cold climate mine; the daily mean peak summer dry bulb and wet bulb temperatures were 18.1° and 15.0° Celsius and daily mean trough winter dry bulb and wet bulb temperatures were -12.9° and -13.5° Celsius. Mean annual dry bulb and wet bulb temperatures were 4.5° and 2.7° Celsius. During the winter months, the air is heated to 3° Celsius at the intakes, to prevent icing. The mine is serviced by an exhaust and intake shaft situated close to one another and ventilated by two exhaust fans providing 650 cubic metres per second of airflow quantity.

At the sensor locations, airflow, dry and wet bulb temperature and pressure are recorded every 3 hours for the entirety of the year.

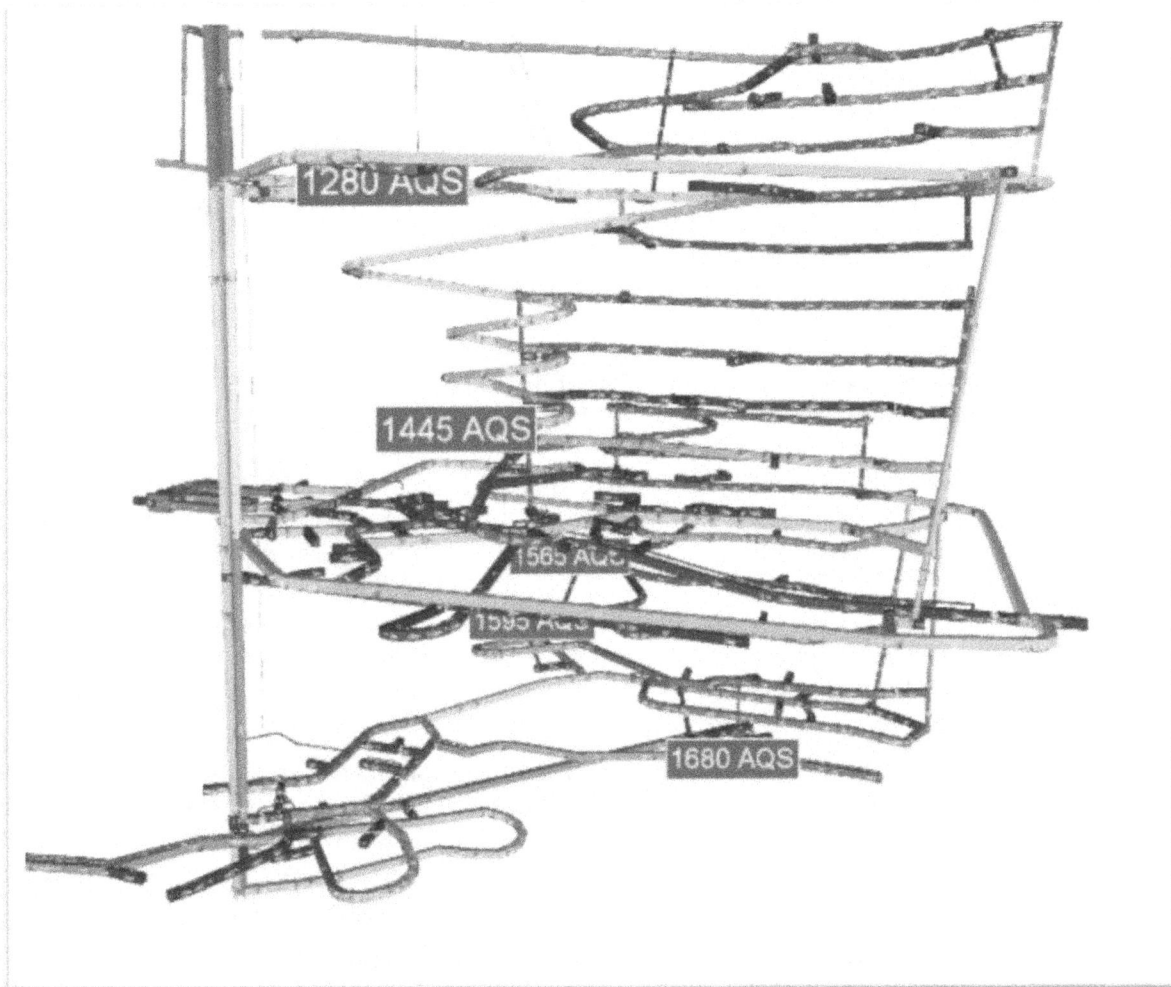

Figure 1 - Ventilation model showing the relevant area of interest.
Airways with sensors are denoted by purple tags.

RESULTS

Comparing the simulation techniques

Figure 2 plots the hourly variation throughout the year of dry bulb and wet bulb temperatures for our case study. The plot gives an idea of the daily and the annual variation of input temperatures to the mine, and of the various maximums and minimums that are smoothed out by the monthly mean. In the mine examined for this case study, heating was applied at the intake by use of a burner, to maintain a minimum inlet temperature of 3°C. Therefore in our model a lower threshold of 3° C was applied to the inlet dry bulb temperature. For the cases where the threshold was applied, the wet bulb temperature was deduced from the existing humidity of the atmospheric air and the assumption of 100% sensible heat from the additional inlet heat source. For burners operating in these temperature conditions, the majority of the water vapour contained in the combustion products will condense, hence allowing this assumption (McPherson, 1993). The resultant variation of inlet dry bulb and wet bulb temperatures can be seen in the later figures 3 and 4.

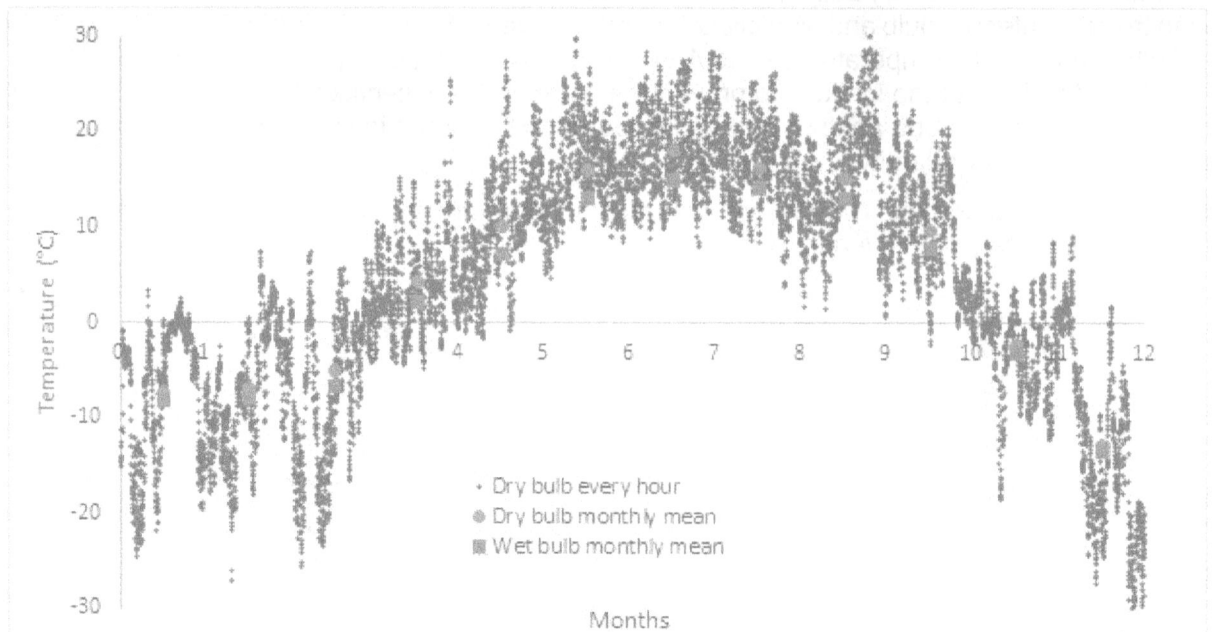

Figure 2 - Daily variation of atmospheric dry bulb temperature throughout the year at the mine site with the monthly mean dry bulb and wet bulb temperatures.

The inlet temperature variations can be applied to the three simulation techniques discussed earlier. Figure 3 plots the heat variation in five locations in the mine, for constant mean temperature input, variable inputs using the traditional Gibson algorithm and the modified flywheel method described earlier. The five airways chosen are at locations progressively further (in terms of shortest air path from the intake) into the mine ((a) 1467 m, (b) 1841 m, (c) 2133 m, (d) 2404 m and (e) 2498 m). The model provided was not a highly calibrated thermodynamic model, therefore some liberties were taking with regard to assumptions of airways age exposure (average 5 years) and rock properties in all tested methods.

Method 1 results (labelled '95th percentile') are obtained from a single heat simulation, with intake air temperatures of 21.8° Celsius dry bulb and 17.5° Celsius wet bulb temperature. These temperatures are at the 95th percentile of the temperatures over the whole year. Taking the 95th percentile temperatures is a common method used to model the mine at the extreme end of the seasonal temperature variation.

Method 2 results (labelled 'Monthly mean') are obtained from 12 separate heat simulations, with the intake temperatures calculated as the monthly mean from the annual temperature variation, which is also plotted on the graphs. Each of these heat simulations is also run using the Gibson algorithm to calculate the rock surface temperature used in the rock strata heat transfer equations. The same assumption regarding constant application of the intake temperatures during the life of the mine applies. The heat simulation run at the peak summer temperatures of July assumes those intake temperatures have been run for the life of the mine. Accordingly, the variation of the temperatures in the selected airways closely follows the variation of the intake temperatures. In each case, the airway temperatures peak at the same time as the intake temperatures. The peak-to-peak dry bulb temperature differences for the five airways are 14.6°, 13.8°, 12.22°, 12.46° and 11.21 Celsius. The reduction of the peak-to-peak temperature difference, or the damping of the temperature variation, is consistent with the increasing dominance of rock heat transfer in relation to the intake temperatures the deeper the airway in the mine and the more rock surface area the airflow must pass.

It should be noted that the temperatures at the 95th percentile exceed the peak monthly mean temperature of July. Accordingly the results of method 1 return temperatures exceeding all the temperatures of method 2. The technique of using the 95th percentile intake temperatures to model the mine heat is intended to focus the engineer on designing the mine ventilation to handle the hotter parts of the year. Figure 3 gives a useful indication of how such a simplification affects the usefulness of method 1 for a whole-of-year mine model. The differences between the 95th percentile temperatures and the peak temperatures of the monthly mean decrease with increasing distance from the air intake (i.e. from figure 3(a) to 3(e)). This reflects the increasing dominance of the rock heat transfer over the intake temperatures in influencing the airway temperatures; the greater the surface area of rock the air must pass, the less the temperature will depend on the intake temperatures. In a much larger mine, this effect would be more noticeable. The mine focused on in this study is small enough that even the furthest airways from the air intake still exhibit a strong dependence on intake temperatures.

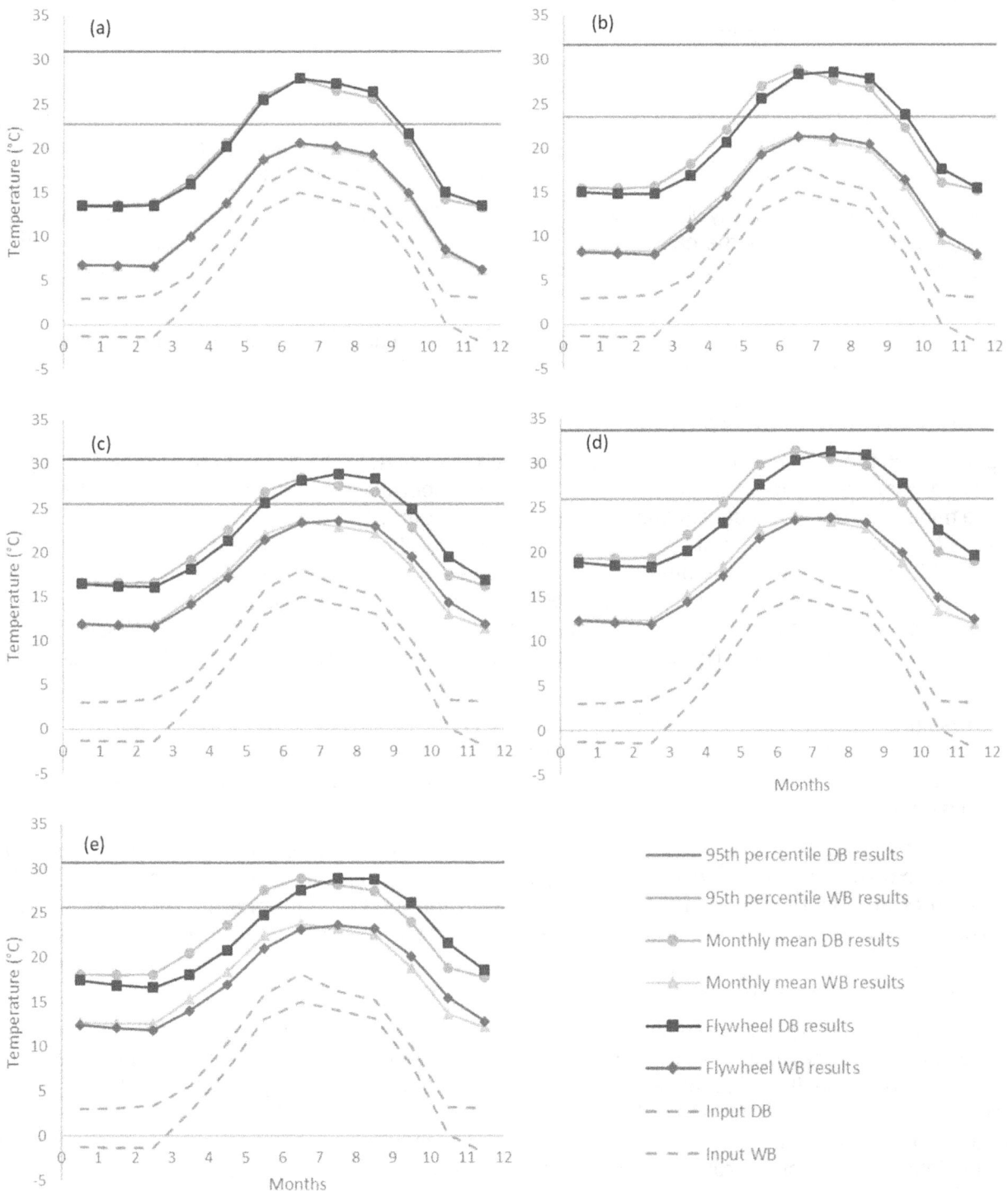

Figure 3 - Annual variation of temperatures in 5 different airways, showing the atmospheric input temperatures (with heating) and the temperature results using the Gibson method with constant temperatures at the 95th percentile of annual temperatures, at temperatures varying according to the monthly mean intake temperatures, and results using the flywheel method; airways are located at distances from the intake of (a) 1467 m, (b) 1841 m, (c) 2133 m, (d) 2404 m and (e) 2498 m.

Method 3 results (labelled 'Flywheel') are obtained using the flywheel method described earlier. The airway of figure 1(a) is located near the bottom of the intake shaft and the results of the flywheel simulation do not vary greatly from the results of the standard heat simulations. At each airway further into the mine, the results begin to progressively differ more. The peak-to-peak temperature differences do not change significantly between the Gibson monthly mean simulations and the flywheel simulations. However, a noticeable phase lag is increasingly apparent the further in the mine the airway is located.

The five airways plotted in figure 3 show phase lags with respect to the intake temperature of 0.1, 1.5, 1.6, 1.7 and 1.9 months; this is the amount of time the peak airway temperature occurs after the peak intake temperature occurs. This is the thermal flywheel effect, whereby the rock acts as a thermal capacitor, delaying the response of the airway temperatures to variations in the intake temperatures. The capacitance effect increases with the greater amount of rock exposure into the mine, so the phase lag should increase with the distance from the air intake.

Figure 3 shows the thermal flywheel effect can be modelled producing the characteristic phase lag. Of note also, or at least for the case discussed here, is that the results obtained using the standard Gibson function, with simulations run on the monthly mean intake temperatures, are still reasonable. Similar temperatures are obtained, with the only difference being the presence of the phase lag.

We have only examined the difference between the two simulation methods. In the next section, the flywheel method will be compared with observed data from the case study mine.

Comparing to measured data

Figure 4, for the same airways as figure 3, plots a comparison of the variation of temperatures obtained from the flywheel method with the observed temperatures taken from underground sensors in the mine. The observed data is taken at intervals of 3 hours. The first thing to note in the observed data is the approximate time of the peak temperature. In all cases, the period of peak temperature is flatter than what the simulations of figure 3 showed. Nonetheless, looking particularly at figure 4(e), a phase lag is observable in the measured data. The 'flat' period of peak observed temperatures is noticeably shifted to one side of the peak July intake temperature. The effect is diminished in the airways closer to the intake. This shows the presence of the thermal flywheel effect phase lag in the mine.

The results of the flywheel method shown in figure 4 compare well in some respects, and not so well in others. For example, the phase lag prediction is reasonably accurate; in all cases the peak temperature from the flywheel simulation is located in time close to the period of peak temperatures in the airway.

However, there is some inaccuracy in the prediction of the winter temperatures. There are several possible explanations of the differences. None of our simulation models take into account the variation of any heat sources in the mine, apart from the intake temperatures, rock heat and the winter-time intake burner. Other possible variations not taken into account are likely changes in fan usage throughout the year; changes in machinery usage; any variations in winter time heating and any variations in airway age not accounted for in the mine.

In terms of validating the flywheel method, the results show some success and some room for improvement. The recreation of the phase lag in the flywheel simulation results is encouraging. It should be noted that the mine model tested in this study is not large. It is possible that the phase lag effect could be better observed in observed data from a larger mine. The lack of accuracy in the winter temperatures is problematic for the flywheel simulation and the Gibson function based-simulations but may stem from inaccurate airway age and rock property selections unique to this model, or possible additional heat sources not being included in the model.

CONCLUSIONS

A method has been proposed for modelling the annual thermal flywheel effect, observable in mines subject to significant seasonal temperature variations. The method was run and compared to a series of standard heat simulations and to observed data from the mine the model is based on. The results from the flywheel simulation method show the characteristic phase lag of the flywheel effect. Comparison with the observed data from the mine shows a reasonable match with the data; notably the phase lag is present for deeper airways, however, the winter-time variation of temperatures was not accurately modelled. The lack of winter-time accuracy was also present in the simulations run with the standard Gibson function-based heat simulations, so it is likely the discrepancy is due to some other modelling inaccuracy.

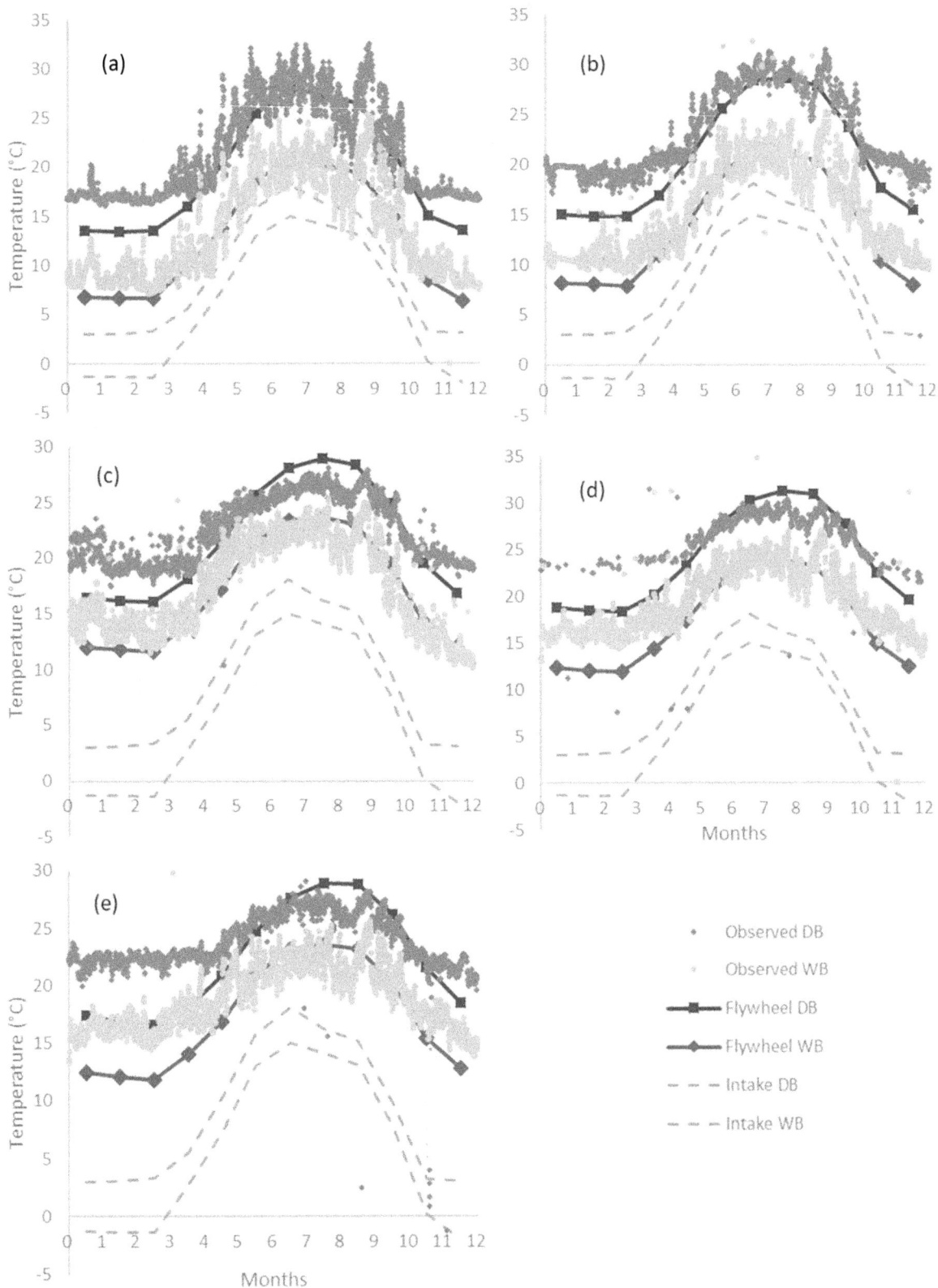

Figure 4 - The variation of temperature in the same airways as Figure 3, comparing the observed data to the data from the flywheel simulation. The intake temperatures are included for reference.

As a first attempt at modelling the thermal flywheel effect in an entire moderately-complex mine model, the method shows some promise. Work is ongoing on improving the method, especially making it more robust for extreme conditions and more consistent across different simulation intervals. It is hoped more large data sets such as shown here may be made available from other mines to further improve the method.

REFERENCES

Danko, G., Bahrami, D., Asante, W. K., Rostami, P., & Grymko, R. (2012). Temperature Variations In Underground Tunnels. *14th United State/North American Mine Ventilation Symposium* (pp. 365-373). 2012: Calizaya & Nelson.

Gibson, K. L. (1976). *The computer simulation of climatic conditions in underground mines.* University of Nottingham: Ph.D. Thesis.

McPherson, M. J. (1993). *Subsurface Ventilation and Environmental Engineering.* Chapman & Hall.

Stewart, C. M. (2014). Practical prediction of blast fume clearance and workplace re-entry times in development headings. *10th International Mine Ventilation Congress*, (pp. 169-176). Sun City.

Todini, E., & Pilati, S. (1987). A gradient method for the analysis of pipe networks. *International Conference on Computer Applications for Water Supply and Distribution.* Leicester, UK.

Effective temperature: is it an effective heat stress management index?

M A Tuck[1] and B Belle[2]

1. Associate Professor, Federation University Australia, Ballarat VIC 3353.
   m.tuck@federation.edu.au
2. Principal Ventilation and Gas Manager, Anglo American Metallurgical Coal, Brisbane QLD 7000.
   bharath.belle@angloamerican.com; UNSW

ABSTRACT

This paper reports on a recent re-examination on use of the basic effective temperature heat stress index in the heat trigger action response plan (TARP) values in coal mines. The paper examines the methodology of calculating normal and basic effective temperature and compares the results against an equation used to develop the UK Health and Safety Executive (HSE) nomogram for Basic Effective Temperature. In addition, the calculated results are also compared against the simple wet-bulb temperature, air cooling power and Thermal Work Limit indices using operational data. As a result of these investigations, a number of flaws in the Effective Temperature method are revealed and allow the question to be asked 'is Effective Temperature an effective heat stress index?' The authors implore the coal mine ventilation engineering community and coal mine regulators to re-visit the current legislative requirements on the use of Effective Temperature parameter against a more practical and proven heat stress parameter such as wet bulb temperature (WBT) as an initial assessment tool which does not require additional calculations.

INTRODUCTION

Heat stress continues to be a safety concern in underground mining. This is due to a number of reasons including increasing depths of mining, increased levels of heat generated from machines, ventilation systems being constrained from providing suitable levels of airflow and others. In order to assess if a thermal environment is safe for miners under normal or modified work arrangements, or to specify stop work conditions a heat stress index is used to guide decisions. In Queensland the Coal Mining regulations specify that Effective Temperature should be used as the heat stress index to make these assessments.

The paper examines the methodology of calculating normal and basic effective temperature and compares the results against an equation used to develop the UK Health and Safety Executive (HSE) nomogram for Basic Effective Temperature. In addition, the calculated results are also compared against the simple wet-bulb temperature, air cooling power and Thermal Work Limit indices using operational data. As a result of these investigations, a number of flaws in the Effective Temperature method are revealed and allow the question to be asked 'is Effective Temperature an effective heat transfer index?'

HEAT STRESS

The body can be regarded as a human heat engine. Heat balance is achieved when the rate of producing heat (the metabolic heat production rate) is equal to the rate at which the body can reject heat mainly through radiation, convection and evaporation. Heat exchange between the lungs and the air inhaled and exhaled is normally less than 5% of the total and therefore usually ignored. Any heat not rejected to the surroundings will cause an increase in body core temperature.

Heat stress is related to the balance between the body and the surrounding thermal environment, the main parameters required to be known when determining acceptable conditions are those associated with the heat production and transfer mechanisms. The main parameters Stewart (1982) involved are:

- Metabolic heat production rates

- Skin surface area (As) (and effects of clothing)

- Dry bulb temperature (t_{db})

- Radiant temperature (t_r)

- Air velocity (V)

- Air pressure (P)

- Air vapour pressure (e_s)

The rate of heat transfer to or from the environment depends on the equilibrium skin temperature t_s and the sweat rate S_r. These in turn depend on the response of the body to the imposed heat stress and the effect of thermoregulation Stewart (1982).

HEAT STRESS INDICES

According to McPherson (1993) heat stress indices can be classified into three types:

1. Single measurements e.g. wet bulb temperature, Kata thermometer and others.

2. Empirical methods such as Effective temperature (Basic and Normal), Corrected Effective Temperature, Kata thermometer, Wet bulb temperature, Wet Bulb Globe temperature and others.

3. Rational indices based on the rational heat balance equation e.g. Air Cooling Power, Thermal work limit and others

EFFECTIVE TEMPERATURE

As stated by Brake (2002)

'The other heat stress index that was critical to many subsequent developments in the field of heat stress and comfort in the 20th Century is the Effective Temperature (ET). This was originally derived in 1923 by ASHVE (the American Society of Heating and Ventilation Engineers, the predecessor to ASHRAE) in private research that was investigating comfort conditions Houghton and Yagloglou (1923). In the early 1900s, the vapour-compression refrigeration system had been invented and found immediate application in movie theatres, where windows could not be opened due to the ingress of light. This created significant heat problems in summer and a series of studies was undertaken to establish the comfort conditions for theatre audiences. A series of rooms each with different temperature, humidity and wind speed was set up, and three near-nude males were asked to move between rooms, giving their instantaneous perception as to which condition was hotter. From this, a series of nomograms were developed for "lines of equal comfort". Even though it was never intended for such a purpose, ET was soon introduced as a heat *stress* index, due to the non-availability of other stress indices at the time. Further experiments by ASHVE were then completed with three subjects wearing normal "office clothing", which lead to a separate nomogram called the "normal" effective temperature scale (the earlier scale then being designated as the "basic" scale). Shortly before World War 2, Bedford (1946) proposed that the same scales be used but with the globe temperature instead of the dry bulb temperature, resulting in the "corrected" effective temperature (CET) scales (in both "basic" and "normal" versions). CET therefore allowed radiant heat loads to be incorporated into the original ET formulations'.

Effective temperature in all forms is a subjective index based on the instantaneous sensations of subjects who were moved between rooms. The idea was to determine combinations of dry bulb temperature, wet bulb temperature and air movement, which gave the same thermal sensation as air conditions in which the air movement was minimal. Thus the effective temperature of an environment is defined as the temperature of a saturated environment with no air movement, which would produce the same instantaneous thermal sensation as the environment under consideration.

BET has been used in mines worldwide for a number of years. The index takes into account the dry and wet bulb temperature of the air and the wind speed .It can be calculated from formulae or read from empirically constructed nomograms. There are two scales, the Basic or American effective temperature was developed for nude men, while the normal scale is applicable to lightly dressed men.

The Basic effective temperature can be calculated using the following equation for air velocities in the range 0.5 - 3.5 m/s Pickering and Tuck (1997):

Basic Effective temperature = BET°C

Dry bulb = td°C

Wet bulb = tw°C

V = air velocity m/s

Works for range of velocities from 0.5 to 3.5 m/s, if V > 3.5 m/s assume V = 3.5 m/s, if V < 0.5 m/s assume V = 0.5 m/s.

$$BET = \frac{4 \times (4.12 - x1) + x2}{1.65176} + 20$$

Where

$$x1 = \frac{8.33 \times (17 \times x3 - (x3 - 1.35) \times (tw - 20))}{(x3 - 1.35) \times (td - tw) + 141.6}$$

$$x2 = \frac{17 \times ((td - tw) \times x3 + 8.33 \times (tw - 20)}{(x3 - 1.35) \times (td - tw) + 141.6}$$

$$x3 = 5.27 + 1.3 \times V - 1.15 \exp(-2V)$$

The Basic effective temperature index is used widely in the UK and Europe. It forms the basis of legislation in Germany where the working shift is reduced if the effective temperature exceeds 28°C. 28°C effective is a commonly applied design condition applied to mine workings.

Effective temperature satisfies the practical requirements of the mining industry, however it does have its limitations, and these will be discussed later in this paper.

THE REQUIREMENTS OF QUEENSLAND LEGISLATION

According to the Queensland Coal Mining Safety and Health Regulation 2017, the following apply:

364 Effective temperature at coal face

The ventilation officer for an underground mine must ensure the wet and dry bulb temperature, and the resultant effective temperature, of the atmosphere at each coal face where mining operations are in progress at the mine are measured and recorded as often as is necessary, having regard to the circumstances at the mine.

Division 2 Heat stress management

369 Managing risk from heat

(1) An underground mine's safety and health management system must provide for ensuring the health of persons in places at the mine in which—

(a) the wet bulb temperature exceeds 27°C; and

(b) persons work or travel.

(2) In developing the part of the safety and health management system mentioned in subsection (1) (the heat stress management provisions), the site senior executive for the mine must—

(a) have regard to any criteria stated in a recognised standard for managing heat; and

(b) comply with section 10, other than section 10(1)(a) and (d)(ii)(C), as if a reference in the section to a standard operating procedure were a reference to the heat stress management provisions.

(3) A person must not work in a place at the mine where the effective temperature exceeds 29.4°C unless the person is—

(a) carrying out the work in an escape or emergency; or

(b) engaged in work designed to reduce the effective temperature; or

(c) a mines rescue member carrying out training or emergency response under procedures developed by an accredited corporation; or

(d) wearing self-contained breathing apparatus and undertaking an emergency response under a standard operating procedure for the mine; or

(e) an ERZ controller carrying out an inspection—

(i) for which a risk assessment has been undertaken to identify the hazards associated with the inspection; and

(ii) under the controls agreed between the ERZ controller and the mine's underground mine manager to manage the risk.

(4) Subsection (3)(e) does not apply to an inspection included in a schedule of inspections mentioned in section 309(4).

370 Calculating effective temperature

An underground mine's safety and health management system must provide for the way of calculating the effective temperature of the atmosphere at the mine.

In summary a heat issue exists under the regulations at a basic effective temperature above 27°C with a stop work condition at a BET of 29.4°C under Queensland legislation. This is similar at the lower end to UK and German experience as described by Webber, Franz, Marx and Schutte (2003) but differs with respect to the stop work condition. No definition of why 29.4°C is specified as a stop work condition is given nor can be found from Queensland government or other sources.

Section 370 of the regulations is also of interest as it does not specify which version of Effective Temperature needs to be applied i.e. Basic, Normal or Corrected. Therefore, ambiguity exist at operations on which formulas were used in the derivation of guided values, let alone their shortcomings.

COMPARISON OF MINE SITE ET CALCULATORS WITH FORMULA CALCULATION

As part of the investigation, three versions of basic effective temperature (BET) calculators (basically .xls sheets) used by three different operations and another version of NET calculator used by a fourth operation were evaluated. These were compared to the BET equations originally specified by Pickering and Tuck (1997). The following observations are made:

- The calculators for the normal scale are based on curve fitting to the original graph/nomogram or BET.

- The results for BET compare well overall with the values calculated from the equations with only minimal difference of less than 0.1 degrees. The only time this differs is in one or two cases where the curve fit equation changes within the data set and the differences in this case are still less than 0.3°C.

- The Normal effective temperature calculator consistently calculates higher values than basic effective temperature as would be expected as this is for clothed individuals. At the inflection point above 35°C ET the NET results are lower than for BET which reflects the graphs for both BET and Normal ET.

- Therefore, it can be reasoned why for the same wet bulb temperature and velocity conditions different mines using the normal and basic scale report and action at different Effective temperatures.

ADVANTAGES AND SHORTCOMINGS OF THE BASIC EFFECTIVE TEMPERATURE INDEX

Basic effective temperature is a simple empirical index needing three inputs to evaluate it, these being:

- Air dry bulb temperature
- Air wet bulb temperature
- Air velocity

It can be simply determined using nomograms such as can be found at (http://www.hse.gov.uk/mining/effectivetempchart.pdf)

However it does have disadvantages:

As described by Brake (2002) ET in all its forms was found to have significant shortcomings particularly for moderate to hard work. Leithhead and Lind (1964) comment:

- …the {ET} scales do not give sufficient weight to the deleterious effects of low air movements in hot and humid conditions.

- …climates of similar severity as judged by rectal temperature, pulse rate, weight loss or tolerance times do not have corresponding values of Effective Temperature.

- …it is clear that there is an inherent error in the construction of the {ET} scales if they are to be used as an index of physiological effect, an error that increases as the severity of environmental conditions increase. This is hardly surprising since the scales were originally devised from instantaneous appreciation and comparison of the sensory warmth of different climates; thus if the error is concerned with, for example, changed physiological circumstances … in which the skin becomes fully wetted, the method by which the scales were devised … is unsuitable for this type of evaluation".

Parsons (1993) also comments on the ET scale as follows:

- "A comprehensive series of studies was conducted on behalf of ASHVE in their Pittsburgh laboratory, USA, which led to the influential effective temperature (ET) index. Incredibly, the studies contained a fundamental experimental error. Three subjects were used and each walked between two chambers and compared different combinations of air temperature and humidity in terms of subjective impressions of warmth. However, subjects gave their immediate impressions which would be largely determined by the effects of transient absorption and evaporation of moisture from skin and clothing. The data may therefore be useful for studies of transient effects but overestimates the effects of humidity when considering steady state conditions which was the aim of the study."'

Brake (2002) also states:

'ET was therefore generally superseded as an index of occupational heat stress by the WBGT from the late 1950s, and ET of all forms was subsequently also abandoned as a comfort index by ASHRAE in 1961.'

Additionally:

- Never really intended for use as a heat stress index
- Developed for a different situation than for mining
- Lacks a work rate component
- Applicable only in the 0.5 to 3.5 m/s air velocity range
- The BET basic effective temperature index does not take into account radiant heat, this can be overcome by replacing the dry bulb measurement with a globe temperature measurement i.e. use the Corrected effective temperature. However this may not overcome the issue.
- Issues at low velocity and low humidities

- Clothing requirements for Queensland coal mining differ from those assumed by BET

BET has been applied to underground coal mining in the UK and Germany and has been used to ensure worker health and safety in both of these countries. As such it can be viewed as a probable effective measure. The question to ask at this point is are conditions in underground coal mines in Queensland the same as those that were experienced in the UK and German mines?

As described in Belle and Biffi (2018) in most countries, WBT is typically used to quantify the heat stress and determine the acceptability of environmental conditions. It has been found by Haldane and several investigators that WBT is the most reliable single instrumental indicator to evaluate the thermal conditions at the workplace (Hinsley, 1949). In Australia and UK, regulations often use effective temperatures that require the measurement of dry bulb temperature (DBT), WBT as well as air velocity conditions. While this paper does not delve into the appropriateness of any indicators, in almost all cases these heat indices are correlated in some form and assist in assessing heat stress. Whillier (1971) suggested that in working places where air velocity is low, additional cooling of workers can be achieved more effectively by increasing air velocity rather than by decreasing the WBT.

ALTERNATIVE HEAT STRESS INDICATORS

In looking at alternatives to Basic effective temperature the approach used by Graves and Graveling (1987) will be employed. This is illustrated in Figure 1 and involves:

1. Initial application of a simple index to determine the effects of hot conditions on workers

2. The simple index is to determine if a problem exists, if no problem exists then no more action is required

3. If the simple index indicates a problem exists/might exist then a more detailed assessment is required

4. The detailed assessment may involve the application of more complex index/indices or direct measurement of the worker(s)

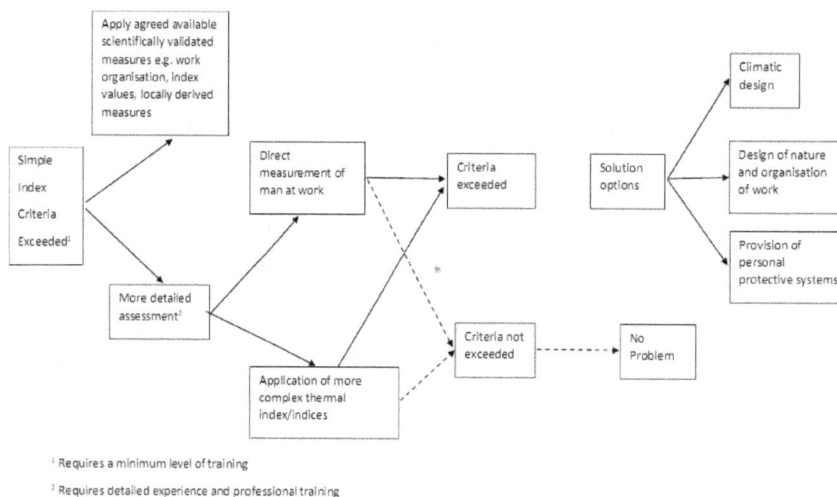

Figure 1 Approach to heat problems. After Graves and Graveling (1987)

Simple Index

The simple index investigated was the wet bulb temperature of the air. The reason why wet bulb temperature was selected was because in hot humid environments the majority of heat transfer is effected by evaporation of sweat where wet bulb is the most important parameter influencing the rate of sweat evaporation. It is recognised that many other factors also affect the human heat balance. Wet bulb is used around the world as an indicator of heat stress despite its limitations as detailed in Belle and Biffi (2018)

The following tables indicate the results of the analysis. The core of the table's shows calculated Basic effective temperatures for air velocities of 0.5, 1.5, 2.5 and 3.5 m/s. Then for defined wet bulb temperatures limits from 27°C to 32°C effective temperatures above 27°C are highlighted in yellow on the tables. Effective temperatures above the current Queensland stop work limit are highlighted in orange. Tables 1 to 4 show some interesting results:

1. For the range of air velocities encountered on coal faces during production wet bulb temperature can be used as a good indicator of heat stress. That is for air velocities above 2.5 m/s and wet bulb up to 30°C. Above a wet bulb of 30°C it would be advisable to use a rational heat stress index to determine the heat stress potential given that other variables are also involved.

2. For longwall development work where are velocities are lower due to auxiliary ventilation being supplied wet bulb temperature is not as good at indicating a heat issue but can still be applied

3. In many circumstances in underground coal mines the wet bulb depression or difference between dry and wet bulb temperature will be in the range of 2 to 4°C depending on the location. Given this using wet bulb temperature as a simple index to specify hot work, modified hot work and stop work conditions is appropriate for air velocities above 1.5 m/s where 27°C wet bulb is used to define the onset of a heat issue for dry bulb temperatures up to 37°C.

4. Modified work conditions at wet bulb temperatures above each of 28, 29, 30 and 31°C are also appropriate. As is a stop work condition above 32°C.

5. Wet bulb is a useful indicator but should also be used in conjunction with a rational index to ensure worker health and safety.

Table 1: Basic Effective temperatures for dry/wet bulb temperatures in the range 25 to 40°C, for air velocity = 0.5 m/s. Limit wet bulb temperature = 27 to 32°C

		Dry Bulb Temperature °C														
		25	26	27	28	29	30	31	32	33	34	35	36	37	38	39	40
Wet Bulb Temperature °C	25	22.6															
	26	23.2	23.8														
	27	23.8	24.4	25.0													
	28	24.3	24.9	25.6	26.2												
	29	24.8	25.4	26.1	26.7	27.4											
	30	25.3	25.9	26.5	27.2	27.9	28.6										
	31	25.8	26.4	27.0	27.6	28.3	29.1	29.8									
	32	26.2	26.8	27.4	28.1	28.7	29.5	30.2	31.0								
	33	26.6	27.2	27.8	28.5	29.1	29.8	30.6	31.4	32.2							
	34	27.0	27.6	28.2	28.8	29.5	30.2	30.9	31.7	32.5	33.4						
	35	27.4	28.0	28.6	29.2	29.9	30.5	31.3	32.0	32.8	33.7	34.6					
	36	27.8	28.3	28.9	29.5	30.2	30.9	31.6	32.3	33.1	34.0	34.9	35.8				
	37	28.1	28.7	29.2	29.9	30.5	31.2	31.9	32.6	33.4	34.2	35.1	36.0	37.0			
	38	28.4	29.0	29.6	30.2	30.8	31.5	32.2	32.9	33.7	34.5	35.3	36.2	37.2	38.2		
	39	28.7	29.3	29.9	30.5	31.1	31.8	32.4	33.2	33.9	34.7	35.6	36.4	37.4	38.4	39.4	
	40	29.0	29.6	30.2	30.8	31.4	32.0	32.7	33.4	34.2	34.9	35.8	36.6	37.5	38.5	39.5	40.6

Table 2: Basic Effective temperatures for dry/wet bulb temperatures in the range 25 to 40°C, for air velocity = 1.5 m/s. Limit wet bulb temperature 27 to 32°C

		Dry Bulb Temperature °C															
		25	26	27	28	29	30	31	32	33	34	35	36	37	38	39	40
Wet Bulb Temperature °C	25	19.8															
	26	20.7	21.2														
	27	21.5	22.1	22.7													
	28	22.3	22.9	23.5	24.1												
	29	23.0	23.6	24.2	24.8	25.5											
	30	23.7	24.3	24.9	25.5	26.2	27.0										
	31	24.3	24.9	25.5	26.1	26.8	27.6	28.4									
	32	24.9	25.5	26.1	26.7	27.4	28.2	29.0	29.8								
	33	25.4	26.0	26.6	27.3	27.9	28.7	29.5	30.3	31.3							
	34	26.0	26.5	27.1	27.8	28.4	29.2	30.0	30.8	31.7	32.7						
	35	26.5	27.0	27.6	28.2	28.9	29.6	30.4	31.2	32.1	33.1	34.1					
	36	26.9	27.5	28.1	28.7	29.4	30.1	30.8	31.6	32.5	33.5	34.5	35.6				
	37	27.3	27.9	28.5	29.1	29.8	30.5	31.2	32.0	32.9	33.8	34.8	35.8	37.0			
	38	27.8	28.3	28.9	29.5	30.2	30.8	31.6	32.4	33.2	34.1	35.1	36.1	37.2	38.4		
	39	28.1	28.7	29.3	29.9	30.5	31.2	31.9	32.7	33.5	34.4	35.3	36.3	37.4	38.6	39.9	
	40	28.5	29.1	29.6	30.2	30.9	31.5	32.2	33.0	33.8	34.7	35.6	36.6	37.6	38.8	40.0	41.3

Table 3: Basic Effective temperatures for dry/wet bulb temperatures in the range 25 to 40°C, for air velocity = 2.5 m/s. Limit wet bulb temperature 27 to 32°C

		Dry Bulb Temperature °C															
		25	26	27	28	29	30	31	32	33	34	35	36	37	38	39	40
Wet Bulb Temperature °C	25	17.5															
	26	18.7	19.1														
	27	19.8	20.2	20.7													
	28	20.7	21.2	21.8	22.4												
	29	21.7	22.2	22.7	23.3	24.0											
	30	22.5	23.0	23.6	24.2	24.9	25.6										
	31	23.2	23.8	24.4	25.0	25.7	26.4	27.2									
	32	24.0	24.5	25.1	25.7	26.4	27.1	28.0	28.9								
	33	24.6	25.2	25.7	26.4	27.1	27.8	28.6	29.5	30.5							
	34	25.2	25.8	26.4	27.0	27.7	28.4	29.2	30.1	31.1	32.1						
	35	25.8	26.3	26.9	27.6	28.2	29.0	29.8	30.6	31.6	32.6	33.7					
	36	26.3	26.9	27.5	28.1	28.8	29.5	30.3	31.1	32.0	33.0	34.2	35.4				
	37	26.8	27.4	28.0	28.6	29.2	30.0	30.7	31.6	32.5	33.5	34.5	35.7	37.0			
	38	27.3	27.8	28.4	29.0	29.7	30.4	31.2	32.0	32.9	33.8	34.9	36.0	37.3	38.6		
	39	27.7	28.3	28.9	29.5	30.1	30.8	31.6	32.4	33.2	34.2	35.2	36.3	37.5	38.8	40.3	
	40	28.2	28.7	29.3	29.9	30.5	31.2	31.9	32.7	33.6	34.5	35.5	36.5	37.7	39.0	40.3	41.9

Table 4: Basic Effective temperatures for dry/wet bulb temperatures in the range 25 to 40°C, for air velocity = 3.5 m/s. Limit wet bulb temperature 27 to 32°C

		Dry Bulb Temperature °C															
		25	26	27	28	29	30	31	32	33	34	35	36	37	38	39	40
Wet Bulb Temperature °C	25	15.3															
	26	16.8	17.1														
	27	18.1	18.5	18.9													
	28	19.3	19.7	20.2	20.7												
	29	20.4	20.8	21.3	21.9	22.5											
	30	21.4	21.9	22.4	23.0	23.6	24.3										
	31	22.3	22.8	23.3	23.9	24.6	25.3	26.1									
	32	23.1	23.6	24.2	24.8	25.5	26.2	27.0	27.9								
	33	23.9	24.4	25.0	25.6	26.3	27.0	27.8	28.7	29.8							
	34	24.6	25.1	25.7	26.3	27.0	27.7	28.5	29.4	30.4	31.6						
	35	25.2	25.8	26.3	27.0	27.6	28.4	29.2	30.1	31.1	32.1	33.4					
	36	25.8	26.4	26.9	27.6	28.2	29.0	29.8	30.6	31.6	32.7	33.9	35.2				
	37	26.4	26.9	27.5	28.1	28.8	29.5	30.3	31.2	32.1	33.1	34.3	35.6	37.0			
	38	26.9	27.4	28.0	28.6	29.3	30.0	30.8	31.6	32.6	33.6	34.7	35.9	37.3	38.8		
	39	27.4	27.9	28.5	29.1	29.8	30.5	31.2	32.1	33.0	33.9	35.0	36.2	37.5	39.0	40.6	
	40	27.9	28.4	29.0	29.6	30.2	30.9	31.7	32.5	33.3	34.3	35.3	36.5	37.7	39.1	40.7	42.4

Heat stress index for detailed analysis (not day to day operations)

If the simple index indicates a more detailed analysis is required a rational index should be applied. Many of these exist and selection needs to be undertaken based on a logical basis with the needs of the mine accounted for. Use of Air Cooling Power (ACP) is appropriate, but others such as

Thermal Work Limit can be applied. Whichever is used there will be a need to account for local differences not accounted for by the algorithms used to calculate the index, such as clothing factors.

One key aspect with any heat stress index is that it must be able to be determined using simple measurements and calculation methodologies, especially if deputies or ventilation technicians are used to measure them and undertake calculations/evaluations underground. As such this study has focused on a simple Air Cooling Power index Wyndham (1974). It is a rational index and so is based on the metabolic heat balance given by:

M=Br+Rad+Con+Evap+Cond+Ac

Where

M = metabolic heat generation

Br = respiratory heat exchange

Rad = radiative heat transfer

Con= convective heat transfer

Evap = evaporative heat transfer

Cond = conductive heat transfer

Ac = heat storage/accumulation in the body

Under hot conditions found in mines the conductive component is very small as is the respiratory heat exchange, as such these are assumed to be zero. To ensure health and safety the core temperature of the body should not be allowed to rise therefore Ac in the equation is also zero. This means that the heat balance equation becomes:

M=Rad+Con+Evap

Equations, as given by Pickering and Tuck (1997), can be written for each of the radiative, convective and evaporative components to determine the right hand side of the equation which represents heat transfer to the environment or ventilation airflow, Wyndham (1974), Gibson (1976).

According to Wyndham (1974) a reasonable assumption is that in hot environments the skin temperature can be assumed to be 35°C and the radiative temperature is equal to the dry bulb temperature, this however may not be correct. It should also be noted that these equations are for very lightly clothed workers so would need to be modified by a clothing factor for Australian conditions. Applying these the same environmental parameters used to determine effective temperature need to be measured at the location. The only other parameter required is barometric pressure, note this can be determined from for example a calibrated VentSim model and does not need to be exactly known as it has only a small influence on the calculated values.

Tables 5 to 8 provide ACP values for the range of dry and wet bulb temperatures from 25°C to 40°C for velocities in 1.0 m/s increments from 0.5 m/s to 3.5 m/s for a barometric pressure of 100 kPa. The figures highlighted in yellow in this table are where the ACP is 175 W/m^2 or less. The figure of 175 W/m^2 is taken from Moreby (2002) and represents a medium work rate.

Table 5: Air cooling powers for the range of dry and wet bulb temperatures from 25°C to 40°C for velocity = 0.5 m/s. Yellow highlighted cells show where ACP <= 175W/m².

		Dry Bulb Temperature °C															
		25	26	27	28	29	30	31	32	33	34	35	36	37	38	39	40
Wet Bulb Temperature °C	25	345															
	26	342	316														
	27	338	312	286													
	28	335	309	282	254												
	29	331	305	279	251	222											
	30	328	302	275	247	218	188										
	31	324	298	271	244	215	185	153									
	32	320	295	268	240	211	181	150	117								
	33	317	291	264	236	207	177	146	113	79							
	34	313	287	260	232	203	173	142	109	76	40						
	35	309	283	257	229	200	170	138	106	72	37	0					
	36	306	280	253	225	196	166	134	102	68	33	-4	-42				
	37	302	276	249	221	192	162	131	98	64	29	-8	-46	-86			
	38	298	272	245	217	188	158	127	94	60	25	-12	-50	-90	-131		
	39	294	268	241	213	184	154	123	90	56	21	-16	-54	-94	-135	-178	
	40	290	264	237	209	180	150	119	86	52	17	-20	-58	-98	-139	-182	-226

Comparing Tables 5 to 8 with Tables 1 to 4 it is evident that the impact of large wet bulb depressions or high dry bulb temperatures impacts to a greater extent with effective temperature than it does with ACP as in Tables 1 to 4 the shading occupies much more of the bottom left hand side of the Tables. If clothing factors and others were taken into account in the calculation of ACP the likely impact is that there would be a shift of the yellow highlighted element of the tables towards the left hand side of the tables.

In a hot humid environment evaporation of sweat is the primary mechanism of body cooling, this to a large extent is controlled by the wet bulb temperature and also the air velocity. The ACP index appears to reflect this better than effective temperature and apart from requiring the atmospheric pressure as an additional input ACP requires the same environmental parameters to be measured as effective temperature does. This also lends weight to using wet bulb temperature alone as an indicator of a heat problem existing and to set initial modified work and stop work rules based purely on wet bulb temperature alone especially given the low wet bulb depression in coal mines.

Table 6: Air cooling powers for the range of dry and wet bulb temperatures from 25°C to 40°C for velocity = 1.5 m/s. Yellow highlighted cells show where ACP <= 175W/m².

		Dry Bulb Temperature °C															
		25	26	27	28	29	30	31	32	33	34	35	36	37	38	39	40
Wet Bulb Temperature °C	25	623															
	26	620	571														
	27	618	568	517													
	28	616	566	514	460												
	29	613	563	512	458	402											
	30	611	561	509	455	399	341										
	31	608	558	506	453	397	339	278									
	32	606	556	504	450	394	336	275	213								
	33	603	553	501	447	391	333	273	210	144							
	34	601	551	499	445	389	331	270	207	142	74						
	35	598	548	496	442	386	328	267	204	139	71	0					
	36	595	545	493	439	383	325	264	202	136	68	-3	-76				
	37	593	543	491	437	381	322	262	199	133	65	-6	-79	-156			
	38	590	540	488	434	378	320	259	196	130	62	-9	-82	-159	-239		
	39	588	538	485	431	375	317	256	193	127	59	-12	-85	-162	-242	-325	
	40	585	535	483	429	372	314	253	190	125	56	-15	-88	-165	-245	-328	-414

Table 7: Air cooling powers for the range of dry and wet bulb temperatures from 25°C to 40°C for velocity = 2.5 m/s. Yellow highlighted cells show where ACP <= 175W/m²

		Dry Bulb Temperature °C														
		25	26	27	28	29	30	31	32	33	34	35	36	37	38	39	40
Wet Bulb Temperature °C	25	829															
	26	828	760														
	27	826	758	688													
	28	824	757	686	613												
	29	823	755	685	612	536											
	30	821	753	683	610	534	455										
	31	819	752	681	608	532	453	371									
	32	818	750	679	606	530	451	369	283								
	33	816	748	678	604	528	449	367	282	193							
	34	814	746	676	603	526	447	365	280	191	98						
	35	813	745	674	601	525	445	363	278	189	96	0					
	36	811	743	672	599	523	443	361	276	187	94	-2	-102				
	37	809	741	670	597	521	442	359	274	185	92	-4	-104	-208			
	38	807	739	669	595	519	440	357	272	183	90	-6	-106	-211	-319		
	39	806	738	667	593	517	438	355	270	180	88	-8	-109	-213	-321	-434	
	40	804	736	665	591	515	436	353	268	178	86	-11	-111	-215	-323	-436	-553

Table 8: Air cooling powers for the range of dry and wet bulb temperatures from 25°C to 40°C for velocity = 3.5 m/s. Yellow highlighted cells show where ACP <= 175W/m².

		Dry Bulb Temperature °C														
		25	26	27	28	29	30	31	32	33	34	35	36	37	38	39	40
Wet Bulb Temperature °C	25	1004															
	26	1003	920														
	27	1002	919	833													
	28	1001	918	832	743												
	29	1000	917	831	742	649											
	30	999	916	830	741	648	551										
	31	998	915	829	740	647	550	449									
	32	997	914	828	738	645	549	448	344								
	33	996	913	827	737	644	547	447	342	234							
	34	995	912	826	736	643	546	446	341	232	119						
	35	994	911	825	735	642	545	444	340	231	118	0					
	36	993	910	824	734	641	544	443	338	229	116	-1	-124				
	37	992	909	823	733	640	543	442	337	228	115	-3	-125	-253			
	38	991	908	822	732	638	541	441	336	227	113	-4	-127	-254	-387		
	39	990	907	821	731	637	540	439	334	225	112	-6	-128	-256	-388	-526	
	40	989	906	819	730	636	539	438	333	224	111	-7	-130	-257	-390	-528	-671

Thus it is possible for a given mine to produce a chart of ACP for a panel or development of interest for specific atmospheric pressures and likely velocity range for a deputy to assess thermal environments underground.

Thermal Work Limit

A variation on air cooling power is provided by the thermal work limit (TWL) index. Again this is a rational index based on the metabolic balance equation like air cooling power. It does however account for some additional elements as described by Brake (2002) including clothing factors and radiant temperature, note radiant temperatures are measured by a globe thermometer.

Time for this project did not allow for the development of a separate calculation spreadsheet for TWL, so an existing web based calculator was used, this can be found at https://www.haad.ae/Safety-in-Heat/Default.aspx?tabid=63

With the calculator used it should be noted that it was not developed specifically for mining. As with the examination of Effective Temperature and ACP, TWL was investigated over the following conditions:

1. Atmospheric pressure 100 kPa

2. Radiant temperature = dry bulb temperature of the air

3. Air velocity range 0.5 to 3.5 m/s in 1.0 m/s increments

4. Dry and wet bulb temperature range 25 to 40°C in 1°C increments

The results for TWL are presented in tables 9 to 12. For Thermal work limit the values in W/m² are not listed just an indication of safe or not safe is given. Values with green shading have a TWL > 140 W/m², values with orange shading 115 ≤ TWL ≤ 140 W/m² and red shading TWL < 115 W/m², the colours are those used in the calculator.

Table 9: Thermal Work Limit for velocity = 0.5 m/s. Values with green shading have a TWL > 140 W/m², values with orange shading 115 ≤ TWL ≤ 140 W/m² and red shading TWL < 115 W/m²

Wet Bulb Temperature °C	Dry Bulb Temperature °C															
	25	26	27	28	29	30	31	32	33	34	35	36	37	38	39	40
25																
26																
27																
28																
29																
30																
31																
32																
33																
34																
35																
36																
37																
38																
39																
40																

DISCUSSION AND CONCLUSIONS

Inspection of tables 1 to 12 shows a number of interesting points:

1. As expected the values for simple ACP and TWL show a reasonable correlation, this is expected as they are both rational formulations of the heat balance equation with some differences due to clothing and other factors.

2. All three indices show a migration of modifies and stop work conditions to the right of the table as air velocity increases. This is to be expected as the cooling power is influenced by velocity. The effect increases markedly at initial increases in velocity and diminishes as further increases occur.

Table 10: Thermal Work Limit for velocity = 1.5 m/s. Values with green shading have a TWL > 140 W/m², values with orange shading 115 ≤ TWL ≤ 140 W/m² and red shading TWL < 115 W/m²

Wet Bulb Temperature °C	Dry Bulb Temperature °C															
	25	26	27	28	29	30	31	32	33	34	35	36	37	38	39	40
25																
26																
27																
28																
29																
30																
31																
32																
33																
34																
35																
36																
37																
38																
39																
40																

Table 11: Thermal Work Limit for velocity = 2.5 m/s. Values with green shading have a TWL > 140 W/m², values with orange shading 115 ≤ TWL ≤ 140 W/m² and red shading TWL < 115 W/m²

Wet Bulb Temperature °C	Dry Bulb Temperature °C															
	25	26	27	28	29	30	31	32	33	34	35	36	37	38	39	40
25																
26																
27																
28																
29																
30																
31																
32																
33																
34																
35																
36																
37																
38																
39																
40																

Table 12: Thermal Work Limit for velocity = 3.5 m/s. Values with green shading have a TWL > 140 W/m², values with orange shading 115 ≤ TWL ≤ 140 W/m² and red shading TWL < 115 W/m²

Wet Bulb Temperature °C	Dry Bulb Temperature °C															
	25	26	27	28	29	30	31	32	33	34	35	36	37	38	39	40
25																
26																
27																
28																
29																
30																
31																
32																
33																
34																
35																
36																
37																
38																
39																
40																

1. Simple ACP and TWL show a reasonably close correlation with wet bulb temperature limits imposed in hot workings such as 28°C for the start of modified work arrangements and 32°C for stop work. In hot and humid conditions it is the wet bulb temperature that has the greatest influence on a human's ability to transfer heat to the environment by sweating, the main form of heat loss in these conditions.

2. Effective temperature is more strongly influenced by higher radiant temperatures or lower humidity air than simple ACP or TWL. This is enhanced at lower air velocities too.

Effective temperature is limited as an effective evaluator of physiological stress as detailed previously in this paper when referring to the literature particularly Leithhead and Lind (1964), Parsons (1993) and Brake (2002).

In addition Effective temperature is limited in the velocity range over which it can be applied. This means that for typical longwall coal faces the air velocities may be in excess of the maximum air velocity specified for use by the index.

No heat stress index is perfect and further research is still required at a basic level to improve the existing indices. This work includes the effect of clothing factors, clothing combinations and other factors, and the efficacy of WBGT. It should also be recognised that each mine site is different and so heat stress indices should be tailored to fit the individual mine sites to enable rapid and accurate determination of heat stress at a site level. Elements to include here may be likely radiant temperature differences from the dry bulb, clothing types and combinations worn on site, likely work rates, acclimatisation and the provision of cooled locations such as air conditioned cabs in light vehicles.

Wet bulb temperature does provide a reasonable correlation for modified and stop work limits set at moderate and high work rates for both ACP and TWL. As such wet bulb could be used as an initial first indicator of if a heat problem exists prior to using more involved indices such as ACP and/or TWL to determining the degree of heat stress risk. This fits in well with the methodology developed previously by Graves and Graveling (1988).

RECOMMENDATIONS

The coal mine regulations for Queensland specify the use of Effective temperature to evaluate the thermal comfort of the underground environment. However the regulations do not specify which effective temperature index needs to be applied, basic, normal and corrected effective temperature all produce different results for the same environmental conditions. The equivalent metalliferous mine regulations for Queensland do not state any particular thermal comfort/heat stress index for use in underground hard rock mines. Effective temperature has been shown in the literature to have a number of failings as a index of heat stress and is no longer a recommended index in the field of occupational hygiene. Given this the following are recommended:

1. Underground coal mines, if required, may seek an exemption from the regulation requiring the use of the effective temperature index on the basis of the flaws in the index and so long as they can show as provided under the coal mine act that a health and safety system, plan and procedure have been developed to ensure the thermal safety of the underground workforce.

2. Mines develop and implement a heat stress management system as per the coal mines act requirements.

3. This study agrees with the approach developed by Graves and Graveling (1988) of a tiered approach to monitoring heat stress and managing heat stress in underground mines. A suggestion for such an approach is:

    a) Apply a simple measurement based approach to indicate the potential for a heat issue. A suggested index to use is the simple wet bulb temperature with 28°C possibly as a first trigger action level although a lower value may be applied

    b) A stop work trigger action may potentially be set at 32°C wet bulb

    c) Once the initial trigger action point has been reached then apply a rational method of heat stress determination, examples include simple air cooling power or thermal work limit. With these work rates can be set to develop modified and stop work conditions which are based on local conditions and work practices at the individual mine site, although these are not easy to estimate.

4. If a rational index is to be applied it should be fine-tuned to the individual site by using local values for clothing and other factors and local values /knowledge of the difference between the radiant temperature and the dry bulb temperature on average. However, such Australian studies do not exist.

5. Use of rational indices also would allow variations at individual work locations to be accounted for such as different measures of heat stress for those working in the general airflow and those working in lower flow areas such as for example a service crew installing cables or pipes close to the roof and walls of roadways where the airflow velocity may be reduced. As an example wet bulb would be applied to the general airflow whilst a rational index applied to the service crew

6. Further research and development of rational indices that reflect differences between mine site is recommended to ensure that evaluations can be made using simple, non-expensive instruments, using non-specialist staff and that appropriate rapid calculation/evaluation techniques such as tables or graphs are provided to those undertaking the measurements.

ACKNOWLEDGEMENTS

The authors would like to acknowledge the support and provision facilities to undertake this work by Federation University Australia. The authors also would like to acknowledge the financial assistance provided by Anglo American Metallurgical Coal

REFERENCES

Bedford T. (1946) Environmental warmth and its measurement. Medical Research Memorandum; No 17, HMSO, London

Belle B. and Biffi M. (2018) Cooling pathways for deep Australian longwall coal mines of the future. International Journal of Mining Science and Technology, Vol 28, No 6, November 2018, pp 849-858

Brake D. J. (2002) The deep body core temperatures, physical fatigue and fluid status of thermally stressed workers and the development of Thermal Work Limit as an index of heat stress. PhD Thesis, Curtin University of Technology.

Gibson K. L. (1976) The computer simulation of climatic conditions in underground mines. PhD Thesis, University of Nottingham.

Graves M. and Graveling R. A. (1988) Notes of guidance for the Health and Safety Executive on working in hot conditions in mining. IOM Research Report TM/88/13S. Edinburgh UK.

Houghton F. C. and Yagloglou C. P. (1923) Determining equal comfort lines. J ASHVE; **29**; pp 165-176

Howes M. J. and Nixon C. A. (1997) Development of procedures for safe working in hot conditions. Proc 6th Int. Mine. Vent. Congress, R. V. Ramani (ed), pp 191-198, SME Littleton CO.

Leithead C. S and Lind A. R. (1964) Heat Stress and Heat disorders. Cassell, London

McPherson M. J. (1993) Subsurface Ventilation and Environmental Engineering. Chapman Hall, London UK

Moreby R. (2002) Sources of heat and heat management in Australian longwall coal mines. Proc NA/9th US Mine vent Symp, E De Souza (ed), Kingston, Ontario, Canada 8-12 June, pp 363-370. Balkema, Lisse NL

Parsons K. C. (1993) Human Thermal Environments. Taylor and Francis.

Pickering A. J. and Tuck M. A. (1997) Heat: Sources, Evaluation, Determination of heat stress and heat stress treatment. Mining Technology, June 1997, **79**, No 910, p 147-156. Doncaster, UK

Queensland Government (2017) Coal Mining Safety and Health Regulation 2017.

Stewart J. M. (1982) Fundamentals of human heat stress. Chapter 20 Environmental Engineering in South African Mines, J Burrows (ed). Pp 495-533. The Mine Ventilation Society of South Africa, Johannesburg, RSA.

Webber R. C. W, Franz R. M, Marx W. M. and Schutte P. C. (2003) A review of local and international heat stress indices, standards and limits with reference to ultra-deep mining. J SAIMM, June 2003, pp 313-324.

Wyndham C. H. (1974) The physiological and psychological effects of heat. Chapter 7 The ventilation of South African Gold Mines, Burrows J (ed), pp 93 – 137. The Mine Ventilation Society of South Africa, Johannesburg.

Methane and Coal Dust Explosions

Ventilation and gas management strategies
for re-entry to Pike River Mine Drift

R Hughes, R Moreby, B Poborowski and J Rowland

ABSTRACT

This paper describes the development and modelling of ventilation and gas management strategies to be employed in the re-entry of the Pike River Mine Drift together with results of nitrogen injection to date. The purpose of re-entry is for forensic examination, possible recovery of bodies and to bring, at least some, degree of closure to the families of the 29 miners who were killed in the Pike River mine explosion on 19th November 2010.

Following the initial and subsequent explosions at the mine, it was finally sealed in 2011. A review of the situation in 2017 led to the establishment of the Pike River Recovery Agency (PRRA) in January 2018. The PRRA is charged with the task of developing and implementing a plan to re-enter the 2.2 km access drift if deemed technically feasible and safe to do so.

At the time of commencing this re-entry planning work, the mine was effectively sealed and inert by virtue of methane concentrations being in excess of 95%. A series of design reviews and risk assessments have led to a force ventilation system being selected combined with application of nitrogen to control atmospheres in the drift and coal mine workings during various re-entry phases.

The recovery plans developed by PRRA involve initial purging with nitrogen followed by re ventilating the drift in stone with fresh air whilst simultaneously maintaining an inert (<5% oxygen) nitrogen/methane atmosphere in the coal mine workings. The current plan is then to re-enter the drift from the portal under force ventilation with forensic examination, involving members of the New Zealand Police force if significant forensic evidence is discovered.

The first phase of nitrogen injection to the drift commenced on 18th December 2018 and was successful. Infrastructure for the next phases of re-entry are currently being installed and tested. Based on monitoring data available so far, it is anticipated that the identified plan will progress to the next phases of re ventilation then re-entry during the first half of 2019.

MINE GEOMETRY

A plan and section of the mine workings is shown in Figure 1. The main features are a 2.2 km long 28 m^2 drift mined up dip to the mine workings. At the time of the explosion, the first goaf had been formed by hydro mining and development was underway for future hydro panels.

Through previous inspection of the mine with robots and downhole cameras it is known that there is a significant fall of ground, of unknown length and severity, at the top of the drift at or about the point that the coal seam was intersected. This fall location determines the limit of proposed re-entry and, prior to nitrogen injection, it was uncertain if gases could pass through it.

The total volume of workings is small at a total of approximately 141 x 10^3 m^3 comprising 66 x 10^3 m^3 for the drift and approximately 75 x 10^3 m^3 for the mine workings. By way of comparison, a single 4km two heading gate road would have a volume of approximately 144 x 10^3 m^3.

With respect to management of the mine atmosphere during periods of changing barometric pressure, the small volume of the mine workings limits the extent of gas migration into or out of the drift or mine workings. This is a consideration for necessary nitrogen injection rates to manage periods of falling barometric pressure and emergency egress strategies in the event of a fan failure.

The location and diameters of existing boreholes together with those being installed for the re-entry plan are shown in Figure 2.

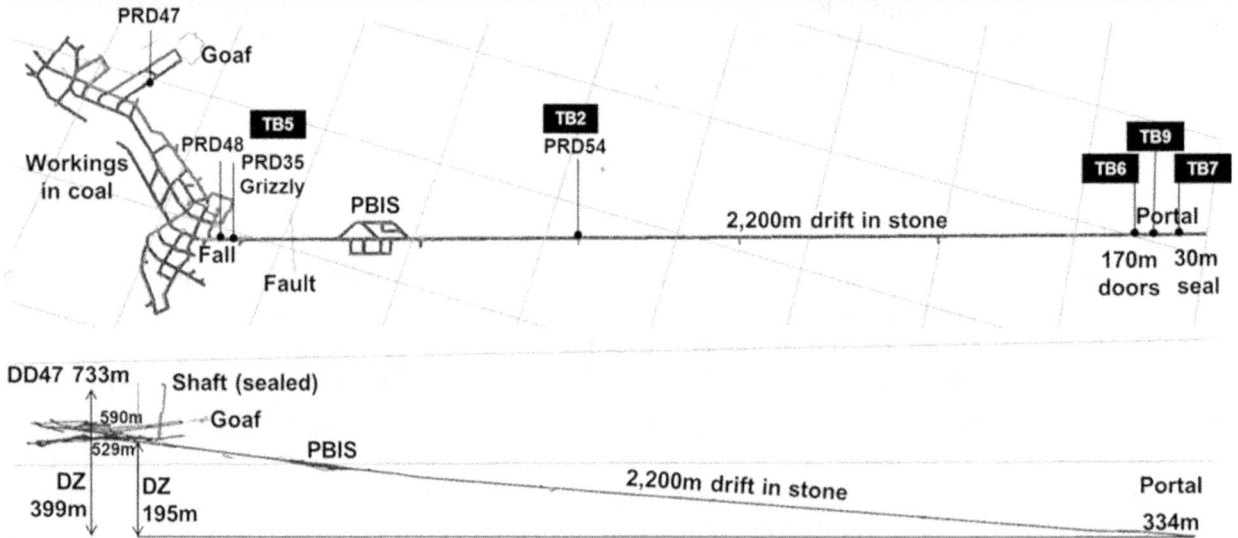

Figure 1 Geometry of Mine Workings

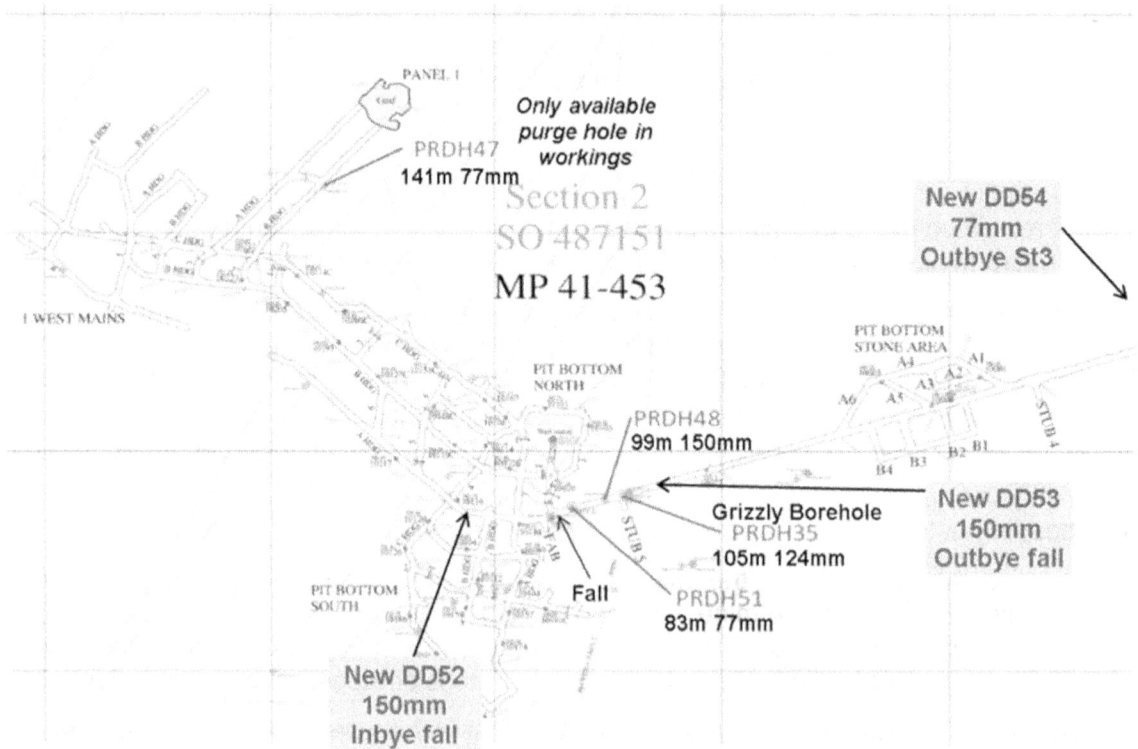

Figure 2 Location and Diameter of Existing and New Boreholes

Some of the existing boreholes are available for monitoring and decanting gas but these with diameters less than 150 mm will limit gas flow rates to below that required. For example, at a differential pressure of 2.0 kPa, a 77 mm ID hole has a capacity of 35l/s air to 48 l/s methane and a 150 mm ID hole has a capacity of 242 l/s air to 325 l/s methane. The main design issue arising from these values is that injecting nitrogen at pressures up to 10bar can be employed at rates limited to about 420 l/s. However, decanting significantly higher rates of gas mixtures under lower pressures during the re ventilation stage requires additional 150 mm ID holes.

The diameter and function of primary holes are as follows;

- PRDH47 has the highest collar elevation and will be used for decanting gas and nitrogen injection. However, flow rates are limited by the diameter of 77 mm.
- PRDH35 is used as a tube bundle monitoring hole then as an exhaust hole..
- PRDH48 will be used initially as a decanting hole then as an exhaust hole.
- PRDH51 is on the rill of the roof fall and will be used as a monitoring hole.
- PRDH52 is being installed for nitrogen injection to the mine workings and over the fall.
- PRDH53 is being installed as an exhaust hole in parallel with PRDH48.
- PRDH54 is being installed as a tube bundle monitoring hole.

It needs to be borne in mind that, although the hole lengths are not long (40 m to 140 m), there are limited drilling sites and all drilling equipment (rigs and compressors) has to be flown to drill site by helicopter.

RISK ASSESSMENT ISSUES AND CONTROLS

A generic list of ventilation related hazards and issues of concern addressed by risk assessment and ventilation and gas management strategies are as follows;

1. Flammable or explosive atmospheres, either steady state or during variations that may occur due to changing barometric pressure or failure of infrastructure.
2. Irrespirable atmospheres due to oxygen depletion or use of nitrogen.
3. Use of diesel equipment (gaseous emissions or fire)
4. Re activation of heating at the fall area, rider seam or in the mine workings.
5. Response time and pedestrian egress during changing conditions (system failures, change in barometric pressure etc).

Applicable regulations are the Health and Safety at Work (Mining Operations and Quarrying Operations) Regulations 2016. Applicable standards are the New Zealand Worksafe ventilation and exposure codes of practice Worksafe, 2014A, 2014B and 2016.

Principal controls and operational practice to be employed are as follows;

1. All parts of the drift and workings accessed during re-entry will be classified ERZ0 or ERZ1 for which relevant regulatory standards applying to equipment and operation will be adhered to.
2. A ventilation rate of 15m³/s or more from the end of force ventilation duct will be available. This is adequate for dilution of residual methane emissions (<65 l/s) and diesel equipment to be employed (133 kW loader and 55 kW Drift Runner).
3. Application of nitrogen (98% nitrogen at up to 420 l/s), initially for purging methane from the drift then for maintaining an inert (<5% oxygen) atmosphere in the coal workings while the drift is re ventilated.
4. Management of potential sources of ignition and electrical conductors in the same manner as in an operating coal mine. In addition, a withdrawal TARP based on weather forecasts of lightning in the area.
5. Provision of self rescuers, CABA and a mobile refuge bay.
6. Gas monitoring systems including;

- Calibrated handheld instruments gas monitoring instruments (e.g. Industrial Scientific MX4 or similar for CH_4, CO, H_2S and O_2)

- Real time telemetric monitoring (TX9165 Sentro 8 Sensor station or similar) to be taken in with re entry team and located at six points in the drift.

- 20 point tube bundle system for CO, CH_4, CO_2 and O_2

- A gas chromatograph on site with back up at the mine rescue station.

7. Only experienced miners appropriately trained in the use of breathing apparatus will enter during re ventilation stages. Current plans provide for police or other non-experienced or untrained miners (but trained in CABA use) to undertake work in stages as made safe by re-entry teams.

8. Pedestrian egress rates (1 m/s) are about an order of magnitude faster than gas expansion rates during periods of falling barometric pressure.

STATUS OF THE MINE PRIOR TO NITROGEN INJECTION

The mine was progressively sealed in 2011, including plugging of the exhaust shaft and construction of a seal at the portal (170 m seal). In 2016 a concreate seal was also installed that the 30 m point. This resulted in the body of the workings and drift becoming filled with methane (>90% CH_4) with only minor ingress of oxygen at the portal seal during periods of rising barometric pressure.

The main design parameters arising from analysis of pressure differentials across the 30m seal and elevated hole collars in 2018 were;

1. The mine is effectively sealed as evidenced by the 30m seal holding pressure differentials up to circa 3.5 kPa during periods of falling barometric pressure and, more recently, up to 9 kPa during nitrogen injection..

2. There is a flow path through the fall as evidenced by collar pressure changes when the seal was partially opened and collar pressure being that of calculated buoyancy pressure (about $(1.2 - 0.7) \times 9.8 \times 399 = 1,955$ Pa). This assumption was subsequently proven to be correct during the initial nitrogen injection phase.

3. The residual methane emission rate is 65l/s or less based on changes that occurred during changes in barometric pressure. However, this was a difficult value to determine with confidence with values obtained during the initial nitrogen phase suggesting the actual value is lower. It remains unknown if the residual methane make is from the goaf, rib emission or both.

4. There is no evidence of residual heat in the mine as evidenced by absence of gas indicators associated with pyrolysis. This finding is consistent with the composition of the mine atmosphere and the time that the mine has been sealed.

An important outcome of previous inspections of the drift by remote cameras is that there is no evidence of significant deterioration of the roof of the drift due to application of the GAG engine. This can be a problem for re-entry to coal mines but, as the drift was driven in metasedimentary gneiss, it does not appear to be as significantly affected by heat and moisture from the GAG engine delivery stream.

RE ENTRY STRATEGY

A number of possible re-entry strategies were evaluated including development of an additional tunnel at the top of the drift and large diameter (>600 mm) holes for ventilation and emergency egress. Through a process of design review and risk assessment the current re-entry strategy is based on a conventional force ventilation system and is summarised as follows;

1. Inject nitrogen into the portal seal and decant methane from elevated borehole collars. The purpose being to use buoyancy pressure to remove methane from the drift. This step is now complete with the drift now purged of methane and replaced with nitrogen.

2. Prior to re-entry, transfer nitrogen injection to the coal workings to hold an inert atmosphere in the body of the mine and over the fall while the drift is re ventilated with the portal fan(s) and venturi exhauster on PRDH53 (possibly also on PRDH48 in parallel). The objective being to hold a positive pressure on the workings compared to that in the drift.

3. Extend the force ventilation duct during re-entry in 10 m to 20 m lengths. Importantly, the duct is to be extended into sections of the drift that already contain a respirable non-explosive atmosphere compliant with ERZ1 conditions.

FORCE VENTILATION SYSTEM

The portal ventilation system installed at the mine comprises 2 x 2 stage 90/90 kW axial fans with one stage being VSD and three stages being DOL, Figure 3. These will deliver force ventilation to the drift portal via a 1.4 m diameter layflat duct.

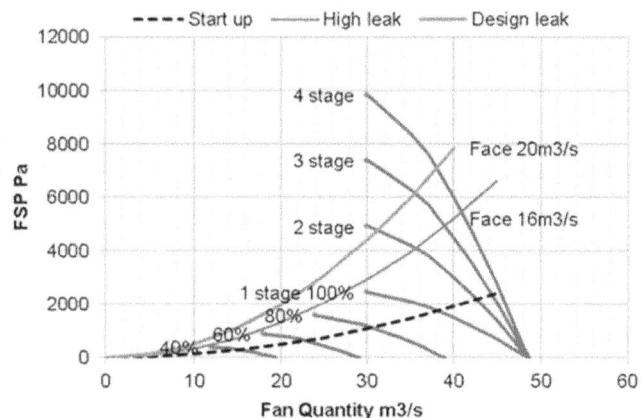

Figure 3 Portal Fans and Characteristic Curves

The main reasons for selecting a conventional force ventilation system for the re-entry phase were as follows;

1. The drift will be purged of methane and excess nitrogen ahead of working locations. Importantly, it is not the design intent for the force ventilation system to flush methane or excess nitrogen hence contaminating the return egress route.

2. The same infrastructure was employed during the original drift development with a measured capacity (20 m³/s at 2.2 km) sufficient for re-entry.

3. It is a conventional method for ventilating long single-entry tunnels worldwide i.e. performance can be calculated with confidence.

4. It is appropriate in this project given the absence of development activities being undertaken and with the absence of a seam gas make in outbye 2km section the drift. That is, potential sources of contamination, other than diesel equipment, are inbye of the working location.

5. It will be easier to install layflat duct, particularly around obstacles, and with fewer joints resulting in lower leakage rates.

6. It provides a positive pressure and forward scouring effect at the discharge point.

7. It provides for rapid and safe degassing of the drift following a fan failure.

8. It reduces the likelihood for the need to install high voltage electrical power underground to supply a smaller inbye auxiliary fan.

Analysis of duct leakage values and fan characteristics provides an inbye delivery quantity of up to 20 m³/s which can be modified to 15 m³/s by speed control or outbye bypass.

NITROGEN SYSTEM

The capacity and characteristics of nitrogen injection plant are summarised in Table 1. The primary nitrogen plant consists of a Memoss membrane generator and three (plus one spare) air compressors and air dryer. The back-up nitrogen plant is of the cryogenic type. Its purpose is to provide a short duration back-up supply of nitrogen when the primary plant is off-line. The plant consists of three ISO liquid nitrogen vessels, containerised vaporiser, liquid gas manifold skid, monitoring skid.

Table 1 Nitrogen Plan Specifications

Parameter	Primary Plant	Back-up Plant
Flow (average):	1500 Nm³/h	750 Nm³/h
Duration:	Continuous supply	24-hrs supply
Pressure (min):	10 bar(g)	9.5 bar(g)
Temperature:	55°C (max)	-4°C(min)
Oxygen % (max):	2% vv	0.05% vv
Oil carry-over (max):	0.03 mg/m³	N/A

Nitrogen will be delivered at the nitrogen compound to the portal area using the combination of 300 mm, 150 mm and 100 mm (nominal) diameter pipes, Figure 4. From the portal area it is distributed to the portal location for seal injection and to bore-hole locations via twin 4.1 km x 77 mm ID pipelines.

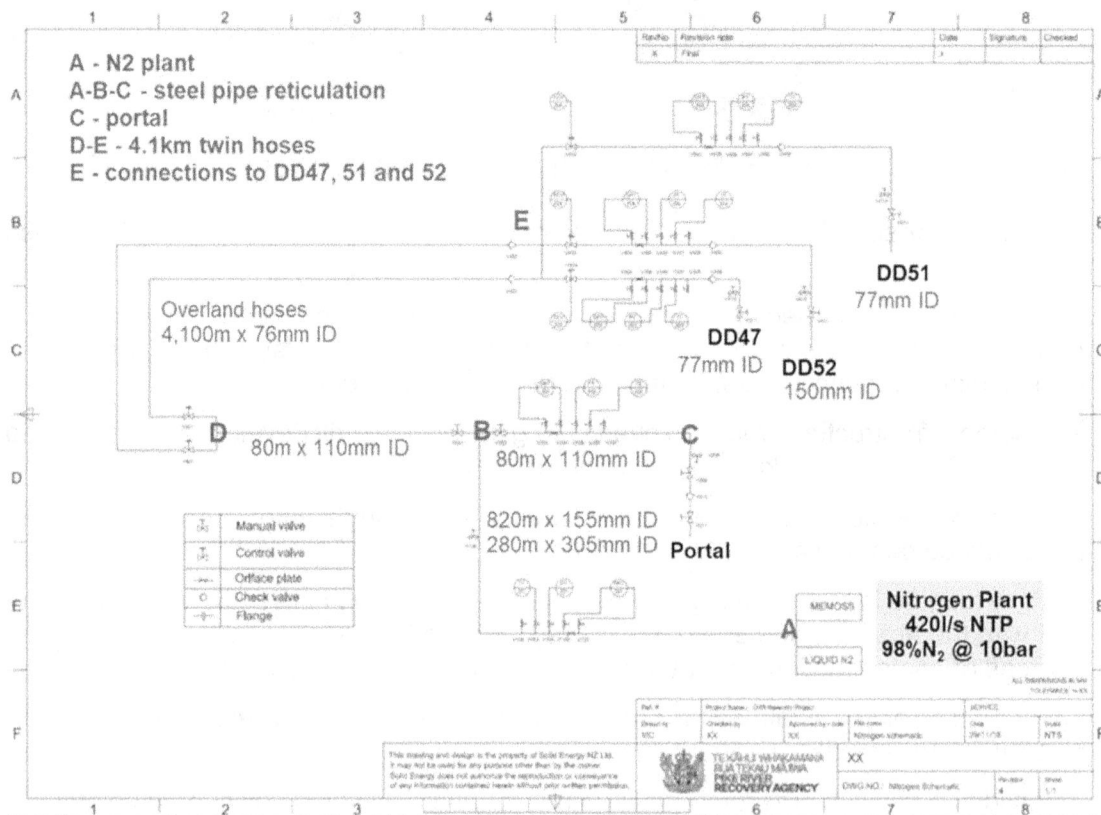

Figure 4 Nitrogen Reticulation System

A significant feature of the reticulation system is the twin 4.1 km 77 mm ID pipeline length from the nitrogen plant area to remote hole collars. This pipe had to lowered in section by helicopter to ground crews who secured and joined it. The need for two rather than one delivery pipeline was confirmed by calculation and Ventsim simulation as described below.

HOLE COLLAR ASSEMBLIES

All decanting, Figure 5, and nitrogen injection holes will be fitted with 150mm flame arrestors and remotely operated valves. An example of a tube bundle monitor hole collar is shown in Figure 6.

Figure 5 Collar Assembly
(Decanting or injecting)

Figure 6 Collar Assembly
(Tube Bundle Points)

Remote monitoring and operation of valves has been employed due to the difficulty of maintaining 24 hour supervision at the holes collars, particularly during bad weather.

A venturi operated by remote compressor will exhaust 800 l/s to 1,000l/s from PRDH53 and, if required, PRDH48 in parallel. This will be applied during the re ventilation stage to promote a respirable non explosive atmosphere ahead of the working location and to ensure that there is a flow of nitrogen from the workings out over the fall.

SYSTEM CHARACTERISTIC CALCULATIONS AND VENTSIM MODEL

Two methods were used to assess nitrogen reticulation system characteristics and resultant purge times. The first used the theory describing frictional losses in gas reticulation systems provided by, amongst others, McPherson, (1993) and Boxho, et al., (2009). In summary, the calculation proceeds as follows;

1. Calculate the "rationale "resistance of the pipe from dimensions and wall roughness factor. This being the physical resistance of the pipe without a correction for density.

2. For a defined gas composition, calculate the gas constant (J/(kg.K)) and density from which the mass flow rate (kg/s) can be calculated for a given gas flow rate (m³/s at specified temperature and pressure).

3. Calculate the change in pressure due to frictional losses using gas laws corrected for compression or expansion of the gas mixture.

The form of the final equation for pressure loss in a horizontal pipe with compression is shown in Equation 1 with that for the rational resistance in Equation 2.

$$\frac{P_1^2 - P_2^2}{2} = M^2.R.T.r_t \quad \text{Eqn.1} \qquad r_t = \frac{64.f.L}{2.\pi^2.d^5} \quad \text{Eqn.2}$$

Where,

P_1 and P_2 = absolute pressures at the start and end of the pipe, (Pa)

M = mass flow rate of the gas mixture, (kg/s)

T = absolute temperature, (K)

r_t = rationale resistance of the pipe, (m⁻⁴)

f = dimensionless friction factor

R = gas constant for the gas mixture, (J/(kg.K))   L = pipe length (m)

d = pipe diameter (m)

The second method was to develop a Ventsim model of the combined gas reticulation and ventilation systems, Moreby, 2019. Due to the low flow rates and high pressure differentials occurring in a nitrogen injection system, compared to those in ventilation circuits, it was found to be necessary to modify Ventsim simulation settings to avoid simulation errors. The custom settings to change are those focused on increasing simulation precision, even if the simulation time increases.

Important note on Ventsim releases – the gas density calculation algorithms in Ventsim for high pressure conditions were updated for revision 5.1.2.2 or later. Earlier revisions of Ventsim may fail to simulate high pressure systems.

For nitrogen injection systems, the modelling strategy was to set an NTP flow or mass flow rate (at a set gas composition) into the plant from surface pressure then set the back pressure by applying fixed pressure drops to the end of pipes where control valves will, in reality, be located, as shown in Figure 7 . A single very high resistance branch is used to provide a theoretical open split for mesh selection.

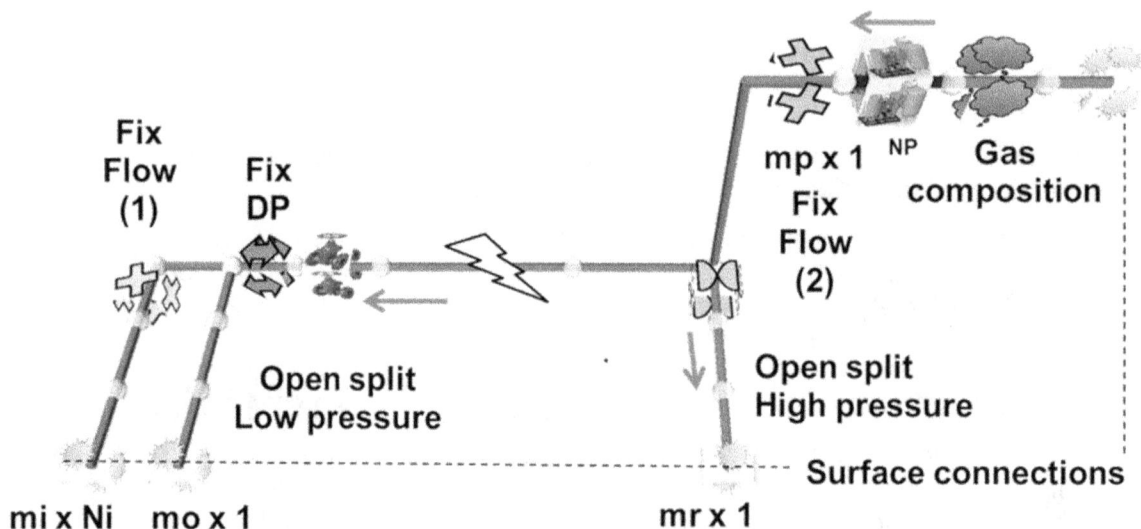

Figure 7 Nitrogen Injection circuit (generic schematic)

The mass balance for the system is then determined by the sum of fixed hole collar flow rates which will be determined by that of the total flow rate provided by the nitrogen plant distributed to outlet points.

This generic model shows nitrogen pipes discharging to atmosphere for the purpose of sizing pipes. However, it was then also combined with the ventilation model to predict inertisation times.

The purpose of the modelling exercise was as follows;

1. To demonstrate that two rather than one 4.1 km pipeline was required to connect the plant to holes into the workings.

2. To estimate what the purge times would be (using dynamic gas simulation)

3. To provide a graphical representation for explanation of the injection strategy to other interested parties.

The model included a normal 3D representation of the mine to scale combined with a schematic representation of the nitrogen reticulation system. In this case, the exit points from the injection holes were connected to the mine's ventilation model using dummy low resistance pipes. Screen shots of the models for three phases of nitrogen injection and then re ventilation are shown in Figure 8.

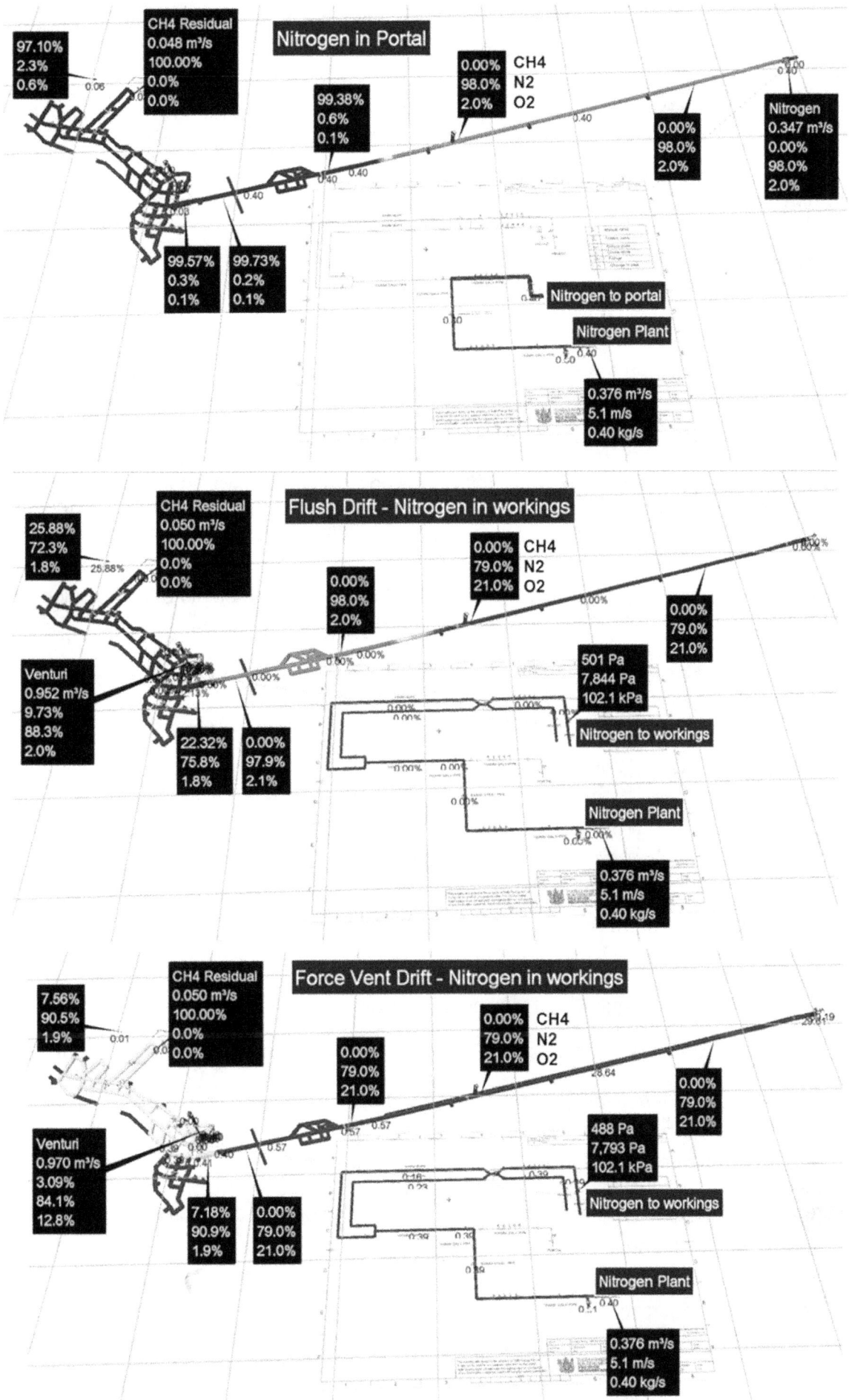

Figure 8 Ventsim Model for the Three Phases of Nitrogen Injection

This model, together with crosschecks with spreadsheet calculations using the calculations described above, provided the following design outcomes;

1. Two 4.1 km 90 mm (76 mmID) nitrogen hoses would be required to deliver full plant capacity to the mine working's injection boreholes at less than 4 bar (400 kPa). This analysis also provided a planning value of approximately 300 l/s NTP through a single pipe should the other fail.

2. Two additional 150 mm boreholes would be required, one for increased injection capacity to the mine workings and another for increased exhaust capacity. These are now PRDH52 and PRDH53.

3. Drift purge times could vary between 2 and 7 days depending on the status of additional holes and degree of connectivity between the drift and workings through a fall of ground at the top of the drift.

The model will be reviewed further once nitrogen is being delivered to the remote boreholes, the main uncertainty being cumulative shock losses due to the nature of the pipeline run.

CURRENT STATUS OF NITROGEN INJECTION

An initial trial of nitrogen injection in to the drift via the portal 30 m seal was started on 18th December 2018. This was followed by full injection at about 400 l/s for 12 hours per day on the 18th January 2019. At this time, PRDH54 had been competed to the drift at about No.3 stub for monitoring and gas was vented from PRDH48 at the top of the drift.

The progress of nitrogen injection was monitored by the tube bundle points at the locations shown in Figure 1 with the effect on methane concentrations in the drift shown in Figure 9.

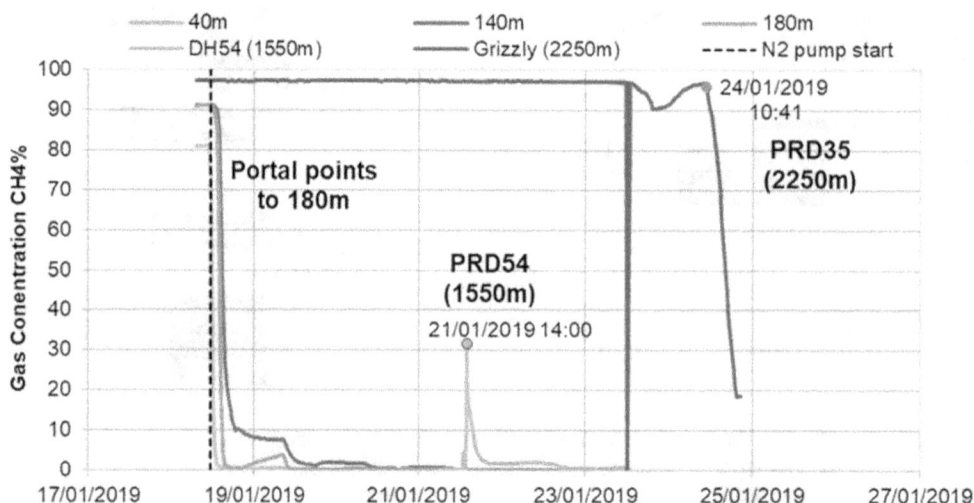

Figure 9 Methane Concentration Reduction with Nitrogen Injection at the Portal

A comparison between times for a single flush (based on nitrogen injection rate and void volume of the drift) and observed flush times is provided in Table 3.

The observed flush time compared to that for a single exchange volume increased further up the drift, for example 3.1 days actual compared to 2.49 days calculated at PRDH54 and 6.0 days actual compared to 3.62 days calculated at PRDH35.. Possible reasons for this increase in time are;

1. Leakage from the 30m seal at a positive pressure of 2.0 to 4.0 kPa

2. Buoyancy of methane resulting in some degree of mixing. This is also evidenced by the time taken for nitrogen to reach T6 (180 m) being shorter than that for T9 (140m). The T9 monitoring point is known to be at a high roof level point.

3. Some degree of goaf atmosphere expansion post injection and as a result of barometric pressure changes also contributing to mixing.

Table 3 Drift Volumes and Actual Flush Times

	T7	T9	T6	PRD54 Approx	PRD35 Approx	
Location	40	140	180	1550	2250	m
XS area	28	28	28	28	28	m2
Volume	1120	3920	5040	43400	63000	m3
N2 rate	403	403	403	403	403	l/s
time per 1 flush	46	162	209	1796	2607	mins
	0.8	2.7	3.5			Hours
	0.03	0.11	0.14	2.49	3.62	Days (12hr/day)
Observed						
Pump on	11:05	11:05	11:05	18/01/19 11:05	18/01/19 11:05	
Flushed time	12:53	18:29	15:11	21/01/19 14:00	24/01/19 10:41	
Actual time	108	444	246	4495	8616	mins
	1.8	7.4	4.1	75	144	hours
	0.1	0.3	0.2	3.1	6.0	days

The nitrogen stream oxygen concentration was shown to be 1.5 to 1.7% which is as expected. It was also noted that low methane and low oxygen concentration were held stable at the outbye monitoring points even with the nitrogen plant being turned off for about 12 hours per day.

By 7th March 2019, nitrogen injection from the drift portal was found to be reporting to PRDH47 proving that there is flow through the fall. At this time, gas concentrations at PRDH47 were 50% CH_4, 0% O_2 and 50% N_2 and at PRDH35, 48 and 51 were <1.5% CH_4, <2.0% O_2 and >97% N_2.

These results are consistent with modelling but are preliminary. The next step is to transfer nitrogen injection to the mine workings in preparation for breaching of the 30 m seal followed by removal of material at the bottom of the drift then staged re ventilation and re-entry.

CONCLUSIONS

To date, the results of risk assessment, peer review, monitoring and engineering design support the position of the PRRA that the Pike River mine drift can be re-entered safely. Clearly, the potential hazards are not trivial but in principal, are no different, to those managed by gassy coal mines worldwide on a daily basis.

With consideration to inertisation and re-entry strategies at other mines, a significant outcome of work to date is the ability to control mine atmospheres with relatively low inertisation rates if the void volume being managed is also small. This may, for example, encourage review of emergency sealing strategies in larger mines to include that of individual panels rather than just the mine as a whole.

Ventsim software can be used to model gas reticulation systems at pressures significantly higher than normal mine ventilation systems. Results are consistent with those from spreadsheet methods based on conventional, and well tested, calculation methods.

The monitoring data now available from this project, together with that from future re –entry phases, will be made available to the mining industry. In this respect, it is hoped that this body of work can form a small legacy in recognition of those who lost their lives in the Pike River mine explosion.

ACKNOWLEDGEMENTS

The PRRA has given permission to use re-entry ventilation and gas management strategies for this paper under its policy of openness and transparency.

REFERENCES

Boxho J, Strassen P, Mucke G, Noack K, Jeger C, Lescher L, Browning E, Dunmore R, Morris L, 2009. Firedamp Drainage. (VGE Verlag GmbH· Essen)

McPherson M, 1993. Subsurface Ventilation and Environmental Engineering. Chapter.12 Methane. Available from https://www.mvsengineering.com/index.php/downloads/publications. [Accessed:1 December 2018].

Moreby R, 2019. Application of Ventsim To Low Pressure Gas Drainage And High Pressure Nitrogen Reticulation Systems. Coal Operators Conference, University of Wollongong, February 2019.

PRRA, 2018. Strategic Intentions. Available from https://www.pikeriverrecovery.govt.nz/documents/document/?ID=188. [Accessed 1 December 2018]

Analysis of methane and coal dust explosion at two stage process using video and image processing

A Ihsan[1], N Priagung Widodo[2], B Sulistianto[2,3] and D Agung Prata[4]

1. Academic Assistant, Institut Teknologi Bandung - Department of Mining Engineering, Bandung 40132. Email: ihsan@mining.itb.ac.id
2. Lecturer, Institut Teknologi Bandung - Department of Mining Engineering, Bandung 40132. Email: agung@mining.itb.ac.id
3. Professor, Institut Teknologi Bandung - Department of Mining Engineering, Bandung 40132. Email: bst@mining.itb.ac.id
4. Researcher, Education and Training Unit for Underground Mine, Sawahlunto 27428. Email: darius_agung@esdm.go.id

ABSTRACT

A dust explosion in an underground coal mine consists of two explosions, namely a primary explosion and a secondary explosion. The primary explosion usually occurs in methane gas, causing waves along the air way and the dispersion of dust. The second explosion occurs due to the concentration of coal dust and methane in the explosive range and is triggered by first explosion. This research was conducted to analyse how the two-stage coal dust explosion process occurs in underground mining by indirect measurement, specifically video and image processing. The purpose of the study is to analyse the effect of methane gas and dust concentration on dust explosibility, characterized by flame speed and flame length. Travel time is obtained from video processing, while distance is obtained from image processing. The experiment was carried out in three conditions: varying methane gas concentrations, varying coal dust concentrations and different positions and areas of coal dust. The experiments lead to the following conclusions. (1) Increased methane concentration will increase the flame speed of a coal dust explosion. In this experiment, the flame speed increased from 1.5 to 2 times the initial condition (methane with concentration 5-6 percent). (2) An increase in dust concentration will increase the flame distance by 1.5 to 2 times compared to without the presence of dust. (3) Coal dust with initial concentrations below lower explosive limit (LEL) will be dangerous and can explode when the dust is mixed with methane gas. This condition often occurs in mine explosions, (4) dust which initially in low concentrations can cause fatalities, the explosibility of coal dust will be higher if it is spread over a wider area and positioned close to the explosion of methane gas.

Keywords: Methane, coal dust, explosion, video processing

INTRODUCTION

Coal Mining in Indonesia

Indonesia is known as one of the top coal-producing countries in the world. Based on International Energy Agency data (2017), Indonesia ranked as the 5th-largest producer of coal in the world with total production of 460.5 million tons in 2016. According to data from the Ministry of Energy and Mineral Resources, around 64% of Indonesian coal is classified in the medium quality category (5,100 – 6,100 Kcal/Kg). Most Indonesian coal is produced in the Kalimantan and Sumatra islands. Coal mining in Indonesia is still dominated by open pit mining, with only a few underground coal mines currently operating in East Kalimantan and West Sumatra.

Based on data from the Agency for the Assessment and Application of Technology - Indonesia (2018), coal production in Indonesia is expected to increase due to the increasing demand for it in domestic industry. It is estimated that total coal production in Indonesia will be around 500 million tons by 2037. The increase in production is certainly an opportunity for coal mining in Indonesia. Indonesia has recently begun to implement more environmentally-friendly mining practices. One of these is to apply underground mining methods, which will minimize land clearing on the surface.

Explosion Risk in Underground Coal Mines

Underground coal mining has several potential high-risk conditions. According to Widodo et al. (based on data from Brnich 2010), there were 11,615 total fatalities in USA coal mines from 1900 to 2008, 95% of which were due to fire and explosion. Fires and explosions are caused by several factors, but in underground coal mines, the main ones are coal dust and methane gas. The major source of coal dust comes from coal extraction activities, while methane is released from the coal when the coal seams are exposed.

In Indonesia, there have been five coal mine explosions in the Sawahlunto mining area in West Sumatra since 1990 (Hendra, 2017). The worst incident occurred in 2009 and resulted in the deaths of 32 mine workers.

BACKGROUND

Explosibility of Methane and Coal Dust

"Methane is produced by bacterial and chemical action on organic material. It evolves during the formation of both coal and petroleum, and is one of the most common strata gases. Methane is not toxic but is particularly dangerous because it is flammable and can form an explosive mixture with air. This has resulted in the deaths of many thousands of miners. Methane: air mixture is sometimes referred to as firedamp" (McPherson, 2012). Methane can explode at a concentration of 5 to 15% by volume. The highest energy occurs at a concentration of 9.8%. A Coward diagram can be used to determine the explosive potential of a methane-air mixture.

McPherson (2012) identified several factors that influence the concentration of methane in ventilation systems for underground coal mines, including: initial gas content of the coal, degree of prior degassing by methane drainage or mine workings, method of mining, thickness of the worked seam and proximity of other seams, coal production rate, panel width (of longwalls) and depth below surface, conveyor speeds, the natural permeability of the strata and, in particular, the dynamic variations in permeability caused by mining and comminution of the coal.

Coal dust is a particle that appears when coal extraction activities. It is classified as fuel in the process of fire or explosion. McPherson (2012) named four factors that influence coal dust explosion:

1. Concentration of the dust and presence of methane. The coal dust can explode at concentrations of 50 g/m^3 to 5000 g/m^3. Coal dust will be more reactive when mixed with methane. The explosive range for coal dust and methane will be wider when both are mixed.

2. Fineness of the dust. The smaller the size of the dust particles, the wider the contact area becomes. The explosibility also increases.

3. Type of coal. The coal that has high volatile matter will have increased explosibility. This means that low rank coal is more dangerous than high rank coal.

4. Strength of initiating source. The stronger the energy ignition, the more powerful the explosion.

Two-Stage Process of Methane and Coal Dust Explosion

Fire will occur if three components are present, namely: oxygen, heat and fuel. These three components are called the fire triangle. If one of the components is not present, fire will not occur. Explosion is a sudden increase of pressure and temperature due to oxidation or other exothermic reaction (Keller, 2014). An explosion will occur if there are fire conditions and two more conditions are added, namely a confined area and the mixture of air and fuel (suspension). Methane gas and dust are two types of fuel that are produced during underground coal mining. Both of these substances have the potential to cause explosions.

Many researchers have analysed the risk of explosion in underground coal mines. According to Kruger (1996) "dust explosion is the uncontrolled exothermic combustion in air of ultra-fine particles of coal in which the resultant aerodynamic disturbance disperses additional coal dust into the air, fuelling the combustion in a self-sustaining process". "The propagation of a coal mine explosion involving coal dust depends on a conducive environment with respect to the following main factors: sufficient heat radiation must be present to ignite unreacted coal particles, the coal dust must be dispersed to form a dust cloud with an explosive concentration and the distribution of the particles must be within the explosive range" (Knoetze, 1993). Cashdollar (1996) developed equipment to directly measure the explosibility of coal dust. The equipment uses 20 litre laboratory chambers, is near-spherical in shape and made of stainless steel. These chambers are used for explosibility measurements such as maximum explosion pressures, maximum rates of pressure rise, minimum explosible concentrations, and inerting effects.

Li et al. (2017) performed an experiment on a semi-confined pipe with a volume of 10 L and a ratio of length and width of 10 to study premixed gasoline-air mixture explosions. From the research Li conclude the results: the explosibility of gasolineair mixtures are different from some other fuels such as hydrogen, methane and LPG, etc. Plessis (2015) studied the development of barriers against methane and coal dust explosions. The equipment is a tunnel made of steel pipe with a length of 200 m and diameter of 2.5 m. In his research, Plessis used ammonium phosphate powder as the suppression material. From the testing, he captured the data of pressure and flame, then plotted graphs with time, distance and maximum readings on the other axes. Plessis' findings clearly illustrate how explosions occur in a coal mine. The dust that is naturally found on the floor can explode due to being lifted and mixed with air, triggered by methane gas that has exploded. These findings are in good agreement with the explanation by Hertzberg (1982) that explosions in coal mines occur through several stages:

1. Growth of a large, flammable methane/air zone near the face that is being mined, resulting from increasing methane emissions as the mining process advances into the fresh seam.

2. Ignition of that flammable volume by some means.

3. Development of a localized methane/air explosion (termed a primary explosion) and flame acceleration outward from the face.

4. Lifting of coal dust accumulations by the flow generated ahead of the accelerating methane/air flame and mixing of that dust with air to create a flammable coal dust/air mixture.

5. Ignition of the dust/air mixture by the methane/air flame.

6. Further turbulent acceleration of the dust flame front which lifts more coal dust, mixing it with air throughout an increasingly lengthening zone ahead of the flame.

7. Propagation of a dust explosion (termed a secondary explosion) throughout the mine.

From Hertzberg's explanation and the experiments conducted by Plessis, it can be understood that the dust explosion in an underground coal mine consists of two explosions, namely a primary explosion and a secondary explosion. The primary explosion usually occurs in methane gas, causing waves along the air way and causing dust dispersion. The second explosion occurs because the concentration of coal dust and methane is in the explosive range and is triggered by the first explosion.

The current research was conducted by indirect measurement, specifically video and image processing, to analyse how the two-stage coal dust explosion process occurs in underground mining. The equipment used in this experiment is relatively simple and affordable. This research aims to analyse the effect of methane gas and dust concentration on dust explosibility.

EXPERIMENT APPARATUS

The test equipment was made from acrylic that is resistant to explosion, with dimensions of 164 cm in length, 19 cm in width and 25 cm in height. There are two main parts: room I, which functions as source of ignition (primary explosion), and room II, which functions as the room for the explosion of coal dust (secondary explosion). The two rooms are separated by thin paper that will be destroyed when an explosion occurs in room I. In room I, there are 2 connector holes through which methane gas is injected and gas samples are taken (measuring the concentration of methane gas in the room). A high-voltage electric trigger is also installed in this room. In room II, coal dust is spread along the tunnel. To record the explosion process, a high-speed camera is placed on the side of the test equipment (Figure 1 and Figure 2).

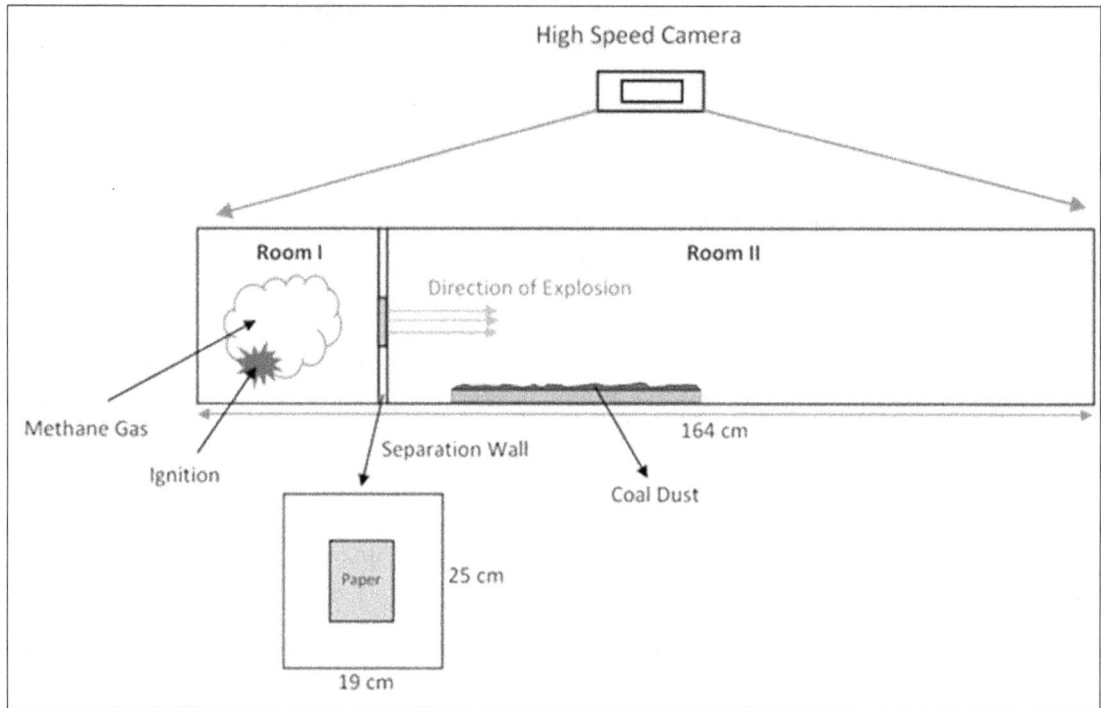

Figure 1 Sketch of Test Equipment

Figure 2 the Test Equipment

EXPERIMENT PROCEDURE

The experiment in this study carries a high risk. Methane gas can explode immediately, therefore the procedure must be done carefully and obey operational standard procedures. The following are the steps in the experiment.

1. Prepare coal dust.

    Coal is reduced and crushed to specific size and then weighed according to the desired concentration. After that, the dust is spread along the tunnel in room II.

2. Create a separation wall between room I and room II.

    The wall is made of thin paper that will be destroyed when exposed to the explosion from room I. The separation wall divides the tunnel into two parts with lengths of 19.5 cm (room I) and 144.5 cm (room II).

3. Prepare methane gas and air mixture.

    Pure methane gas (99.95%) is slowly injected into room I. At the same time, an infrared gas analyser is also connected with room I to determine the concentration of methane gas in the room. After the desired methane concentration is achieved, the connector pipe and analyser are both removed.

4. Put a high-speed camera to the side of test equipment, about 2 meters away. The camera is in standby for recording.

5. Trigger the methane gas explosion using sparks from high-voltage electricity. The trigger equipment is installed in room I and is made of explosive- and fire-resistant material.

EXPERIMENTAL SETUP

Experiments in this study were carried out with three conditions, namely:

1. Several different concentrations of methane gas in room I and the same concentration of coal dust in room II (see Table 1). This experiment aims to analyse the effect of methane concentration on explosions.

Table 1 Concentrations of Methane Gas Used in the Study

Experiment	Methane Concentration (percent of volume)	Dust Concentration (gram/m³)
1	5-6	67
2	6-7	67
3	7-8	67
4	8-9	67

2. The same concentration of methane gas in room I, and several different concentrations of coal dust in room II (see Table 2). This experiment aims to analyse the effect of coal dust concentration on explosions.

Table 2 Concentrations of Coal Dust Used in the Study

Experiment	Dust Concentration (gram/m³)	Methane Concentration (percent of volume)
1	0	9-10
2	27	9-10
3	40	9-10
4	53	9-10
5	67	9-10

3. The position of the coal dust is varied: located at a higher position and spread over a longer area (see Figure 3). Dust concentrations in both experiments were the same, namely 67 g/m³. This experiment aims to analyse the effect of the position of coal dust on explosions.

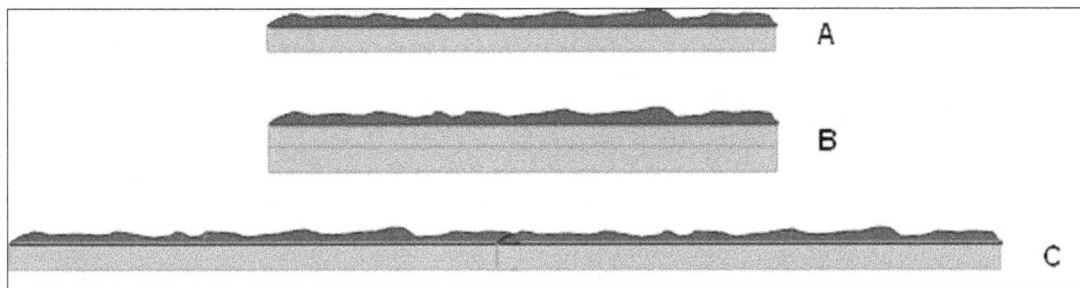

Figure 3 (A): Coal Dust in its Original Position, (B): Coal Dust at Two Times Higher than its Original Position, (C): Coal Dust at Two Times Longer than its Original Position

The coal used in this experiment is taken from Sumatra island. The average size of the coal dust is 45 microns (passed 325 mesh screen). Table 3 shows the proximate test results of the coal.

Table 3 Proximate Analyses of Coal Used in the Experiments

Parameter	Value
Moisture (%), ar	6.39
Volatility (%), db	32.20
Fixed carbon (%), db	41.89
Ash (%), db	25.90
Gross caloric value (Kcal Kg-1), db	5753.57

ANALYSIS

To analyse the explosibility of coal dust, two main parameters are used, namely speed and distance of flame propagation. These two parameters are obtained by processing the video and images from a high-speed camera. Travel time is obtained from video processing, while distance is obtained from image processing. To process the images, the authors used the trial version of Digimizer software. Figure 4 shows one example of image processing using the software.

Explosion Due To Methane Concentration

Table 4 presents the time and distance data for each experiment taken from the video and image processing.

Table 4 Time and Distance for Each Methane Concentration

5-6 Percent		
Image	Time (s)	Distance (m)
1	0.78	0.00
2	0.82	0.44
3	0.84	0.53
4	0.85	0.68
5	0.87	0.91

6-7 Percent		
Image	Time (s)	Distance (m)
1	0.93	0.00
2	0.98	0.37
3	0.99	0.69
4	1.01	0.97
5	1.03	1.05
6	1.04	1.11

7-8 Percent		
Image	Time (s)	Distance (m)
1	0.74	0.00
2	0.78	0.43
3	0.80	0.74
4	0.81	1.01

8-9 Percent		
Image	Time (s)	Distance (m)
1	0.38	0.00
2	0.40	0.56
3	0.41	0.72
4	0.43	0.95
5	0.45	1.00

Figure 5 displays the result of the plot between the flame distance and flame speed for the four different concentrations.

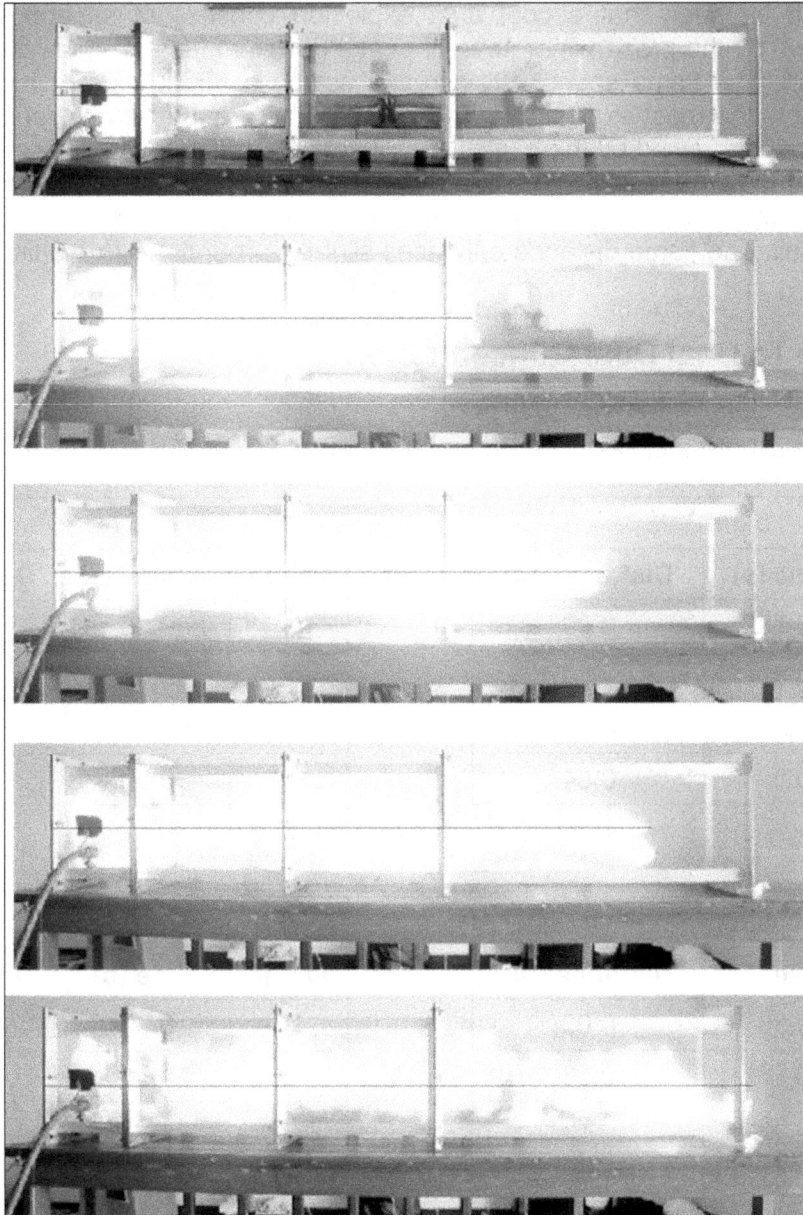

Figure 4 Example of Image Processing Using Digimizer Software

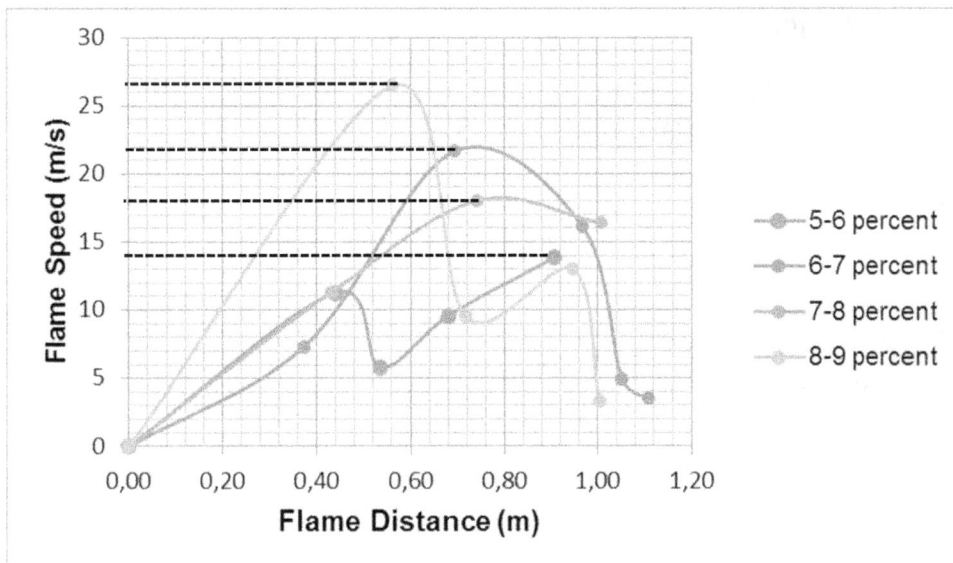

Figure 5 Plot of Flame Distance and Flame Speed for Each Methane Concentration

In Figure 5 it can be seen that a coal dust explosion consists of several explosions which propagate as shown by the presence of several peaks (the flame speed goes back up after going down) on the chart. The flame speed will increase as the concentration of methane increases. Logically, methane is the most explosive at a concentration of about 9%. When the concentration gets higher, the wave produced by the explosion will get stronger, which will also blow up coal dust and finally increase the degree of dispersion. In this experiment, the flame speed rise ranged from 1.5 to 2 times from the initial condition (methane with concentration 5-6 percent). Flame distance has a different result. The flame distance only rises slightly, around 1.1 to 1.2 times from the initial condition.

Explosion Due To Coal Dust Concentration

Table 5 presents the time and distance data for each dust concentration.

Table 5 Time and Distance Parameter for Each Coal Dust Concentration

0 g/m³				27 g/m³		
Image	Time (s)	Distance (m)		Image	Time (s)	Distance (m)
1	0.78	0.00		1	0.45	0.00
2	0.82	0.59		2	0.47	0.49
3	0.84	0.79		3	0.49	0.78
4	0.85	0.84		4	0.50	1.07
				5	0.52	1.23

40 g/m³				53 g/m³		
Image	Time (s)	Distance (m)		Image	Time (s)	Distance (m)
1	0.56	0.00		1	0.99	0.00
2	0.58	0.56		2	1.01	0.49
3	0.59	0.89		3	1.03	0.56
4	0.61	1.06		4	1.04	0.65
5	0.63	1.10				

67 g/m³		
Image	Time (s)	Distance (m)
1	0.38	0.00
2	0.40	0.56
3	0.41	0.72
4	0.43	0.95
5	0.45	1.00

The plots between the flame distance and flame speed are shown in Figure 6.

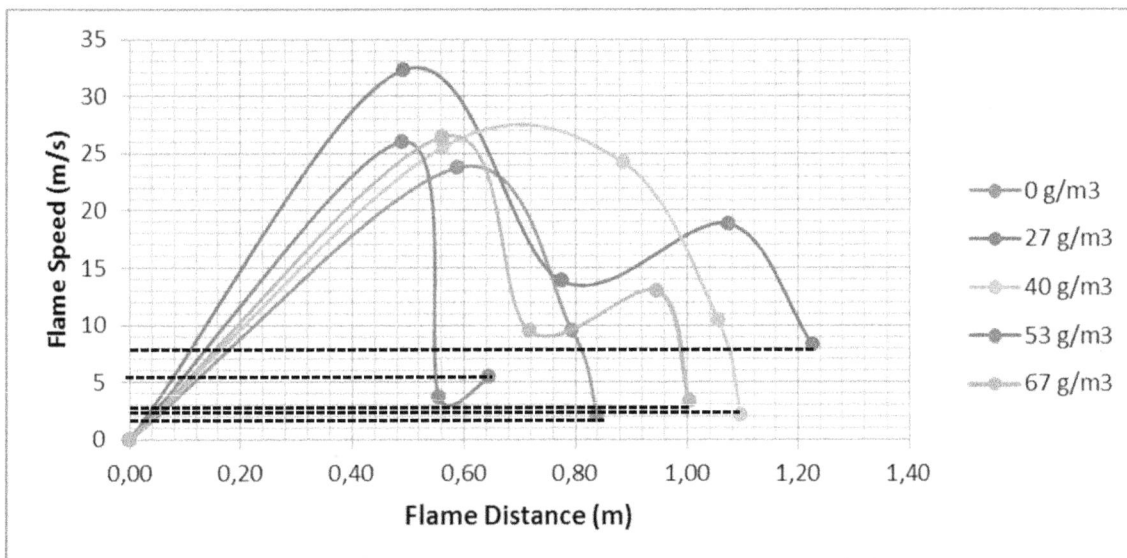

Figure 6 Plot of Flame Distance and Flame Speed for Each Coal Dust Concentration

The test results show that an increase in the dust concentration results in a increased flame distance. But the pattern is inconsistent. The dispersion degree is very influential in each explosion. Low concentrations tend to be more spread in the tunnel, so the flame distance is longer. The increase in flame distance ranges from 1.5 to 2 times compared to explosions without the presence of coal dust, while the flame speed does not significantly increase. The experimental results also showed that coal dust could explode below its LEL, 60 g/m^3 (Hartman, 2012). The authors think this happened because methane gas from room I had flowed into room II and mixed with coal dust shortly before it exploded. This condition causes the dust to be pushed into the explosive zone (Figure 7).

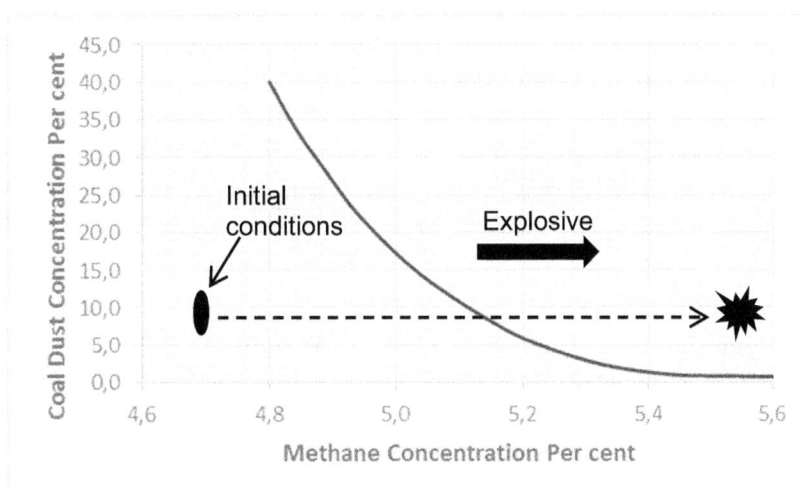

Figure 7 Plot of the Reduction in the Lower Flammability Limit of Coal Dust
In the Presence of Methane (McPherson, 2012)

Explosion Due To Coal Dust Position

The third experiment aims to analyse how the position coal dust influences the explosion (see Figure 3). The results are shown in Figure 8.

In Figure 8 it can be seen that at position B (higher), the explosion has a higher flame speed, which is about twice that of condition A. This happens because the coal dust is closer and directly exposed to the fire propagation. Condition C (longer) produced a longer flame distance, which is around 1.7 times of the condition A. This happens because the explosion propagates along the coal dust that is spread in the tunnel.

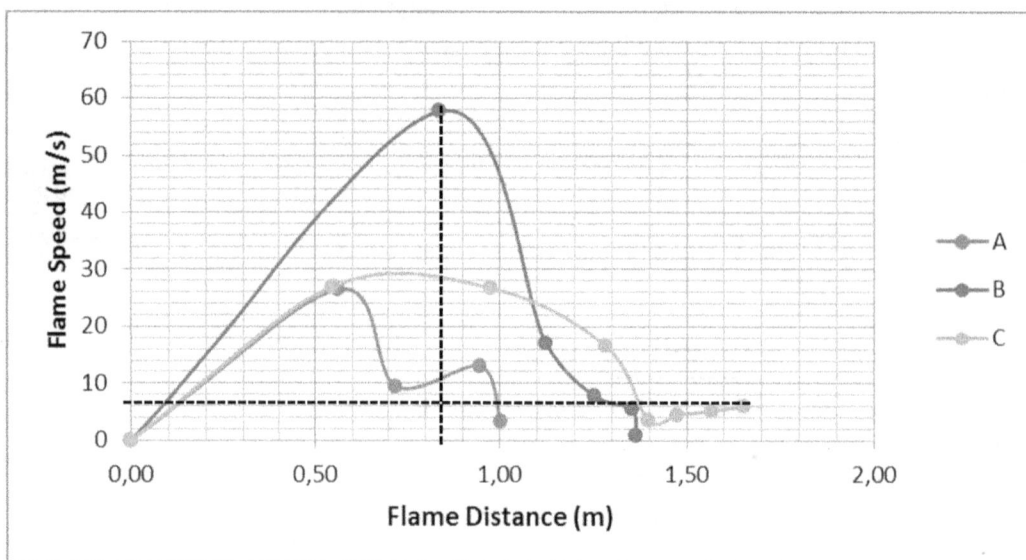

Figure 8 Plot of Flame Distance and Flame Speed for Each Coal Dust Positition

CONCLUSIONS

There have been many studies on methane and coal explosions in tunnels. In the current study, the authors tries to explain the explosive process with visualization and image processing. The results of the experiments lead to the following conclusions:

1. The explosion of coal dust in a tunnel occurs several times and propagates. This occurs because the coal dust is initially lifted and spread by the waves from the explosion of methane gas.
2. Higher methane concentration will increase the flame speed of the coal dust explosion. In this experiment, the flame speed increased from 1.5 to 2 times compared to the initial condition (methane with concentration 5-6 percent). Higher dust concentration will increase the flame distance by 1.5 to 2 times compared to without dust.
3. Coal dust with initial concentrations below LEL is dangerous and can explode when the dust is mixed with methane gas. This is a common condition in which fatal mine explosions occur even though the dust concentration is low.
4. The explosibility of coal dust will be higher if it is spread over a wider area and its position is close to the explosion of methane gas. The increase in flame distance is proportional to the wider spread area, while the flame speed is proportional to the distance from the methane explosion.
5. The indirect measurement methods used in this study, specifically video and image processing, offer good accuracy.

ACKNOWLEDGEMENTS

The authors gratefully acknowledge a 2018 research grant from Institut Teknologi Bandung, "Program Penelitian, Pengabdian kepada Masyarakat, dan Inovasi (P3MI) ITB", for its financial support of this research. The authors would also like to acknowledge the Education and Training Unit for Underground Mining- Sawahlunto for supporting the research facilities.

REFERENCES

Agency for the Assessment and Application of Technology – Indonesia, 2018. Indonesia Energy Outlook 2018. Jakarta.

Brnich, M.J., Kowalski-Trakofler, K.M. and Brune, J., 2010. Underground coal mine disasters 1900-2010: Events, responses, and a look to the future. Extracting the science: A century of mining research, pp.363-372.

Cashdollar, K.L., 1996. Coal dust explosibility. Journal of loss prevention in the process industries, 9(1), pp.65-76.

Du Plessis, J.J.L., 2015. Active explosion barrier performance against methane and coal dust explosions. International Journal of Coal Science & Technology, 2(4), pp.261-268.

Hartman, H.L., Mutmansky, J.M., Ramani, R.V. and Wang, Y.J., 2012. Mine ventilation and air conditioning. John Wiley & Sons.

Hendra, Y., 2017. Riwayat ledakan tambang batubara Sawahlunto. Padangkita.com. Retrieved from http://padangkita.com/riwayat-ledakan-tambang-batubara-sawahlunto/

Hertzberg, M., 1982. Inhibition and extinction of coal dust and methane explosions (Vol. 8708). US Dept. of the Interior, Bureau of Mines.

International Energy Agency, 2017. Coal information: Overview. Paris.

Keller, J.O., Gresho, M., Harris, A. and Tchouvelev, A.V., 2014. What is an explosion? International Journal of Hydrogen Energy, 39(35), pp.20426-20433.

Knoetze, T.P., Kessler, I.I.M. and Brandt, M.P., 1993. The explosibility of South African coals as determined in a 40-litre explosion vessel. Journal of the Southern African Institute of Mining and Metallurgy, 93(8), pp.203-206.

Kruger, R.A., du Plessis, J.J.L. and Vassard, P.S., 1996. The potential of fly ash for the control of underground coal dust explosions. Contract Report for Ash Resources (Pty) Ltd, CSIR Division of Mining Technology.

Li, G., Du, Y., Wang, S., Qi, S., Zhang, P. and Chen, W., 2017. Large eddy simulation and experimental study on vented gasoline-air mixture explosions in a semi-confined obstructed pipe. Journal of Hazardous Materials, 339, pp.131-142.

McPherson, M.J., 2012. Subsurface ventilation and environmental engineering. Springer Science & Business Media.

Widodo, N.P., Sulistianto, B. and Ihsan, A., 2018. Analysis of explosion risk factor potential on coal reclaim tunnel facilities by modified analytical hierarchy process. International Journal of Coal Science & Technology, 5(3), pp.339-357.

Ventilation Planning and Management

Performance analysis of surface goaf hole gas drainage systems and use of CO triggers for spontaneous combustion risk management in Australian coal mines

B Belle[1,2] and G Si[2]

1. Anglo American Coal, Brisbane, Qld 4001, Australia
2. School of Minerals and Energy Resources Engineering, University of New South Wales, Sydney, NSW 2052, Australia

ABSTRACT

This paper focuses on assessing gas drainage performance from vertical surface goaf holes in an operating mine with high specific gas emission rates (~20 m^3/t). Field monitoring results of gas flow rate and gas composition collected from individual gas drainage holes were analysed. The major challenge of implementing the full potential of goaf drainage system is the perceived risk of spontaneous combustion (sponcom) in goaf using *arbitrary trigger set points*. Although the provenance on the introduction and reasons for the use of CO is unknown, the paper highlights that the current use of CO based trigger action response plan (TARP) for sponcom management leads to the premature closure of goaf holes, which largely constrains goaf hole gas drainage performance and has a significant bearing on longwall gas management. The CO concentration in gas captured by goaf holes is found to be positively correlated with O_2 concentration and negatively correlated with CH_4. Building upon abundant field measurement data, goaf gas profiles for CO, O_2, CH_4 and CO_2 concentration were established, which suggest that the increase trend of CO level behind the face is a normal behaviour of goaf closure and would recover to trivial concentration after 300 m deep into the goaf. The paper provides the basis to eliminate or review the use of CO triggers in current surface goaf gas management TARP levels of longwall panels, which has detrimental effect on longwall TG gas management for explosion prevention.

INTRODUCTION

Coal has been a primary source of global energy production and development for the past two centuries and is expected to continue into the near future. Today, coal supplies 40% of global electricity, and in some countries such as South Africa and China, it supplies over 90 % and 70% respectively of the electricity generating source. In the case of Australia, its association with coal can be traced back to as early as 1797. Methane is a major gaseous constituent of the coal seam. It gets released during mining and can result in unsafe working conditions.

Methane is explosive in the range of 5% to 15% in air, its transport, collection, or use within this range, or indeed within a factor of safety of at least 2.5 times the lower explosive limit (2.0%) and at least two times the upper limit (30%), is generally considered unacceptable because of the inherent explosion risks. Methane is contained under pressure within the micropores and fractures of coalbeds and in the adjacent strata and gets released into the mine atmosphere during mining (Moore, Deul and Kissell, 1976). The removal of methane by conventional ventilation is particularly difficult in Longwall (LW) mining of gassy coalbeds, which are prone to dangerous outbursts. In Australia, typically, pre-drainage of working seam would entail management of the outburst risks by ensuring the gas content values below 7 m^3/t as one of the rule-of-thumb outburst risk indicators. A combination of pre- and post-drainage using advanced, Surface-based In-Seam (SIS), Medium Radius Drilling (MRD) and Underground In-Seam (UIS) directional drilling techniques are utilised in Australia (Belle, 2017).

Goaf is a broken and loose, highly permeable ground where coal has been extracted by coal mining and the roof has been allowed to collapse, thus fracturing and de-stressing strata above and, to a lesser extent, below the seam being worked. In a typical U ventilated longwall coal mine, the goaf gas is generated from the broken and left over coal as well as the liberation from the upper and lower coal seams adjacent to the working seam. When these goaf gases are not managed, they will enter into the general body of the ventilation return air by means of goaf fringe on the tailgate side of the longwall. It is important to note that with limited longwall ventilation dilution capacity, ensuring adequate and timely goaf drainage is critical. In multi-seam longwall mining, goaf drainage plays a fundamental and critical role in managing the gas levels. The volume of gas in the post mining depends on the thickness of upper coal seams, gas content, rate of longwall retreat, gas purity, magnitude gas domain. Generally, the volume of goaf gas is significantly higher than the pre-drainage of the working seams. Typically, goaf drainage holes used are as follows, viz., horizontal boreholes, cross-measure boreholes and vertical surface goaf holes. For longwall retreat mining, surface vertical goaf boreholes are drilled from the surface into the upper limits of the goaf, collecting goaf gas from the de-stressed goaf area. Vertical goaf holes are the most appropriate, efficient and safe methodology where there is surface access to an operating longwall mine. Where there was a failure of surface goaf holes, attempts have been made to access the goaf space using 203 mm (8 in) drainage pipes through perimeter roadway seals (Belle, 2017) under controlled Trigger Action Response Plan (TARP) values for methane purity and oxygen levels for safe and effective longwall mining.

The provenance of the introduction of CO (carbon monoxide) as a trigger for sponcom management is unknown or not well documented. Anecdotal discussions suggest that they are an indication of sponcom activity in the active and sealed longwall goaf. It is measured and documented the presence of elevated levels of oxygen behind the goaf as a result of the used ventilation system. The presence of this oxygen under very humid and hot conditions enable the left-over broken coal behind the longwall chocks leading early coal oxidation. If not monitored and controlled, this condition may turn itself into elevated levels of heating leading to sponcom fire. Therefore, to prevent potential sponcom hazards, underground seal gas compositions are measured for sponcom indicator gases.

The current goaf gas drainage practice in Queensland mines also monitor the gas composition in particular CO levels in individual goaf holes to stop the goaf hole operation. Although CO level in ventilation air acts as a robust trigger value for sponcom control in underground ventilation system, it is unclear whether CO level in drained goaf gas also would provide an accurate trigger for sponcom incidents in the goaf. Building upon the analysis of gas drainage performance from goaf holes at a high production gassy mine, this paper reviews the performance of goaf drainage holes and their gas composition to manage the tailgate gas levels as well as reviews the relevance of the current CO-based TARP for sponcom management in goaf holes.

GOAF HOLE GAS DRAINAGE AND SPONCOM TARPS

This section of the paper provides the background to the goaf hole spacing and their performance in managing the methane gas levels in a high production longwall operating mine. In the recent panels, in order to maintain ventilation and gas regulatory compliance, a disciplined data collection system was established, which includes total goaf gas flow and composition from each of the goaf holes, underground longwall real-time gas, ventilation airflow, barometric and production. These have enabled operators to understand the goaf gas dynamics, location of gas zones and anticipate controls and optimise future drainage programs. The results and analyses carried out on three longwall panels (LW6, LW7, and LW8) are based on this data collection system, with particular reference to goaf gas flow, gas composition and use of CO levels as a trigger for sponcom management. Typically, the manual goaf hole gas composition parameters are used to slow down or shut-off individual goaf hole operation.

This paper will focus on analysing data collected from a representative panel LW7 along with additional two LW panels. There are in total 92 goaf holes in the LW7 panel. Among these goaf holes, 71 were drilled a 30-40 m away from the tailgate roadway with ~50 m spacing, while 21 holes were drilled 50 m away from the maingate (MG) with ~150 m spacing (Balusu, 2016). From inbye to outbye, the goaf holes at the Tailgate (TG) side were named as 7-01 to 7-71, and the goaf holes at the MG side were named as 7-MG01 to 7-MG21. Typical goaf holes would have been fully cased in the top portion and the bottom 48 m of the holes were cased with slotted pipes to allow gas flow into the pipes. The diameter of these goaf holes was 250 mm (10") and their length was 330 m with bottom hole completion depth at 10 m above the coal seam. The first tailgate goaf hole was approximately 25 m from the start-up line.

GOAF HOLE PERFORMANCE ANALYSIS

Overall Performance

The following paragraphs provide analyses of LW7 goaf hole performance analyses. A large variation of cumulative flow volume in individual goaf holes can be observed across the entire LW7 panel (Figure 1). Goaf hole 7-20 and 7-MG19 achieved more than 5×10^6 m^3 methane production while in goaf holes such as 7-04 and 7-58, the cumulative methane flow volume was less than 5×10^4. Average methane flow rate and methane concentration of drained gas in individual goaf holes are shown in Figure 2, which also suggests a large variation of the purity and efficiency of goaf gas drainage in different holes. A general increase trend of methane purity can be observed as goaf holes move from inbye to outbye at both TG and MG sides of this panel. Gas drained from tailgate goaf holes ranging from 7-01 to 7-19 at the start-up stage shows low flow rate and methane purity due to incomplete goaf cavity formation and lower Specific Gas Emission (SGE) levels along the inbye portion of the panel. On the other hand, methane flow rate in goaf holes 7-47 to 7-63 is relatively low, but they have much higher methane purity. The lower flow rate could be attributed to hole blockage and failures.

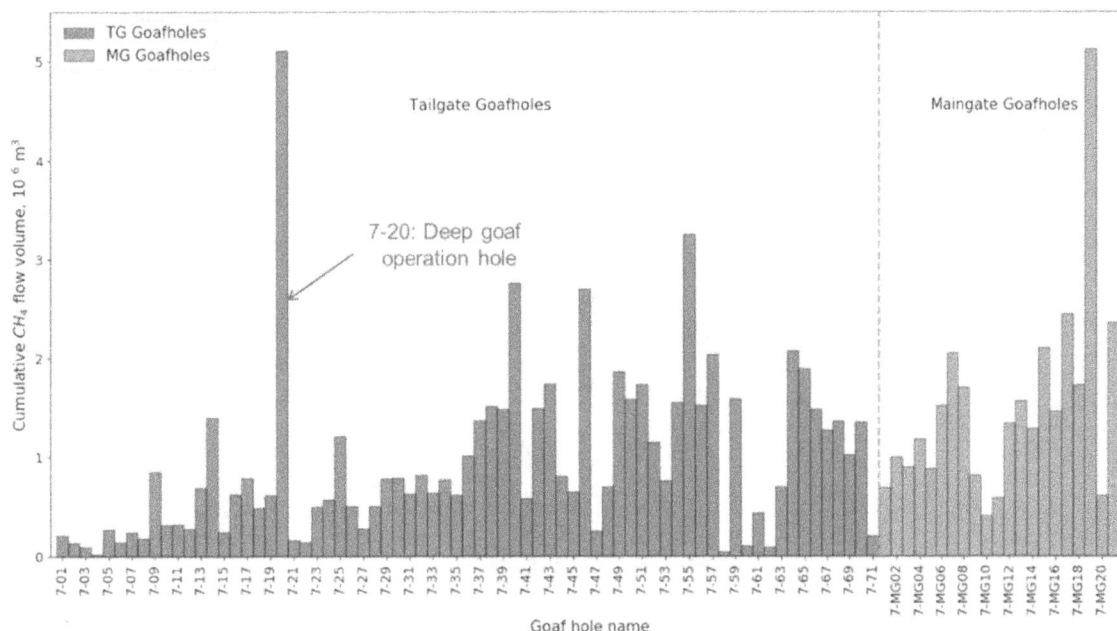

Figure 1 Cumulative flow volume in individual goaf holes from inbye to outbye at LW7.

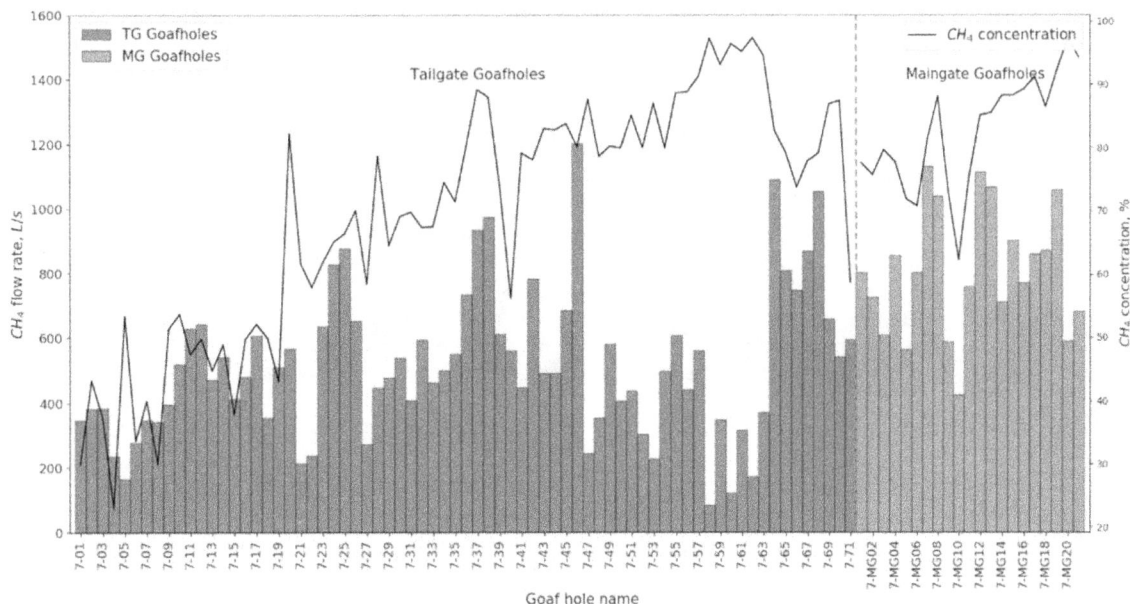

Figure 2 Average methane flow rate in individual goaf holes.

Goaf holes were opened sequentially along with the progress of longwall face retreat. There is no standard prescribed procedure to shut-off these holes, but CO is seen to be the main reason for premature shut-off. The average active production time for each goaf hole was approximately 21 days as shown in Figure 3. Note that a certain number of goaf holes with extremely short active days, particularly 7-01 to 7-19, were result from the high concentration of CO observed (>60 ppm) in those goaf holes. This led to the premature closure of those holes due to the concern of sponcom activity and the lower CO trigger values used in the goaf hole TARPs. Although flow rates were much lower as observed in goaf holes from 7-47 to 7-63, a much longer active/production days (>40 days) was achieved in these holes due to high methane purity and low CO presence. Note that two goaf holes (7-20 and 7-MG19), which achieved the best gas drainage performance, accredited to the long production period, stable methane flow rate (>400 l/s) and high methane purity (~90%).

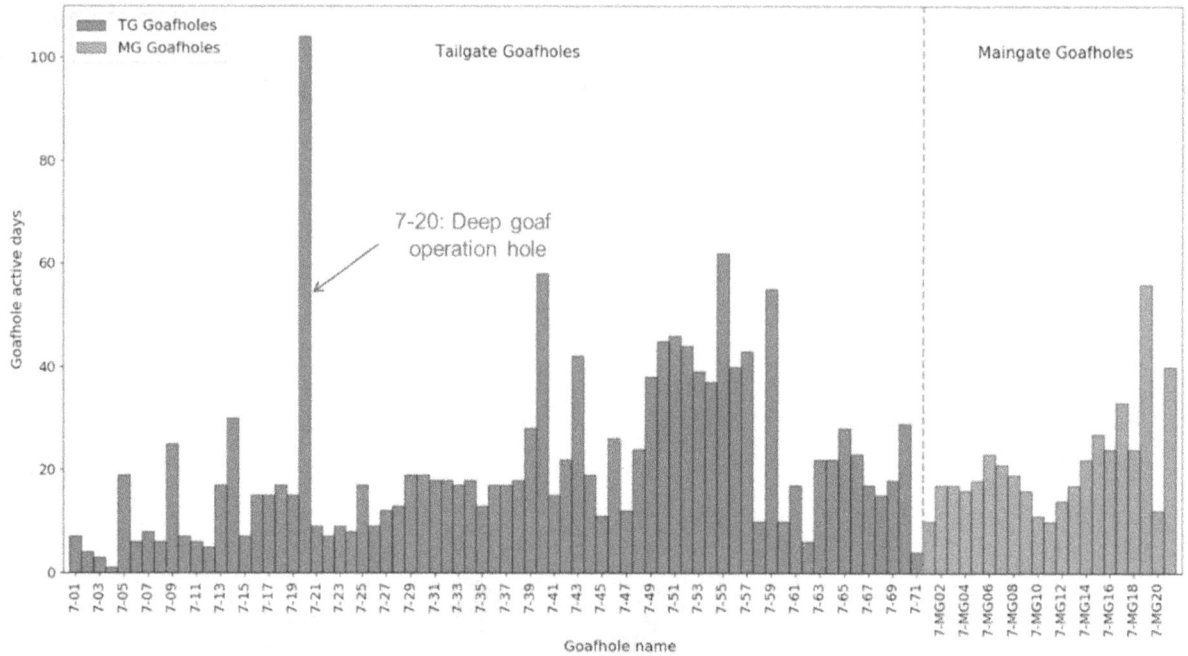

Figure 3 The active production time for individual goaf holes

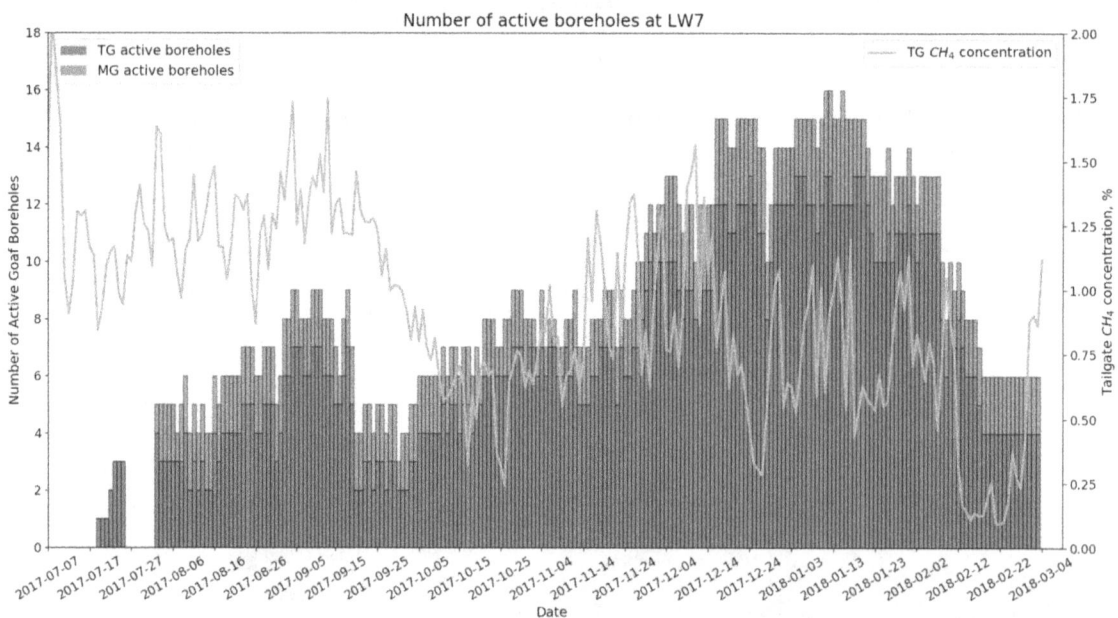

Figure 4 Number of active goaf holes at each longwall production day

The number of active goaf holes in each coal production day during the lifespan of LW7 panel is shown in Figure 4. As explained in the previous figure, in the start-up stage between 17 July to 17 Sep 2017, the low life span of goaf holes results in less active holes (~6) and higher CH_4

concentration observed in the tailgate (~1.25%). On the other hand, there were 14 active goaf holes in average between 2 Dec 17 to 2 Feb 18, which significantly reduced CH_4 concentration in tailgate to ~0.75%. This further demonstrates that with increased size of the goaf gas reservoir, it is important to increase the number of deployable goaf holes to manage the TG gas levels.

Histograms of average methane flow rate and average methane purity in each goaf hole are shown in Figure 5. The average methane flow rate in over three-quarters of goaf holes is higher than 400 l/s and nearly 65% of goaf holes presented methane purity higher than 70%. In general, compared with the TG side, goaf holes drilled at the MG side have better gas drainage performance in terms of higher methane flow rate and methane purity. This can be attributed due to MG holes having access to larger source of gas reservoir from adjacent undrained panels.

Methane production curves for a number of goaf holes are shown in Figure 6, where side by side (linear neighbouring) goaf holes were plotted. In general, neighbouring goaf holes presented similar gas production trend. Among all goaf holes, a sharp increase of methane production rate was observed within the first five production days. Peak flow rate normally achieved at the fifth production day and then was followed by a gradual decline until the 20th day. If the goaf hole was still active, a long and stable tail with around 400 l/s methane flow rate would be generally observed. The long and stable tails from these methane production curves suggest most of the goaf holes in LW7 were prematurely switched off using either CO TARP or the need to deploy the well-head for the next outbye hole, were seriously under-performing and there is significant potential to increase goaf hole performance by extending their drainage lead time and thus assisting the management of longwall TG gas levels.

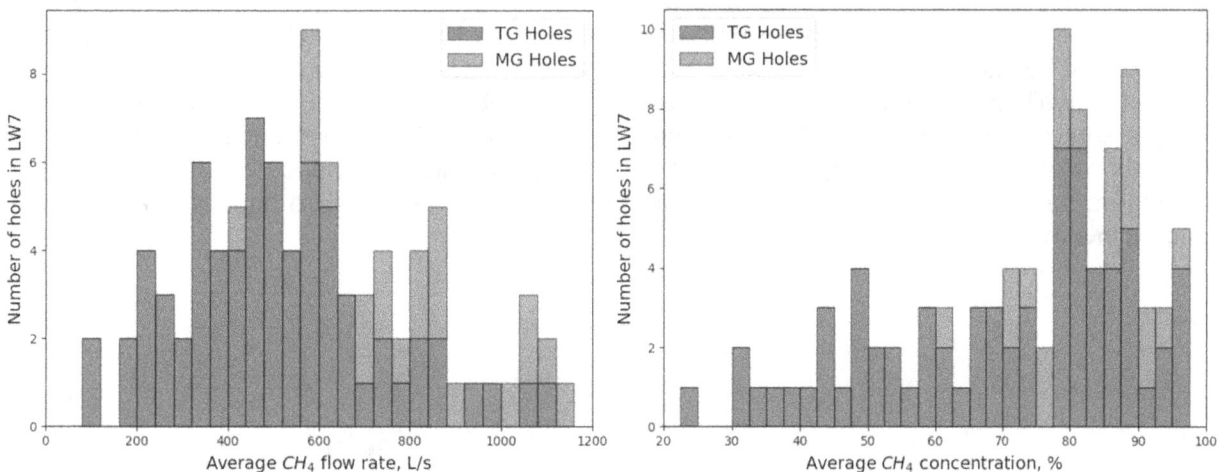

Figure 5 Histograms of number of active goaf holes as the variation of average methane flow rate and concentration

Goaf Hole Gas Composition Analysis

The following paragraphs provide an analyses of gas composition of monitored goaf holes. For each goaf hole, the daily captured gas composition was manually monitored over its production period. Heatmaps illustrating the concentration of CO, O_2, and CH_4 for all the tailgate and MG side goaf holes are shown in Figure 7. Each small colour-coded block in these heatmaps represents gas level for a specific goaf hole at one production day. Non-active goaf holes or production days are not coloured. Thus, for each goaf hole, the horizontal length of coloured blocks represents the active operation time (days) of that hole. Similarly, at each production date, the vertical length of coloured blocks represents the number of active holes on that specific day.

As expected, a strong correlation between CO, O_2, and CH_4 levels can be observed, i.e., O_2 percentage is positively correlated with CO and negatively correlated with CH_4. As expected, at the start-up stage of the panel, a high percentage of O_2 was repeatedly monitored in corresponding surface goaf holes, in particular at the TG side, which coincided with the higher levels of CO and consequently the early termination of these holes as a result of TARPs. Higher levels of CO and O_2 measured during the goaf hole start-up is merely a reflection of goaf gas composition immediately behind the longwall face of coal oxidation. It may be possible that the higher velocity of goaf gas flowing past the upper seams may also be contributing to CO generation. There is a misperception that the CO levels recorded is an indicator of elevated levels of sponcom activity in the active goaf. However, based on the recent sponcom incidents of coal mines in Australia, it is suggested that the measured CO levels are an order of upto 1000 ppm higher than the levels recorded from the active goaf drainage holes. Similarly, the CO levels recorded in the Goonyella longwall seam is higher than the German Creek seam longwall. Note that the goaf hole 7-20 was shut-off at the early stage due to high CO but reopened after two months and only a negligible amount of CO was found after re-opening the goaf hole. From mid of October 2017, O_2 level was largely reduced along with a notable improvement in CH_4 purity and production time. During the panel completion stage (early February 2017), another CO increase trend was observed in outbye goaf holes 7-64 to 7-71. Note that high CO or low CH_4 blocks tend to cluster at neighbouring goaf holes or consecutive production dates, which indicates the spatial and temporal continuity of drained gas quality.

The average CO, O_2 and CH_4 concentration over the lifespan of individual holes are shown in contour plots in Figure 8, where the correlation between three gases can also be observed. Goaf gas drainage data collected from LW6 and LW8 were also used here to generate these contours. Note that tailgate goaf holes close to the face start-line (within 800 m outbye) tend to have high CO, high O_2 and low CH_4, which reflects the challenging CO condition during the start-up period and they are certainly not reflective of heightened spontaneous combustion activity. Also, there is no correlation between underground bag samples from sealed areas and goaf hole CO levels. Similar gas composition was observed in goaf holes next to each other, which suggests the gas drainage performance for goaf holes are also spatially correlated.

Compared with the delayed response of CO level, the presence of O_2 or the drop of CH_4 purity suggest that the longwall retreat gas reservoir area is of low rate of gas emission and goaf hole is operating just behind the longwall faceline collecting the oxygen rich goaf gas. It is to be noted that in these longwall panels, there is no large volume of coal gets left in the longwall goaf. Based on the goaf gas analyses of historic 17 LW panels, the use of CO levels from goaf holes have not provided any relationship between underground sponcom activity. On the other hand, the use of lower level CO triggers based on coal gas evolution tests have resulted in premature shut-off the goaf holes resulting in elevated TG gas levels.

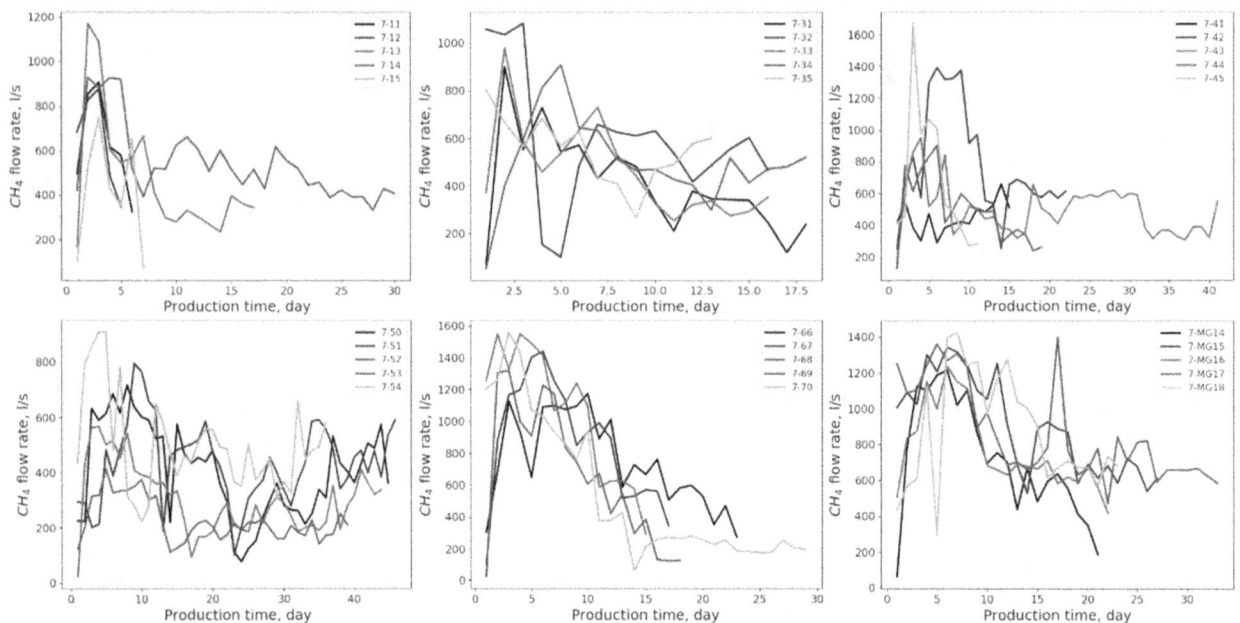

Figure 6 Methane production curves in side by side neighbouring goaf holes

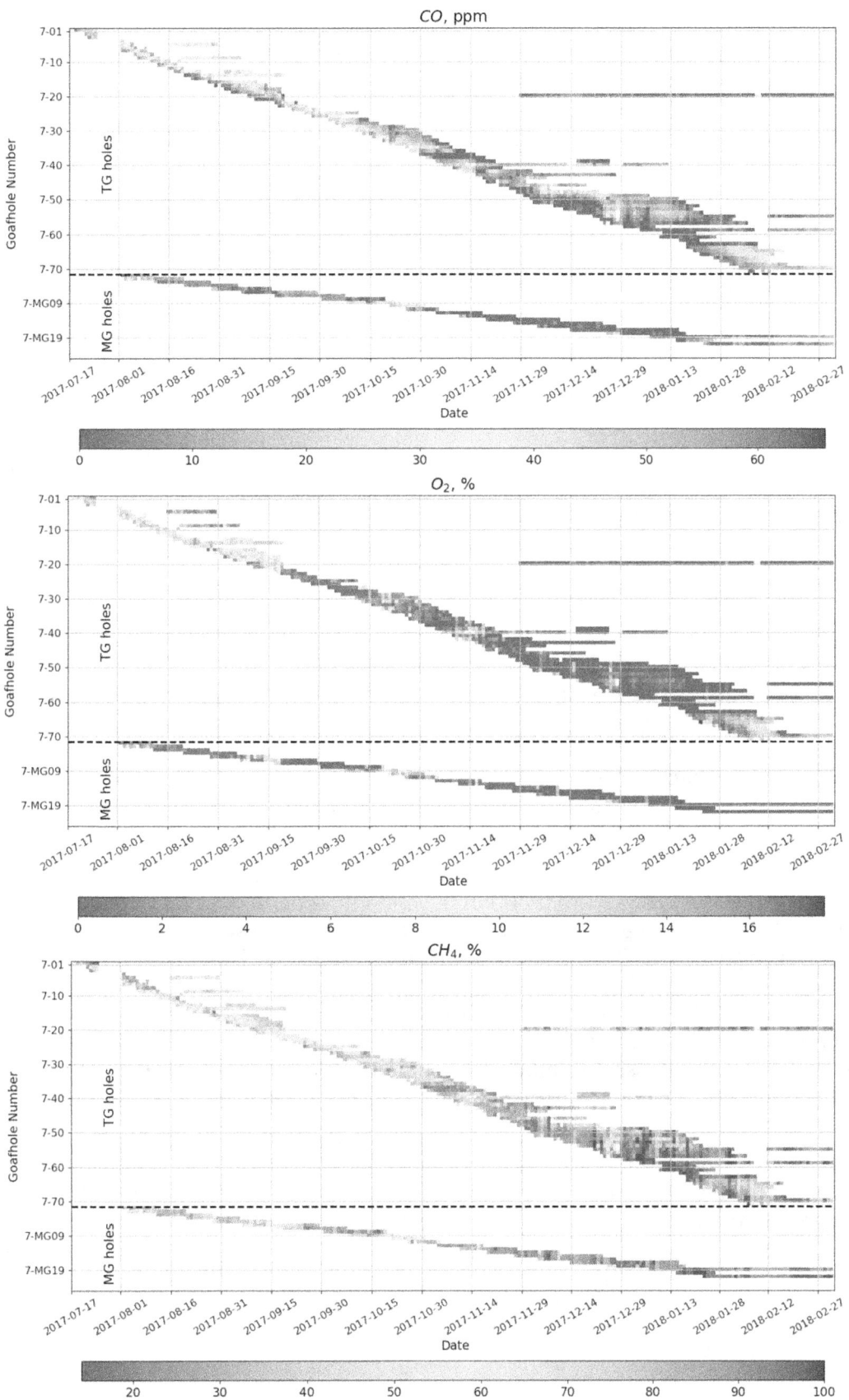

Figure 7 Composition of drained gas in all goaf holes during the LW7 retreating period.

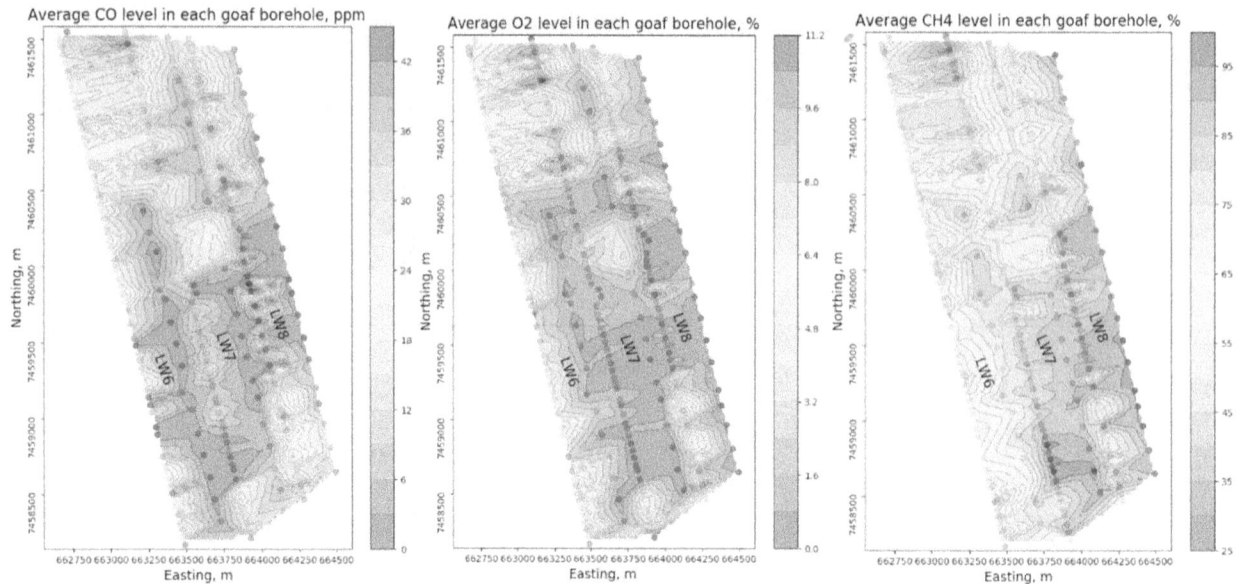

Figure 8 Contours of average gas levels of individual goaf holes during each production period in LW6, LW7 and LW8.

Goaf Gas Profiles

The gas composition measurement results from goaf holes at different distances pertinent to the longwall face during the start-up period (July to September 2017) and completion period (Dec 2017 to Feb 2018) are plotted in Figure 9 and Figure 10, respectively. A clear trend of gas profiles can be observed in the goaf during these periods, which are consistent with the field observations. As discussed earlier, the observation of high CO level led to the premature shut-off of goaf holes, which largely limited their degasification effect. However, according to goaf gas profiles, the presence of high CO level is most likely due to the natural behaviour of goaf closure, which is characterised by an active zone close to the face and an inert zone far deeper in the goaf.

From the gas profiles in Figure 9 and Figure 10, it can be concluded that the active zone is around 300 m wide and the inert zone is from 300 m beyond in the deep goaf and these would vary from mine to mine. As moving deeper into the goaf, the active zone shows an increase trend of CO and CO_2, a relatively high level of O_2, and a mix of CH_4 level response. The gradual increase of CO level in the active zone is a reflection of normal coal oxidation in an oxygen-rich and high moisture and hot goaf environment. On the other hand, in the inert zone, a general declining trend of CO and O_2 can be observed, together with an increase of CH_4 and CO_2 concentration in deeper goaf. This indicates the continued emission of methane from upper and lower destressed coal seams in the deep goaf and depletion of O_2, which leads to the progressive decline of the CO level.

The identification of the active zone and its behaviour is of significant importance for the goaf gas drainage plan since it defines the 'normal' trend of gas behaviour in the goaf. The observed relatively high CO level in goaf holes may not be a sufficient shut-off trigger if these holes are in the active zone (ranging from 100 to 300 m behind the face). Furthermore, it is imperative to extend the degasification period of goaf holes since once these holes are 300 m behind the face (in the inert zone), they can produce a much purer methane flow with no association to sponcom activity.

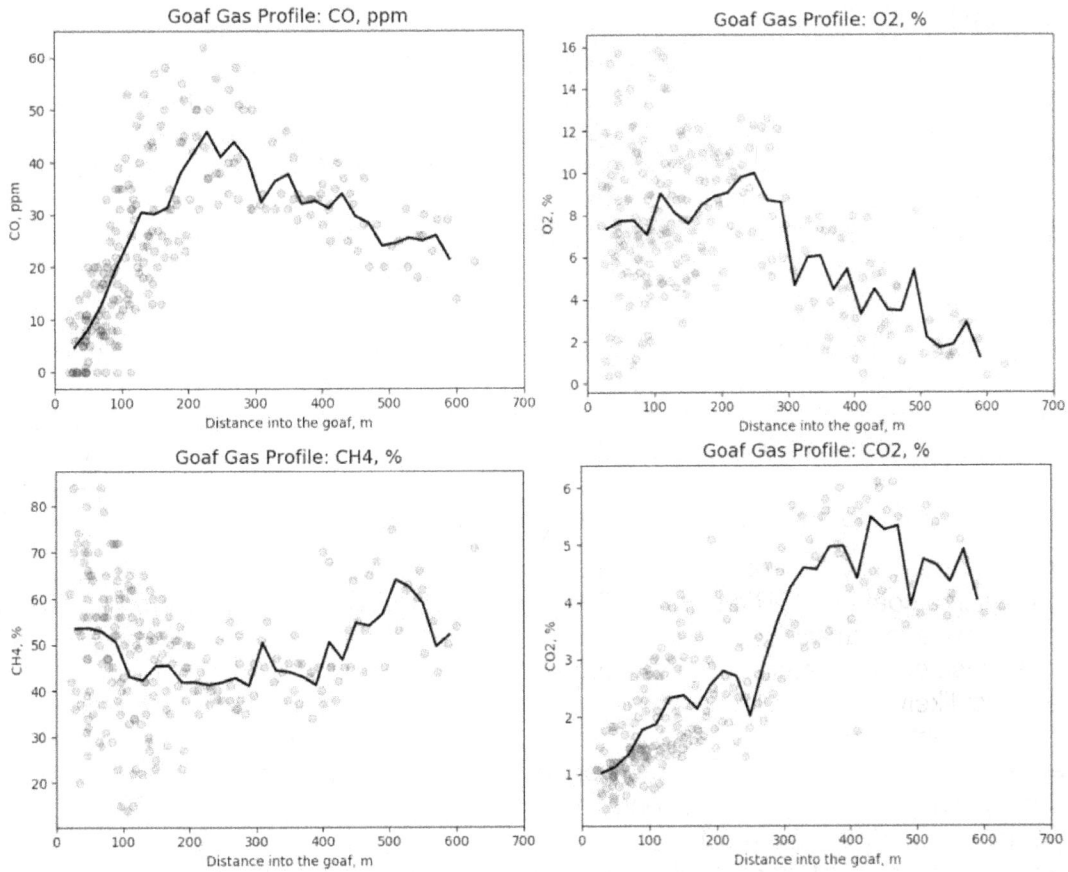

Figure 9 Goaf gas profiles from goaf holes at different distances pertinent to the longwall face during the start-up period (17/07/2017 to 17/09/2017).

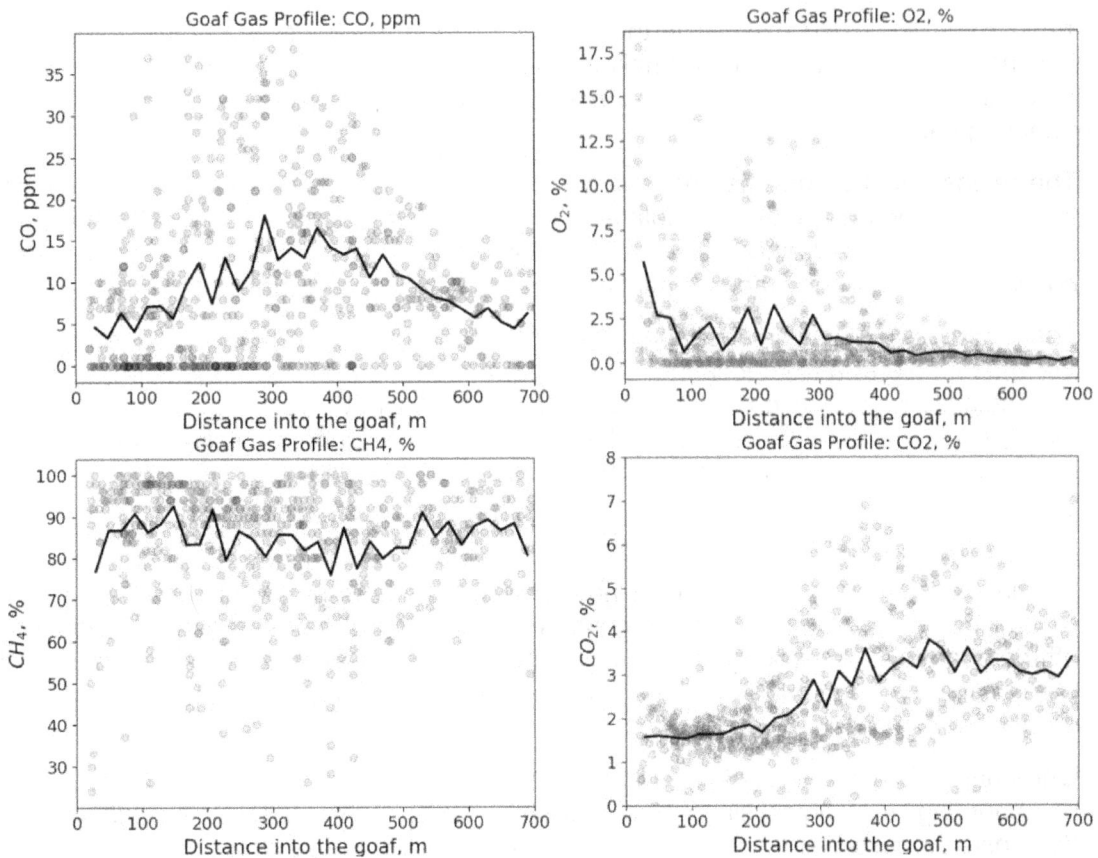

Figure 10 Goaf gas profiles from goaf holes at different distances pertinent to the longwall face during the completion period (12/12/2017 to 12/02/2018).

DISCUSSION

The current operational time for most goaf holes is too short with an average gas production period of 21 days with some holes operated for less than five days. All methane production curves in Figure 6 suggest the existence of a long tail with ~400 l/s methane flow rate and over 90% methane purity. A number of goaf holes with long production periods (7-20 and 7-MG19) also confirm that there is a large potential to increase the production period to 60-100 days without compromising methane quality or the presence of spontaneous combustion risk. In addition, as noted above, a goaf hole with longer production period also indicates that the face would be able to move further away from that hole and result in less volume of O_2 ingress.

Goaf holes drilled in the panel start-up region and completion region observed high level of CO concentration. The high CO level triggered the premature closure of goaf holes, particularly in the start-up region due to the TARP. These problematic goaf holes had relatively low methane flow rate at low methane purity. Goaf holes with low flow rate but high methane purity were also observed, such as 7-47 to 7-63, and these holes tend to have long active time. Goaf holes at the side generally have higher flow rate and higher methane purity compared with goaf holes at the tailgate side.

High CO concentration is a direct result of ventilation air migrating behind the retreating longwall and further ingress into goaf holes. Rich air (O_2) flow observed in goaf holes provides an environment for the slow coal oxidation as well as heat accumulation around the goaf hole vicinity. This leads to a delayed response of CO release and consequently observed in captured gas. The air sucked in goaf holes was most likely from ventilation air leakage while fresh air passed through the longwall face. It is reasonable to anticipate that as the face moves away from a goaf hole, the amount of ventilation air migrate to that goaf hole will be much lower. This can be validated by goaf gas profiles (Figure 9 and Figure 10), where CO/O_2 level dropped and CH_4/CO_2 level increased as the face moved 300 m away from that hole. While the reason behind the introduction of CO level trigger to switch-off the goaf holes is not known, it has certainly affected negatively on the optimum goaf hole operation to minimise the TG gas levels. Therefore, it is recommended to re-consider the usage of CO level in TARPs to switch off the goaf holes or consider the possibility to measure goaf gas temperature as an indicator of the elevated level of coal oxidation.

The current CO-based TARP resulted in the premature closure of a large number of goaf holes in the study mine. Findings from this data analyses suggest it is arguable to not simply use CO as the TARP for identifying spontaneous combustion activity to shut-off goaf holes. A number of reasons are summarised below:

- The release of CO from the coal oxidation process is a delayed response, which requires time for CO percentage to accumulate to the trigger level. Compared with CO, the consistent presence of O_2 in drained gas can be observed earlier and should also be taken into consideration regarding the potential elevated levels of coal oxidation process.

- The progressive longwall retreating causes dynamic changes to CO, O_2, CO_2 and CH_4 levels. For a goaf hole at a fixed location, even though high CO has been observed soon after the goaf hole starts production, the rapid advance of the face will soon isolate the goaf hole from the O_2 source (ventilation air) behind the longwall face. A general declining trend of CO level as the face moves away from that goaf hole can be anticipated.

- The current CO-based TARP focuses on a single threshold value rather than the trend over a consecutive period. As noted in this analysis, the accumulation of CO normally shows an upward trend given that sufficient O_2-rich environment lasts for a long period. A single record of maximum CO value may rise the concern but should not be treated as the trigger to shut-off the goaf hole. Temporary increase of CO may be caused by a slow-down or stoppage of the longwall face, and the CO level will decline to an acceptable range once normal face production resumes.

- Once high CO has been detected in a goaf hole, it is useful to examine the CO level in its neighbouring goaf holes since CO release is found to be spatially correlated. A robust diagnosis process be placed underground using a tube-bundle and bag sample monitoring regime rather than using goaf hole gas composition to evacuate the mine for spontaneous combustion risk.

CONCLUSIONS AND RECOMMENDATIONS

This paper analysed gas drainage performance from vertical surface goaf holes in an operating mine with high specific gas emission rates (~20 m³/t). Field manual monitoring results of daily goaf gas flow rate and gas composition collected from individual goaf holes were analysed. The major challenge of implementing the full potential of goaf drainage system is the perceived risk of sponcom in a goaf using *arbitrary CO trigger set points, without any scientific basis or empirical evidence.* Based on the analysis of extensive dataset from the study mine, this analysis suggests that the following approaches can be taken to maximise goaf hole gas drainage performance without compromising underground sponcom risk management:

- The start-up period and completion period of a longwall panel showed a general trend of CO increase, which indicates the presence of conducive environment for coal oxidation. The deployment and operation strategy of goaf holes in these two periods should be carefully reviewed with specific reference to the use of CO trigger level.

- Optimum control of goaf gas drainage rate based on the distance between the goaf hole and face can be applied, whereby goaf holes are operated in low flow rate when they are close to the face but high flow rates when they are away from the face.

- From an explosion management perspective, the key trigger gases to be monitored are oxygen and methane rather than the use of CO as a trigger to control the goaf hole operation. It is strongly recommended that no goaf holes should be operated at methane concentrations below 30% regardless of the oxygen concentration to foolproof the probability of explosion risks. If there is a consistent presence of O_2 were observed in a number of neighbouring goaf holes, temporary shut-in of these holes should be applied as per the leading practice. Reopening of these holes can only be considered when the face is at least 300 m away from these holes, and their gas composition should be reviewed to ensure the O_2 concentration has dropped to an acceptable level.

- The analyses of data as well as empirical evidence clearly suggested that the use of CO gas as a trigger in goaf hole to identify underground sponcom activity could not be established and use of CO levels based on goaf hole gas composition for mine evacuation is flawed.

- While the investigation into the origin and introduction of CO trigger in TARP could not be established, the paper provides adequate background and the basis to eliminate or revise the use of CO triggers in current surface goaf gas management TARPs. The continued mis-use of CO as a trigger for underground sponcom activity has a detrimental effect on longwall TG gas management for explosion prevention due to early goaf hole termination despite higher levels of methane purity.

- Longwall operations must continue to deploy and monitor underground goaf environment using appropriate tube bundle monitoring and bag sample regime, real-time longwall tailgate airflow and CO monitoring for early indication of sponcom related activities.

REFERENCES

Balusu,R., (2016). Personal communications, CSIRO, Australia.

Belle, B., (2017) *Optimal goaf hole spacing in high production gassy Australian longwall mines- operational experiences,* in Proceedings Australian Mine Ventilation Conference, Brisbane, 28-30, August.

Moore, T.D., Deul, M. and Kissell, F.N., (1976). *Longwall gob degasification with surface ventilation boreholes above the Lower Kittanning Coalbed* (Vol. 8195). US Department of the Interior, Bureau of Mines.

Ventilation of waste and orepasses, tramming routes and drawpoints

D J Brake[1]

1. FAusIMM(CP), Director, Mine Ventilation Australia, Brisbane QLD 4017.
   Email: rick.brake@mvaust.com.au Adjunct Assoc Professor, Resources Engineering, Monash University, Melbourne

ABSTRACT

Many modern metalliferous mines continue to use ore and waste passes to transfer broken material vertically through the operation. Handling broken material produces dust and any open vertical connections between levels provide opportunities for short-circuiting, unwanted low flows ("dead spots"), flow reversals and recirculation. As loaders and trucks have become more productive by using larger buckets and trays and can handle a larger size fraction, their loading and dumping rate is much faster, so orepasses have also become physically larger in diameter. Loaders are also now frequently used in teleremote and even autonomous operation. Reducing mine capital costs has meant vertical development for exhaust raises or other connections to ventilate orepasses has often been removed from mine designs. Ventilation controls for large headings and passes are very expensive and more easily damaged by heavy equipment impact and blast overpressures than previous smaller controls. In some cases, finger raises into the pass system have also been removed. The number of passes used for a given production rate has been reduced either because loader productivity is much higher than in the past, or to lower capital costs, and this has resulted in even more pressure to use multi-level tipping into one pass from separate levels to increase pass utilisation and productivity. At the bottom of passes, traditional rail haulage has been replaced by loader or truck haulage, and chutes at the bottom of passes have been removed due to cost and delays to commencement of production from that pass. All of these changes have resulted in much greater difficulty in providing satisfactory ventilation outcomes for mines using ore or waste passes. This paper discusses the impact of these changes on available controls for orepass ventilation and tramming routes, as well as crusher and conveyor ventilation, and offers some additional solutions for these systems.

INTRODUCTION

Definitions

In this paper, the following terminology and definitions have been used.

- *Orepass, Finger, Tipple, Orepass Access* and *Tramming Drive* (See Figs. 1 and 2).

- *Chute* is the (optional) device at the bottom of a pass that allows controlled loading of material from the pass into a rail car or truck or crusher

- *LHD* for "loader", "bogger" or "digger"

- *Mucking* for "loading", "bogging" or "digging"

- *Tramming* is the process of carrying the broken material between the mucking and tipping points

- *Drawpoint (brow)* is the "brow" where the LHD mucks the broken material from a stope

- *Drawpoint access* is the horizontal connection between the drawpoint and the tramming drive

- *Multi-level tipping* occurs when connections into an orepass are simultaneously open on two or more levels, usually to allow concurrent tipping into the pass on those levels

- *Single-level tipping* occurs when tipping only occurs into an orepass on one level at any time, and all other connections into that pass are sealed off

The paper uses the term "orepass" but the comments are equally applicable to waste passes. In addition, the scenarios are usually described in terms of an LHD but are usually equally applicable to trucks.

Material is usually tipped into an orepass via a *finger*. This is for safety and operating reasons. On the top tipping level, material might be tipped directly into the pass (not via a finger), but this is also undesirable. For safety reasons, the tipple is usually located at the end of a short access off to the side of the main travel way (the tramming route). This allows the use of tipple covers and safety barricades at or before the tipple, and also ensures the LHD operator has good visibility of the tipple during the approach and the LHD is not "bent" when it is tipping.

There are many important safety and productivity issues relating to the design and operation of orepasses that are not related to ventilation. There have been many fatalities associated with orepass operation, including falling into the pass, being struck by rocks coming out of a pass, being drowned by mud rushes out of passes, caught by falling rock bringing down hangups in passes, refurbishing passes and their linings etc. For these and other reasons, good practice is that any mine using orepasses should have an "Orepass Management Plan". For a description of general orepass design issues other than ventilation, refer to Hadjigeorgiou et al, 2005, Hadjigeorgiou and Stacey, 2013 and Queens University, undated.

In modern mining where stopes have flat bottoms and mucking is via remote or teleremote LHD, an important source of dust and other ventilation problems can be short-circuiting through open stopes particularly when the stope rills. This dust is best controlled by installation of temporary ventilation controls on the upper open levels of the stope such as Nixon flaps and water sprays.

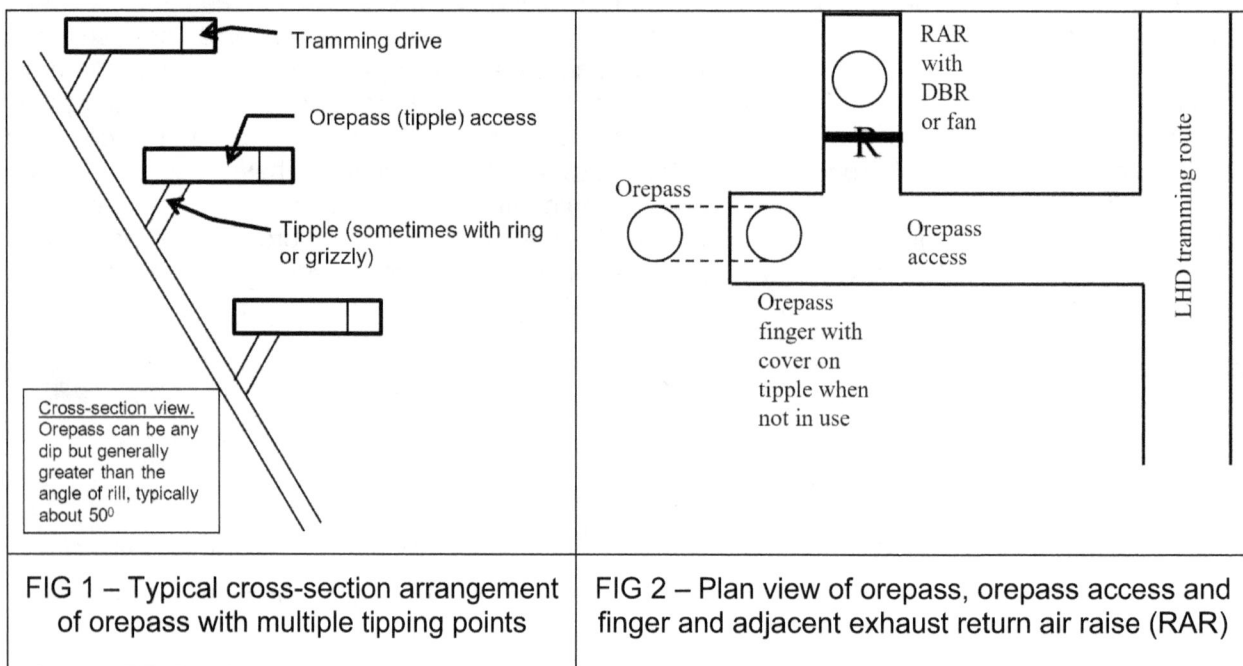

FIG 1 – Typical cross-section arrangement of orepass with multiple tipping points	FIG 2 – Plan view of orepass, orepass access and finger and adjacent exhaust return air raise (RAR)

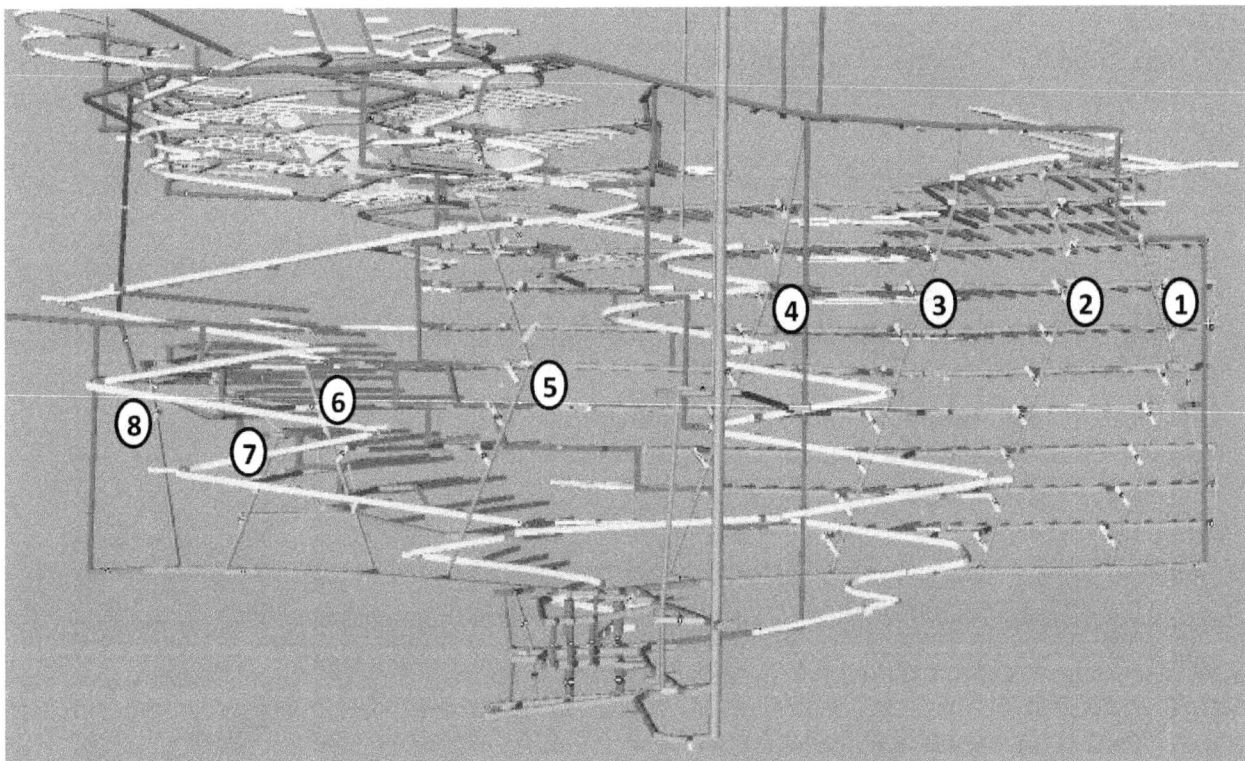

FIG 3 – This operation has 8 orepass systems, most with multi-level tipping, to achieve 3.5 Mtpa

History of orepass ventilation and original design concepts and 'rules'

Prior to the advent of diesel loaders, mine productivity was low. Blastholes were small diameter and short in length so the burden and spacing was small (and generally so was fragmentation). Orepasses were loaded by scraper, compressed air-powered rocker shovels or other low productivity equipment. Due to the small buckets and good fragmentation, passes were small in diameter. If there were multiple tipping points into a pass (from different levels), passes often had wooden cover doors to stop short-circuiting of air, and chains (which opened and closed slowly) controlled the flow between levels within the pass so that the maximum distance the muck had to fall (and hence the piston effect produced by falling muck) was modest. Muck was usually also wet. Where tipples were near access drives that carried fresh air, a dust scrubber on the tipple might be used to incast the tipple (when the cover door was open) to keep fugitive dust levels low. The rule of thumb was to downcast the tipple by 0.5 m/s (e.g. if the tipple had a 10 m² opening, then it would incast 5 m³/s which was then scrubbed and the scrubbed air returned back out to the tipple access) but this guideline was to capture dust particles with the assistance of gravity; higher capture speeds were required for coarser dust or for vertical upwards capture (usually about 1 m/s). For all these reasons, where there was only a single level tipping into a pass, orepass ventilation was usually not very important.

By contrast, modern mines have large loaders (or trucks) with large buckets (or trays). They load quickly and when they dump, a large plug of material is dropped almost instantly into the pass creating a significant piston effect and dust problems both on the tipping level and any levels below the tipping level.

VENTILATION OF OREPASSES AND TIPPLES

As mucking, tramming and tipping operations produce dust, the ventilation engineer *must* have a strategy for dust control at each location: the drawpoint and drawpoint access, the tramming drive, the orepass access, the tipple, the loading point at the bottom of the pass and the tramming route at the bottom of the pass. Where the orepass is open on more than one level at a time, then a ventilation strategy for every connection into and out of the pass is required.

Sources of dust in orepass operations

There are four main sources of ventilation problems relating to orepasses:

- Dust produced at the actual tipping point due to the fine dry material becoming airborne as it falls from the LHD bucket or truck tray to the ground or into the top of the tipple etc. This dust can make visibility difficult for the LHD operator and can get into the tramming drive and disrupt downwind working areas,

- Dust coming out of the tipple on one level ("belching") due to tipping operations into the pass from other levels and then disrupting downwind working areas,

- Short-circuiting of air due to the pass being open (and incasting) on one level and outcasting on another level. Even if there are no tipping activities, this can result in low flows, flow reversals or even recirculation of auxiliary fans on one or both levels. In turn this can result in temperature problems, gas or DPM issues, and slow clearance times after blasting,

- Dust produced when pulling from the pass at the bottom and tramming the material via rail, truck or LHD to either a crusher or to the surface.

It is also necessary to distinguish between a *production* situation (where an LHD is tipping into a pass for an entire shift and usually for many shifts in a row) and a *development* situation (where the LHD is only tipping at most for a couple of hours per shift, often less frequently if there are multiple faces in different locations). The reason to make this distinction is that the cumulative dose of dust or other contaminants for persons exposed is much higher when the exposure is continuous than when it is "only" a few hours per shift. In addition, it may be possible to temporarily cease downwind activities of a development LHD without major impact on the operation, but it may not be possible to cease activities downwind of a production LHD.

Orepass pressures

The mechanism by which pressures rise and fall in orepasses and the factors influencing this have been analysed by many researchers including Linhart and Pollack (1980) and McPherson and Pearson (1997). Various control strategies were tested at the Freeport mine, particularly trialling a recirculation system for air in the pass and a pressure relief system, with varying success (Casten et al, 1999; Calizaya F and Duckworth K (2004)). Critical factors were found to be the size of the pass and its verticality, as well as the rate (tonnes per hour) at which material is fed into the pass. Continuous, high-rate tipping (e.g. from a conveyor) into a vertical pass is particularly problematic as air pressures in the pass of up to 100 kPa have been recorded along with significant and hazardous "belching".

A comprehensive study including measurements and development of predictive equations was conducted by Moss et al (2005). Moss found that peak pressures in the pass in that particular application were about 3 kPa. He also noted that as the tipped material falls down the pass, positive pressures are generated below the falling material and negative pressures above the falling material.

The pressure rise in a pass below the plug of falling material travels at the speed of sound (about 300 m/s) so the increase in pressure is almost instantaneous along the full length of the raise (for typical orepasses of up to a few hundred metres in length). If the pass is unvented, then the increase in pressure below the falling plug has a corresponding zone of negative pressure above the plug, which results in the typical cracking failure due to flexing of the walls that seal off orepass connections. In typical orepass situations, the plug of falling rock attains a steady-state speed of about 15 m/s with the main drag (limit on speed) coming from the interaction of the falling material and the walls of the pass.

Dust controls and circuit design

Water sprays and scrubbers

In all cases, fugitive dust emissions will be reduced if the material is thoroughly wet prior to tipping. This is particularly important given that water sprays are only useful to prevent dust entering the air; once the dust is in the air, then even an "atomised" (very fine droplets) water spray system is of little use in removing dust from the air. Water sprays at the tipping point are less effective than sprays at the point where the material is picked up in the LHD bucket (e.g. the drawpoint); water that gets into the pass can create material handling problems and safety issues (mudrushes) at the chute or at the bottom of the pass or plugging of (blockages in) the pass. Where water is used, it is important that the sprays at the tipple only come on when the material is actually tipped into the pass (and then turned off).

Whilst dust scrubbers are a potential way to remove dust from the air, in practice the amount of dust produced during tipping in modern operations usually overwhelms most dust filtration systems. Where scrubbers are used (e.g. crusher excavations), they tend to be large and expensive. They need to be able to handle high humidity and potential condensation (NIOSH, 2003). They are impractical in terms of routine orepass dust control in mines (Moss et al, 2005).

Orepass exhaust systems

In an ideal situation, an orepass would only be tipped into on one level at any time (single-level tipping), although this includes two LHDs tipping into the one pass on the same level. All other levels would be sealed off. This prevents any dust being pushed out by the air compression in the pass onto the lower level. With single-level tipping, an orepass exhaust may not be needed for the pass itself. However, the orepass *access* (from the tramming drive to the tipple) usually needs ventilation and unless this is done with auxiliary ventilation (e.g. ducted exhaust into a return airway), then it is easier and more reliable to put in some exhaust ventilation on the orepass.

Single-level tipping can only be achieved if there is one orepass allocated to each producing level and these passes are within an efficient tramming distance of the stope drawpoints. Separate development passes for surplus waste would also be required (or ore and waste would need to be batched through the one pass) and careful scheduling would be needed to not "double book" one pass to two concurrent stopes on two different levels (which will force the operation into multi-level tipping into that pass). However, even where such provisions have been made, multi-level tipping is still likely at least on some occasions.

Where an LHD is tipping infrequently into the top of a pass (e.g. development mucking), and especially if there are no downwind activities or these can be relocated for the few hours when mucking is in effect, then it may not be essential to have an exhaust at that tipping point. However, where the orepass access is longer than the LHD, then there remains the issue that the LHD itself needs to be ventilated due to its own exhaust gases, DPM and heat in the access especially when the access is very much longer than the LHD.

Where the LHD is tipping constantly into a pass (e.g. production mucking), or into a crusher, then an exhaust at the back of the tipple (or at least somewhere in the tipple access) is needed to ventilate both the access to the tipple and the tipple itself to remove any dust directly into the return preventing it from entering the tramming drive.

Where possible, production mucking from high tonnage drawpoints should be set up in its own "dedicated" ventilation circuit (providing "one pass" ventilation) so that no person is working downwind of the LHD tramming route. Whilst such a dedicated circuit is also desirable for development mucking, it is usually the case that air from a development end is relatively clean except when mucking is in progress (a few hours per shift or even less) and so can then be reused in some other workplace underground. In some mining methods, the amount of time spent by a production loader in any given portion of its tramming route is also small in which case the situation may become similar to that of a development loader with respect to contamination of downwind workplaces.

A tipple exhaust is particularly important where two or more LHDs are tipping into the one pass (e.g. from two different drawpoints) especially if there is more than one access to the tipping point or in circumstances where the trams are very short as the amount of dust produced at the tipping point is much higher than long trams with a single LHD. In both these cases, strong exhaust flows are needed to keep the tipple area clear for visibility and safety. As a guideline the minimum airflow would be about 1 m/s per access point to the tipple or 0.05 m^3/s per kW for the diesels in that ventilation split, whichever is the greater. Certainly the "100% air change time" in the access should be faster than the cycle time of the LHD.

Where a grizzly and rockbreaker are used over the tipple, then the rockbreaker operator should either remotely control the rockbreaker or be inside an air-conditioned cabin with filtered air.

Multi-level tipping should not be used at all in some circumstances, such as where the dust is highly toxic, or in uranium mines, due to the time-related growth of radon daughters and radioactive nuclides being attached to dust particles.

Ventilation controls on passes

Tipples into the orepass can be sealed off on non-tipping levels (Figs. 2 to 8) by:

- A brick wall or other ventilation control such as a door or rubber flaps across the orepass access. However, these need to be able to cope with the pressure of the air wanting to escape from the pass. Less effective than a wall is a ventilation door or strips of rubber flap or a curtain hanging in the orepass access. Where flaps are used, the individual vertical strips should "overlap" and be in a suitable steel frame to reduce leakage across the back and walls. Ideally the flaps should be at least far enough back from the tipple that the volume of air that is expelled from the tipple is not pushed past the flaps, assuming there is some form of exhaust on the orepass or at the tipple. The tipple access needs to be sufficiently long that the LHD is fully inside the flaps or door when tipping.

- A cover or cap or plug on the orepass tipple. These need to be robust, easy to lift on and off (or self-powered) and provide a good seal, as well as handling the positive and negative overpressures in the pass. Concrete plugs can work but fibreglass plugs are not sufficiently robust. Where high tipping productivity is not essential, then it is possible to raise and lower the orepass cover for each bucket tipped otherwise the cover is generally removed when tipping starts and replaced when tipping is completed. In most cases operators detest orepass covers as they reduce productivity and operators have even been known to deliberately sabotage them. In addition, even steel tipple cover doors often have insufficient weight to not "belch" dust when material is tipped into an upper level of the pass.

- Keeping the pass full above the level of that finger. However, this is usually not practical as the "live capacity" of an orepass finger is often very small, so that the interactions between "pulling" the pass at its bottom and filling it at the finger make this impractical. In addition, if the finger is left full of muck which has reactive mineralogy, the muck can set like concrete.

- Where passes are close together, two or more connections between the passes can allow the air to "recirculate" between the two passes when material is tipped into either of the passes as shown in Fig. 4 from Yourt (1969).

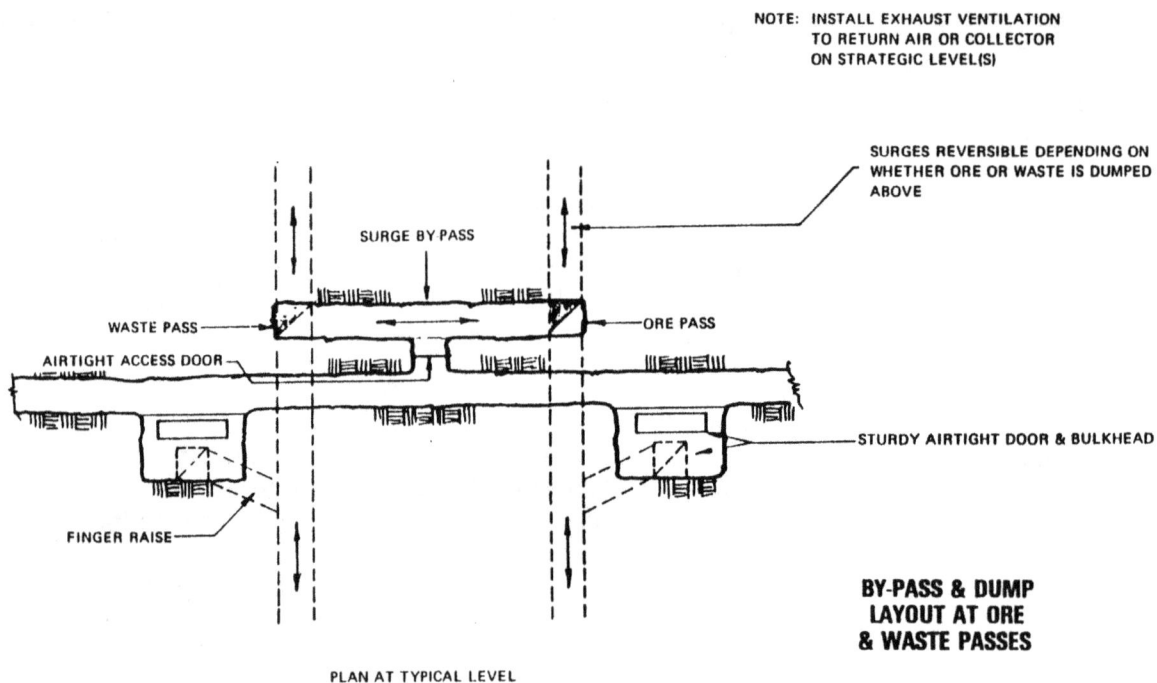

FIG 4 – Interconnected passes (on minimum of 2 levels) provides exhaust buffer capacity irrespective of which pass is being tipped into. Note the doors on each tipple into each finger raise.

Required exhaust flows

The minimum target airflow in the pass would be about 15 m³/s as this is the typical value needed to maintain 1 m/s inside a 3.5 x 3.5 m (or 3.5 m Φ) orepass as well as being sufficient to achieve a minimum wind speed of 0.5 m/s in the 5.5 x 5.5 m orepass access and sufficient for a 300 kW diesel loader at 0.05 m³/s per kW. Where exhaust is applied directly to the pass using circuit fans in a wall on the pass, it is important that at least one connection into the pass is kept open (or the fans turned off) to avoid putting the fans into stall.

Where multi-level tipping is employed, then the amount of air with dust being expelled on the lower open level will depend on how much air is "displaced" in the pass between the upper and lower levels. If these two tipping levels are close vertically, a smaller amount of dusty air will be expelled; if the levels are much further apart, then a much larger plug of air will be displaced. For example, if the orepass is 2.4 m Φ (4.5 m²) and there is 60 m vertically between the two open tipples, then the "swept" volume of the pass between the levels, when material is dropped into the upper tipple, and which can potentially be pushed out the lower tipple, would be 4.5x60=270 m³ (more if the pass is inclined). This assumes no wear in the pass and some other simplifications. If the tipple access is 5 x 5 m (25 m²), then 270/25=11 m of the orepass access on the lower level (ignoring any billowing effects) could be filled with dusty air when material is tipped on the upper level. In practice, the situation is not as simple as this, but this 11 m of orepass access would be an absolute minimum on the lower level that could be contaminated.

A further check is to examine the clearance time between tipping: if the LHD has a 3 minute tramming cycle, then ideally the 270 m³ of air potentially expelled from the pass would be sucked into the pass exhaust in 3 minutes which is a rate of about only about 1.5 m³/s. Obviously this would only rarely be the limiting design criteria.

In summary, to keep the orepass access clear and prevent dusty orepass air entering the tramming drive in multi-level tipping, experience indicates that taking in approximately 15 m³/s per open tipple is a prudent design value. Such a flow should also provide sufficient exhaust to ventilate the orepass access. So an orepass with 3 open tipples at 15 m³/s per level would require 45 m³/s as the target orepass exhaust. In practice, higher flows may be required as the air will be pulled preferentially from the upper tipple (assuming top exhaust), unless a system is used to achieve the design flows at each level. Providing 15 to 45 m³/s for an orepass may sound straightforward, but with the increasing cost of ventilated air in many mines, providing 15 to 45 m³/s per orepass system is a considerable ongoing operating cost, as well as the capital cost to provide the orepass exhaust ventilation connections and fan power.

Top, mid and bottom exhaust of passes

In terms of the ventilation connections into an orepass to provide exhaust for the pass, often the easiest and most robust solution is to "top out" the orepass on a dedicated return air level. The orepass exhaust is then obtained by putting circuit fans (the most reliable system) or a regulator (a less reliable system) on top of the pass on the return air level.

In some cases (e.g. block caving) a top exhaust is not possible in which case a mid-pass exhaust can potentially be used. A mid-pass exhaust can even be preferable to a top exhaust providing the mid-pass offtake cannot become blocked with material falling down the pass. In some cases, a new level that is currently not active (e.g. is developed ahead of time) or an old level that is worked out can be used as the top- or mid-level exhaust. Careful scheduling may allow the pass exhaust to move location as levels become active or inactive.

A third type of orepass exhaust is where there is a separate return air raise in the orepass access (Fig. 3). This RAR can either exhaust the access (and hence also the displaced dusty air) by a regulator or by circuit fans. Two circuit fans are ideal as this provides the ability to increase the exhaust from that individual tipping point as required, as well as providing some redundancy if one fan fails.

A variation on this theme is to set up an exhaust duct at the orepass so that the dusty air from the tipple is ducted away to a safe location, e.g. a return air raise or return airway. However, the duct may need to be large which is difficult if using rigid duct with a fan on the suction end. It is easier if a blowing fan is used with the fan located at the tipple. Mining a small chamber above the tipping point to act as a capture hood as the dust billows upwards and to protect the fan can help with this system. Any such orepass duct strategy assumes there will be no secondary blasting ("popping") of oversize material (e.g. material too big to pass into the tipple) near the duct.

If the lower level only has an LHD tipping into the pass, and the LHD has an effective sealed air-conditioned cabin, then the dust is more of a nuisance issue than a respiratory health issue, in which case it may be acceptable. Alternately, if there are fresh air raises on the level, then even if the travel way or tramming route on the level is dusty, it may still be possible to feed fresh air to other activities on the level, providing satisfactory working conditions.

Damage to orepass exhaust ventilation controls

Secondary blasting near the tipple must be avoided if there are ventilation controls or fans in the area (or a ducted system is in use as discussed above). However, orepass covers and ventilation controls in multi-level tipping remain subject to damage and it is essential that spares are held can be and easily installed.

Stope blasting can result in high overpressures on production levels and adjacent orepass connections. With modern mine automation, it is desirable for ventilation doors and fans in orepass accesses to be remotely controlled so they can easily be opened/turned off before blasting in the area and closed/turned back on after blasting.

MODELLING, THE VENTILATION CONTROL PLAN AND THE OREPASS MANAGEMENT PLAN

It is essential that the ventilation engineer models the impact of multi-level tipping (or orepasses being open on two levels at the same time, even if there is no tipping); this is true even if the mine planners or operators at feasibility study stage assure him that multi-level tipping will never happen. There are two reasons to model a multi-level tipping situation on each pass:

- Irrespective of anyone's assurances, multi-level tipping will almost always happen so understanding the impacts will help the ventilation engineer to assess contingency plans,

- Modelling of the impacts will help the ventilation engineer explain to the planners and operators what the impacts will be on the operation which may result in changes to the design or schedule.

The procedures and controls associated with orepass operation, and especially multi-level tipping, must be included in the mine's **Ventilation Control Plan** and also the **Orepass Management Plan**.

| FIG 5 – Grizzly on top of a pass | FIG 6 – Rockbreaker on top of a tipple |

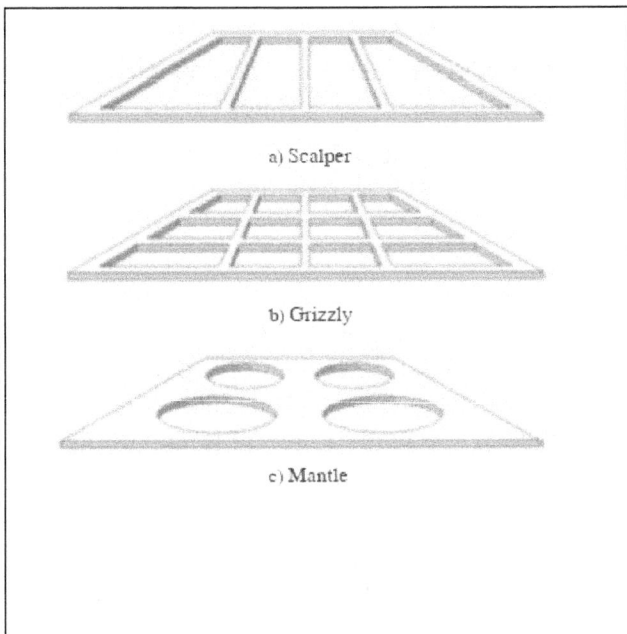

FIG 7 – Tipple screening systems
(Hadjigeorgiou et al, 2005)

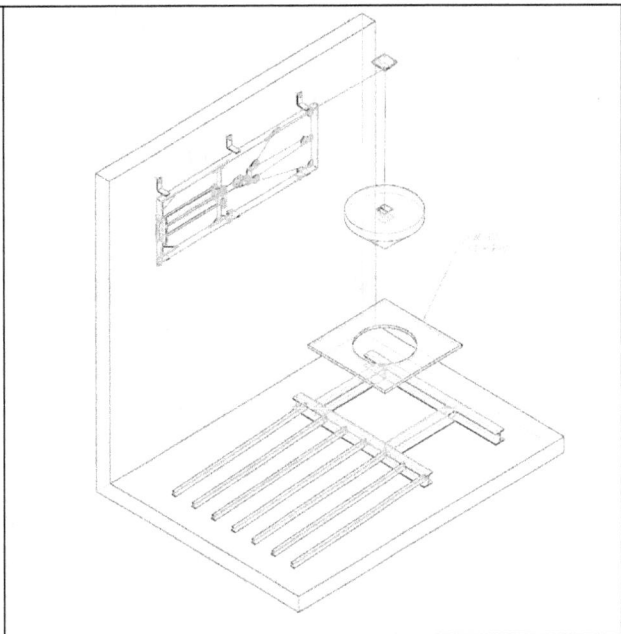

FIG 8 – Cone type orepass plug

FIG 9 – Hinged tipple door

FIG 10 – Plug installed and removed by LHD

FIG 11 – Hinged tipple door (Rock and Roll Mining, undated)

PULLING FROM THE BOTTOM OF THE PASS

Pulling the chute or mucking the bottom of the pass also produces dust. Even tipping into the pass will produce dust coming out of the bottom unless a minimum level of muck is kept in the pass.

Where possible, operation of chutes or mucking of the pass should be conducted remotely or autonomously. Flowthrough ventilation should be set up to take fresh air past the chute or pass and into a return.

VENTILATION OF TRAMMING DRIVES

Production loading and the tramming drive at the bottom of a pass (e.g. to the crusher) should have its own ventilation circuit, i.e. there should be no activities working downwind of a dust-producing production activity. Alternately other activities in blind headings in the same circuit should be fed fresh air directly by duct from a fresh air raise. In most cases, it is best for the air to go counterflow to the loaded LHD (Fig. 9 method A).

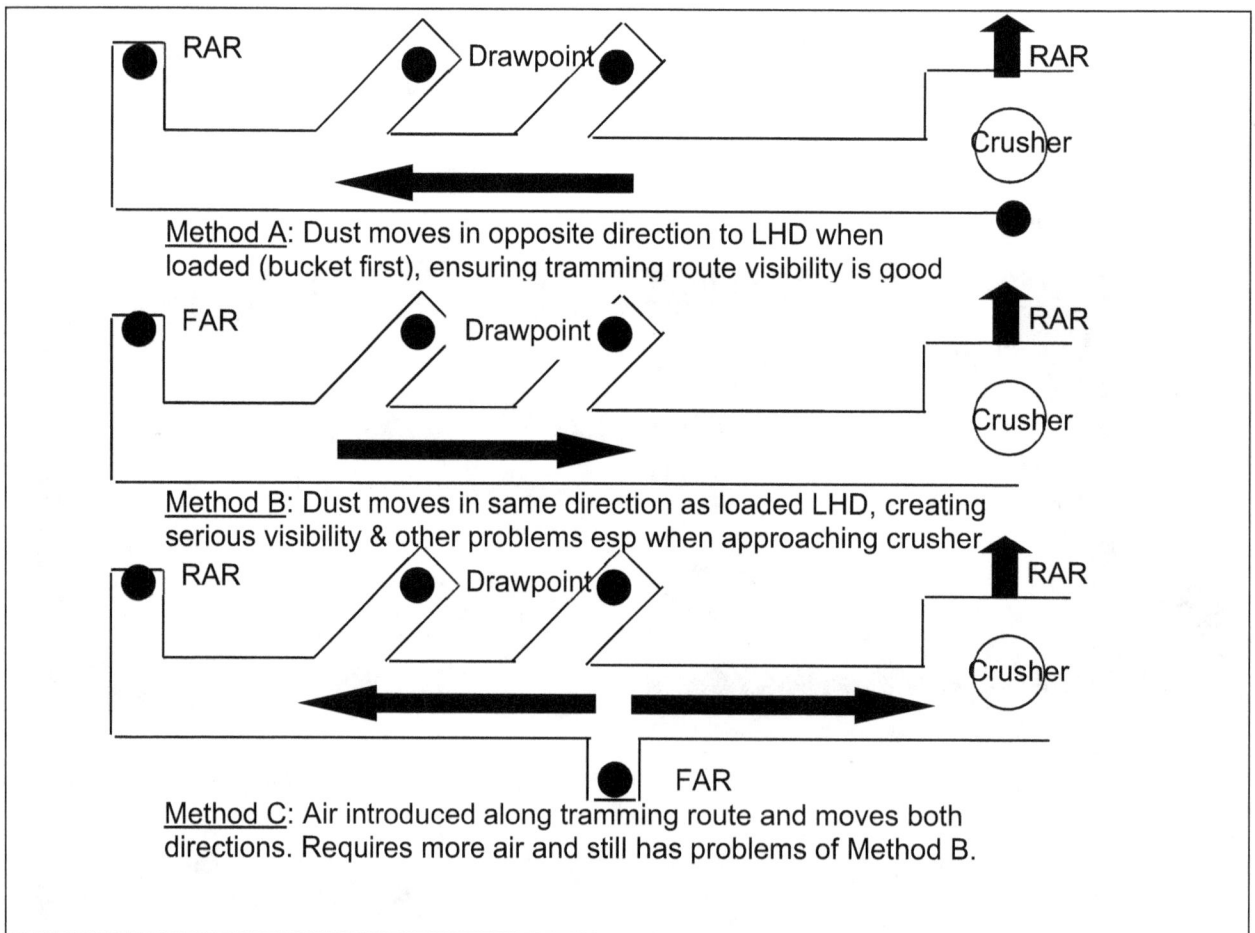

Method A: Dust moves in opposite direction to LHD when loaded (bucket first), ensuring tramming route visibility is good

Method B: Dust moves in same direction as loaded LHD, creating serious visibility & other problems esp when approaching crusher

Method C: Air introduced along tramming route and moves both directions. Requires more air and still has problems of Method B.

FIG 12 – Ventilation strategies for feeding a crusher

CRUSHER VENTILATION

Most of the discussion about dumping into an orepass tipple also applies to dumping into the crusher.

The crusher chamber is often a large opening both in plan view and in height. Keeping a reasonable wind speed through the area is difficult due to the large cross-sectional area. Where possible, a dust extraction system at the place where most of the dust is generated (e.g. the jaws of the crusher) should be combined with an overall exhaust from the crusher chamber. Any dust extraction system must take into account the abrasiveness of the dust (including on fan impellers and motors), plus potential dust build-up in ducting (or the return airway) and dust toxicity (and hence the frequency and arrangements for dust removal). Where the crusher has an operations control room, this room should be fed filtered, air-conditioned air. During maintenance days, the ventilation flows through the crusher may need to be increased to clear dust or reduce temperatures for maintenance workers; therefore, it can be desirable to have a ventilation system that has a "high" and "low" setting—low being for unmanned operation and high when there is maintenance in progress. In all such cases, any circuit fans must be fitted with self-closing dampers.

Comprehensive design recommendations for crusher and transfer point dust extraction ventilation are given in NIOSH (2012), MAC (1980), Yourt (1969) and ACGIH (2001). Some specific points noted by Yourt regarding take-off points for dust extraction from crushers or transfer points include:

- The base of the take-off cone should be large enough so the wind speed at the base is less than 2.5 m/s,

- The take-off should be at least 1.8 m from the path of the falling rock to avoid picking up larger particles,

- Any ductwork carrying dust should have a wind speed greater than 15 to 18 m/s to avoid deposition of dust in the duct.

CONVEYOR DRIVE

Conveyors can produce dust especially if there are multiple transfer points in the conveyor system. For dust minimisation, conveyors should keep wind speeds to a maximum relative wind speed of 4 to 6 m/s. Hence if the conveyor is moving at 3 m/s in one direction and the airflow is in the other direction, then the maximum wind speed would be 1 to 3 m/s. This means the carrying capacity of a conveyor drive should be kept low.

Following the ALARA principle, to aid with keeping both dust and products of combustion from a conveyor fire polluting the mine workings, conveyors should ideally use a "neutral intake system", meaning the conveyor has its own supply of fresh air but this fresh air is not then used for any other activities in the mine—it is sent directly to the exhaust. Conveyor drives can also be used as full exhaust or full intakes. Where used as full intakes, a thorough risk assessment/HAZOP analysis should be conducted regarding fire protection, fire detection and fire extinguishment.

A key problem with conveyors is where there is a service road parallel to the conveyor for conveyor access and maintenance, which often means multiple connections between the service drift and the conveyor. In some cases, it is desirable for the service drift to carry substantial flows of fresh air (e.g. if the service drift connects to the hoisting shaft, which acts as a major mine intake), but if the conveyor is to be a neutral intake, then it will only carry sufficient air for its own purposes (typically about 1 m/s). This means there needs to be ventilation controls (usually also acting as smoke or fire doors) at every connection between the service drift and the conveyor drift. These doors are difficult to maintain and keep closed. The problem is aggravated if the conveyor is at a steep angle (say 1 in 5, to save development) but the service drift is at a more standard 1 in 7 inclination, as the accesses connecting the service and conveyor drifts can then be off-horizontal, making ventilation doors even more problematic as ventilation doors on an inclined airway are especially difficult to install and maintain. Where practical, the best solution is to also set up the service drift as a neutral intake, as this means ventilation doors are not required between the conveyor and service drifts. In addition, in this case perhaps only every 2nd or 3rd connection between the conveyor and service drifts needs a door, with the others having more secure walls.

The tail end of the conveyor drift needs permanent ventilation as it is often a frequent work area and this is best achieved by a small return air raise at the end of the drift. To ensure there is always the minimum required flow, a circuit fan should be installed rather than a regulator. Alternately, a permanent rigid ducted system can be used if the tail end is a blind heading. The tail end also needs to be set up with a second egress or suitable rated refuge chamber.

Conveyors are higher risk environments for fire. Fire protection for conveyor belts is a separate issue. However, in general, there should be automatic fire detection and suppression on at least the drive end of the belt (where slippage is most likely), but preferably also on the tail end of the belt. Other controls such as using fire resistant belt should also be investigated.

GOOD ENGINEERING PRINCIPLES

Finally, as with other engineering designs, the ventilation of orepasses, tipples, conveyors and tramming drives should explicitly take into account:

- The Hierarchy of Controls ensuring stronger more reliable controls are used where practicable (Business Queensland, 2018).

- The ALARA Principle to provide a solution that is not only safe in an absolute sense, but also where the risk is as low as can reasonably be achieved (DNRME, 1999 s26).

It should be noted that while ventilation controls are arguably the most important in terms of controlling dust in the materials handling system, other non-ventilation related controls are also important including:

- PPE and procedures to cover especially dusty activities, such as flowing off floors with compressed air blowpipes,

- Air-conditioned cabins and their dust filtration systems (NIOSH, 2008; Cecala et al, 2005),

- Suitable systems to clean heavily dust-laden clothing (Pollock, 2006 and NIOSH, 2004),

- Use of well-designed water spray or dust suppression systems to reduce dust entering the air. This is application not only for cutting machines such as roadheaders and continuous miners, but also for drawpoints, tramming routes and, if accompanied by strict controls, dump points (NIOSH, 2012).

CONCLUSIONS

Materials handling in mines produces dust at drawpoints, tramming routes, dump points, crushers, conveyor transfer points and along the conveyor belt. Dust production in orepass systems using multi-level tipping is especially problematic with no easy solutions available. However, good design and planning based on a sound understanding of the issues and what can potentially work, along with what will definitely not work, when combined with strong support from management and a degree of operational discipline, can result in an acceptable outcome for all stakeholders.

REFERENCES

ACGIH, 2001. Industrial ventilation: a manual of recommended practice. 24th ed. Cincinnati OH: American Conference of Governmental Industrial Hygienists.

Anon, 1964. Standard practices and drawings for dumps at Sullivan mine. The Consolidated Mining and Smelting Company of Canada, 11 pp. (unpublished).

Business Queensland (Qld Govt), 2018 [online]. Controlling the risk of dust exposure to workers in mines. Available from https://www.business.qld.gov.au/industries/mining-energy-water/resources/safety-health/mining/hazards/dust/control

Beus M J, Iverson S R andStewart B M, Design analysis of underground mine ore passes: current research approaches. 100th CIM Conference (Montreal, Quebec, Canada, May 3-7, 1998); 8 pp

Calizaya F and Duckworth K, 2004. Study of Pressure build-up in long ore passes using computational fluid dynamics (CFD). SME annual meeting, Denver Colorado pp 1-5

Casten T, Calizaya F and Pearson N, 1999. Results of the orepass recirculation system at the Grasberg Mine, P.T. Freeport Indonesia. Mining Engineering. April. Pp 48-51

Cecala A B, Organiscak J A, Zimmer J A, Heitbrink W A, Moyer E S, Schmitz M, Ahrenholtz E, Coppock C C and Andrews E H, 2005. Reducing enclosed cab drill operator's respirable dust exposure with effective filtration and pressurization techniques. J Occup Environ Hyg 2005 Jan;2(1):54-63

Hadjigeorgiou J and Stacey T R, 2013. The absence of strategy in orepass planning, design, and management. The Journal of The South African Institute of Mining and Metallurgy,

Hadjigeorgiou J, Lessard J F and Mercier-Langevin F, 2005. Orepass practice in Canadian Mines. The Journal of The South African Institute of Mining and Metallurgy, Vol 105 (Dec). pp. 809-816.

Linhart, J and Pollack, R, 1979. Influence of mine ventilation by falling material – solids and water – during sinking and filling in of shafts or during actions for extinguishing mine fires in shafts, in Proceedings Second International Mine Ventilation Congress (ed: P Mousset-Jones), pp 510-517 (Society of Mining Engineers of American Institute of Mining, Metallurgical and Petroleum Engineers: New York).

MAC, 1980. Design guidelines for dust control at mine shafts and surface operations. 3rd ed. Ottawa, Canada. Mining Association of Canada

Maree, J A, 2011. Orepass best practices at South Deep. The Journal of The South African Institute of Mining and Metallurgy, Vol 111 (Apr). pp. 257-272.

McPherson, M J and Pearson, N, 1997. The airblast problem in the ore passes of the Grasberg mine, PT Freeport Indonesia. Proc Sixth International Mine Ventilation Congress (ed: R V Ramani), pp 113-117 (The Society for Mining, Metallurgy and Exploration Inc: Littleton).

Moss E A, Sheer T J, Rose J H and Dumka M, 2005. Measurements and Modelling of Pressure Surges in Orepasses. Proc 8th Int mine vent congress, Bris, 6-8 Jul (ed: S Gillies). AusIMM. Pp 391-398

NIOSH, 2004. Technology News No. 509. A new method to clean dust from soiled work clothes. CDC. 2 pp.

NIOSH, 2008. RI9677 Key design factors of enclosed cab dust filtration systems. CDC. 51 pp.

NIOSH, 2012. RI9689 Dust control handbook for industrial minerals, mining and processing. CDC. 314 pp.

Pollock D E, Cecala A B, Zimmer J A, O'Brien A D and Howell J L, 2006. A new method of clean dust from soiled work clothes, in Proc 11th US/Nth Am Mine Ventilation Symposium 2006, Mutmansky and Ramani (eds), Taylor & Francis, London.

Queens University, undated. Ore and waste passes [online]. Available from http://minewiki.engineering.queensu.ca/mediawiki/index.php/Ore_and_waste_passes. [Accessed 17 Feb 2019].

DNRME, 1999. Queensland Mining and Quarrying Safety and Health Act 1999. Queensland Department of Natural Resources, Mines and Energy

Rock and Roll Mining, undated. Ore Pass Doors & Lifting Devices [online]. Available from http://www.rocknrollmining.com/doors-lifts. [Accessed 17 Feb 2019].

Sjoberg A, Mureithi E, Stacey T R, Ockendon J R, Fitt A D and Lacey A A, 2005. Piston effect due to rock collapse. Proceedings of the Mathematics in Industry Study Group 2005, University of the Witwatersrand, Johannesburg. Editor DP Mason.

Yourt G R, 1969. Design principles for dust control at mine crushing and screen operations. Can Min J 90(10):65-70.

Ventilation planning considerations for the Carrapateena Sub Level Cave

K Manns[1], J Holtzhausen[2], A Mooney[3] and L van den Berg[4]

1. MAusIMM, Senior Ventilation Engineer, BBE Consulting (Australasia), Suite 6, 89 Winton Road, Joondalup, Perth, WA, 6027. Email: kmanns@bbegroup.com.au
2. MAusIMM, Senior Engineer – Ventilation Carrapateena, OZ Minerals, 2 Hamra Drive, Adelaide Airport, SA, 5950. Email: Johannes.holtzhausen@ozminerals.com
3. MAusIMM, Project Director - Carrapateena Expansion, OZ Minerals, 2 Hamra Drive, Adelaide Airport, SA, 5950. Email: Andrew.mooney@ozminerals.com
4. MAusIMM, Principal Ventilation Engineer, BBE Consulting (Australasia), Suite 6, 89 Winton Road, Joondalup, Perth, WA, 6027. Email: lvandenberg@bbegroup.com.au

ABSTRACT

OZ Minerals is an Australian based modern mining company with a focus on copper. Their Carrapateena Project, located in South Australia is one of Australia's largest undeveloped copper deposits. Construction of Carrapateena is underway and commissioning is scheduled for Q4 2019 after which the project will ramp up to steady state production.

The project will be an underground sub-level caving (SLC) operation, with an estimated mine life of 20 years. The planned ore handling comprises of ore-passes feeding underground crushers and loaded onto a conveyer belt that transports the ore up the decline to the surface.

The underground workings will extend to a depth of 1.4km with a steep virgin rock temperature (VRT) gradient and future mine cooling is also required. In addition, the orebody contains very low levels of uranium, which requires suitable ventilation strategies to manage possible natural occurring radon emanation from strata. In keeping with the philosophy of designing a modern mine, ventilation on demand (VoD) is a central part of the ventilation strategy.

This paper explores the feasibility and on-going operational ventilation planning for the Carrapateena Project. Although natural occurring radon emanation from strata is common in many hard rock mines there are very few SLC examples and therefore this paper provides a unique perspective on ventilation planning. The paper also presents the feasibility and subsequent implementation planning from an owner's and consultant's perspective.

INTRODUCTION

OZ Minerals is a copper-focused, global, modern mining company based in South Australia and listed on the Australian Securities Exchange. OZ Minerals has a growth strategy focused on creating value for all stakeholders and owns and operates the copper-gold mine at Prominent Hill, are developing one of Australia's largest copper-gold resources at Carrapateena and have assets in Brazil. OZ Minerals genuinely cares about their people, our environment, the heritage of the people we work with and the communities they operate in. OZ Minerals' strategy is anchored by stakeholder value and is underpinned by their How We Work Together principles. It outlines what they focus on, how they work, how they create value and how they deliver.

Carrapateena is a copper-gold project located approximately 160km north of Port Augusta in South Australia's highly prospective Gawler Craton. The project is located on Pernatty Station and its supporting infrastructure is located within Oakden Hills Station. The Kokatha people are the traditional owners of the land.

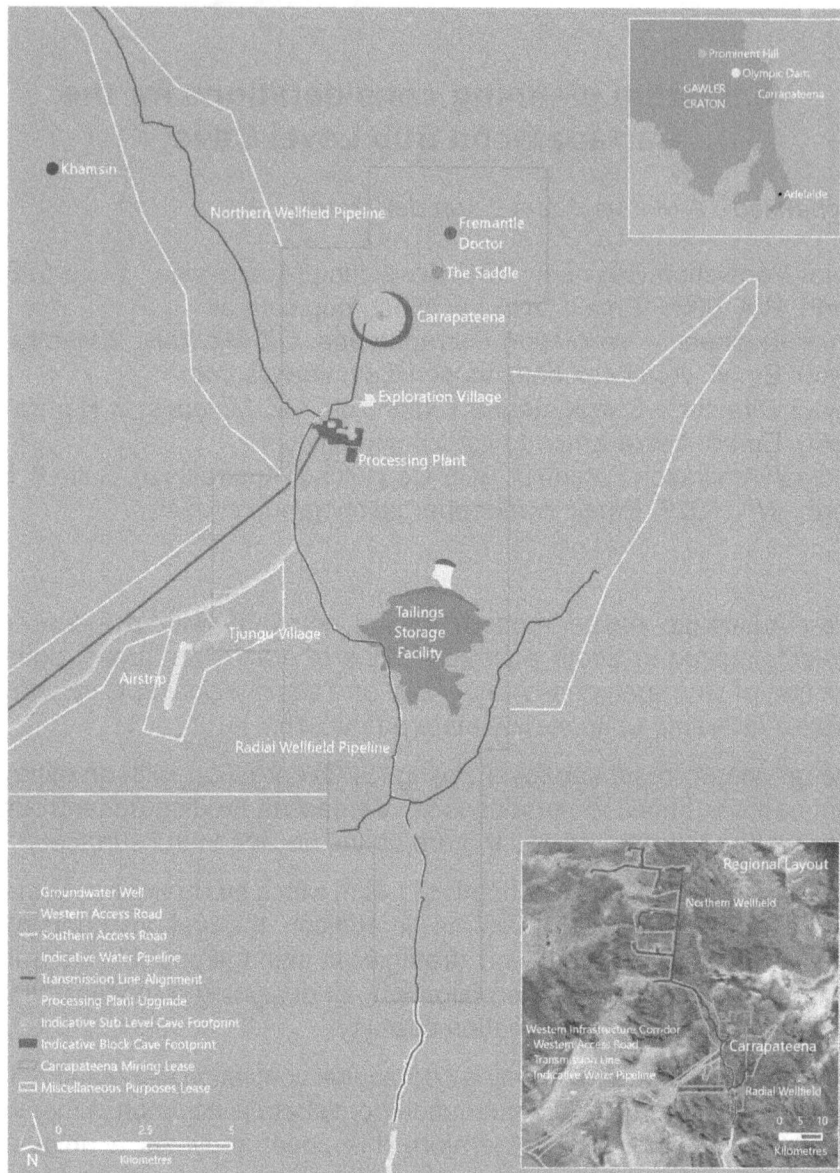

FIG 1 – Carrapateena Location Map

BACKGROUND

The Carrapateena project will be an underground sub-level caving (SLC) operation, with an estimated mine life of 20 years (OZ Minerals 2017). A recently completed scoping study confirmed the potential to transition the initial SLC mine into a future Block Cave (BC) mine without impacting on the SLC production (OZ Minerals 2019).

The mineralised orebody occurs at a depth of approximately 515 m below ground level (mBGL), extending 950 m to 1,465 mBGL with maximum virgin rock temperature (VRT) of 66°Cwb. The ore body has uranium grade and will experience low levels of radon emanation. An important aspect of SLC is that both development and production miners work inside the mineralised area. Ventilating workplaces in a SLC operation can be complex and the overriding principle is to ensure that the mine ventilation system delivers fresh air to each of the workplaces and that all contaminated air is exhausted from the mine without going through downstream workplaces. Every mine ventilation system has their primary drivers, in this case it is the interaction between the number of activities over the number of levels, the low levels of radon and the heat management due to the level of VRT. The secondary ventilation strategy has been defined with an automated ventilation-on-demand (VoD) system in mind. This VoD includes variable speed secondary fans, automated level exhaust louvres, automated in duct dampers and air quality monitoring stations. The mine ventilation planning is described in this paper, which presents a unique perspective on both the SLC and Block Cave designs.

FEATURES OF THE MINE

Initial SLC Mine

Carrapateena will be mined using a sub-level cave mining method. Using this method, the deposit is blasted in 25 metre vertical sections and then collected by loaders. From there it will go into the ore pass, then to the crusher and then onto the conveyor belt to the surface. A separate decline to surface provides people and services access.

The mine will typically have four producing levels and two levels in development. Each level comprises of a cluster of SLC ore drives referred to as 'crows feet' (see Figure 2). On average, a level will have four of these 'crows feet' active with various mining activities cycling through each.

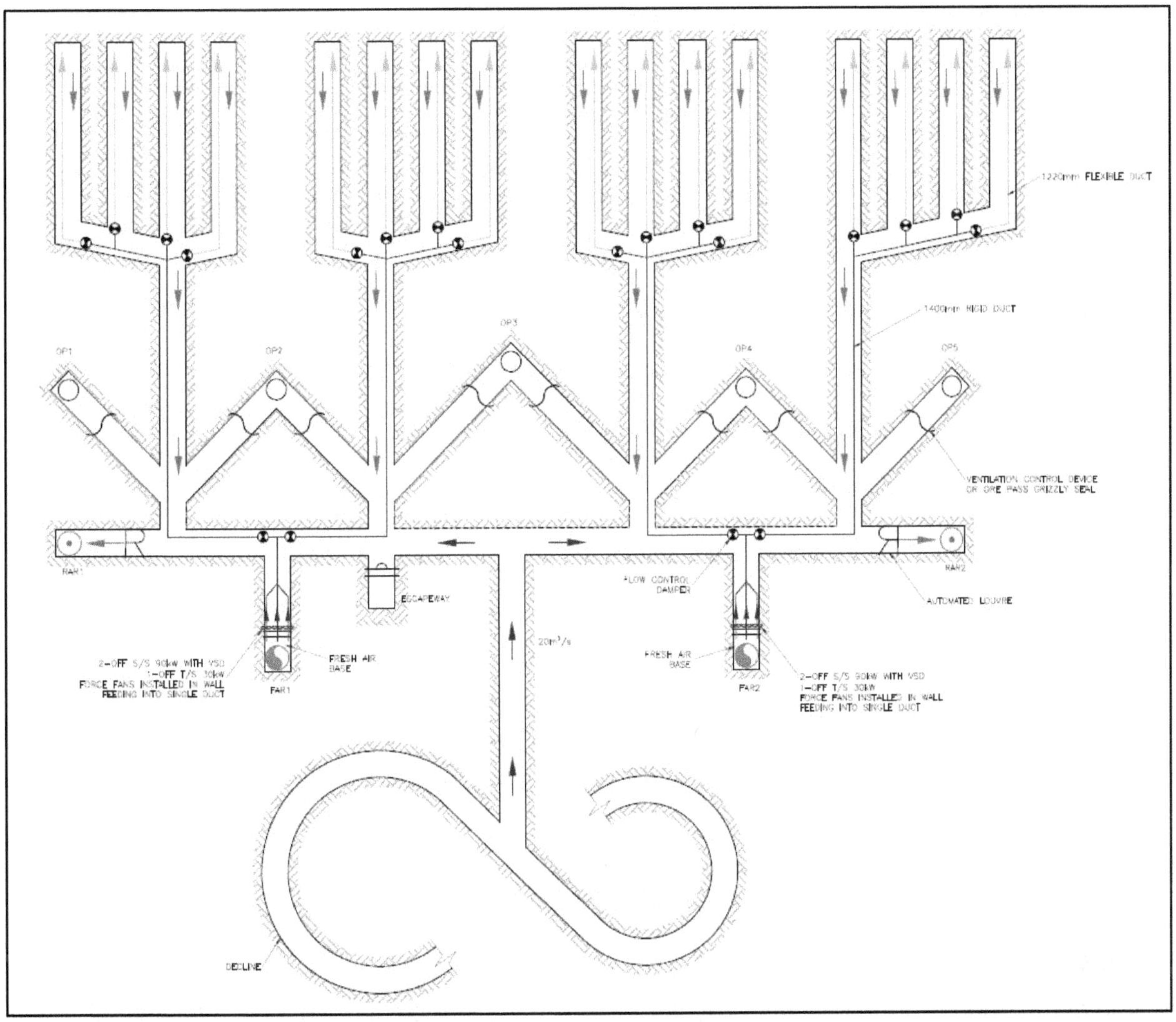

FIG 2 – Typical SLC Level Layout

Block Cave

A recently completed scoping study confirmed the feasibility of a block cave extension to the initial SLC (OZ Minerals 2019). The conceptual design includes an Apex, Undercut and Extraction level. The extraction level comprises of 14 drives with ore transported via loader on the level to a crusher station which feeds the conveyor decline (as an extension of the SLC).

RADON CONSIDERATIONS

Above the orebody, there is non-mineralised cover sequence material (containing less than 5 ppm U). The mineralisation itself lies within a basement host rock, which is also non- mineralised, but with a higher natural uranium content of approximately 20 ppm. For reference, the world average uranium content of soils is 2.8 ppm (World Nuclear Association, 2016). The Carrapateena project falls into the category of being classified as a Naturally Occurring Radioactive Material (NORM) deposit. OZ

Minerals does not intend to produce uranium from the deposit. Compared to some other deposits, the one at Carrapateena has comparably low levels of radon emanation. Figure 3 shows the measured radon emanation rates against those published in literature (vd Linde; 2014). Despite the emanation rates being comparatively low it is recognised that the elevated uranium concentrations bring a small additional risk from radiation and there is a need to pro-actively plan for a ventilation system that can manage any associated risk.

A conservative ventilation design approach was followed, which can be adapted once mining reaches the ore horizon and actual field data on underground radon behaviour becomes evident. The radon is only expected to be a factor once the ore body is reached, with no uranium grade anticipated in the surrounding waste rock. Radon management ventilation principles have only been planned for the production levels and not infrastructure development such as the decline, crusher areas, workshops etc. Current decline development has not encountered any detectable levels of radiation.

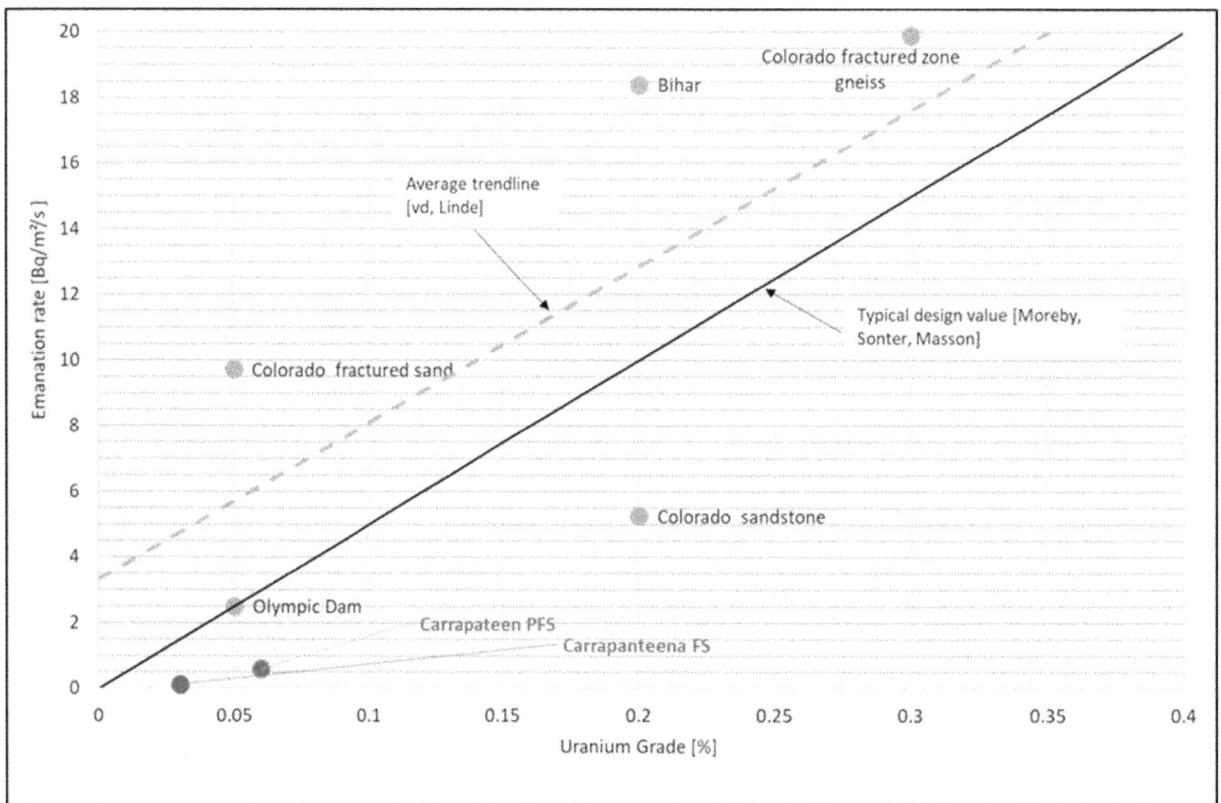

FIG 3 – Radon Emanation Rates

Ventilation Strategy

The role of the underground ventilation system is to ensure sufficient fresh air is available at all times in all underground workplaces, thereby controlling airborne contaminants such as dust, blasting fumes, heat and the decay products of radon. This requires careful consideration of the both the primary ventilation strategy for the overall mine and secondary ventilation strategy which supplies fresh air via ducting to production areas.

Design of the underground ventilation system considered possible leakage of radon bearing air into the levels via the fractures in the cave zone. A number of primary ventilation configurations to manage this were developed including a primary force system, primary push-pull system and a hybrid primary ventilation strategy. Ultimately, a hybrid ventilation strategy was selected for Carrapateena, which incorporates secondary ventilation fans and ducts as an integral part of the primary circuit.

The primary ventilation circuit consists of a network of large diameter surface raise bores. Centrifugal fans installed on surface will exhaust contaminated air from the mine through a range of large diameter shafts into the workings (i.e. VR1, VR3). Fresh air will be drawn into the mine via large diameter fresh air raises/shafts (i.e. VR2, VR4). The hybrid system allows fresh air to be distributed directly to the workplaces from the FARE by means of the secondary fans and duct. The overall mine operates as a negative pressure ventilation system.

The secondary ventilation system will take fresh air from the primary circuit and distribute it to working areas using secondary ventilation fans and ducting (i.e. a "positive pressure" system that minimises the potential for drawing contaminated air in through the caved material zone). The fans deliver fresh air to the working face, with the "contaminated" air flowing back along the drives to the exhaust by the primary ventilation system. The decline is a source of fresh air and will be developed through inert cover material. This would be achieved by interlocking the surface fan and underground secondary fan controls to ensure a slight positive pressure on the level [see Figure 4].

It should be recognised that managing leakage through the cave zone will only become a factor at depth when the cave breaks through to surface. Therefore, the installation of doors to "pressurise" the level is considered a fall-back position. The initial level set-up will exclude the level access doors, with a determination to be made in the future as to the need for these doors, once some operational experience is gained.

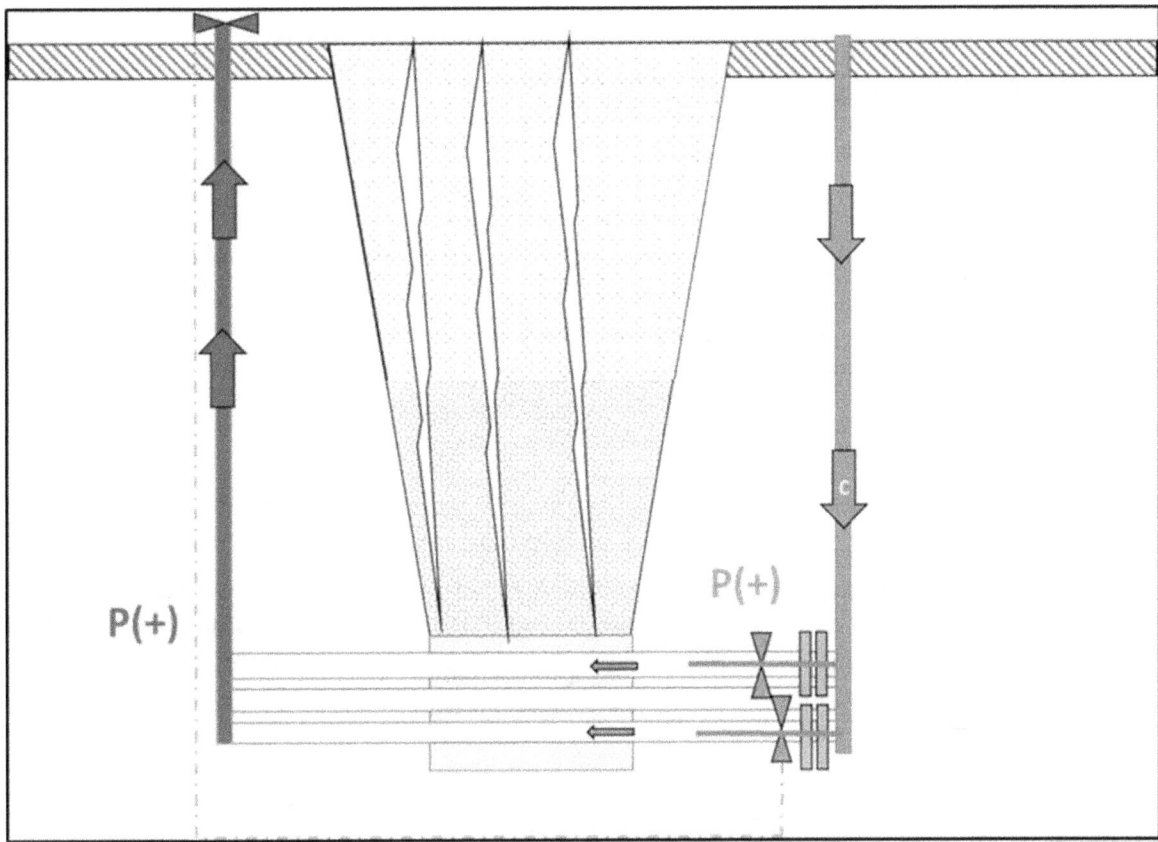

FIG 4 – Hybrid Primary Ventilation Strategy

Production Level Ventilation

Before intersecting ore at the first production level, the ventilation circuit will be established to the point where flow and contaminants will be restricted to the level. To manage airborne contaminants on production levels, dedicated intake and return air raises will be in place.

Radon modelling suggested that to maintain RnDP concentrations in a non-working end below $2\mu J/m^3$ a continuous flow of only 2 m³/s is required. This shows that a large flow rate is not necessarily required in drives to manage radon. The adopted approach for volume planning was 3m³/s per drive continuous ventilation as a minimum. This will then allow enough flow to adopt a flushing strategy at a later stage, should it be necessary.

Figure 5 shows the results from radon modelling undertaken on a typical mine level with a "worst case" uranium grade. It is impractical to expect to maintain such a low airflow as 3m³/s in a drive. Furthermore, ventilating all available drives with the airflow requirement for activity (which is about

15m³/s minimum for a loader and to maintaining a velocity above 0.5m/s for heat management) and utilising no reuse will result in excessive primary ventilation infrastructure and cost.

The mine ventilation system was therefore designed with ventilation on demand (VoD) principles and technology in mind. The VoD system would control the airflow to available but inactive headings. The following VoD features have been incorporated in the design:

- Variable speed secondary fans (Dual speed fans or VSDs)

- Duct control valves

- Automated louvres

- Air monitoring stations

Monitoring systems will be installed to assist with the control and monitoring of the ventilation circuit. The system will include control room software interface, allowing alarm set points to be entered, and continuous monitoring of the ventilation circuit. Remote management of the circuit will be supported by local and manual control options which will be defined by the Production Plan. Real time data on flow rates and contaminants will allow for the proactive management of the ventilation circuit and confidence in physical working conditions present underground.

General Radon Management

The following underlying principles to manage radon were applied to the ventilation design:

- The primary ventilation circuits are designed to ensure air contaminated by production activity is immediately directed to the exhaust on each level and no re-use occurs.

- All return air that has been in contact with the ore body travels via dedicated exhaust raises and drives before being exhausted to surface.

- Primary ventilation and other infrastructure are developed in areas that contain little or no uranium grade.

- Construction of the ventilation circuit in this manner has minimised the residence time for any dust generated during mining and the dust laden air is directed to return as far as practical.

   o Ore passes must be single-level tipping to prevent dust roll back when ore is tipped from a higher level, with controls on ore pass accesses to stop air flowing in or out.

   o Each ore pass is connected at the top level directly to return and is regulated to ensure a minimum of 20m³/s is drawn into the ore pass on the open tipping level for dust management.

   o Air from crusher and transfer stations is channelled directly to return using dedicated drives or ducted arrangements.

   o Sprinkler systems are to be mounted at crusher stations, conveyor transfers, the backs of the declines and intake airways where high velocity is experienced. General road and wall watering done using a water truck. Watering down development headings prior to mucking if required.

- No workers will be located permanently at the location in an airway with the oldest air, and hence the highest RnDP concentration.

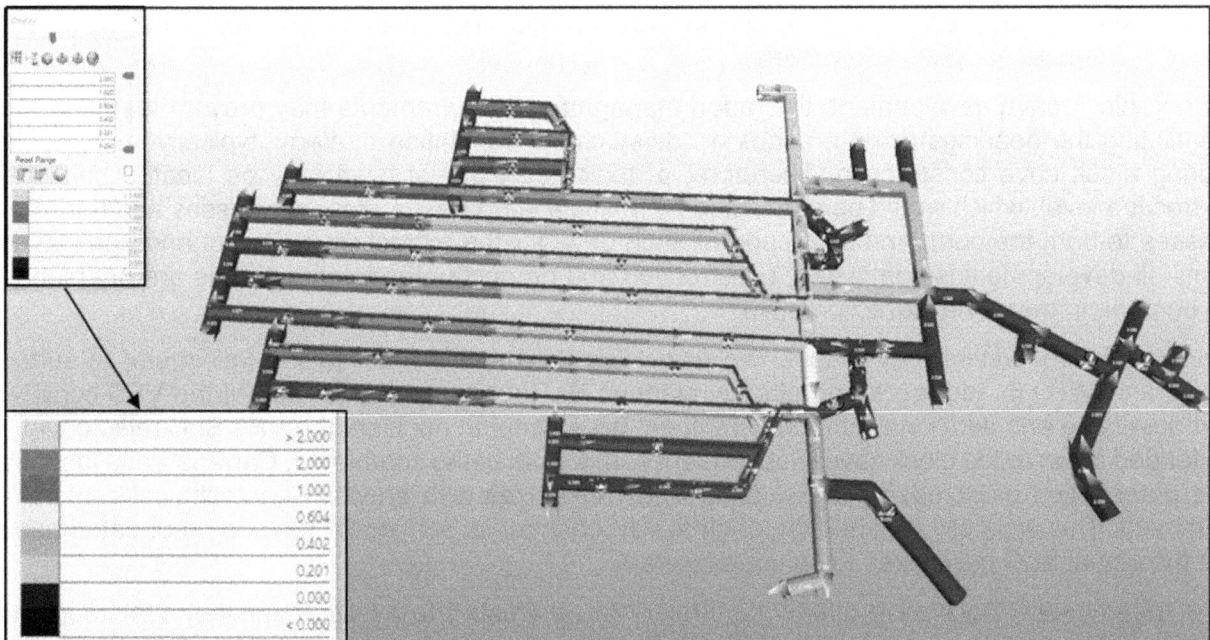

FIG 5 – Radon Modelling

PLANNED PRIMARY INFRASTRUCTURE

The planned capacity for primary airflow is 1180m³/s (at surface density). The planned primary exhaust infrastructure comprises two Φ5.5m surface return raises (VR1 and VR3) and one Φ4.1m return raise to independently ventilate the conveyor (VR5) to manage radon and fume risk in the unlikely event of a fire. Individual raise sizes were based on geotechnical constraints due to expected poor ground. Allowance was made for three surface fan stations comprising high-pressure centrifugal fans.

The primary intake system consists of two Φ5.0m surface fresh air raises (VR2 and VR4), a conveyor decline and an access decline. The fresh air raises are also located outside the cave influence zone and connected by 6mx6m ventilation drives. A stepped 5.5mx5.5m raise along the access decline will be used for developing the decline. In the ultimate steady state, this raise will also be used as a chilled air intake to inject cool air along the decline.

Mine cooling was planned to manage heat but will only be required in future when the critical depth of about 700 mBGL is approached. The VR2 and VR4 fresh air raises are planned to include allowance for future surface bulk air coolers comprising two 4MWBAC rated cross flow cooling tower modules. This gives a total duty of 16MWBAC that will be phased in over time. The planned refrigeration strategy will comprise of a centrally located R134a refrigeration plant distributing chilled water to the raises.

TRANSITIONING FROM AN SLC TO A BLOCK CAVE

A Scoping Study was completed to investigate possible transition from an SLC to Block Cave operation (OZ Minerals 2019). The block cave mine could be able to take over from the initial SLC mine, which is currently under construction, and will be reviewed further during the Carrapateena Expansion Pre-Feasibility Study. Radon management remains an important consideration for the ventilation design and a currently acceptable approach had been followed in this regard, for the Scoping Study. Some strategies to manage the risk of radon are:

Minimise flow through the cave

- Manage connections to the ventilation system
- Barricade connections to the cave where possible

Manage pressure gradient

- Manage pressure gradient across the cave to discharge radon to non-working areas (not the Undercut and Extraction levels)

During block cave development, the radon management requirements may prevent the mine from ventilating the headings using a series or "daisy-chain" ventilation strategy, typically implemented during block cave construction. Therefore, a dedicated exhaust level may be located below the extraction level, which would be connected by multiple short raises, fresh-air-passes and return-air-passes to both the north and south end of each drive for the extraction, undercut and apex levels. Until all development is completed, construction and production bogging activities are not planned to be done in the same extraction drive.

The ventilation system may include two new 4.8m finished diameter raises (minimum) to surface (VR6 and VR7) in addition to the already planned SLC raises and extension of the VR5 conveyor exhaust raise and additional surface centrifugal fan station. In addition, the VR3 SLC raise could be extended down to the block cave. Subsequent to this work being completed, Carrapateena identified the potentiation of increasing the existing SLC raise sizes with previous geotechnical constraints removed. This work still includes the original size constraints and would have to be updated based on the actual SLC raise sizes to be implemented.

The steady state block cave airflow requirement is estimated to be approximately 850m³/s, initial development would be supported by 490m³/s from a new dedicated raise set. The peak construction and development airflow requirement would be 1000m³/s which could be served by the dedicated block cave raise set, as well as the additional capacity from the SLC raises as the SLC production ramps down.

The 16MWR cooling capacity initially estimated for the SLC is deemed to be enough for the potential block cave mine. However, cooling demands will peak during construction and development, and it is proposed that provision be made for an additional 8MWR rental unit.

CONCLUDING REMARKS

The ventilation strategy for a SLC mine with radon considerations is described. The approach to managing radon will rely on a hybrid push pull secondary ventilation strategy combined with a conventional exhaust primary ventilation system. The hybrid secondary system allows for production levels to be pressurised to manage radon from the cave. High VRT gradients and depths mean the mine will require cooling as a 700 mBGL depth horizon is reached. The primary airflow capacity planned for initially is circa 1180m³/s with 16MWR refrigeration.

Transitioning from a SLC to a BC in the future without impacting the SLC operation may require additional surface raises. Radon management during development and construction of the potential block cave will need particular consideration, and a flexible system that allows intake and exhaust connections to be interchanged may be required. The steady state block cave airflow requirement is estimated to be 850m³/s, initial development will be supported by 490m³/s from a new dedicated raise set.

REFERENCES

OZ Minerals 2017, Carrapateena Feasibility Study Update, OZ Minerals, Adelaide, viewed 24 May 2019, http://minerals.statedevelopment.sa.gov.au/__data/assets/pdf_file/0007/296458/20170824_OZ_Minerals_Carrapa teena_Feasibility_Study_Update.pdf

OZ Minerals 2019, Carrapateena Expansion Scoping Study, OZ Minerals, Adelaide, viewed 24 May 2019, https://www.ozminerals.com/uploads/media/190306_ASX_Release_Carrapateena_Block_Cave_Expansion_Scop ing_Study.pdf

Vd Linde, A; Ramsden, R; Rose, H.J.M. (2014) Radiation in Mines. Ventilation and Occupational Environment Engineering in Mines3 3rd Editiond edited by Prof. J.J.L du Plessis.

World Nuclear Association, 2014, World Nuclear Association website: (http://www.world- nuclear.org/information-library/nuclear-fuel-cycle/mining-of-uranium/uranium-mining-overview.aspx) accessed: 211116.

Calibrating model airway size and resistance with survey asbuilt data

F Michelin[1], C Stewart[2], M D Griffith[3] and T Andreatidis[4]

1. Senior Software Applications Engineer, Howden VentSim, Brisbane, QLD 4101.
   Email: florian.michelin@howden.com
2. Operations Manager, Howden VentSim, Brisbane, QLD 4101.
   Email: craig.stewart@howden.com
3. Senior Software Applications Engineer, Howden VentSim, Brisbane, QLD 4101.
   Email: martin.griffith@howden.com
4. Senior Ventilation Engineer, Ernest Henry Mine, QLD

ABSTRACT

Airway size is one of the key factors to calculate correct resistance for ventilation models in the absence of a pressure quantity survey. Use of traditional estimation methods such as design size or time-consuming underground spot measurements can produce poor airflow simulation results, particularly if there are frequent size variations in tunnels. The merits of using an algorithm to automatically convert three-dimensional survey information to correct airway size is examined, and the effect on model accuracy is observed.

INTRODUCTION

To simulate the pressure and flow in a mine it is critical to use accurate resistances. Atkinsons equation is commonly used in mine ventilation to calculate the relationship between flow, pressure and resistance (Equation 1) and airway resistance can be calculated using the Atkinson resistance formula (Equation 2)

$$\Delta P = \frac{\rho_{actual}}{\rho_{ref}} RQ^2 \qquad \text{Equation 1}$$

Where R = resistance

Q = airflow quantity

P = pressure loss

ρ_{actual} = current air density

ρ_{ref} = air density when the resistance was measured

$$R = \frac{kLS}{A^3} \qquad \text{Equation 2}$$

Where k = Atkinson's friction factor

L = length of airway

S = perimeter of airway

A = cross sectional area of airway

The resistance equation shows the power coefficient of the airway area results in the largest influence on resistance given measurement variation in all factors. Resistance in ventilation modelling is either estimated using measurements from a pressure and quantity survey (using Equation 1 to calculate resistance) or estimated by measurements of the area and perimeter of an airway together with an estimate of the friction factor (Equation 2). Due to the time-consuming nature of underground pressure quantity surveys (either barometric or trailing tube), it is the experience of the authors that the latter method is more commonly used for Australian mines.

A common method of entering data in ventilation modelling software is to enter a standard profile with a height and width, or to directly enter the perimeter and area of the tunnel. Ventilation models commonly use assumed design dimension data to estimate airway size, however potential over or under blasting can significantly affect the final size of each airway. This can lead to discrepancies between actual airway data and modelling assumptions. Most mines now have access to three-dimensional survey data which can be used to accurately estimate the true size of each airway in the model. While matching the airway to the shape is possible by manually entering dimension data for each airway, or by visually adjusting the airway size and shape until it graphically matches the imported reference, the process can be inaccurate as well as time consuming. Extracting the size and exact position of each airway automatically would improve accuracy and productivity. This study observes potential methods, and the effect on accuracy of matching airway size to the survey data.

ADJUSTING THE PROFILE SIZE

Methodology

The strongest influence of airway size is found in the high airflow areas where pressure loss inaccuracies due to incorrect airway size will be significant. To simplify the study, only the high airflow airways are considered, limiting the profile adjustment to the decline and the exhaust airways downstream of the main fans. Other high flow areas include the hoisting shaft, but in this case, additional infrastructure in the shaft will have a more significant effect on the resistance than the size.

An algorithm was developed in VentSim™ Design to use 3D survey data to adjust the size and position of each airways. The survey data is a group of 3D polygons forming the true shapes of the mine excavations. For each airway, the proposed algorithm calculates the intersection of the survey data with the plane normal to the direction of the existing airway and located at the middle of the centreline of the airway. In order to avoid intersecting with other tunnels, only survey points less than 20m from the airway are considered. The intersection between the plan and the survey polygons will give the cross section of each airway. The cross-sectional polygon can then be measured to obtain the maximum width and height of the airway as well as the area and perimeter.

Due to the likely irregularity of the shape, using the maximum width and height as measurements for a standard airway profile (like square or arched for example) may not give an accurate enough estimate of the area or perimeter of the airway. Using this method, the measured area of the survey versus the calculated area of the height and width using a standard profile was on average 5% smaller.

Because of the importance of the area in calculating the resistance, the width and height are then calibrated and scaled to obtain less than 0.1% difference between the survey area and the adjusted airway area. Doing so provided an adjusted airway perimeter within 1% of the measured perimeter. The standard airway arched shape (normally used in VentSim and common for most hard rock underground mines) is kept for simplicity as the airways are frequently irregular and slicing only one position does not provide enough information to calculate an average shape.

FIG 1 – 3D survey data imported in VentSim™ Design

Airway Size Adjustment

The table below summarises the adjustments made from the design sizes generally used in the ventilation model representing the mine, to actual sizes measured from the survey data. The most common sizes used were 6.5m width by 6m height, and 7m width by 6m height; both arched in shape. These sizes are used extensively in the primary ventilation circuit, especially in the ramp where large trucks are required to travel.

TABLE 1 – Average change in size for 7m by 6m airways

	Width	Height	Area	Perimeter
Original	7 m	6 m	39.4 m²	24.2 m
Average Change	+ 0.48 m	+ 0.28 m	+ 4.83 m²	+ 1.41 m
Percentage Change	+ 6.9 %	+ 4.6 %	+ 12.3 %	+ 5.8 %

TABLE 2 – Average change in size for 6.5m by 6m airways

	Width	Height	Area	Perimeter
Original	6.5 m	6 m	36.6 m²	20.5 m
Average Change	+0.4 m	+0.01 m	+3.1 m²	+1.1 m
Percentage Change	+ 6.4 %	+ 0.2 %	+ 8.5 %	+ 17.4 %

Table 1 and Table 2 demonstrate an overbreak increase in area of between 8 and 12%, and a perimeter increase of between 6 and 17%. Using the Atkinson's resistance equation, the adjusted resistance shown in Equation 3 can be calculated from the original resistance ($R_{original}$), the percentage change in area ($A_\%$) and the percentage change in perimeter ($P_\%$).

$$R_{adjusted} = \frac{P_\%}{A_\%{}^3} R_{original} \quad \text{Equation 3}$$

The adjusted resistance calculated from the true areas averages 25% lower than the assumed resistance for the 7m by 6m airways. For a given pressure, the flow would therefore be 15% more than in the base case model. For the 6.5 m by 6 m, the adjusted resistance is 8% lower than the original resistance, which would result in a flow increase of around 5% given similar pressure.

Shape comparison

Figure 2 represents the adjusted model size from original model shapes (see table 1). While the survey data (in blue) can be seen to be highly irregular, the adjusted shape (in orange) maintains the original model shape and correctly approximates the area and perimeter of the survey shape.

Figure 3 represents a worst case difference between the original model and adjusted model shape. The survey area is 45% larger than the original area and the perimeter is 22% larger. In this case the adjusted model resistance is 60% lower than the original model resistance.

DISCUSSION

Cause of size differences

The first factor affecting the discrepancies of the airway size is the inaccuracy of the blasting process compared to the plan, where drilling inaccuracy, poor ground and blast damage causes an increase in excavation size. This is exacerbated in larger excavation as show in Figure 4, where the decline was purposely excavated wider in the turns to allow trucks to go around corners more easily, leading to overbreak to both the wall and roof.

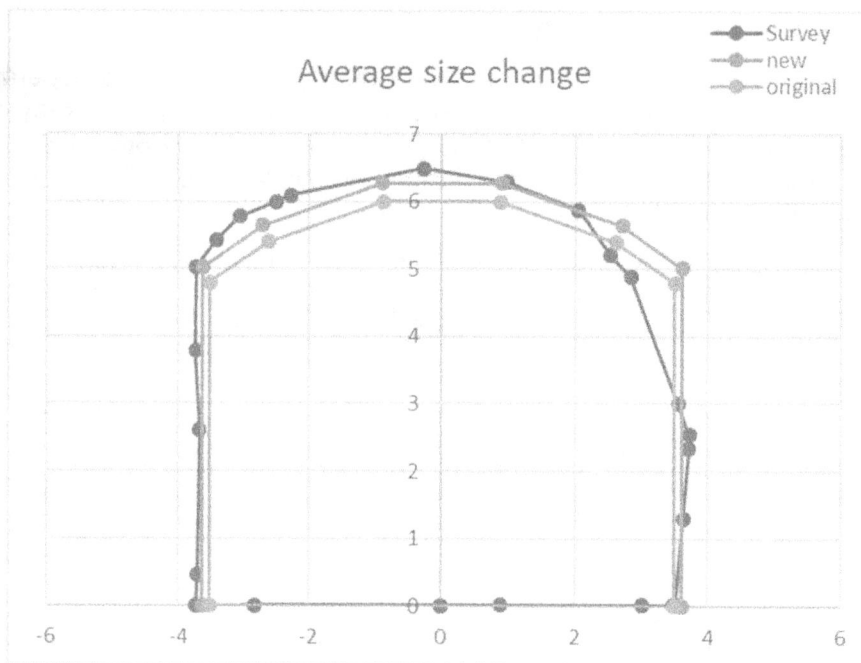

FIG 2 – Average airway profile

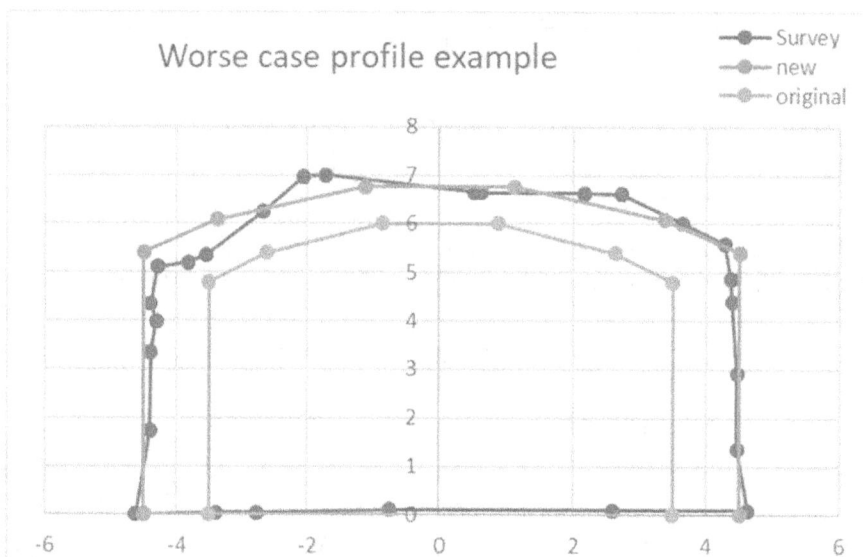

FIG 3 – Worst case profile example

Effect on the model

The modified model was compared to the original model where airflow measurement is made. On average the flow has increased by 41%. Considering only a small but significant part of the model was converted, the results are consistent with the theoretical change.

However, when comparing to the ventilation survey data, the modified model matches poorly with the airflow survey. The original model had an average error of a 6% underprediction, however after adjustment this difference has increased to 27% overprediction.

To examine the reasons for this, other factors affecting resistance must be considered. The most likely candidates are incorrect friction factors, and shock loss estimation. If Atkinson's resistance equation is used to calibrate friction factors from measured sizes, the factors calculated demonstrate the friction factors used in the model are significantly underestimated. Therefore it is likely that the friction factors calibrated for the model compensate for the incorrect airways size used, nullifying the effect on resistance. In addition, shock loss factors caused by sudden airway size or direction changes are often not considered in the ventilation model, and the under sizing of true airway size also compensates for this. The model used in this study includes some automatically calculated shock loss along the decline, which will account for size changes and direction changes.

Some survey positions have improved however, such as the ramp, which has been entirely converted. The top of the ramp survey was 161.5 m3/s and 139.2 m3/s in the original model hence a 14% difference. The simulation result was 155.9 m3/s which is a 3% difference compared to the survey data.

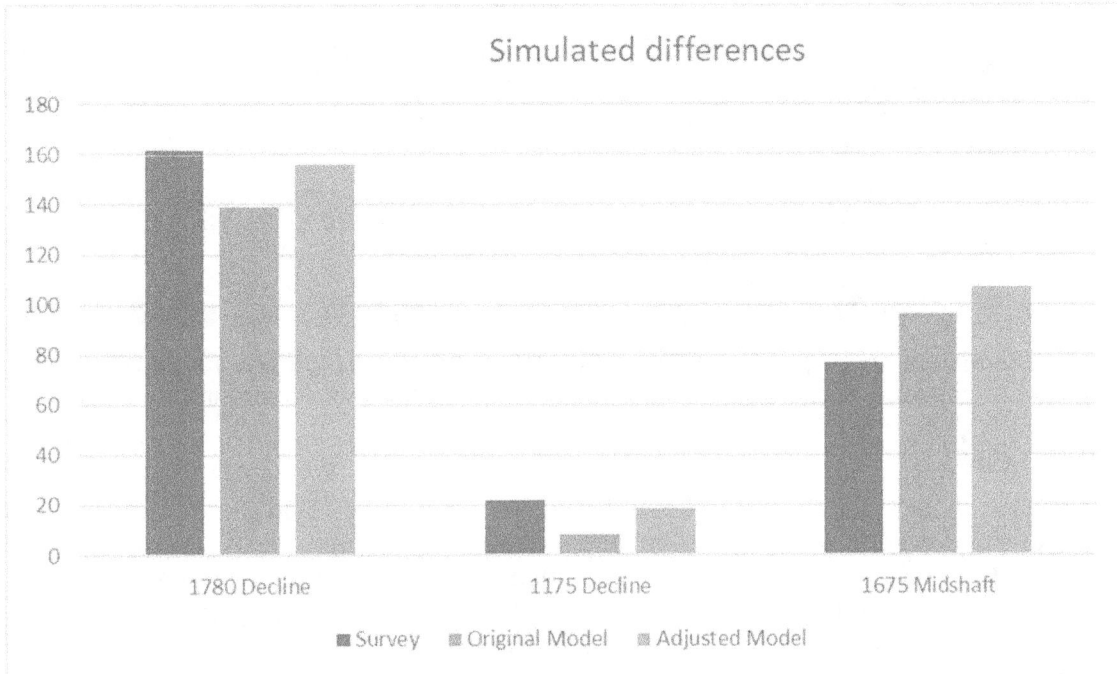

FIG 4 – An example of airway simulated flow differences (m³/s) at various points in the mine

Does size matter?

If underestimation of airway size is compensated for and nullified by underestimation of friction factors and shock losses in a model, it may not affect airflow model accuracy, since resistance calculated is still correct. However, problems may arise on several other fronts. For accurate simulation, friction factors would need to be calculated for every size, as the overbreak amount and airway size effects on resistance are non-linear. Airway sizes are also used for other simulation functions. For example, heat and moisture transfer from rock to airflow is highly dependent on the exposed area and perimeter of tunnels. In addition, correct airway size allows air velocity to be accurately calculated, which is particularly important for accurate timing in blast fume dispersion clearance, or fire modelling for smoke or gas transport through a model. Using underestimated design size, travel time of smoke or fire fumes through a mine may be too rapid.

Limitations of the algorithm

A fully automatic algorithm may occasionally produce very large cross-sectional areas, for example at the intersection of several tunnels. Applying this size to a length of airways as shown in Figure 5 gives an unreasonable estimation of actual sizes unless it is only applied to a very short length. The frequency of cross-sectional slicing of the survey data can be increased to limit this issue, however this may create an unnecessarily complex model which may simulate more slowly and take up more computer processing and memory. Likewise, if the frequency of slicing becomes too sparse, accuracy can also suffer particularly if airway sizes change regularly.

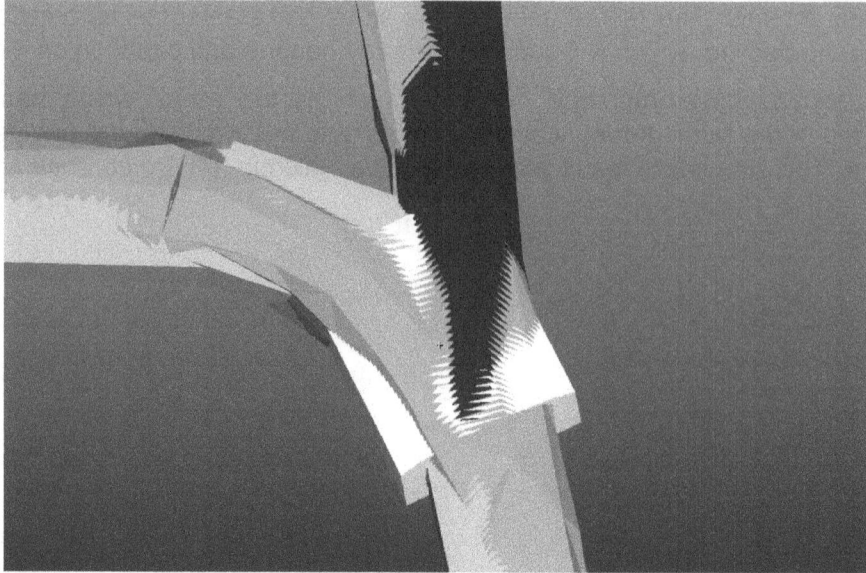

FIG 5 – Example of an inaccurate airway. The selected airway (white) larger than in a typical model and not matching the shape (brown), other airways are in blue.

CONCLUSIONS

Many ventilation models observed (including the model used in this study) are built using planned design dimensions which do not represent the true size and shape of the airways. This study observed significant differences between the sizes used in the model and the survey measurements, which had a large influence on the simulation result. The inaccuracy however was mostly compensated for by using friction factors which are too low but correct the resistance for the error in airway size estimation. This indicates that the existing model, if already accurate in terms of airflow, may not benefit from better matching of airway sizes to survey data for airflow simulation.

However, the size of the airway does not only have an impact on the resistance of the airways. The area will impact the velocity of the airflow which can have a critical impact on fire and contaminant simulation. The total surface of the exposed wall will also have an impact on correct heat and moisture simulation.

For new models the benefits of correct actual airway sizes and factors will result in more accurate simulation, providing friction factors are also calibrated to actual measurements. Further work on this method could lead to more accurate models without performing a time consuming PQ survey; however a PQ survey would still be required to achieve accurate airflow modelling.

ACKNOWLEDGEMENTS

This paper would not have been possible without help from Ernest Henry Mine who provided the model, survey and measurement data necessary for this study.

REFERENCES

Hall, C. J. (1981). *Mine Ventilation Engineering.* New York: Society of Mining Engineering of The American Institue of Mining, Metallurgical and Petroleum Engineers, Inc. New York.

McPherson, M. (1993). *Subsurface Ventilation and Environmnetal Engineering.* Chapman & Hall.

Rowland, J. (2010). Survey Execution to Build a Ventilation Model, Australian Style. *13th United States/North American Mine Ventilation Symposium.* Sudbury, Canada: Hardcastle and McKinnon (Eds.).

Rowland, J. A. (2011). Ventilation Surveys and Modelling - Execution and Suggested Outputs. *11th Underground Coal Operators Conference.* Wollongong.

Accuracy and confidence prediction in ventilation models

C M Stewart[1] and M D Griffith[2]

1. Operations Manager, Howden VentSim, Brisbane, QLD 4101. Email: craig.stewart@howden.com
2. Senior Software Applications Engineer, Howden VentSim, Brisbane, QLD 4101. Email: martin.griffith@howden.com

ABSTRACT

Ventilation models simulate underground pressures and flows based on data entered by the user, however it is not common knowledge the accuracy of the results can vary widely through different parts of a model regardless of the accuracy of the data entered. Depending on the structure of the ventilation system and the effect of natural ventilation pressures the results could vary significantly with only small input parameter changes. Modelled regions of a mine often cause considerable concern with ventilation professionals when simulated data does not match measured data or critical ventilation areas fail to meet expectations after design and implementation. This paper examines the causes of variability and inaccuracy in modelling results and where to expect lower confidence results in a ventilation system. Methods to improve confidence through more robust ventilation design and contingency are also discussed.

INTRODUCTION

Ventilation modelling has become a critical tool in most modern mine ventilation designs and relies on the mining professional to accurately design and input data into the modelling software to obtain accurate results. When done correctly, models can reliably confirm design assumptions and provide confidence in achieving planned airflow and environmental conditions.

The most critical input data for airflow modelling is the resistance of the model airways which includes assumptions of friction and shock losses. For existing mines, airway resistance data can be validated with actual resistance measurements in a mine, however measured data can also have error variability due to the survey methods used and constantly changing mine conditions during surveys. For future ventilation planning, assumptions must be made based on previous experience or theoretical data which will typically be less accurate than measured data. The accuracy of input data will have a direct impact on the accuracy of simulation results, however the impact of resistance errors can vary significantly, depending on the location in the model and the construction design of the interacting airways. The resultant variability in airflow may be much more than what the resistance variability would initially suggest.

In critical areas, where airflow or temperature must precisely be engineered, these variations can cause significant problems with health, safety, and lost productivity from reduced work output and delays.

This paper shows variations are greatly impacted by ventilation design structure, the accumulation and interaction of small errors, and the variable pressure of natural ventilation. Analysis of the key causes of large variability in ventilation model results will be discussed, and contingency plans raised about what can be done to identify and limit the variability in critical areas.

ANALYSIS OF VARIABILITY

The most common causes of variability of modelling airflow results compared to actual results are errors in resistance estimation and natural ventilation pressure changes (Griffith & Stewart 2019). As described in Equation 1, resistance errors can be caused by poor estimations of airway size, friction factor or shock loss. For example, variations between design and actual resistance can occur when the mining process causes overbreak and results in larger than expected airway sizes. If the engineer fails to survey the correct size and assumes design sizes, the resulting simulation may be incorrect.

Natural ventilation pressure variation is more difficult to identify. It is caused by heat changes in the mine (from mining activity, underground rock strata or surface temperature changes) creating pockets of more buoyant air in the mine, changing the balance of airflow at different times. Although well below the pressures normally applied by ventilation fans, small changes in pressure *balance* particularly in areas outside of the direct influence of fans can cause large changes of airflow without significantly affecting overall mine fan pressures or total airflow. This variability may cause unexpected changes in ventilation, particularly away from the influence of the primary ventilation circuit.

Resistance sensitivity

Atkinson's resistance equation for an airway (Equation 1) is dependent on the airway length, L, perimeter, *per*, cross-sectional area, A, and the friction factor, k (McPherson, 1993):

$$R = kL\frac{per}{A^3}. \qquad (1)$$

Where

R = Resistance (Ns^2/m^8)

k = Friction Factor (kg/m^3)

L = Length (m)

A = Area (m^2)

per = Perimeter (m)

Resistance can therefore be subject to the following errors in measurement or estimation including;

- Poor or incorrectly assumed measurement of size or friction factors underground.

- Deformation of the surrounding airway host rock causing size changes over time.

- Partial blockage of airways with pipes, ducts, conveyors or other obstructions that are not included in the model.

- Poorly considered shock losses around bends or contractions in the mine.

- Passing vehicles and machinery causing temporary blockages or a piston effect.

The relationship between pressure, airflow and resistance is defined by Atkinson's equation (Equation 2) or it's variation (Equation 3).

$$P = RQ^2 \qquad\qquad (2)$$

$$Q = \sqrt{\frac{P}{R}} \qquad\qquad (3)$$

$$\frac{1}{\sqrt{Rt}} = \frac{1}{\sqrt{R1}} + \frac{1}{\sqrt{R2}} + \frac{1}{\sqrt{Rn....}} \qquad (4)$$

Where

P = Pressure (Pa)

Q = Quantity (m^3/s)

Rt=Combined resistance of parallel airways (Ns^2/m^8)

In the case of a single airway leading to a pressure source, if resistance (R) varies and pressure (P) remains the same, then airflow (Q) according to Atkinsons equation should theoretically vary by the square root of the resistance change percentage (Equation 3). For example, a 10% change in resistance should only cause a 3% change in airflow if fan pressure remains constant. However, as most mines are driven by ventilation fans defined by pressure/quantity curves, the change in airflow may be more than this.

In the case of multiple parallel airways ventilated by the same fan pressure, a variation in resistance applied to either parallel airway will result in a change only in that airflow, with the combined flow changing by a lesser percentage than a single airway (Equation 4). If multiple random changes are made to multiple parallel airways, then the combined resistance change will always be less than a single airway.

However, in the case of connecting airways between multiple parallel airways a much less predictable result may occur as demonstrated in Figure 1. If two parallel airways with identical resistance (Airways A and B) exist with a connecting airway (Airway C) at the same point along the airways, then no flow will occur between the connecting airway. If the resistance is increased in Airway A relative to Airway B then the connecting airway will flow towards Airway B, and vice versa. In summary the flow in the connecting airway may change magnitude and direction as the airflow tries to find the path of least resistance. It is this configuration that causes the greatest uncertainty in model design.

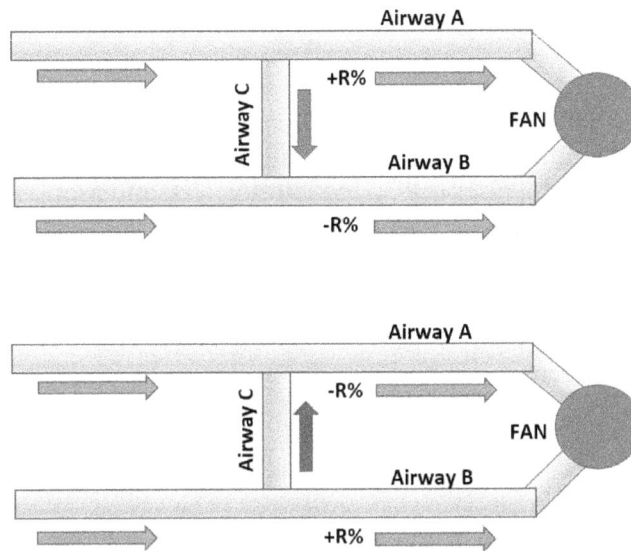

Figure 1 Connections Between Parallel Airways

Methodology of Resistance Sensitivity

Small errors in resistance estimation are unavoidable, therefore a proposed method of examining the effect of resistance errors is to deliberately apply a series of randomised errors to produce multiple variations of the model. Like the Monte Carlo approach of statistical analysis, a method can be applied where the resistance of each airway is modified by applying a random factor, ε, of maximum size, ε_{max} as described in equation (5). Every simulation will result in slightly different airflows depending on the variations in resistances in the model.

$$(1 - \epsilon_{max}) \leq \epsilon \leq (1 + \epsilon_{max}), \qquad (5)$$

An algorithm can be derived as follows;

- A variation factor of 1/3 or 33% provides a reasonable random variation to each airway resistance in the model.

- A significant number of iterations of variable simulations can be performed to provide a statistically significant average. One hundred (100) iterations is suggested, however larger models may need to be satisfied with fewer iterations due to simulation time constraints.

- The simulation results will provide a mean airflow $\bar{Q}$, the airflow standard deviation σ_Q and the frequency of airway airflow directions changes for each airway.

- Further statistical properties can be determined including the mean standard error, $\frac{\sigma_Q}{\sqrt{n}}$, the deviation of the mean airflow from the current simulated airflow, Q, and the airflow quantity confidence, C.

- The quantity confidence, expressed as a percentage, is given by equation (6) which is useful in highlighting any airways with large deviations from the mean, but will also highlight low flow airways with small, but relatively large, deviations.

$$C = 100 \times \left(1 - \left|\frac{(Q - \bar{Q})}{\bar{Q}}\right|\right). \tag{6}$$

It is important to note the method cannot estimate the likely actual variation of resistance or airflow in a real mine. Instead, the method is testing the resilience of the airway flows in the model to change because of variations in resistance. If the airflows do not change any more than the amount predicted by Atkinson's equation (Equation 2 and 3), then the confidence of the model airflow matching the inputted resistance is considered 100%. If, however the airflow variation is greater than expected by the simple variation in resistance, then it is likely the structural configuration of the model (in particular airway connections between parallel airways) is likely causing the variability and therefore the airflow magnitude and direction should be treated with more caution.

Heat sensitivity

A second cause of variability of airflow in model predictions is the distribution of heat and natural ventilation pressures throughout a model. If natural ventilation pressure is considered (in most cases it should be considered) then the simulation and distribution of temperature and air density is an important part of the calculation. Figure 2 demonstrates the effect on temperature changes and natural ventilation pressures causing airflow changes between what would be otherwise stagnant airway connections between parallel airways.

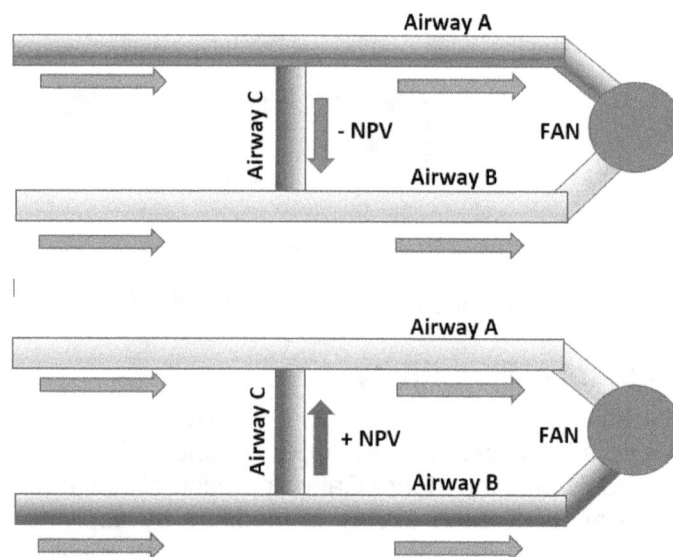

Figure 2 Variable Temperatures (shown as Red and Blue) cause natural ventilation pressures and flow changes between parallel airways

Most ventilation modelling is performed using steady state analysis. In this case, the distribution of heat through a model is predetermined and considered constant for an indefinite period. A simplified approach to calculating the effect of natural ventilation pressures is to use prior airflow simulations to predict heat flow distribution through a mine, and then use the heat flow distribution to predict natural ventilation effects which influence subsequent airflow simulations. This coupled approach will provide different steady state solutions every iteration which may or may not eventually converge to a steady solution.

A more sophisticated approach is to couple the flow and heat distribution into the energy balance equations used to solve airflow distribution (Danko 2017), however any method using a steady state approach is only valid for a single moment in time and real mines will be constantly disturbed by subtle changes in heat distribution due to changing surface conditions or moving heat sources (for example machinery) underground.

The first simplified approach is more likely to provide an indication of the stability of a model due to changing heat distribution for each iteration being considered for subsequent iterations. While this approach does not simulate actual likely variance of heat throughout a mine (because the ultimate steady state solution between each iteration would be interrupted before completion in a real mine), is does provide an indication of the possibility of airflow variance due to changing heat distribution. Another approach would be to simply change the temperatures (and therefore air density) randomly throughout the model (much as the resistance was changed randomly in the previous section), however this would require a more sophisticated distribution mechanism to consider the initial heat distribution and how the heat flowing from one random change would flow into downstream areas.

Methodology of Heat Distribution Sensitivity

Using the simplified iterative steady state method described above;

1. A model with no temperature variation is run to provide initial airflow distribution

2. A thermodynamic heat simulation is run to distribute heat along the pathways determined by the airflow simulation.

3. The air density and natural ventilation pressure is calculated for the new heat distribution in every airway.

4. Another airflow simulation is performed taking into consideration the temperature and density changes causing new natural ventilation pressures.

5. The iteration loops back to Step 2 and repeats a statistically significant number of times.

Similar to the resistance sensitivity method, the mean airflow, $\bar{Q}$, the airflow standard deviation, σ, and the percentage of samples in which the airflow direction changed, X, for each airway can be recorded. In addition, the mean dry-bulb temperature, $\bar{T}$, and the standard deviation, σ_T can be recorded, as well as the temperature mean standard error, $\frac{\sigma_T}{\sqrt{n}}$, and the deviation of the mean dry-bulb temperature from the currently simulated dry-bulb temperature are calculated. Sections of the mine that do not converge to a solution after repeated heat and airflow simulations are detected by observing airways with low airflow and direction confidences.

RESULTS

The proposed methods and algorithms were tested in VentSim Design to provide visual feedback of results. These methods have now been incorporated into tools in the software for users to test their own designs.

Resistance sensitivity

Figure 3 shows a metalliferous mine design with fresh air entering the mine via a ramp and shaft system on the right-hand side. Whilst most areas on the design show confidence of 100%, where resistance variations will not affect airflow any more significantly than defined by Atkinsons equation (Equations 2 and 3), several areas show significantly less confidence where airways branching off the main ramps show in yellow and green, suggesting a confidence level of 90% or below. A deeper airway in the model design traversing across the design shows a blue colour with a confidence level of only 50%.

This does not mean the airflow results in the real mine could vary by this amount. However, it does indicate the model results in these areas are being easily influenced by small resistance changes locally and elsewhere in the model. Any errors in data input and assumptions for airway friction factors or sizes are more likely to have a strong effect on these airways. Conversely, areas of full confidence are less likely to be affected and can be considered more reliable predictors of actual airflow.

Airways with low quantity confidence will often also have low direction confidence. An airway prone to airflow reversal can cause numerous safety and productivity issues as the expected direction of gases, blasting fumes and heat cannot be guaranteed. The resistance sensitivity analysis will often highlight connecting airways between parallel zones sourcing air to or from a common pressure source as low confidence, whereas airways leading from or to major fans will show as high confidence.

Figure 3 Airflow quantity confidence (%). Cooler (green and blue) indicate higher variability.

Heat sensitivity

The same model was simulated for heat sensitivity with the results shown in Figure 4.

Figure 4 Heat sensitivity effects. Cooler colours (green and blue)
show increased variability and reduced confidence

Considerable variability is demonstrated in this case with large sections of the upper main ramp on the right (away from the main fans and near multiple other surface connections) showing poor airflow confidence. Areas under the direct influence of a fan show good confidence with little susceptibility to natural ventilation pressure variations. The low confidence areas also showed a high risk of airflow reversal, a significant concern if a guaranteed ventilation direction is required for safety or productivity purposes.

DISCUSSION

The example model presented in Figures 3 and 4 shows a poorly designed ventilation system, producing variable results in some areas from the sensitivity analysis. The model consists of many interconnected parallel pathway connections to primary ventilation fans and to surface connections, a configuration previously discussed as being more likely to cause variation in predicted airflow with subtle changes to resistance and heat distribution. The author notes there are many mines designed in this manner, and normal steady state simulation results may therefore be misleading as they perhaps incorrectly infer adequate airflow and correct directions to all critical areas.

The larger variations appear more likely to occur in regions a long distance from the direct influence of a main ventilation fan, or between parallel ventilation circuits driven by the same fan pressures. If primary airflows and directions can be easily influenced by small changes to airway sizes, resistance or heat, or by even parking or moving obstructions, such as vehicles, through these zones, then the ventilation cannot be considered robust and may require redesign in critical areas.

Engineers and ventilation officers are often perturbed about why some areas of their ventilation models perform poorly against measured results. Sensitivity analysis can also assist in identifying these areas and help allay (or confirm) fears of model inaccuracy or measurement mistakes. The same variability factors and sensitivity mean that it is common for the sensitive areas to show a high variability in actual measured results as well, with ventilation surveys on one date being substantially different to surveys taken on other dates. Trying to improve a model that performs poorly in a non-critical area shown to have high sensitivity could therefore be dismissed as a waste of effort and engineers could focus on more important parts of a model or ventilation design.

Critical areas that should not be exposed to the possibility of airflow variation (production zones, workshops, crushers, explosive magazines or fuel stores) should be checked more closely for variability. If the analysis shows the areas are sensitive, there may be a need for additional ventilation controls or engineering redesign to ensure guaranteed airflow. The causes of variation in models and real mines can often be overcome by utilizing fans dedicated to ventilating only one region, or by adjustable regulators that can be quickly changed to respond if undesirable ventilation variations are detected.

CONCLUSIONS

Sensitivity analysis using variable resistance and heat inputs may be a valuable method to determine the reliability and robustness of a ventilation design toward a given application. If strong variations are shown likely to occur in critical regions, ventilation professionals can take proactive steps to prevent them by improving ventilation design. Model sensitivity analysis will assist engineers in developing better and more robust designs that ensure safe conditions and that meet regulatory and productivity requirements in all circumstances.

REFERENCES

McPherson, M. J. (1993). *Subsurface Ventilation and Environmental Engineering.* Chapman & Hall.

Danko, C. , Barahmi, D., Stewart, C. (2017). Computational Energy Dynamics Solver for Mine Ventilation Simulations. *16th North American Mine Ventilation Symposium.* Golden, Colorado USA.

Stewart, C. M. (2014). Practical prediction of blast fume clearance and workplace re-entry times in development headings. *10th International Mine Ventilation Congress,* (pp. 169-176). Sun City.

Todini, E., & Pilati, S. (1987). A gradient method for the analysis of pipe networks. *International Conference on Computer Applications for Water Supply and Distribution.* Leicester, UK.

Griffith, M. D., & Stewart, C. (2019). Sensitivity Analysis of Ventilation Models - Where Not To Trust Your Simulation. *North American Mine Ventilation Symposium,* Montreal Canada.

Selection guide when considering rental or fixed mine cooling plants

L van den Berg[1], J Raubenheimer[2,] H Mohle[3] and S Bluhm[4]

1. Principal Mechanical Engineer/MD, BBE Consulting, Perth Western Australia, 6027. lvandenberg@bbegroup.com.au
2. Senior Ventilation Consultant, BBE Consulting, Perth Western Australia, 6027. jraubenheimer@bbegroup.com.au
3. Principal Mining Engineer, BBE Consulting, Perth Western Australia, 6027. hmohle@bbegroup.com.au
4. CEO, BBE Group, Johannesburg South Africa. sbluhm@bbe.co.za

ABSTRACT

Australian mines are increasingly getting deeper and these operations are starting to encounter heat issues. At critical depth the heat cannot be managed by ventilation alone and refrigerated air cooling becomes necessary as part of the primary ventilation system.

Historically, the first Australian refrigeration systems comprised of fixed plant typically Ammonia. Examples include Mt Isa operations, CSA mine, Argyle, Tanami and Telfer amongst others. In the last decade or so mine refrigeration trends such as rental and fixed plant R134a systems have become more prevalent. Implementation of rental R134a units has become an attractive option especially for operations with small refrigerated air cooling requirements.

When evaluating refrigeration strategies for an underground mine, project teams often need to trade off fixed plant and rental units. A typical evaluation strategy with key considerations is described. The main advantages, disadvantages and pitfalls of both approaches are discussed. The main differences including design, technical performance, economics, safety, system expandability, implementation and contracting approach is explored. The paper provides guidance to select between either strategy or in some cases a combined hybrid strategy. The factors and parameters to determine the most feasible option including life of mine and deployment period are also investigated.

With the increasing depth of the Australian underground mines fit for purpose refrigeration strategies will become more important and a selection guideline will assist with determining a 'best value' solution for the asset.

INTRODUCTION

Heat is a very important consideration in the mine ventilation design and planning process. Underground operations in both the coal and hard rock industry experience heat from hot humid ambient conditions, high virgin rock temperatures, auto compression (depth) and mining equipment (diesel/electrical). Failure to effectively plan for possible heat load in mines can lead to loss in productivity, production delays and increased accident rates. Failure to plan for the potential heat impact often leads to 'crisis management approach' when it comes to the implementation of mine cooling system. This could result in suboptimum systems being implemented which costs the mining operation in the long run.

Mine cooling systems have been implemented in the underground mining industry in various forms over many years. These technologies are well established yet highly specialised for mining applications. Traditionally when a mine requires cooling the mine would construct a refrigeration system to manage heat by cooling the air either on surface or underground. However, over the past decade or so, mines have also used rental units for refrigeration machines, cooling towers and bulk air coolers (BAC's). This has become a practical consideration for Austrian mines with financial constraints due to remoteness and geological constraints impacting the confidence in the future ore body (i.e. very short planning horizons).

With the exception of Australia, it is very unusual for mines to not acquire this type of plant by outright purchase (although there are some exceptions generally related shaft sinking and development elsewhere in the world). The approach of using this rental equipment is dominant in Australia but not in other leading mining countries such as South Africa, United States or Canada where the use of rental equipment is less frequent.

TYPICAL FIXED COOLING PLANT

Typical fixed cooling plant comprise of custom build air coolers sized for the amount of air that needs to be cooled. These air coolers are served by various different types of refrigeration machine technologies including ammonia machines, R134a machines and absorption refrigeration machines. Mine cooling systems at Australian mines include Mt Isa operations, CSA mine, Argyle, Tanami and Telfer amongst others.

The Mt. Isa 25MW$_R$ ammonia refrigeration fixed plant was commissioned in 2000 as one of the first surface mine cooling plants in Australia (Brake, 2002). To the author's knowledge the Mt. Isa cooling system is still one of the largest in Australia. The Gwalia mine site in Western Australia implemented a fixed absorption refrigeration plant which uses waste heat from the power generators to produce chilled water (Broodryk, 2015). Fixed plant cooling systems are not unique to hard rock mines and are used in Australian coal mines that experience hot ambient conditions such as the ammonia plant at Moranbah North (Belle, 2013).

There are a number of examples of fixed plant R134a refrigeration such as the 4MW$_R$ plant in Figure 1 at a mine site in Western Australia. These cooling plants are configured with custom build refrigerating machines or complete factory assembled units from leading OEM's such as York, Carrier, Trane or Shmardt.

The typical features of a fixed cooling plant are:

- Thermal design safety factors and material selection on heat exchanger equipment to minimise the performance impact from aggressive mine environments.

- Modular sizes of up to 12MW$_R$ for a single size refrigeration machine available.

- Custom construction with optimised thermo-process design to suit cooling requirements is possible:

    o BAC's specifically sized and selected for the airflow quantity an ambient design condition.

    o Implementation of two stage spray systems and larger contact area to enhance heat transfer leading to lower chilled water circulation and BAC fan pressures.

    o Lead-lag refrigeration arrangements possible enabling lead machines to operate at higher chilled water off temperatures with improved power performance i.e. higher coefficient of performance (COP).

- Demand control by using variable speed drives for refrigeration machines and air coolers allowing power draw to be minimised during part load operation.

- Installation constructed with constructability, maintainability and safety principles in mind often to strict client standards and requirements.

- Fully instrumented and integrated to allow real time control, monitoring of equipment and importantly allowing system thermal performance to be verified.

- Ability to chill air to temperature as cold as 5°Cwb or even super chill air to 2°Cwb if combined with technologies such as ice melt systems (Wilsons, 2003; Branch, 2015).

- Energy management systems such as thermal storage systems (Branch, 2015) and (du Plessis, 2015).

FIG 1 – Typical Fixed Plant (R134a)

WHY BUY FIXED PLANT?

For hot mines a cooling plant is an integral part of the mine's ventilation infrastructure just like the primary fans or a gas drainage system in coal mines. The majority of mines will establish permanent ventilation infrastructure by purchasing plant that will service the operation throughout its life. Most underground mine heat loads increase with depth and once the cooling threshold is reached it is likely that the mine will continue to require cooling throughout its operating life. The decision to buy a permanent cooling installation to manage heat is perhaps a fairly obvious choice for mines that have:

- Large heat loads (depth, high VRT) where cooling requirements can only be effectively achieved with custom built cooling systems.

- Long design life for example large mines that employ mass mining methods such as block caving that often have 10+ year mine life.

- Existing permanent cooling installation requiring an upgrade to serve the long term - upgrade will normally be in the form of additional permanent modules.

- The view to exploit strategies that include supporting infrastructure not necessarily available for rent for example

    o   water storage dams and ice systems for load control and energy management,

    o   chilled service water reticulations,

    o   absorption systems to exploit waste heat.

- Locations where rental systems are not offered or rental companies do not have in-country support.

- Requirement for continuous cooling through the year (24hr x365days) for example ultra-deep mines or mines in climates where the ambient wet-bulb is high throughout the year. Rental plant cost is directly linked to operating time and is therefore higher if the plant is used for longer periods.

TYPICAL RENTAL COOLING PLANT

Typical rental plant is made of a standardised fleet of containerised modules including air handling units (AHU's); refrigeration machines and condenser cooling towers. Equipment is typically containerised for ease of mobilisation and transport as is appropriate for a rental business model. These modular units available in the fleet for the components that is used to meet the overall cooling duty. Most reputable rental companies such as Aggreko or CAT use R134a refrigeration machines in their fleet. These cooling plants are configured with e factory assembled units also from leading OEM's mainly Carrier or Trane.

The typical features of a fixed cooling plant are:

- Compact modular standard design with containerised equipment.

- Modular refrigeration machine sizes limited to about 1.5MWR

- Modular air coolers with maximum 20-25m³/s capacity.

- Typically, fixed speed refrigeration and air cooling units.

- Often bare minimum instruments to protect equipment only making performance assessments challenging.

- Air can practically only be chilled to about 8°C wet-bulb.

FIG 2 – Typical Rental Plant (R134a)

WHY RENT PLANT

There are circumstances where renting plant should be considered as part of the mine cooling strategy. Rental systems offer a quick "off-the-shelf" solution that can be implemented in a very short period of time - this is their biggest advantage.

Because components are normally limited to 'off-the-shelf' HVAC type units as offered. The rental companies often pre-purchase the components which are generally supplied ex the rental company's store yard. A rental system will thus not necessarily be an optimised system (this would only be limited the rentals company ability to supply the equipment ex their store yard). Because of this, the rental systems will invariably be sub-optimal in mine refrigeration systems and may not be suitable at all for larger systems. Nevertheless, they offer some strategic advantages under very specific circumstances. The rental system approach could perhaps be considered for mines that:

- A need to or wish to defer capital for strategic reasons;

- Require an interim solution in between planned major ventilation upgrades; major cooling system upgrades or during the construction of a permanent installation.

- Require cooling for only a short period (e.g. mines designed only for a couple of years or mines that will experience heat issues only toward the end of the life-of-mine).

- Projects that are in exploration stages e.g. exploration declines or shafts where there is no guarantee that the ultimate mine will be built.

- The cooling requirement is temporary in nature:

- Experience heat challenges only during the initial mine development or production build-up phases of the project but not during steady state mining.

- Encounter sudden unexpected 'temporary' heat loads that need to be managed in the interim (e.g. unexpected inflow of hot ground water that can ultimately be managed by dewatering strategies).

ECONOMIC CONSIDERATIONS

The above discussion addressed the strategic thinking that goes into evaluating purchase versus rental of cooling plant and, as pointed out, there are specific circumstances where the choice between the two systems will be clear. However, there will also be 'grey areas' where the strategic value is not immediately evident. Under these circumstances an economic evaluation between purchasing and renting the plant should be considered.

A permanent system requires a capital investment for the purchase, delivery and installation of equipment as well as supporting civil, structural and electrical works. The operating cost primarily includes the cost of power to run the system. While, there are also some costs associated with make-up water [for water cooled systems] and maintenance cost these are generally much lower than the power cost.

A rental system on the other hand requires less capital in terms of equipment delivery and installation. The cost of the supporting electrical infrastructure for the same size plant would be similar however civil works would generally be temporary in nature and hence lower cost. The operating costs will include the rental cost (with maintenance generally included) as well as absorbed power. The lower capital cost and higher operating cost associated with rental systems is often attractive from a pure capital deferment point of view.

Payback Analysis

For an economic assessment the payback period between Permanent and Rental options needs to be determined. Figure 3 shows the total owning cost for a typical trade-off comparison between buying Permanent Fixed plant and Rental plant for 12MW$_R$ rated air cooler. The analysis is based on the parameters in Table 1.

This example is for an Australian mine site with about eight months of rental needs which is representative for either a deep mine (1.5km) or a mine in more tropical regions such as North Queensland where high ambient wet-bulbs are sustained for longer periods.

TABLE 1 – Economic Assessment Inputs

Description	Permanent Plant	Temporary Rental Plant
Plant Cooling Capacity	12 MW$_R$	12 MW$_R$
Refrigeration plant - Capital cost [$1,200/kW]	$14.4 Million	-
Plant Rental [35 weeks operational]	-	$2.1 Million
Plant Rental [17 weeks standby]	-	$0.3 Million
Rental plant – Upfront costs [Commissioning etc]	-	$0.9 Million
Plant Coefficient of Performance [COP]	4.6	2.8
Plant absorbed power [Total]	2.6 MW$_E$	4.3MW$_E$
Maintenance cost	$0.3 Million/year	-
Recoverable costs [~18%]	-$2.5 Million	-
Power cost per year	$0.11/kWhr	$0.11/kWhr
Interest rate	6%	6%
Total Owning Cost [TOC]	$28.6 Million	$45.0 Million

A cost comparison between rental and purchase options indicates a breakeven point at approximately 4 years after start of plant operations. For this example, the LOM is 12-years and it will clearly be more cost effective to purchase this equipment rather than to rent it over the life of the project. The difference in total owning cost is $16.4million more for the rental plant compared to purchasing a plant over the same period.

However, if this same mine only had five years operation left a rental option makes a lot of economic sense. Another reason rental may be favoured is if the operation is CAPEX constrained or the business has very aggressive financial payback metrics, say three years.

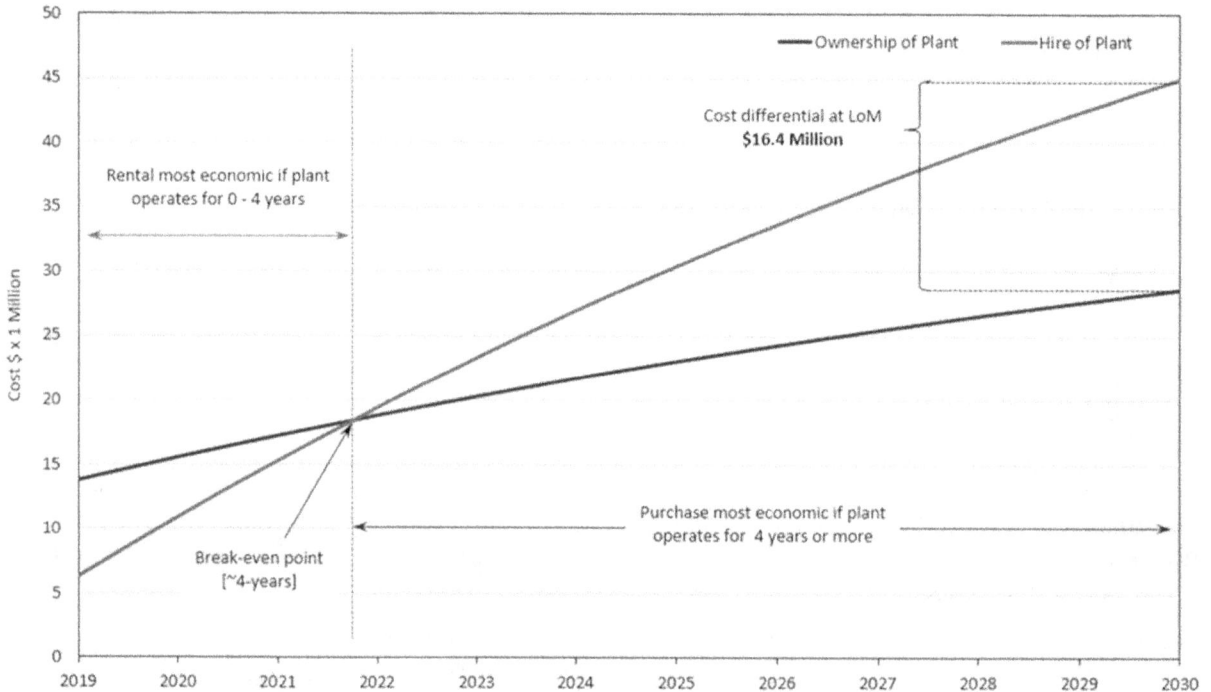

FIG 3 – Economic Sensitivity to Input Parameters

Economic Sensitivity

Economic assessments of this nature relies on a number of input parameters both of a technical and economical nature. The relative sensitivity between these different parameters are critical in interpreting the economic assessments. The results of a typical case study reflecting the delta in the NPV of two cases is presented as a tornado graph in Figure 4. A what-if scenario was on various input variables by adjusting them either -20% and +20%:

- Power cost (Power)
- Water cost (Water)
- Operating Months (Op Months)
- Efficiency of Permanent plant as a ratio from BAC to Rated (PermBACtoR)
- Efficiency of Rental plant as a ratio from BAC to Rated (RentBACtoR)
- Efficiency of UG plant as a ratio from BAC to Rated (UG_BACtoR)
- Permanent plant chillers COP (PCCOP)
- Rental plant chillers COP (HPCOP)
- UG plant chillers COP (UGCOP)
- Make-up water requirement for Permanent plant (Perm_H2O_Mup)
- Make-up water requirement for Rental plant (Rent_H2O_Mup)
- Permanent Plant, fixed plant maintenance cost as a % of CAPEX (FPmaintPER)

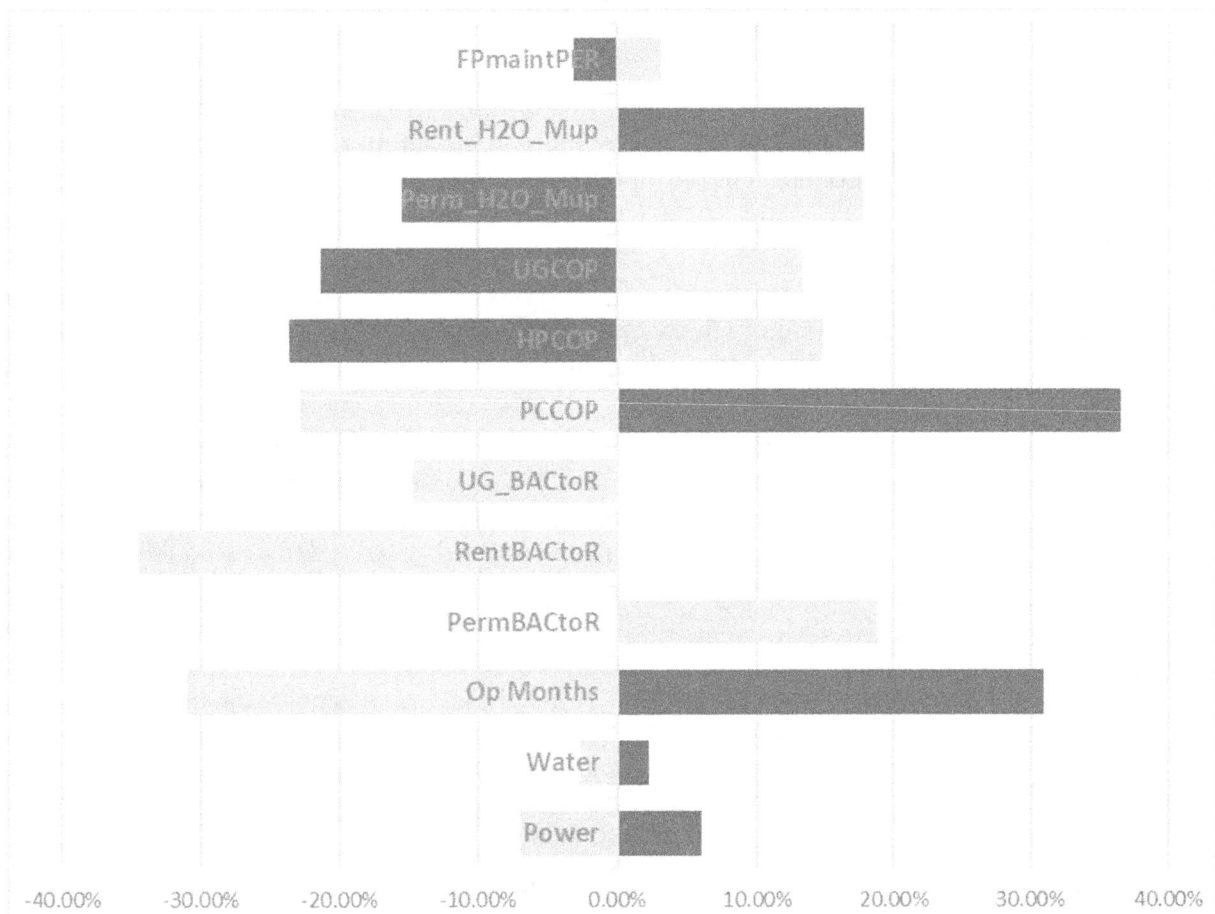

FIG 4 – Economic Sensitivity to Input Parameters

The increase in operating months favours a permanent plant case. The model sensitivity for a 10% increase (3 weeks) in operating months the delta in the NPV improves by more than 15% in favour of a permanent plant. For mines that deepens over time the impact of the ambient temperatures becomes less and thus the summer to winter differential plays less of a role. This is the same for mines in tropical regions where seasonal ambient wet bulb temperatures do not fluctuate significantly. This will inevitably result in the refrigeration system being operated for more months during the year.

Another important observation is that the outcome is more sensitive to efficiency of the BAC for the Rental plant compared to the Permanent plant. The outcome differential is sensitive by a factor of 1.8. In other words, should efficiencies be over-stated by the same amount the effect is more pronounced for the rental plant compared to the permanent plant. Another conclusion from the sensitivity in general is that the delta in the NPV for the two cases in fact has more downside to the rental plant efficiency and COP than it has upside.

From the sensitivity it was concluded that economic assessment of this nature is positively biased to the performance data provided by the vendor. It is therefore critical that mine sites who make decisions in favour of rental plant on these types of economic trade off's negotiate performance clauses to ensure stated performance criteria are met. Another conclusion is that mine sites with cooling demands that is likely to increase with depth needs to factor the increased operating months as mines deepen. Further mine sites in tropical regions with longer term cooling demands will have much shorter payback periods on permanent plant compared to more mild climates where cooling is only required for a few months in the year.

Perils of 'Rolling' Planning Cycles

Many Australian mines have forward planning cycles with a three to five year life of mine. This means that each year planning is based on a limited operational period which falls within the typical economic payback period of a rental plant. This naturally favours the rental approach to cooling. If these mines then operate for this limited period and shut down the adoption of a rental system was the correct call.

In reality a large majority of these mines have a rolling three to five year life. Which means the operation continuous expands and the 'true' life of mine is in fact much longer. The reality is that financial decision making can often only be made on the planning period and if these corporate constraints exists then naturally (and correctly so) rental is an attractive option within these constraints.

Nevertheless, mine with a clear understanding that there is a high likelihood of future extension of the mine plan beyond the rolling three to five year periods need to reflect on the implications of this approach which could be somewhat short sighted and limiting. To illustrate consider an example based on an operating mine in Australia.

The mine has permanent ammonia plant however required additional 4MWR cooling with increasing depth. In the context of limited available CAPEX, an economic payback benchmark of less than 5-years a rental plant is selected based on the economics. This decision making process is repeated for a number of cycles until the installed rental capacity grows to 10MWR. Meanwhile a longer life of min and, increased depth means that and additional cooling demand of another 8MWR needs to be installed.

The implication is that the CAPEX needs to provide for 18MWR with 10MWR to replace the installed rental capacity and 8MWR additional capacity. Not only is this a significant once off CAPEX expenditure that would be difficult to have approved but it also means that once rental is installed with growing cooling demand the economic trade-off is increasing biased to rental. The effect is that mines that technically need permanent plant due to increasing depth may find themselves locked into a rental strategy based on the initial decision to follow this strategy.

Mines that has a view of having permanent plant in the future to meet cooling demands need to reflect on whether the 'Deferral of CAPEX' does not limit their ability to 'Phase-in CAPEX over time'.

OTHER CONSIDERATIONS

There are various technical and other considerations when selecting between rental and permanent plant that needs to be factored. A select view is discussed briefly in the following section:

System Performance

It was stated previously that economic assessments is sensitive to the cooling system performance. Thermal performance is critical for both Permanent and Rental plant in order for the system to meet cooling demands and minimise the mine owners' indirect costs (power and water supply). As with any thermal system performance reduces over time due to fouling of the heat exchange components. For a Permanent plant sustaining the performance of the plant is managed by selecting appropriate thermal design safety factors and reliance on regular maintenance by the mine's maintenance department.

For rental the selection of thermal safety factors and maintenance is done by the rental supply company. Belle, 2013 observes that *'advantages of hiring mine cooling systems over rental may be realised only by structuring hiring contracts carefully, where payment is clearly linked to the achievement of measurable operational parameters''*. It is important to realise that most rental companies offer the rental of the gear along with the maintenance and operation oversight of that gear.

Rental companies generally do not rent chill air to the mine and despite many operations correctly insisting on contract terms that states performance based on a chilled air quantity at a set temperature. This type of agreement, to the authors knowledge, is not accepted by most rental providers which means performance can become a contentious issue.

In the case of permanent plant, the system is normally designed with sufficient instrumentation to measure performance. This is not necessarily the case for rental plant and is important that mines specifically specify minimum instrumentation requirements to allow accurate real time performance assessments. In addition, the provision of instrument ports to allow verification of field instruments by hand held instrumentation needs to be included. As a minimum the following is required airflow off individual BAC's, air temperature off BAC, individual 'BAC water flow, water onto BAC, water off BAC, chiller evaporator entering and leaving water temperature, chiller condenser entering and leaving water temperature.

<u>Health and Safety Standards</u>

In the authors experience rental plant are rarely installed to the same standard in terms of HSE when compared to permanent plant. Permeant plant is typically subject to thorough constructability reviews, risk assessments and Safety in Design reviews. Although similar principles may apply to Rental installations the outcomes are rarely the same. This is evident on the amount of safety related devises and features in permanent plant such as access platforms, ladderways, pipe crossings, cable racks/trays amongst others evident with Permanent plant.

The same level of engineering is only evident on rental systems if specified. In most instances operations accept rental systems "off the shelve 'from suppliers whereas this is not the case for Permanent plant. The point is not that Rental plant is unsafe but rather that these factors are not always accounted for in economic assessments which leads to unforeseen costs after implementation if these plants are subject to safety audits.

CONCLUSION

With the increasing depth of the Australian underground mines refrigeration becomes increasingly important for operations. Both permanent fixed plant which is purchased and rental refrigeration strategies are operational on mine sites with heat issues,

Project teams often need to trade off fixed plant and rental units as part of decision making and justification process. Both strategies have an application in the mining industry but there are clear differentiators as to which strategy offers the best value for the operation. The paper provided guidance to select between either strategy.

REFERENCES

Belle, B (2013). Cooling Pathways for Deep Australian Longwall Coal Mines of the Future. The Australian Mine Ventilation Conference Proceedings. Adelaide, Australia. 2013.

Brake; D. J (2002). The R67 refrigeration plant at Enterprise mine, Mount Isa - the world's largest bulk air cooler. The AusIMM Underground Operators Conference, Townsville, Queensland, Australia.

Branch, A.R; Wilson R.W and Poe, T.S. (2015) Novel Use of Ice Thermal Storage for Energy Management on a Mine Refrigeration Plant. The Australian Mine Ventilation Conference Proceedings. Sydney, Australia. 2015.

Broodryk A; de Vries, J; Kyselica P; McLean, K (2015). The Design, Installation and Commissioning of an Absorption Refrigeration System at Gwalia Gold Mine in Western Australia. The Australian Mine Ventilation Conference Proceedings. Sydney, Australia. 2015.

Du Plessis, J (2015). Strategy and Tactics Implemented to Achieve Energy-efficient Ventilation and Cooling of Mines. The Australian Mine Ventilation Conference Proceedings. Sydney, Australia. 2015.

Wilson, R et al, 2003. Surface bulk air cooler concepts producing ultra-cold air and utilising ice thermal storage, in Proceedings Managing the Basics Conference (mine Ventilation Society of South Africa: Pretoria).

Study on the effect of seasonal air-flow on high temperature environment in deep mines and its prevention technology

X Yi[1], W Yu[2], L Ma[3], L Zou[4] and G Wei[5]

1. Doctor, College of Safety Science and Engineering, Xi'an University of Science and Technology, Xi'an Shaanxi Province , PR China, 710054. Email:
2. Ms., College of Safety Science and Engineering, Xi'an University of Science and Technology, Xi'an Shaanxi Province , PR China, 710054. Email: yuwencong1995@163.com
3. Professor, College of Safety Science and Engineering, Xi'an University of Science and Technology, Xi'an Shaanxi Province , PR China, 710054. Email: mal@xust.edu.cn
4. Mr., College of Safety Science and Engineering, Xi'an University of Science and Technology, Xi'an Shaanxi Province , PR China, 710054. Email: 15594803880@163.com
5. Doctor, College of Safety Science and Engineering, Xi'an University of Science and Technology, Xi'an Shaanxi Province , PR China, 710054. Email: wgm20180326@163.com

ABSTRACT

Seasonal air and surrounding rock temperature accounts for severe heat damage in deep mine, which seriously affected safety and efficient production of coal mining. Especially in summer, the high ground temperature intensifies heat damage at underground work sites. To study the effect of seasonal air and surrounding rock on high temperature environment in deep mines, we established the model of the surrounding rock temperature field in the unsteady wind temperature. A method for calculating the heat-adjusting layer by difference method in the unsteady inlet wind temperature was proposed. Meanwhile, comparative analysis of cooling methods has been conducted. We found a cooling method and system suitable for seasonal heat damage in mines. According to the calculation, the temperature distribution of the heat-adjusting layer after cooled was predicted. The results show that as the ventilation time increased, the radius of the heat-adjusting layer continuously expanded and formed a permanent one. Hence, the heating effect of surrounding rocks located in roadway on the wind flow became smaller. The pithead high air volume non-power air heat exchange system was also designed. It was determined that the design could meet the system cooling requirements by simulating the ventilation state in air leakage condition.

INTRODUCTION

With the reduction of the shallow coal resources, the mining depth is constantly increased (He and Guo, 2013). Correspondingly, the problem of mine heat damage is highlighted. Mine heat damage is the sixth major disaster of coal mine production and listed as the four major occupational hazards of coal mines with dust, noise, and toxic gases which seriously threatens the health of miners. The heat damage deteriorates the environment in the mine, which poses a huge challenge to the design and layout of mining ventilation and cooling system. Therefore, investigating the formation, prediction and cooling methods of mine heat hazards are of great significance for improving the working environment, reducing the accident rate, and improving labor efficiency and safety.

The heat transfer between the surrounding rocks and the wind flow is an important factor affecting the mine temperature. Therefore, mine wind temperature prediction is crucial for controlling and preventing heat damage. On the basis of analyzing the heat exchange between surrounding rock and airflow, Van Heerden (Heerden, 1951) and Knig (Knig, 1952) solved the temperature field of surrounding rock thermal coil in ideal conditions. The development of computer technology has enriched the research prediction. Biccard Jeppe (Jeppe, 1939) predicted the wind flow temperature in deep mines and formed the basic idea of wind temperature prediction and calculation. Uchino, K.(Uchino and Inoue, 1986; (Uchino and Inoue, 1990) corrected the calculation of the wind flow temperature and humidity under complicated boundary conditions. Moloney,K.W. (Moloney, Lownde, Stockes and Hargrave, 1997) applied the computational fluid dynamics principle to obtain the distribution law of wind flow in the roadway. G. Danko (Danko and Bahrami,2008) adopted CLIMSIM and MULTIFLUX to simulate the heat and mass transfer of the roadway wall.

Adopting effective treatment methods to prevent the heat damage is of great significance for ensuring safe and efficient mining. Refrigeration is an important method to control mine heat damage and can be traced back to more than one hundred years ago. In 1915, the Morro Velno metal mine in Brazil built the world's first mine air conditioning system: the first installation of air condition and the establishment of a ground centralized refrigeration station in mine. Adopting a suitable ventilation layout will reduce some of heat damage and play an important role in improving the thermal environment of the mining area. Chorowski, M et al. (Chorowski, Gizicki and Reszewski, 2012) studied the cogeneration cooling technology of copper mines in South Africa. The energy demand of copper mines were analyzed in details, and a reasonable technology scheme were proposed. The artificial refrigeration technology can be divided into air cooling system, thermoelectric glycol cooling system, ice cooling system, water cooling system and mine water inrush cooling system. Due to the differences between mines and cooling systems, it is necessary to select appropriate cooling method which could reduce the heat damage and ensure a comfort environment.

To study the effect of seasonal air and surrounding rock temperature on high temperature environment in deep mines, we simulated the temperature variations of mine main ventilation circuit and surrounding rock as the normal production in coal mines. In this paper, the prediction model of the heat-adjusting layer and wind temperature in unsteady wind temperature was established. The effect of pithead wind temperature to the underground were simulated. Based on the prediction of wind temperature, the cooling method and system suitable for seasonal heat damage of mine were determined.

EFFECT OF PITHEAD AIR TEMPERATURE ON THE UNDERGROUND

The numerical calculations were based on Zhaolou coal mine of JuYe mining area, located in Shandong Province of China. The section of ventilation route from its ground inlet to 3302 coal mining face was taken as the scenario. Due to the difference between roadways, the thermal parameters of the wind flow are constantly changing, which results in complicated heat-humid exchange process between the surrounding rock and the wind flow. Therefore, a simplified analysis of the roadway is required, assuming that the shaft is circular, the rock mass excavated is homogenous and isotropic, the heat transfer conditions are constant, and the heat radiated from the surrounding rock is all transmitted to the wind flow.

Prediction model

According to the assumptions, the temperature field distribution of surrounding rock in the tunnel can be solved with the one-dimensional unsteady heat conduction equation.

(1) Heat conduction in surrounding rocks

The heat transfer from the rock original temperature to the cooling by wind flow is the most typical heat conduction phenomenon in the well. The original temperature distribution of the surrounding rock obeys the Fourier heat conduction differential equation, as shown in Eq. (1).

$$q = -\lambda(gradt) \tag{1}$$

where q, λ and gradt present heat fluxes, thermal conductivity and temperature gradient of surrounding rocks, respectively.

(2) Unsteady temperature field of surrounding rock exposed to cool wind flow

Based on the assumptions and the heat and mass transfer theory above, the prediction model of the surrounding temperature field of the circular drying roadway was established.

$$\frac{\partial^2 t}{\partial r^2} + \frac{1}{r}\frac{\partial t}{\partial r} - \frac{1}{a}\frac{\partial t}{\partial \tau} = 0 \tag{2}$$

The numerical calculation method can be used to solve the partial differential equation, and the temperature value at any position and any time in the surrounding rock can be obtained.

Influence of seasonal air-flow on underground wind temperature

The air temperature and the temperature of the rock in the heat-adjusting layer have similar periods of change, and the temperature decreased as the radial distance increased. The original surrounding rock temperature and airflow temperature determine the radius of heat-adjusting layer. Higher temperature of the surrounding rocks and wind flow means smaller radius of the heat-adjusting layer with lower thermal regulation. The increase depth of mining leads to the reduction of thermal regulation. Hence, the seasonal temperature rise appears in the working face.

According to the meteorological parameters of Zhaolou coal mine, a three-year surface wind temperature variation was shown Fig.1. The surface humidity is averaged with 50%.

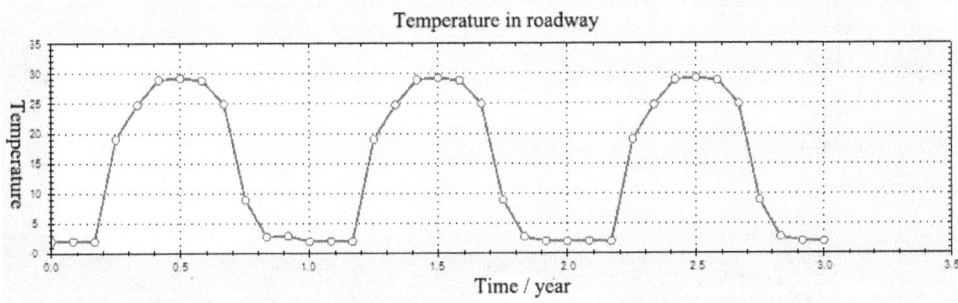

FIG 1– Three-year surface wind temperature variation of Zhaolou coal mine

Based on the data above, only the surrounding rock heat dissipation was considered in the heat source. The temperature and humidity prediction results of underground affected by wind flow was displayed in Fig.2.

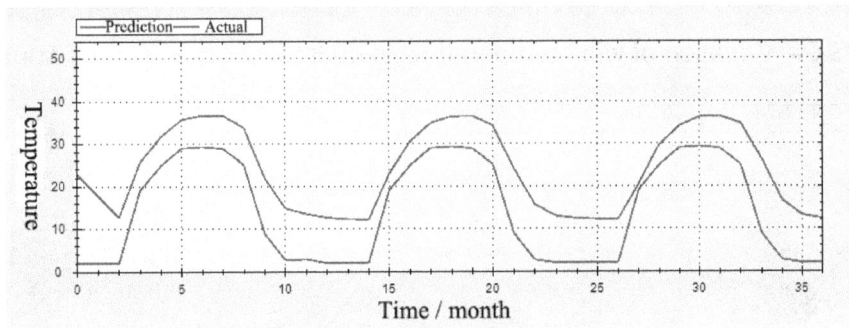

FIG 2 – Change of underground temperature

The data of red line in Fig. 2 was chosen as the inlet air temperature of the next roadway. By analogy, the results of wind temperature on the routes ware obtained and shown in Fig. 3 to Fig. 6.

The prediction demonstrated that the change of air temperature obviously causes that of the underground. Therefore, full air volume cooling can reduce the underground wind flow temperature of Zhaolou coal mine.

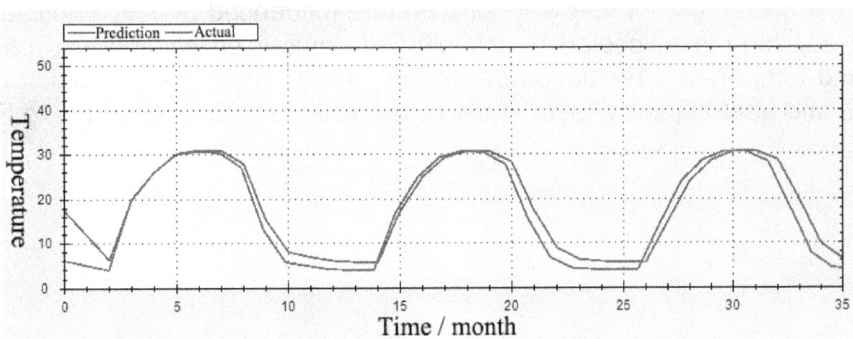

FIG 3 – Variation of wind temperature at outlet 1 of track 2 in the south

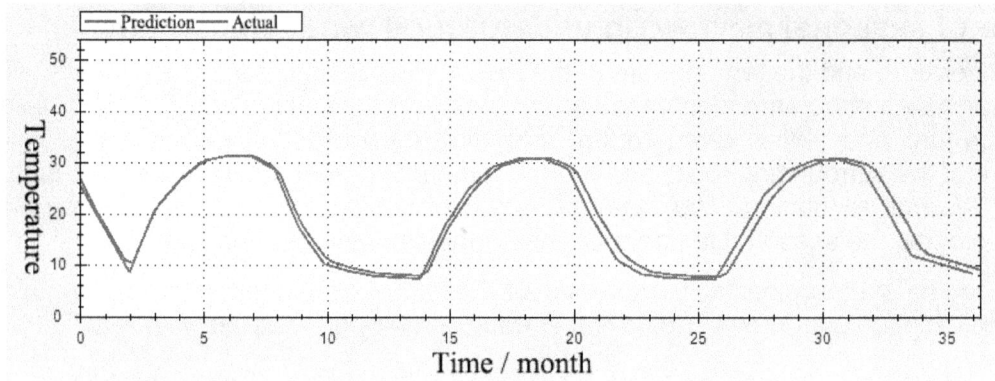

FIG 4 - Variation of wind temperature at outlet 2 of track 2 in the south

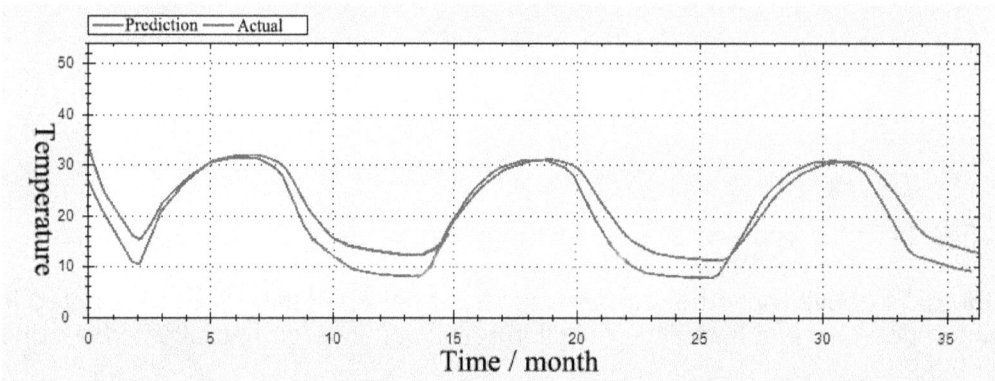

FIG 5 – Variation of wind temperature at outlet 3 of track 2 in the south

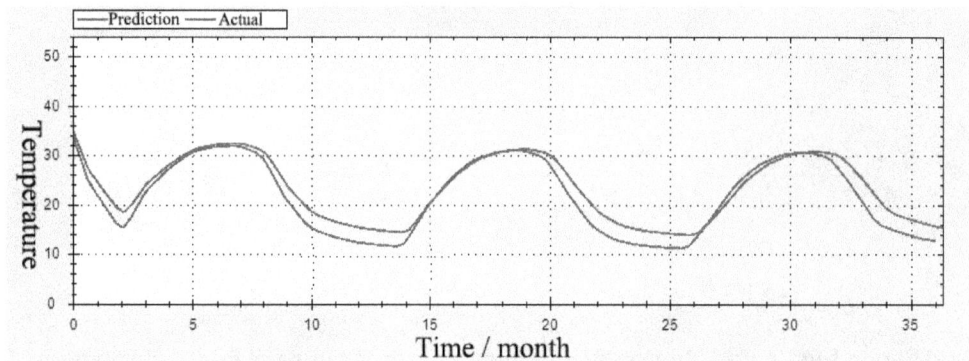

FIG 6 – Variation of wind temperature at transport outlet of 3302 working face

Change of heat-adjusting layer after concentrated cooling at the pithead

In order to learn the distribution of roadway temperature influenced by fully ventilation volume, the variation of heat-adjusting layer cooled by the pithead centralized cooling was calculated. Fig. 7 describes the wind temperature prediction of auxiliary shaft. Then, the prediction of underground wind temperature and heat-adjusting layer in the roadway are exhibited in Fig. 8 to Fig. 10.

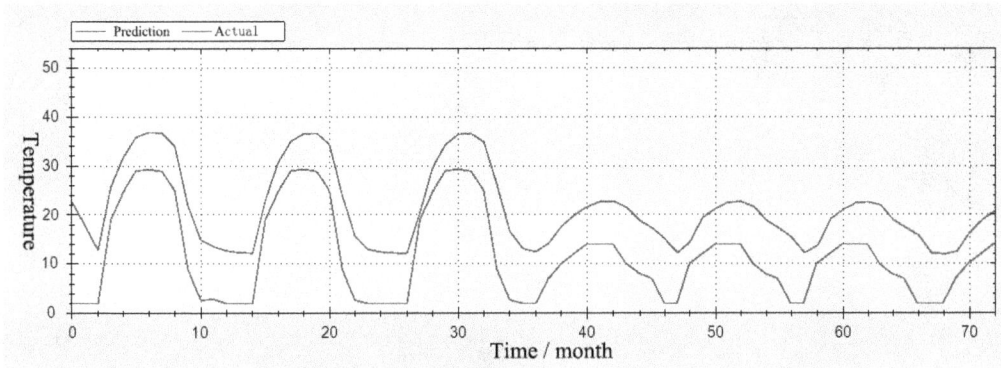

FIG 7 – Prediction of underground wind temperature cooled in three years

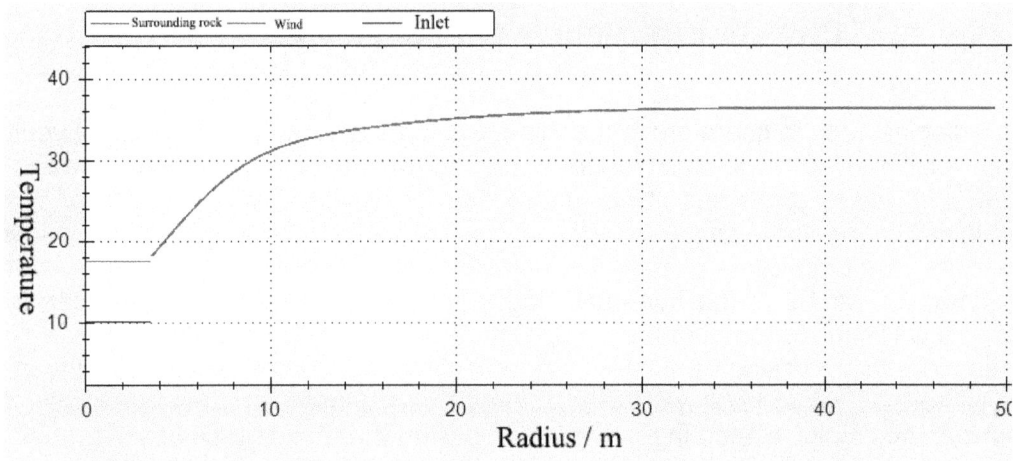

FIG 8 – Heat-adjusting layer and wind temperature after one month of ground concentrated cooling

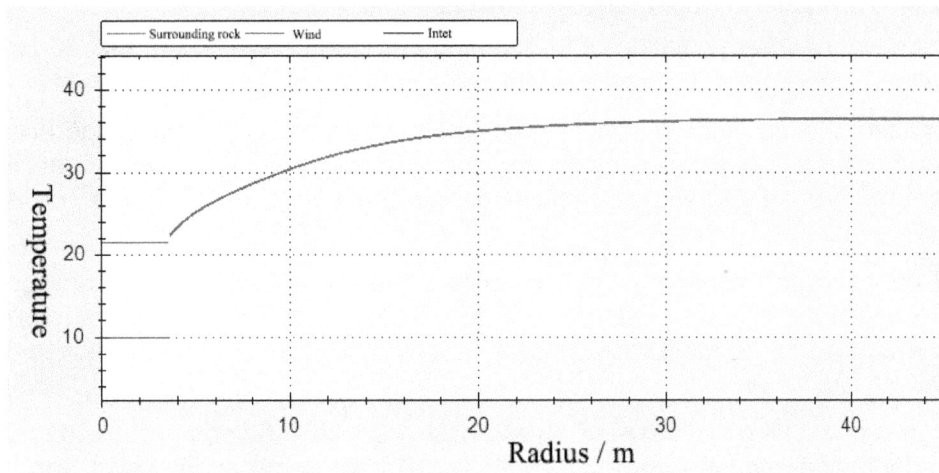

FIG 9 –Heat-adjusting layer and wind temperature after six months of ground concentrated cooling

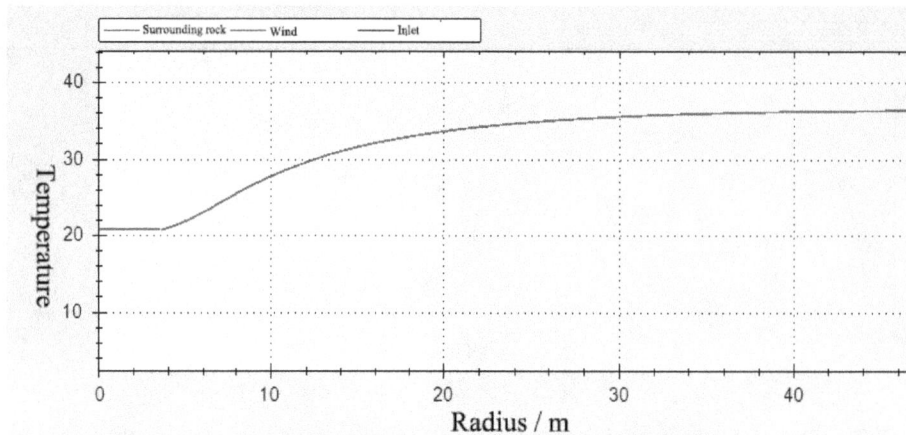

FIG 10 – Heat-adjusting layer and wind temperature after
three years of ground concentrated cooling

Although the original rock temperature of the mine was high, after a certain period of ventilation, the surrounding rock cooling zone can reach a certain radius. The distance between the high temperature surrounding rocks and the center of roadway limited the effect of wind flow heating. After centralized cooling at the pithead, the seasonal high-temperature peaks were weakened, and the temperature distribution law of the heat-adjusting layer was changed. With the increase of ventilation time, the radius of the heat-adjusting layer expanded and formed a permanent one. Meanwhile, the wall surface temperature, temperature gradient and heat dissipation of surrounding rocks decreased with increasing ventilation time, and the decrease margin was smaller and smaller. The range of heat-adjusting layer expanded in depth and formed a permanent one. The heating effect of surrounding rocks located in roadway on the wind flow became smaller.

MINE SEASONAL THERMAL HAZARD PREVENTION TECHNOLOGY

Comparative analysis of cooling methods

According to the arrangement position of the cooling unit, the cooling systems can be divided into ground centralized, underground concentrated, underground moving and combined.

Ground centralized cooling method transports cold water produced by the refrigeration unit to the hot place, and then transfers the cooling to the air flow through the heat exchanger. It can reduce the temperature of inlet air, but with low efficiency and poor economy. This method is suitable for mines with severe seasonal damage.

Underground concentrated cooling system circulates the chilled water to the air cooler, and the cooling air mixed with the high temperature air in the roadway so as to reduce the temperature. The process can be applied in the mine cooling for a long time, and is suitable for mines with local high temperature.

Underground moving cooling system need to establish a cooling station at required location. It is an efficient and economical measure for mines with partial heat damage. However, the management and safety of equipment need to be considered, and the discharge of condensation heat is difficult.

Combined cooling system are equipped with cooling stations on the ground and underground. With the help of ground cooling system, the condensing heat in the well is cooled by the ground cooling system, which can increase the return temperature of the primary coolant and reduce the cold loss. The system is complex and difficult to manage, and suitable for mines with severe heat damage.

Design of large air volume non-power heat transfer at pithead

Constitution of non-power heat transfer system

Non-powered air coolers used in large air volume refrigeration. The power source of the device is the negative pressure formed by the main ventilator at the pithead, which could meet the explosion-proof requirements. Fig.11 exhibits the elevational view of the air cooler structure and a cross-sectional view taken along line A-A.

The heat exchanger design software is used for calculation. The air volume of each group of heat exchangers is designed to be 105m³/h, and the cooling capacity is 740kW. 20 units were arranged in series with a total cooling capacity of 14940 kW. The plan can meet the requirements of air heat and humidity treatment. Each of the cooler is designed a size of 4800mm×4000mm×620mm and consists of six heat exchangers. The size of heat exchanger is 1800mm×1178mm ×600mm.

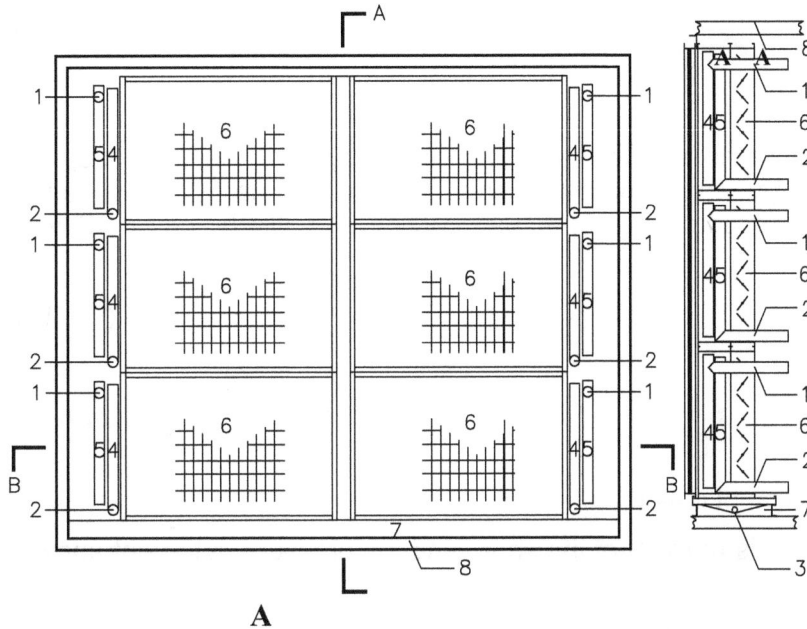

1-chilled water inlet pipe; 2-chilled water outlet pipe; 3-condensate drain pipe; 4-heat exchanger water tank; 5-heat exchanger water tank; 6-heat exchange coil; 7-condenser tray; 8-outer frame

FIG 11 – Large air volume non-power heat transfer at pithead

Auxiliary shaft ventilation simulation under air leakage

The model was constructed in the range of 100m×12m×8m. The arrangement of heat exchangers is described in Fig. 12. A total of 20 heat exchangers and air inlet are arranged on the side of the structure wall. The size of the air outlet is 3m×2m with the heat of 0.5m from the ground and the gap of 1.1m. The air volume of each air supply port is 50,000m³/h with the speed of 2.31m/s. The supplement temperature is 19°C. The total air volume of cold air is 1,000,000m³/h. The air volume entering through the doors is 200,000m³/h with 4.61m/s and 34.8°C.

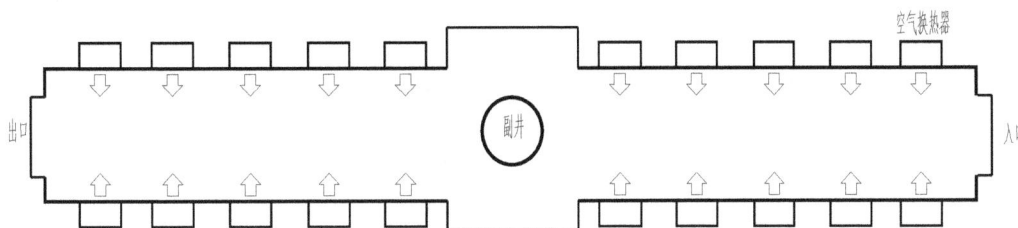

FIG 12 – Arrangement of heat exchangers

GAMBIT was adopted to establish building models and generate mesh. The number of mesh is 50,400 and 8400 in auxiliary shaft and roadway, respectively. The air in the building is incompressible, and thermal properties of the outdoor air are: t=34.1°C, ϕ=60%, ρ=1.125 kg/m^3, C=1.005 $kJ/(kg \cdot K)$, λ=0.0259W/(m·K), a=2.14×10-7 m2/s. After treated by the heat transfers, its properties became: t=19°C ; ϕ=95%, P=1.1833 kg/m^3, C=1.005 kJ/(kg·K), λ =0.0259W/(m·K), a=2.14×10-7 m2/s.

The boundary conditions were set as follows.

(1) Initial condition

Air is a steady flow in the periods of normal ventilation and obeys the Eq. (3).

$$\frac{\partial \varphi}{\partial t} = 0 \tag{3}$$

(2) Boundary condition of inlet

The inlet boundary distribution was specified according to the wind speed, ambient temperature, turbulent flow energy and dynamic energy dissipation rate. Wind speed and environmental parameters were set according to the actual situation. The ENKE and ENDS were calculated by Eq. 4 and Eq. 5.

$$K_m = 0.05V^2{}_{in} \tag{4}$$

$$\varepsilon = \frac{c_\mu^{3/4} k^{3/2}}{ky} , \eta_t = c_\mu \rho k^2 / \varepsilon , \rho\mu L/\eta_t =100{\sim}1000 \tag{5}$$

(3) Boundary condition of outlet

The exit of the model is the inlet of the shaft, and the pressure at the entrance of the shaft was set to P=0.

(4) Boundary condition of wall

The roughness was set as 0.7. The speed value is 0. The wall temperature is calculated with a definite value.

Simulated result

(1) In summer

All air inlets open in summer. The air volume entering the heat exchangers is 83,300 m³/h. It's speed is 2.49 m/s and temperature is 19.9□. Total cold air volume is 1,000,000 m³/h.

The result of summer is shown in Fig. 13 to Fig. 16. 1,000,000 m³/h cooled air volume mixed with 200,000 m³/h air, and the temperature of mixed air was 21□ which could meet the cooling requirement in summer.

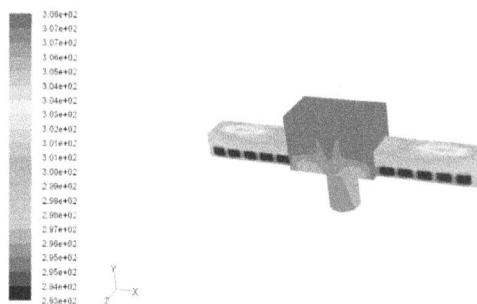

FIG 13 – Three-dimensional (3D) temperature field distribution of wellhead in summer

FIG 14 – Longitudinal section temperature field distribution at pithead

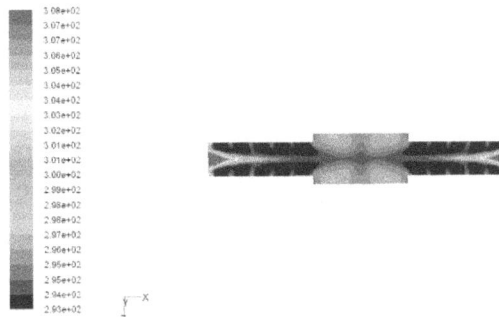

FIG 15 – Cross-sectional temperature field distribution at inlet

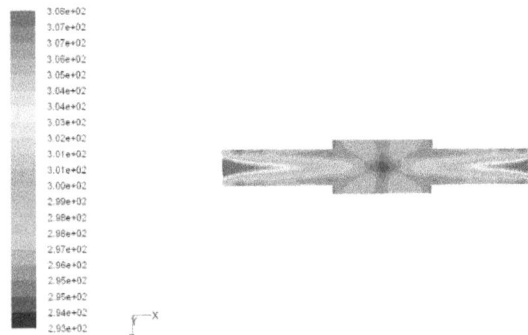

FIG 16 – Cross-sectional temperature field distribution at height of 0.25m

(2) In winter

12 air inlets open in winter. The air volume entering the heat exchangers was 600,000 m³/h. It's speed is 2.31 m/s and temperature is 35□. The air volume entering the doorway is 600,000 m³/h with the speed of 10.85 m/s and the temperature of -12□.

The result of winter is shown in Fig. 17 to Fig. 20. 600,000 m³/h heated air volume mixed with 600,000 m³/h air, and 12 inlets near the two doors were opened. The result indicated that the temperature of mixed air was higher than 10□ and that of pithead could maintain above 2□.

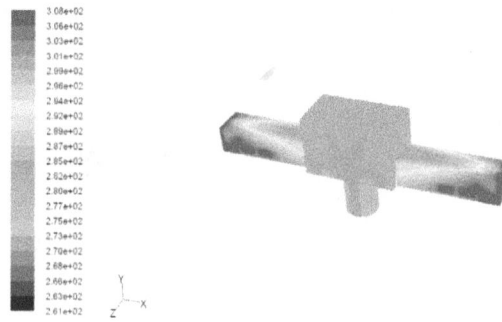

FIG 17 – 3D temperature field distribution of wellhead in winter

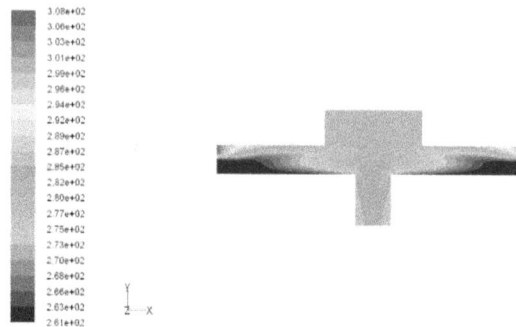

FIG 18 – Longitudinal section temperature field distribution at pithead

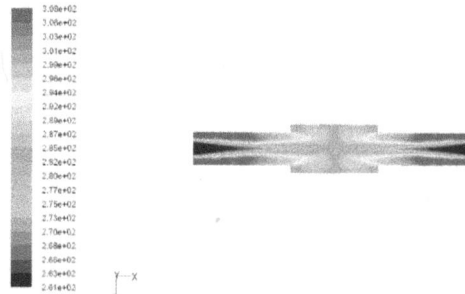

FIG 19 – Cross-sectional temperature field distribution at inlet

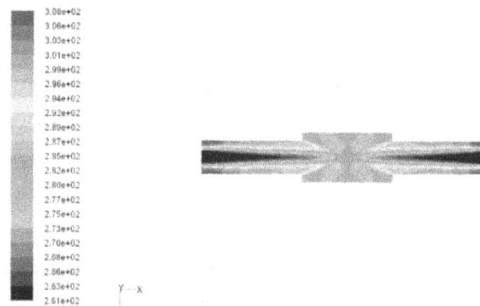

FIG 20 – Cross-sectional temperature field distribution at height of 0.25m

CONCLUSIONS

The model of the surrounding rock temperature field in the unsteady wind temperature was established. A method for calculating the heat-adjusting layer by difference method in the unsteady inlet wind temperature was proposed. The temperature distribution of the heat-adjusting layer after cooled was predicted. The results show that as the ventilation time increased, the radius of the heat-adjusting layer continuously expanded and formed a permanent one. Hence, the heating effect of surrounding rocks located in roadway on the wind flow became smaller.

Comparative analysis of cooling methods has been conducted. We found a cooling method and system suitable for seasonal heat damage in mines. The seasonal heat damage of the mine was treated by the full air volume ventilation and cooling method at the pithead. The pithead high air volume non-power air heat exchange system was also designed. It was determined that the design could meet the system cooling requirements by simulating the ventilation state in air leakage condition.

ACKNOWLEDGEMENTS

This work was supported by National Key R&D Program of China (No. 2016YFC0800100).

REFERENCES

Biccard, JCW, 1939. The estimation of ventilation aer temperatures in deep mines in Journal of the Chemical Metallurgccal and Mining Society of South Africa.

Chorowski, M, Gizicki, W and Reszewski S, 2012. Air condition system for copper mine based on triseneration system, in Journal of the Mine Ventilation Society of South Africa; 65: 20-24.

Danko, G and Bahrami, D, 2008. Application of MULTIFLUX for air, heat and moisture flow simulations, in Wallace. 12th U.S./North American Mine Ventilation Symposium 2008. Nevada: Nevada University Press 267-274.

He, M and Guo, Y, 2013. Deep rock mass thermodynamic effect and temperature control measures in Chinese Journal of Rock Mechanics and Engineering 32: 2378-2393.

Heerden,CV, 1951. A Problem of Unsteady Heat Flow in Connection with the Cooling of Collieries Preceedings of the General Discussion on heat flow. Inst.Mech.Eng 283-285.

Knig H, 1952. Mathematische Unteruchungen über das Grubenklima Bergbau Archiv 13.

Moloney,KW, Lowndes,IS, Stockes, MR, and Hargrave, G, 1997. Studies on alternative methods of ventilation using computational fluid dynamics, scale and full scale gallery tests, in Proceedings of the 6th International Mine Ventilation Congress, 497-503.

Uchino K, and Inoue M, 1986. New practical method for calculation of air temperature and humidity along wet roadway-The influence of moisture on the underground environment in mines (2nd Report), in Journal of the Mining and Metallurgical Institute of Japan, 102: 353-357.

Uchino K, and Inoue M, 1990. Improved practical method for calculation of air temperature and humidity along wet roadway-The influence of moisture on the underground environment in mines (3rd Report), in Journal of the Mining and Materials Processing Institute of Japan, 106: 7-12.

www.ingramcontent.com/pod-product-compliance
Lightning Source LLC
Chambersburg PA
CBHW061103210326
41597CB00021B/3966